AGROCHEMICALS FROM NATURAL PRODUCTS

BOOKS IN SOILS, PLANTS, AND THE ENVIRONMENT

Soil Biochemistry, Volume 1, edited by A. D. McLaren and G. H. Peterson
Soil Biochemistry, Volume 2, edited by A. D. McLaren and J. Skujiņš
Soil Biochemistry, Volume 3, edited by E. A. Paul and A. D. McLaren
Soil Biochemistry, Volume 4, edited by E. A. Paul and A. D. McLaren
Soil Biochemistry, Volume 5, edited by E. A. Paul and J. N. Ladd
Soil Biochemistry, Volume 6, edited by Jean-Marc Bollag and G. Stotzky
Soil Biochemistry, Volume 7, edited by G. Stotzky and Jean-Marc Bollag
Soil Biochemistry, Volume 8, edited by Jean-Marc Bollag and G. Stotzky

Organic Chemicals in the Soil Environment, Volumes 1 and 2, edited by C. A. I. Goring and J. W. Hamaker
Humic Substances in the Environment, M. Schnitzer and S. U. Khan
Microbial Life in the Soil: An Introduction, T. Hattori
Principles of Soil Chemistry, Kim H. Tan
Soil Analysis: Instrumental Techniques and Related Procedures, edited by Keith A. Smith
Soil Reclamation Processes: Microbiological Analyses and Applications, edited by Robert L. Tate III and Donald A. Klein
Symbiotic Nitrogen Fixation Technology, edited by Gerald H. Elkan
Soil-Water Interactions: Mechanisms and Applications, Shingo Iwata and Toshio Tabuchi with Benno P. Warkentin
Soil Analysis: Modern Instrumental Techniques, Second Edition, edited by Keith A. Smith
Soil Analysis: Physical Methods, edited by Keith A. Smith and Chris E. Mullins
Growth and Mineral Nutrition of Field Crops, N. K. Fageria, V. C. Baligar, and Charles Allan Jones
Semiarid Lands and Deserts: Soil Resource and Reclamation, edited by J. Skujiņš
Plant Roots: The Hidden Half, edited by Yoav Waisel, Amram Eshel, and Uzi Kafkafi
Plant Biochemical Regulators, edited by Harold W. Gausman
Maximizing Crop Yields, N. K. Fageria
Transgenic Plants: Fundamentals and Applications, edited by Andrew Hiatt
Soil Microbial Ecology: Applications in Agricultural and Environmental Management, edited by F. Blaine Metting, Jr.
Principles of Soil Chemistry: Second Edition, Kim H. Tan

AGROCHEMICALS FROM NATURAL PRODUCTS

edited by

C. R. A. GODFREY

Zeneca Agrochemicals
Jealott's Hill Research Station
Bracknell, Berkshire, England

Marcel Dekker, Inc.   New York•Basel•Hong Kong

Library of Congress Cataloging-in-Publication Data

Agrochemicals from natural products / edited by C. R. A. Godfrey.
     p.   cm. — (Books in soils, plants, and the environment)
    Includes bibliographical references and index.
    ISBN 0-8247-9553-9 (acid-free paper)
    1. Natural pesticides. I. Godfrey, C. R. A. II. Series.
    SB951.145.N37A37 ~~1994~~ 1995
    668'.65—dc20   94-35411
                                                         CIP

The publisher offers discounts on this book when ordered in bulk quantities. For more information, write to Special Sales/Professional Marketing at the address below.

This book is printed on acid-free paper.

Marcel Dekker, Inc.
270 Madison Avenue, New York, New York 10016

Current printing (last digit):
10 9 8 7 6 5 4 3 2 1

PRINTED IN THE UNITED STATES OF AMERICA

Preface

It has now been over 20 years since publication of the book *Naturally Occurring Insecticides* (M. Jacobson and D. G. Crosby, eds., Marcel Dekker, Inc., New York, 1971), which has since become a classic in the area of applied natural products chemistry. In the intervening years, the importance of natural products as insecticides has been accompanied by a steady growth in other agrochemical sectors. It is the intention of this book to present a thorough account of the current status of each of the most important areas. The main emphasis of the book is on the discovery and agricultural applications of both natural products and synthetic compounds derived from natural products. Where appropriate, information is presented on topics such as isolation, structural studies, biological activity, structure–activity relationships, toxicology, resistance, and commercial data. It is anticipated that the contents of this book will appeal to a wide readership, including scientists working in the agrochemicals industry, agriculturists, synthetic chemists, biologists, and students.

The book consists of nine chapters written by scientists from both sides of the Atlantic. The first chapter, on insecticides, explores the structural diversity of recently discovered insecticidal natural products and is intended to bring the reader up to date on the most significant technical developments that have taken place in this field since the work of Jacobson and Crosby. Several classes of insecticide have reached sufficient prominence to merit chapters of their own. The first of these presents a general overview of the pyrethroids, with much of the emphasis on structure–activity relationships. These compounds arguably represent the most successful example of the development of synthetic agrochemicals from natural products. The juvenoids are analogues of the natural juvenile hormones that are designed to disrupt the endocrine system of insects. Although offering potential advantages in terms of species selectivity, their initial promise has not yet materialized and only a small number have reached the marketplace. The chapter on avermectins and milbemycins charts the development of these complex macrolides, which have been found to exhibit both insecticidal and anthelmintic activity. Some have found wide application in agriculture, thus complementing the diverse range of vitamins, antibiotics, anticoccidials, and animal growth promoters currently in use as animal health products.

In contrast to these relatively mature areas, and despite much continuing research, few highly successful examples of natural products have been developed in the remaining agrochemical markets. The chapter on herbicides and plant growth regulators cites the herbicidal natural product phosphinothricin as the most significant recent development in this area. The primary emphasis in this chapter is on the development of screening techniques for the detection of herbicidal substances produced by nature, which may lead to important discoveries in the future. The use of natural products per se as fungicides and bactericides is declining in comparison with modern synthetic agrochemicals. The focus of this chapter is on natural products (such as pyrrolnitrin and the β-methoxyacrylates) that have been used as leads in the discovery of new, synthetic fungicides. The penultimate chapter describes the impact of natural products on the development of rodenticides. A number of acutely toxic natural products have been used in agricultural situations for many years for the control of rodent pests, such as rats and mice. However, since the 1940s, superior anticoagulant compounds have been developed from the natural product dicoumarin that are more effective in use and safer to nontarget species. Last, but by no means least, the chapter on biological control agents (BCAs) reviews the current practical uses of BCAs in agriculture and highlights those whose mode of action is at least partially connected with the production of natural products. This is a growing area that, although relatively unexploited at the present time, appears to hold much promise for the future.

I would like to take this opportunity to thank each of the authors who have contributed to the preparation of this book. I would also like to thank my colleagues at Zeneca Agrochemicals (in particular, Brian Baldwin, John Clough, Chris Urch, and Paul Worthington) for their continued interest and encouragement and my wife Jayne for her forbearance and support.

C. R. A. Godfrey

Contents

CONTENTS

Contributors

R. W. Addor Agricultural Research Division, American Cyanamid Company, Princeton, New Jersey

A. Clare Elliott, Ph.D. Department of Chemistry, Zeneca Agrochemicals, Jealott's Hill Research Station, Bracknell, Berkshire, England

C. R. A. Godfrey, Ph.D. Department of Chemistry, Zeneca Agrochemicals, Jealott's Hill Research Station, Bracknell, Berkshire, England

Clive A. Henrick, Ph.D. Director, Chemical and Analytical Research, Research Division, Sandoz Agro, Inc., Palo Alto, California

Gabe I. Kornis, Ph.D. Senior Research Scientist, Upjohn Laboratories, Upjohn Company, Kalamazoo, Michigan

Margaret A. Miller-Wideman, M.S. Research Group Leader II, Department of Medicinal Chemistry, G. D. Searle, St. Louis, Missouri

Jane Louise Faull, Ph.D. Department of Biology, Birkbeck College, University of London, London, England

Keith A. Powell, Ph.D. Biotechnology Section Manager, Zeneca Agrochemicals, Jealott's Hill Research Station, Bracknell, Berkshire, England

Richard James Stonard, Ph.D.   Director of Research, The Agricultural Group, Monsanto Company, St. Louis, Missouri

David J. Wadsworth, Ph.D.   Chemistry Group, Animal Health Division, Ciba-Geigy Limited, Basel, Switzerland

1

Insecticides

R. W. Addor

American Cyanamid Company,
Princeton, New Jersey

I. INTRODUCTION

With the exclusion of the pyrethroids, avermectins, and insect hormones, this chapter addresses the ever-broadening subject of insecticidal natural products. An effort has been made to update parts of the earlier outstanding volume edited by Jacobsen and Crosby [1] in which significant new information has been reported, the isobutylamides being a case in hand. Rotenone and the rotenoids [2] have not been re-addressed; additional background on their occurrence, synthesis, and biosynthesis has been provided by Crombie [3]. The earlier piece on the piericidins by Yoshida and Takahashi [4] was updated and expanded [5]. Where older compound types have been included, it is usually because they have recently served as models for the synthesis of new structures.

Even with the newer structures, emphasis has been placed on those compounds isolated from any source in the natural world, other than insect venoms, that have served as models for new work. New compounds for which only some index of repellency, growth inhibition, or insect feeding deterency is reported have usually been passed over. An overview of compounds that generally fall into these categories and their sources can be found in reports by Kubo and Nakanishi [6] and by Klocke and co-workers [7] and references therein. A summary of these activities for preparations derived from a wide range of species from important plant families has been provided by Jacobson [8a]. Recent reports on the isolation of a series of steroidal insect growth inhibitors from the species *Petunia* is also instructive [8b]. Of course, potent antifeedant action, as in the case of azadirachtin, can lead to death at one stage or another in the insects life cycle, and efforts to capitalize on the outstanding activities of this compound are reviewed here. Plant-derived materials with molluscicidal activity have been reviewed [9].

One area of research worthy of greater discussion than is appropriate in this review is that dealing with the so-called "cage" convulsants. These compounds, of which a series of 1,4-disubstituted-2,6,7-trioxabicyclo[2.2.2]octanes are representative, behave like picrotoxinin, a component of the seeds of *Anamirta cocculus* L., in their ability to block receptors modulated by γ-aminobutyric acid [10]. However, synthesis studies have been based more on the chemistry of known antagonists than on the structure of the natural product, picrotoxinin, itself. Reports on recent work provide background [11,12]; extension of the work to an interesting series of 1,3-dithiane insecticides has been reported [13,14].

II. TERPENOIDS

The liminoids, or tetranortriterpenoids, of which azadiractin is a member, represent the upper end of the array of plant-produced terpenoids which discourage predation by insects. At the lower end, such relatively simple monoterpenes as limonene and myrcene play a role in protecting plants that produce them [15]. Coats and co-workers have reported recent studies aimed at gaining a better understanding of the toxic effects of monoterpenes on insects [16]. Much work has been expended on discerning the importance of a variety of monoterpenes, as well as a series of more complex terpenoid aldehydes, e.g., gossypol, which impart resistance to insects by varieties of glanded cotton [17]. A number of polyhalogenated monoterpenes, isolated from marine algae, have demonstrated moderate insecticidal activities [18, and references therein]. Pulegone-1,2-epoxide (*1*, Figure 1) is reportedly the major insecticidal component of the medicinal plant *Lippia stoechadoifolia* [19]. Little insecticidal data were provided, but the observation of a neurotoxic action was backed up by studies showing **1** to be an inhibitor of acetylcholinesterase [20a]. The activity of the essential oils derived from 11 Greek aromatic plants belonging to the Lamiaceae family to eggs, larvae, and adults of *Drosophila auraria* has been reported [20b]. Among

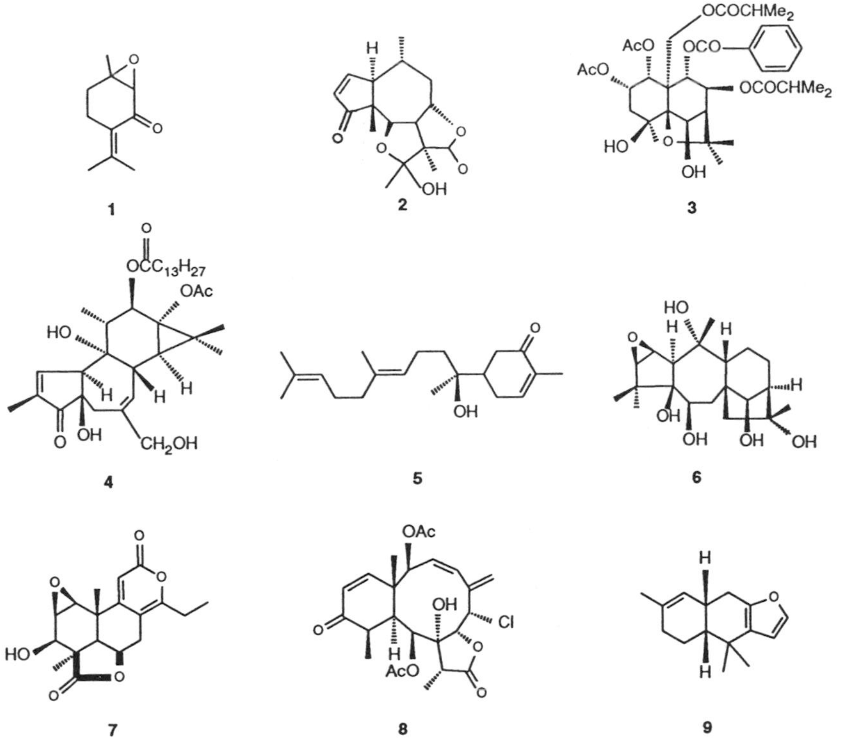

Figure 1 Insecticidal terpenoids.

the most active was that from *Mentha pulegium* L. which includes pulegone as the major constituent [20b].

In terms of insect control, the mass of literature dealing with the higher terpenoids refers mostly to antifeeding and growth-retarding observations. A review by Picman summarizes the effects of insect feeding, development, and survival for over 50 sesquiterpene lactones [21]. Her summary of the extensive studies on the feeding deterency of these compounds confirms the general observation of the importance of the presence of α-cyclopentenone and/or α-methylene-γ-lactone functionalities as electrophilic acceptors for promoting this activity. In this regard, a study of the mode of antifeedant activity of tenulin (2, Figure 1) is representative [22]. On the other hand, the sesquiterpene polyol ester angulatin A (3) is claimed to be strongly antifeedant (and insecticidal) against a variety of insects [23]. Angulatin A and close antifeedant analogs [24,25] are derived from the root bark of *Celastrus angulatus* Max. used in China to protect plants from insects.

To gain a further perspective on the range of work done with naturally occurring terpenoid antifeedants, reference should be made to the review by Mabry and Gill [15], the report on insect antifeedants from tropical plants by Kubo and Nakanishi [6], the tabulation concerning diterpenes with activity against insects included in a paper by Cooper-Driver and Le Quesne [26], the paper on terpenoids from the genus *Artemisia* by Duke, Paul, and Lee [27], and the report by Rodriguez on insect feeding deterrents from arid land plants [28].

There are some recent reports of insecticidal terpenoids of interest. Of 20 esters representing the biogenetically related tigliane, daphnane, ingenane, and lathyrane series of diterpenes, only the tigliane diterpene ester 4 (Figure 1) isolated from the seeds of *Croton tiglium* L. (Euphorbiaceae) and two closely related esters elicited both growth inhibitory and insecticidal activities in diet against newly hatched larvae of *Pectinophora gossypiella* (100% kill at 20 ppm for 4) [29]. Ester 4 also gave 100% kill of *Culex pipiens* larvae at 0.6 ppm but was ineffective against *Oncopeltus fasciatus* and *Tribolium confusum*. Esters of this type are also well known for their vesicant and tumor-producing properties [29,30].

From the leaves and twigs of the medicinal plant *Croton linearis* Jacq. was isolated, as a yellow oil, the novel diterpene of structure 5 [31]. The topical LD_{50} to the adult weevil *Cylas formicarius elegantulus*, a serious pest on sweet potatoes, was 0.32 μg/insect. Symptoms of poisoning were typical of neurotoxic insecticides.

By using antifeeding activity against two insect species for assay, three known grayanoid diterpenes were isolated from the dried flowers of the Chinese insecticide plant *Rhododendron molle* [32]. Of these, rhodojaponin III (6) was the most active antifeedant. In diet, it was also toxic to newly hatched *Spodoptera frugiperda* with an LD_{50} of 8.8 ppm, the same order of magnitude as malathion. Against *Heliothis virescens*, it was nontoxic and failed to retard growth.

Conifers of the *Podocarpus* species are known to be resistant to many insects [33,34 and references therein]. Among the components isolated from various parts of such representative species as *P. gracilior* [33] and *P. nagi* [34] are a series of norditerpene dilactones referred to as nagilactones. These compounds are strong feeding deterrents leading to reduction in larval growth and pupation. The most effective of these is nagilactone D (7, Figure 1), which at 166 ppm in artificial diet prevented all larvae of *Heliothis virescens* from reaching the third instar. Evidence points to effects on food digestion and absorption and on mouth part sensory receptors as the cause of reduced food consumption [34]. *Podarcopus* resistance is also favored by the presence of other identified compounds having growth and ecdysis inhibitory properties [34].

The search for biologically active compounds from marine sources has included, in some instances, insecticidal evaluation. Some 12 briaran diterpenes from coelenterates *Briareum polyanthes* and *Ptilosarcus gurneyi* have been isolated and identified [35]. Of these, five were tested against the tobacco hornworm, *Manduca sexta*. An electrophilic site, as included in the cyclohexenone part of compound *8* (Figure 1), apparently contributes to the toxicity; all five compounds caused a loss in weight gain. Among several sesquiterpenes isolated from the sponge *Dysidea etheria* and the nudibranch *Hypselodoris zebra*, furodysinin (*9*) was the most toxic to the grasshopper *Melanoplus sanguinipes* [36]. Although nontoxic to the tobacco hornworm at 250 ppm in diet, it did produce difficulty in moulting.

III. AZADIRACHTIN

Some 64 triterpenoids have been reported from the seeds, wood, bark, leaves, and fruit of the neem tree, *Azadirachta indica* A. Juss, family Meliaceae [37]. These include azadirachtin A (Figure 2), first isolated by Butterworth and Morgan [38], and the group of liminoids most closely associated with the antifeedant and growth-retarding properties of neem extracts. A review of the isolation and identification of these compounds is provided [37].

Rembold has summarized the evidence supporting azadirachtin's ability to interfere with neuroendocrine control of metamorphosis in susceptible insects [39,40]. Retention of azadirachtin in insects is extremely low and yet sufficient to initiate many endocrine and, thus, morphological and behavioral effects. Significantly, it is this inhibitory activity, as measured against the Mexican bean beetle, *Epilachna varivestis*, which occurs in the 1–10-ppm range that is emphasized. Under these conditions, growth is arrested at the larval stage. Within a few days, larvae refuse to eat even untreated leaves and die of starvation. Other stages are also severely affected and ultimately die. These effects are in contrast to quick kill, which requires concentration in excess of 1000 ppm and the strong antifeedant activity noted in the 10–100-ppm range.

The ability of azadirachtin to function at hormonal concentrations and to produce ecdysone-type effects in susceptible insects had led to the idea that it is, in fact, acting as an ecdysteroid. However, Rembold presents some experimental evidence to the contrary [39,40]. He also rejects structural homology based on the trans A/B ring junction of azadirachtin as opposed to the cis configuration for the ecdysteroids.

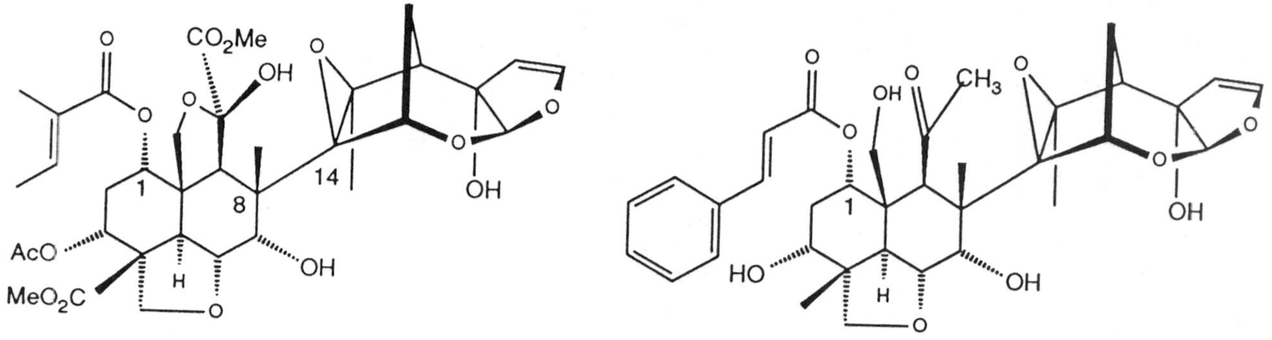

Figure 2 Structures of (a) azadirachtin (from *Azadirachta indica*) and (b) 1-cinnamoylmelianolone (from *Melia azedirach*). (From Refs 42e and 43.)

By contrast, Lee et al. present considerable evidence linking similarities of activity for azadirachtin with phytoecdysones and synthetic ecdysone agonists [41]. They emphasize (Figure 3) structural similarities between deesterified azadiracthin and ecdysterone, especially in regard to the placement of the important hydroxyl group in the 3-position of the A ring and the oxygenation pattern in the B and D rings. Their analysis suggests that despite the differences at the A/B ring junction for these compounds the hydroxyl groups in the 3 position, axial for azadirchtin and equatorial for the ecdysteroids, occupy the same

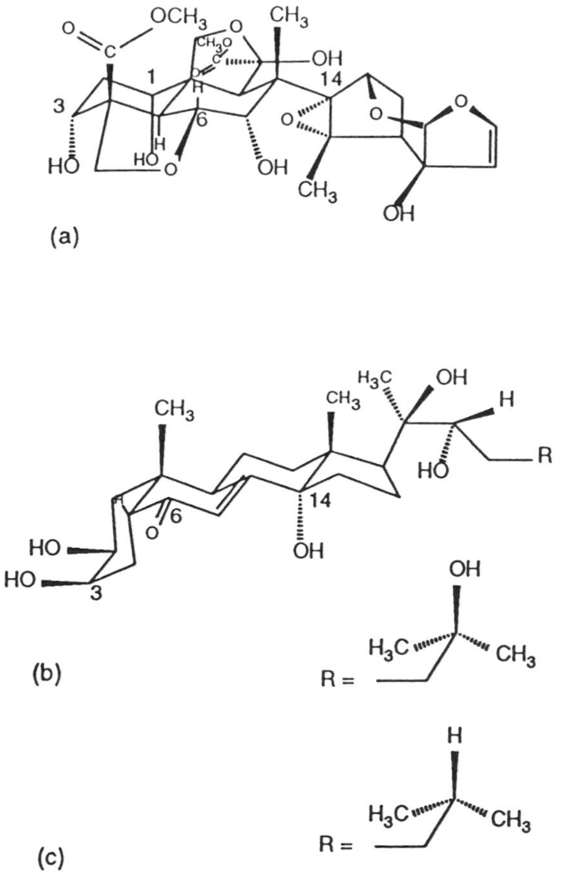

Figure 3 Stereostructural relationships between (a) azadirachtin, skeleton, deesterified (1,3-dideacylate), (b) ecdysterone (the insect molting hormone) (20-hydroxecdysone) and (c) the phytoecdysone, ponasterone A. (Reprinted with permission from Ref. 41. Copyright 1991).

relative space with regard to the rest of the molecule. Thus, either carbon framework may be able to hold the C-3 hydroxyl group in space such that an interaction with an appropriate receptor can occur [41].

By examining the key structural features contributing to potent growth inhibitory effects on *Epilachna varivestis* for six closely related azadirachtins and eight derivatives, Rembold has proposed a minimal structure shown in Figure 4 [39,40]. He adds the following statements:

1. The decalin ring must be substituted at positions 1 and 3 by hydroxy groups, either one or both of which are esterified or both are unsubstituted. The two free hydroxy groups yield the highest biological activity.
2. An epoxy group must be present in correct steric distance from the two hydroxy groups. Additional effects of a hydroxy group at position 7 and/or a ketal function in position 21 are possible.
3. The reduced dihydrofuran ring or a side chain, again in correct steric distance from the epoxy group, increases the growth regulating activity of the azadirachtin analog.

A less complicated structure may evolve from the current efforts of Ley and co-workers. In a convergent approach to the synthesis of azadirachtin, routes to the hydroxy tetrahydrofuran and the decalin fragments, which will ultimately be joined through the C(8)–C(14) bond, have been demonstrated [42a, 42b]. These synthesis schemes promise not only a route to new azadirachtin analogs, but the fragments themselves have proven of interest. Indeed, several components modeled on the hydroxy tetrahydrofuran segment have shown good antifeedant activity against lepidopterous larvae [42c, 42d].

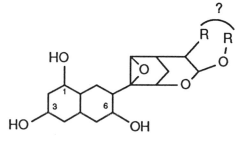

Figure 4 Proposal for a reduced bioactive azadirachtin structure. (Reprinted with permission from Ref. 40. Copyright 1989).

Recent research has revealed, contrary to earlier reports, that the related chinaberry tree (*Melia azedarach*) does not produce azadirachtin [41]. However, from the insecticidal methanolic extract of chinaberry fruits, the new triterpenoid called 1-cinnamoylmelianolone (Figure 2) was identified [41, 43]. This and a number of other natural and derivatized tetranortriterpenoids closely associated with azadirachtin are also potent growth inhibitors and larvicides when fed in diet to first instar larvae of *Heliothis virescens* and *Spodoptera frugiperda* [41].

The development of a formulated neem seed extract for insecticidal use has been described by Larson [44]. Through control of pH and the use of appropriate emulsifiers and stabilizers, a formulation containing 3000 ppm azadirachtin with reasonable shelf life has been achieved and designated "Margosan-O." This formulation is being marketed for use in greenhouses, nurseries, forests, and homes, and several other neem-based products are being developed [45]. Dietary studies required to register Margosan-O in the United States on crops are in progress [45]. Preliminary reports indicate that Margosan-O is nontoxic orally to rats, mallard ducks, and bobwhite quail [45].

A review of the use of various neem extracts and formulations against pests on vegetables and fruit trees has been published [46]. Lepidopterous larvae, a most important group of harmful insects in warm climates, were main target species. Good results were reported especially in application of neem seed kernel extracts.

Numerous other liminoids isolated from Meliaceae and Rutaceae have been evaluated as antifeedants and growth inhibitors against several insect species. In these studies none has shown the exceptional activity of azadirachtin; cedrelone and anthothecol (Figure 5) exemplified the best [47, 48].

Figure 5 Cedrelone (*R*=H); anthothecol (*R*=Me).

IV. ALKALOIDS

Alkaloids are generally defined as those naturally occurring organic compounds that incorporate a nitrogen atom as part of a heterocyclic ring [49, 50]. The term "protoalkaloid" has been used to include nitrogen-containing compounds, e.g., colchicine and mescaline, beyond that definition [49]. In addition to those known for their insecticidal activity, e.g., nicotine and sabadilla, many alkaloids found in plants have been recognized for their ability to prevent or reduce predation by all manner of herbivores [49]. The repellency to insect feeding by a number of well-known alkaloids, including the plant sources from which they come and the insect species deterred, has been reviewed [50a and references therein]. A role for β-carbolines, present in a variety of plants, as antifeedants and possible toxicants has been suggested based on results of studies that incorporated harman and harmine in the diet of the beet armyworm (*Spodoptera exigia*) [50b].

A. Nicotine

As an insecticide, nicotine (*10*, Figure 6) has probably been the most well known and widely used of the alkaloids. An excellent review of nicotine's history, source, and utilization along with information on the major attendant alkaloids nornicotine and anabasine was provided by Schmeltz [51].

Recently, the cuticular components of several species of Nicotiana were shown to include a series of *N*-acylnornicotines of type *11* [52]. Topical application of 200 µg/larve of these acylated nornicotine mixtures gave 100% kill of first instar tobacco hornworms (*Manduca sexta* M.) in 48 hr. This contrasts with nicotine, which gave only 7% mortality at 500 µg after 96 hr. However, no activity was observed in similar tests with the tobacco budworm or the cabbage looper. Further work on the effect of chain length and hydroxylation toward activity is being pursued [52].

Despite nicotine's relatively uncomplicated structure and the knowledge that it competes with the chemically simple neurotransmitter acetylcholine, past efforts to synthesize better insecticides based on this information have been unsuccessful [53, 54]. However, some years ago, "in quest of new classes of insecticides acting by heretofore unrecognized mechanisms," a group of Shell scientists prepared and tested a broad array of heterocyclic compounds incorporating a nitromethylene substituent [55]. Among the best of these was tetrahydro-2-(nitromethylene)-2H-1,3-thiazine (*12*, Figure 6). This compound proved highly active against a variety of insects and was particularly effective

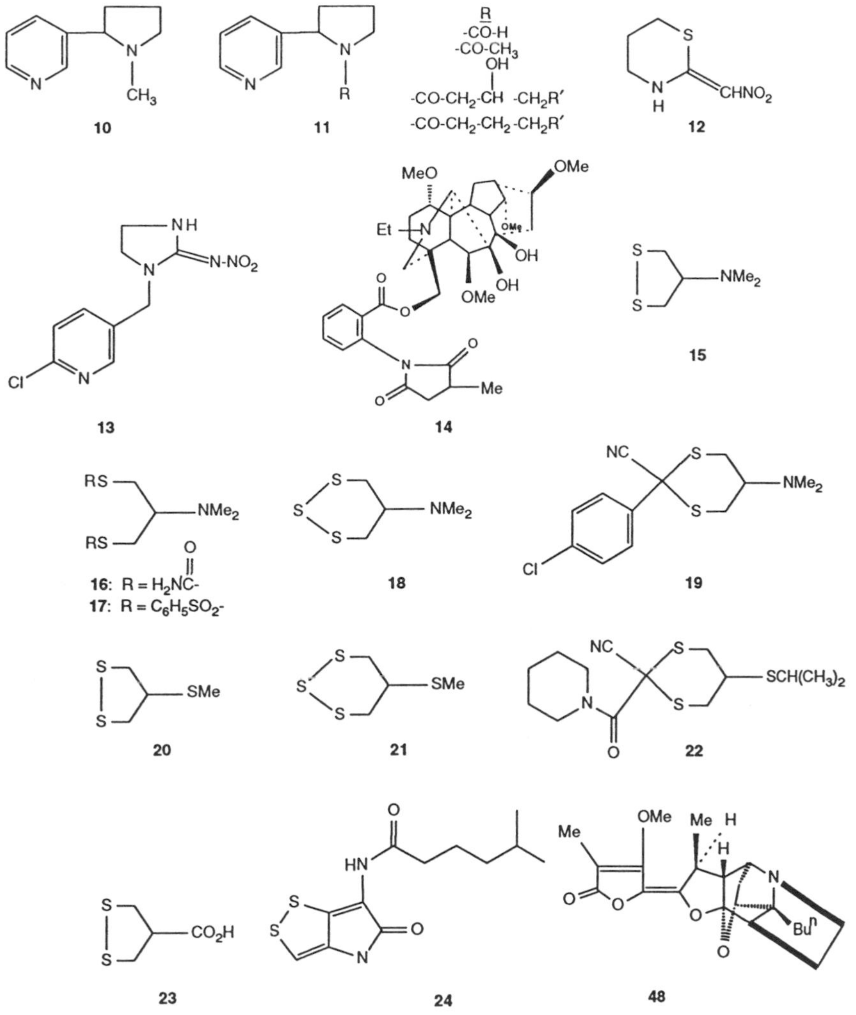

Figure 6 Insecticidal alkaloids.

against lepidopterous larvae. Acute mammalian toxicity was also quite acceptable, being far better than that of nicotine [55]. Some time later, the "unrecognized mechanism" of action was shown, in fact, to be as an agonist at the nicotinic acetylcholine receptor [56, 57].

The major drawback to the further development of these new nitromethylene compounds was an extreme photochemical instability [55]. Extensive synthesis efforts by the Shell group, and later by Bayer scientists in collaboration with a Japanese group, were undertaken in trying to find a solution. Most of this work is documented in the patent literature. Eventually from these efforts the new structurally related N-nitroimino analog, imidacloprid (13) evolved for commercialization [58a]. Imidacloprid is effective against a wide range of insects and is particularly active as a systemic insecticide against sucking insects on many crops [58a]. Mammalian toxicity is low [58a]. Imidacloprid, thus, appears to be the new improved "nicotine" long sought by many people. As with tetrahydro-2-(nitromethylene)-2H-1,3-thiazine, a representative of these newer compounds has been shown to function as a nicotinic acetylcoholine receptor agonist [58b]. A recently reported 2-nitromethylene-thiazoline analog of imidacloprid is highly active against lepidopterous species [58c].

B.   Methyllycaconitine

Recent work with extracts of *Delphinium* seeds, known to be insecticidal, has revealed that the major active toxin is methyllycaconitine (14, Figure 6) [59, 60]. Using a housefly head homogenate, methyllycaconitine was shown to be a potent inhibitor of α-bungarotoxin binding with a $K_{inh} = 2.5 \times 10^{-10}$ M. Nicotine was 10,000-fold less potent in this competition for the nicotinic acetylcholine receptors. Unlike nicotine, methyllycaconitine showed activity against lepidopterous larvae. The LC_{50} for the southern armyworm (*Spodoptera eridania*) fed treated lima bean leaves was 308 ppm; the number for nicotine was much greater than 1000 ppm. Lycoctonine, which lacks the aromatic ester function, was inactive against the armyworm and 1000-fold less effective in blocking binding of α-bungarotoxin to the insect cholinergic receptor.

Based on the methyllycaconitine finding, Pelletier and Ross undertook the preparation of a series of esters and ethers of structurally related alkaloids, i.e., delphisine, neoline, delphine, lycoctonine, aconitine, delphonine, and N-deacetyllappaconitine [61, 62, 63]. No reports of *in vitro* or *in vivo* testing have been published.

There is a recent report on the isolation of the alkaloids responsible for the antifeedant properties of Colorado larkspur (*Delphinium gereri*) [64].

C.   Nereistoxin

The compound 4-dimethylamino-1,2-dithiolane (15, Figure 6), now known as nereistoxin, was isolated from the marine annelid *Lumbriconereis heteropoda* Marenz in 1934 when flies were noted to die after feeding on dead bodies of the annelid [65, 66]. The subsequent structural determination work and synthesis has been reviewed by Konishi [66].

Using the chemistry developed in the synthesis of nereistoxin, Konishi undertook the preparation of a large number of derivatives and analogs of 4-(N,N-dimethylamino)-1,3-propanedithiol (dihydronereistoxin) [66]. Those compounds in which the thiol groups were alkylated, or in some other way prevented from regenerating the parent thiol *in vivo*, were inactive. A variety of compounds in which the SH group was converted to SCN, SCOCH$_3$, or any group that could generate dihydronereistoxin in the insect exhibited insecticidal activity. Synthesis work also established that dimethylamino was the best of a number of amino groups in the 4-position for providing high insecticidal activity. Nereistoxin and active relatives were found to paralyze adult American cockroaches by blocking action in the central nervous system [67]. More recent studies have shown that nereistoxin acts as a partial agonist on nicotinic acetylcholine receptors at low concentrations, and as an inhibitor of the receptor channel at higher concentrations [68].

From the synthesis program, two compounds, the *bis*-amide *16* and, later, the *bis*-benzensulfonyl derivative *17* were developed for commercial application by the Takeda Chemical group. The amide *16*, cartap, proved effective in the field on the rice stem borer (*Chilo suppressalis*) and a number of other insects [66, 69]. The benzenethiosulfonate *17*, bensultap, is especially effective against the Colorado potato beetle (*Leptinotarsa decemlineata*) [70]. Both compounds proved less toxic to mice than nereistoxin (66).

A third product, thiocyclam (*18*), has been developed by Sandoz and appears to have an activity spectrum similar to the Takeda compounds [71].

By reacting bensultap under basic conditions with a series of active methylene compounds, e.g. *p*-chlorophenylacetonitrile, a large number of dithianes, of which *19* (Figure 6) is an example, were prepared by the Takeda chemists [72]. Dithiane (*19*) and other compounds in the series showed good activity against several insect species, e.g., fifth instar rice stem borers by topical application at 50 µg/g and third instar common cutworms (*Spodoptera litura*) on soybean seedlings sprayed at 500 ppm. Reversion to dihydronereistoxin *in vivo* seems highly probable with these compounds.

Of interest relative to the nereistoxin development was the discovery of 4-methylthio-1,2-dithiolane (*20*), and 5-methylthio-1,2,3-trithiolane (*21*) as components that contribute to the ability of the freshwater plant *Chara globularis* to inhibit photosynthesis in surrounding plants [73]. The dithiolane, known as charatoxin, was later shown to be insecticidal, although, by topical application, it was about tenfold less active than nereistoxin against the housefly ($LD_{50} = 2$ µg/fly) and half as active against the weevil *Sitophilus granarius* ($LD_{50} = 10$ µg/weevil) [74]. Synthesis work established that insecticidal activity was retained with 2-methylthio-1,3-propanedithiol or with a derivative, such as the bis-thioacetate, which presumably regenerates the dithiol *in vivo* [74]. Somewhat surprisingly, the bis-thioacetates in which the methylthio group was replaced by methoxy or ethyl (but not phenyl) groups retained some insecticidal activity [74]. Charatoxin and some of its precursors were less potent than nereistoxin in their actions on nicotinic acetylcholine receptors and appeared to bind to a low-affinity noncompetitive blocker site [68].

The Takeda group recently synthesized a series of 4-alkylthio-1,2-dithiolanes [75]. Of these, 4-ethylthio, 4-isopropylthio, and 4-*n*-octylthio-1,2-dithiolanes gave 100% kill of small brown planthoppers (*Laodelphax strialellus Fallen*) at 50 ppm on rice seedlings. However, this replacement of an alkylthio group for the dimethylamino group of nereistoxin gave substantially lower activity against larvae of the rice stem borer and the common cutworm [75].

The Takeda people also prepared a series of 5-alkylthio-1,3-dithianes for insecticidal evaluation and, as in the case of the 5-dialkyl-amino-1,3-dithianes, active compounds incorporated electron-withdrawing groups at the 2-position [76]. Activity was improved by replacing a methylthio by an isopropylthio or a *t*-butylthio group in the 5-position. Among the best compounds, giving 100% mortality of third instar northern house mosquitos (*Culex pipiens molestus* Forskee) at 5 ppm and third instar small brown planthoppers on rice seedlings at 50 ppm was 2-cyano-2-piperidinocarbonyl-5-isopropylthio-1,3-dithiolane (22, Figure 6) [76].

Of interest relative to the discovery of nereistoxin and charatoxin was the recognition of asparagusic acid (23, Figure 6) as a nematocidal constituent in the roots of asparagus [77].

Somewhat more removed but suggested by the position of the nitrogen atom relative to the sulfur atoms are the recently described xenorabdins [78]. This series, members of the pyrrothine family of antibiotics, were obtained from cultures of *Xenorhabdus* spp., bacteria symbiotically associated with insect-pathogenic nematodes. One of these, 24 (Figure 6), in a larval feeding assay on surface-treated artificial diet against the Australian native budworm (*Heliothis punctigera*), gave an LC_{50} of 60 µg/cm^2 and considerable weight loss for survivors [78].

D. Sabadilla

Sabadilla, long known for its insecticidal properties, is the dried powdered ripe seeds of *Schoencaulon officinale* A. Gray also known as *Sabadilla officinarum* Brant and *Veratrum sabadilla* Retz [79]. The major insecticidal components of sabadilla are veratridine (*26*), the 3-veratroyl, and cevadine (*27*), the 3-angenoyl esters of veracevine (*25*, Figure 7) [79]. Veratridine is one of several alkaloids, including aconitine and grayanotoxin, identified as neurotoxins that affect sodium ion channels in excitable membranes [80 and references therein].

An extensive study on the effect of varying the nature of the acyl group attached to the 3-position of veracevine on insecticidal activity and toxicity to mice has been reported [80]. Of a series of benzoyl esters, the 3,5-dimethoxybenzoyl derivative (*28*) was most active. The LD_{50} by topical application was 1.5 and 0.51 µg/g to houseflies (pretreated with piperonyl butoxide) and milkweed bugs (*Oncopeltus fasciatus*), respectively. Veratridine gave values of 6.6 and 2.6 µg/g, respectively, in the same test. In the aliphatic ester series, cevadine remained the most active with an LD_{50} of 10 µg/g against houseflies and 0.51 µg/g

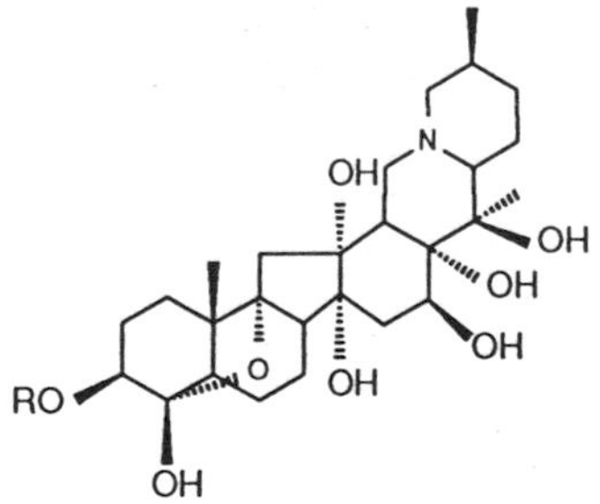

25: R = H   veracevine
26: R = 3,4-(MeO)$_2$PhCO   veratridine
27: R = Me   cevadine

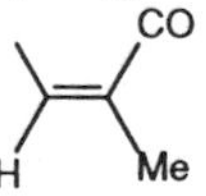

28: R = 3,5-(MeO)$_2$PhCO

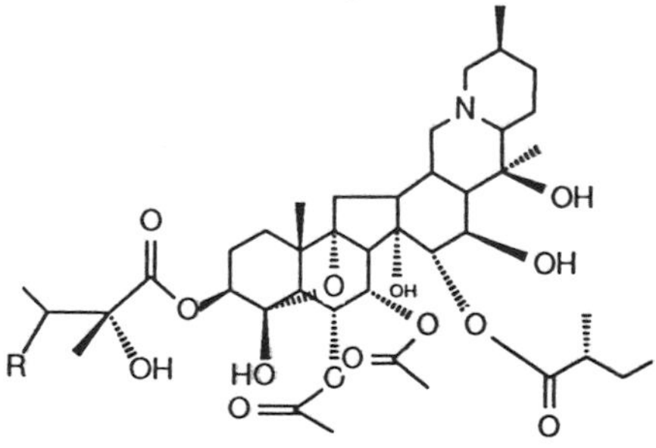

29: R = H   Protoveratrine A
30: R = OH   Protoveratrine B

Figure 7   Insecticidal *Veratrum* alkaloids. (From Ref. 80.)

for milkweed bugs. Veratridine, cevadine, and the ester *28* were toxic to mice at levels of 9.0, 5.8, and 7.5 mg/kg, respectively, by intraperitoneal administration. Veracevine itself was nontoxic to the insects and mice at the upper levels used in the study [80].

The closely related protovaratrines A (*29*) and B (*30*), which are components of "hellebore," the dried rhyzomes of *Varatrum album* (79), were also examined in the veracevine ester study [80]. Only protovaratrine A proved highly insecticidal with an LD_{50} of 1.6 µg/g for the housefly (with PB) and 2.2 µg/g for the milkweed bug. It was also the most toxic to mice with an LD_{50} of 0.20 mg/kg.

E. Ryania

Since the earlier review by Crosby [81], considerable work has been accomplished in isolating, identifying, and establishing the relative insecticidal activities of the compounds of ryania, the powder obtained by grinding the dried plant *Ryania speciosia* Vahl (Flacourtiaceae). In a recent study of ryania, the Casida group, utilizing radial TLC and preparative HPLC, isolated 10 of 11 compounds previously identified and provided an HPLC and ^{1}H NMR procedure for monitoring different lots of powders [82]. Ryanodine (*31*) and dehydroryanodine (*32*, Figure 8) comprised over 80% of two ryania preparations analyzed and were among the most active of the components by injection into adult house-

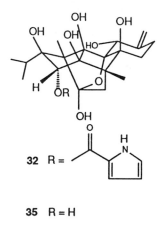

Figure 8 Ryanoids.

flies, pretreated with piperonyl butoxide, and much the more active in diet using first instar flour beetles (*Tribolium castaneum*; LD_{50} of 0.10 ppm for ryanodine). They were also the most toxic by intraperitoneal application to mice. Of interest, the naturally occurring pyridine-3-carboxylate analog (*33*) of ryanodine showed little activity [82]. The toxic action against the flour beetles and mice for the 10 compounds paralleled their ability to bind to the calcium release channel protein of muscle sarcoplasmic reticulum, possibly preventing channel closing, and a mode of action which defines ryanodine's function as a muscle poison [83, 84].

In other studies, which included a number of ryanodine derivatives and degradation products, ryanodol (*34*) and didehydroryanodol (*35*) were shown to have good knockdown activity when injected into houseflies and cockroaches even though toxicities to mice were low and there was little binding to the ryanodine receptor [83–85].

A total synthesis of ryanadol has been reported [86].

F. Pyrroles

The compound dioxapyrrolomycin (*36*, Figure 9) was isolated and identified from fermentation broths of *Streptomyces* strains in quick succession by three different groups. In the case of Meiji Seika Kaisha [87] and SS Pharmaceutical scientists [88], the isolation was directed using antibiotic screens. Dioxapyrrolomycin represents one of a series of anti-

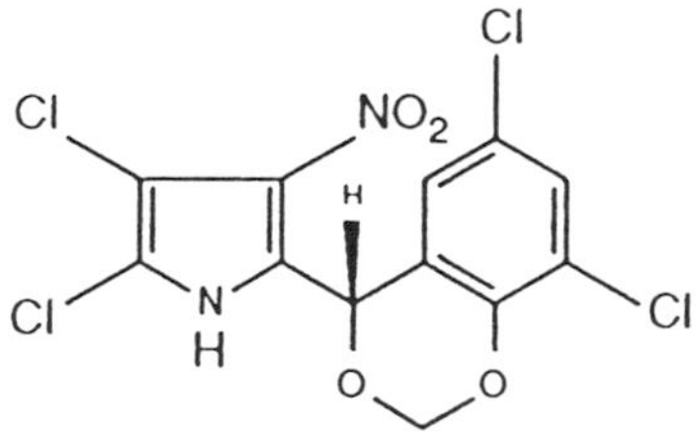

Figure 9 Dioxapyrrolomycin (*36*): Pesticidal activity: (Reprinted with permission from Ref. 91. Copyright 1990.)

Species	LC_{50} (ppm*)
Southern armyworm (*Spodoptera eridania*)	40
Tobacco budworm (*Heliothis virescens*)	32
Two-spotted mite (*Tetranychus urticae*)	10
Western potato leafhopper (*Empoasca abrupta*)	>100

*Leaf-dip assay.

biotic halogen-containing nitropyrroles labeled pyrrolomycins coming from a number of studies reported by the Meiji Seika group [89]. At American Cyanamid, the initial interest was insecticidal activity, which was noted for the crude isolates, and this screen directed the isolation of *36* [90]. In pure form, *36* demonstrated the activity shown in Figure 9 [91].

Dioxapyrrolomycin also proved toxic to mice with an oral LD_{50} of 14 mg/kg [91]. Despite this finding and the recognition that the toxic action of *36* derived from its ability to function as an uncoupler of oxidative phosphorylation, synthesis work was initiated in search of a safer pyrrole more effective against insects [91].

Some of the important structures resulting from this effort are shown in Table 1. Insofar as the major electron-withdrawing group on the pyrrole ring was concerned, cyano was shown to be generally equal to or better than nitro. The presence of an electronegatively substituted phenyl ring in place of the more complex substituent in the 2-position of *36* was found to give a very effective series of novel insecticides. Indeed, compounds such as *37*, *39*, and *40* showed caterpillar activity superior to the "lead" *36*, and *39* proved an effective acaricide as well. When the halogen in the 5-position of the pyrrole ring was replaced by a CF_3 group to give *41*, the result was a potent broad-spectrum insecticide/acaricide.

Although pyrrole (*37*) was nonphytotoxic, the more broadly active analogs such as *39* and *41* produced unacceptable plant damage. As a possible remedy, an examination of N-derivatization was undertaken [91]; such a strategy had been used earlier to circumvent similar problems with a series of insecticidal benzimidazoles [92]. Results of a segment of this work are summarized in Table 2. When pyrrole *37* (Table 1) was converted to the *N*-methyl (*42*) or to the N-ethoxymethyl (*43*) derivative, there was no observable change in spectrum or level of activity. When an α-methyl group was introduced into the N-alkoxymethyl substituent to give, e.g., *44* the level of lepidopteran activity was retained and the spectrum was broadened to include the mites and leafhoppers. Ethoxymethylation of the phytotoxic pyrrole *39* produced a compound (*45*) that proved safe to grass species but remained damaging to broadleaves. Also, mite activity noted for *39* was lost, whereas leafhopper activity was gained. Ethoxymethylation of *40* gave, in *46*, an improved broad-spectrum pesticide.

When the highly phytotoxic pyrrole *41* was ethoxymethylated, the plant-safe broadly active compound *47* resulted [91, 93]. Despite its potent activity against insects and mites, *47* proved only moderately toxic with an oral LD_{50} of 626 mg/kg for combined male and female rats

Table 1 Insecticidal Activity for Selected Pyrroles

	Army Worm[a]	Bud Worm[b]	Mite[c]	Leaf Hopper[d]
37	+++ (4.7)	++ (15)	0	0
38	+++ (8.2)	++ (63)	0	++
39	+++ (1.0)	+++ (9.4)	+++ (7.0)	+
40	+++	++	±	0
41	+++ (3.5)	+++ (3.6)	+++ (2.9)	++ 4.0

a: *Spodoptera eridania*; b: *Heliothis virescens*; c: *Tetranychus urticae*; d: *Empoasca abrupta*. Numbers in parentheses are LC_{50} values in ppm. (Reprinted in part with permission from Ref. 91. Copyright 1992.)

[94a]. Dermal toxicity to rabbits exceeded 2000 mg/kg and it was nonmutagenic in the modified Ames and the Chinese hampster ovary tests [94a]. The ability of insects to convert 47, a prodrug, to 41, a potent uncoupler of oxidative phosphorylation, has been demonstrated [94b].

Table 2 Insecticidal Activity for Derivatized Pyrroles

	Army Worm[a]	Bud Worm[a]	Mite[a]	Leaf Hopper[a]
42	+++	++	0	0
43	+++	++	0	0
44	+++ (3.6)	++± (14)	++	++ (3.5)
45	+++ (2.7)	+++ (10)	0	+++ (2.0)
46	+++ (3.6)	++± (7.3)	+++ (6.0)	++ (2.2)
47	+++ (2.6)	+++ (7.5)	++++ (1.6)	++± (0.92)

a: See footnote, Table 1. (Reprinted in part with permission from Ref. 91. Copyright 1992.)

By leaf-dip assay on cotton, 47 was as active as the pyrethroid cypermethrin against third instar tobacco budworms and about 1.4-fold less active against first instars [95]. By contact exposure on glass, it was tenfold less active. Of considerable importance was the demonstration of a lack of resistance of two strains of pyrethroid-resistant budworms to the effectiveness of 47 [95].

Results of early field tests with 47, designated as AC 303,630, have been published [96]. Necessary efficacy, toxicological, and environmental studies to support registration of AC 303,630 for use on a variety of crops and ornamentals are being advanced [96]. The synthesis of AC 303,630 has been described by Kuhn and co-workers [97].

G. Stemofoline

Extracts of the roots of Stemonaceae plants have been used for the control of agricultural insects in China [98]. Stemofoline, an alkaloid isolated from the leaves and stems of *Stemona japonica* Miq., shown by Irie and co-workers to have structure 48 (Figure 6), has been found to be the major insecticidal component [99]. When fed in artificial diet at 100 ppb to fourth instar silkworm larvae (*Bombyx mori* L), there were immediate toxic symptoms and death within 24 hr. Conversely, stemofoline was inactive at 100 ppm in artificial diet fed to cabbage armyworms (*Mamestra brassicae*).

V. UNSATURATED AMIDES

A. Background

In his 1971 review [100], Jacobson described a number of plant-derived unsaturated isobutylamides isolated from the families Compositae and Rutaceae, for example, pellitorine and affinin (49 and 50, Figure 10), and their insecticidal properties. As a group, they were characterized as pungent compounds which promote numbness and salivation when tasted. Although preparations of ground plant material containing the purported isobutylamide toxicants had found limited use in some areas for insect control, no isolated or synthesized examples were in use [100]. Chemical stability of the pure isolates was reportedly poor; even affinin, described as "by far the most active and stable of the natural isobutylamides thus far isolated and identified," showed a marked tendency to polymerize although it could be kept at 5°C in hydrocarbon solvent [100]. In addition to work establishing methods for the synthesis

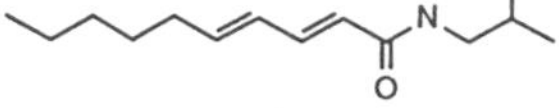

49 Pellitorine

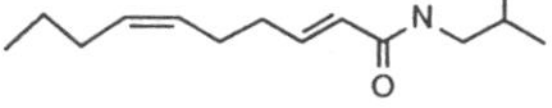

50 Affinin

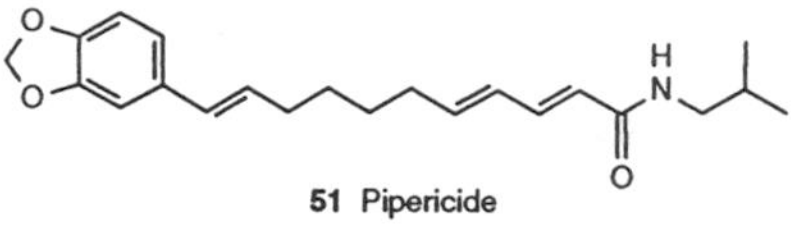

51 Pipericide

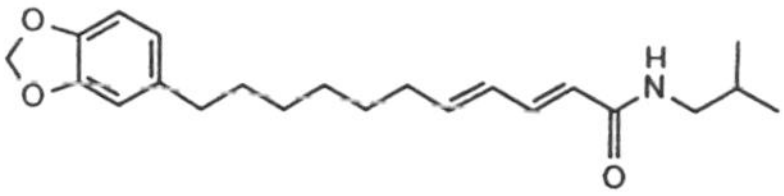

52 Dihydropipericide

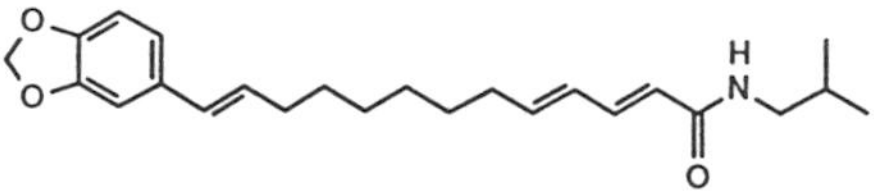

53 Guineensine

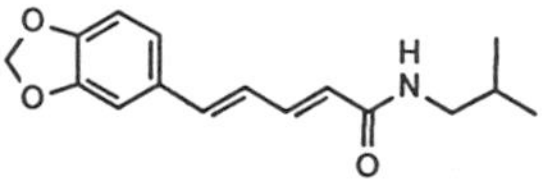

54 Piperine

Figure 10 Insecticidal unsaturated amides.

of these compounds, much of it coming from the laboratory of L. Crombie, some considerable effort had also been made to prepare and test synthetic analogs. These are tabulated in the Jacobson review [100].

Recently, as a prerequisite to undertaking a synthesis program based on what they describe as insecticidal lipophilic amides, Elliott and co-workers developed a broader definition of the inclusive structure [101a]. These were to be naturally occurring amides in which the carbonyl is attached to a chain of at least three acyclic carbons, and the nitrogen to a carbon of a simple group containing up to 12 carbons. In addition, "a structural relationship should be discernible with the amides found in Piperaceae" [101a]. This latter reference is to the work of Miyakado and co-workers who, using the adzuki bean weevil (*C. chinensis*) as a guide, isolated the three isobutylamides, pipericide, dihydropipericide, and guineesine, from the fruit of the black pepper, *Piper nigrum* (*51, 52, 53,* Figure 10) [102, 103]. These, rather than piperine (*54,* Figure 10) reported earlier [104], proved to be the insecticidally active components. Results, including comparisons with pellitorine and pyrethrins, are shown in Table 3. The KT_{50} number in Table 3 is the time that 50% of the weevils remain immobilized. Inclusion of the Piperaceae amides within the definition eliminates amides of long-chain saturated acids which do have mosquito larvacidal activity but undoubtedly work by other mechanisms [105].

The search by the Elliott group turned up 172 amides (structures not included) that fit their definition [101a]. Twenty-eight were reported

Table 3 Insecticidal Activity and Knockdown Time of *Piperaceae*
Amides to Adult *C. chinensis* (Male) on Topical Application

Compounds	LD$_{50}$ (μg/insect)[a]	KT$_{50}$ (min)[b]
pipercide (51)	0.56	11.8
dihydropipercide (52)	0.23	30.0
guineensine (53)	0.36	20.5
51 52 53, 1:1:1	0.11	6.0
pellitorine (49)	6.46	>60.0
piperine (54)	>10.0	>60.0
pyrethrins	0.10	1.0[c]

[a]Mortalities were evaluated after 24 hr.
[b]0.1 μg/insect was applied.
[c]0.05 μg/insect was applied.
(Reprinted with permission from Ref. 102. Copyright 1989.)

to have at least a trace of insecticidal activity and, of these, 23 were
isobutylamides and 5 were piperidides [106]. A recent review by Greger
provides the structures of over 50 purely olefinic and some eighty acetylenic amides along with their plant sources [107]. The acetylenic compounds generally included olefinic bonds as well. This broad collection
of amides are derived mostly from various combinations of 14 amines
with 85 acids.

B. Synthesis/Activity

Serious efforts have been made recently to ascertain just what structural
features contribute to activity of the insecticidal amides. By systemati-
cally varying carbon chain length, changing the position and degree of
unsaturation, replacing methylene groups by oxygen, incorporating
unsaturation into aromatic systems, varying the structure of the amide
nitrogen substituent, and other devices, significant progress has been
made in finding new active analogs. Key contributors have been Elliott
and co-workers at Rothamsted [101, 106], Blade and co-workers at the
Wellcome Research Laboratory [108] and Miyakado and co-workers at
the Sumitomo Chemical Company [102, 103]. Useful recent synthetic
methodology has come from Crombie's laboratory [109, 110].

In order to avoid some of the instability associated with many of
the polyenic natural amides, the Elliott group turned to partial replacement by aromatic systems [101a]. Isobutylamide *55* (Figure 11), which
had been reported earlier [111], was prepared and shown to have some

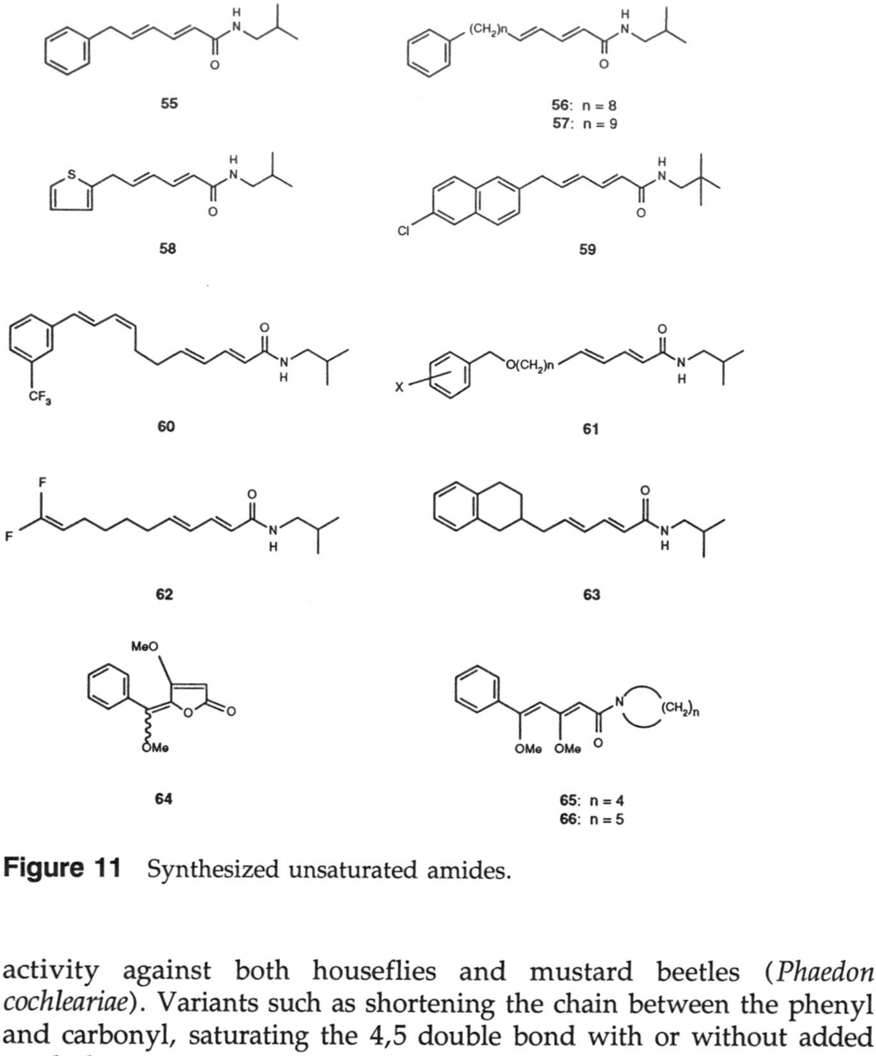

Figure 11 Synthesized unsaturated amides.

activity against both houseflies and mustard beetles (*Phaedon cochleariae*). Variants such as shortening the chain between the phenyl and carbonyl, saturating the 4,5 double bond with or without added methylene groups, or replacing the terminal benzyl by a methyl, methylvinyl, phenoxy, benzoyl, phenoxymethyl, or phenylamino-methyl group gave compounds with no activity [101a, 101b]. Addition of methyl groups individually at C-2 to C-6 of 55 destroyed activity [101b].

Using isobutylamide 55 as the standard, the effects of 47 variants of the structure of the nitrogen substituent on activity were examined [101e]. Among the best were those shown in Figure 12. In general, loss of branching or inclusion of more than five or six carbons in the alkyl group led to reduced activity. A pair of amides prepared from resolved

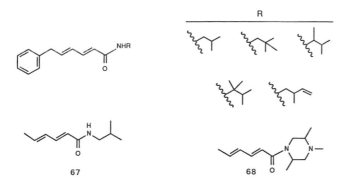

Figure 12 Nitrogen substituents.

1,2-dimethylpropylamine showed no improvement in activity over that prepared from the racemic amine [112]. A methyl group on nitrogen in place of the hydrogen destroyed activity. More limited studies by Miyakado [102] and Blade [108] and co-workers using other synthesized amides support the conclusion that simple alkyl groups of the type in Figure 12 provide the best activity. Of interest to note, however, is the lack of activity to houseflies and mustard beetles reported for the isobutylamide of (2E,4E)-hexa-2,4-dienoic acid (*67*, Figure 12) [101a], whereas nigrarillin (*68*), the amide from 3,4,6-trimethylpiperidine, provided fast knockdown of silkworm larvae by topical application and kill on feeding [113]. Introducing electron-withdrawing groups, e.g., alkyne, CN, or CF$_3$, at the carbon atom adjacent to nitrogen reduced activity [108]. A propesticide was prepared by replacing the amide hydrogen with groups such as phenylthio or ethoxycarbonyl, but these gave no advantage over the parent compounds [108]. A thioamide equivalent was likewise prepared but showed reduced activity [108].

Further work based on *55* as the model established that extending the alkylene chain length to give, for example, *56* and *57* (Figure 11), could lead to a small improvement in activity against mustard beetle, whereas housefly activity was lost [101c]. Small effects were also noted with changes in chain length for analogs terminated by methyl vinyl groups. Extended amides terminated by ethynyl, butadienyl, methoxyiminomethyl, and phenoxy groups offered no improvement over the phenyl series. Synthesized pellitorine and pipericide were only weakly active on these test species [101c].

Except for halogen substitution, there was a loss of activity when a variety of substituents were introduced on the phenyl ring of *55* [101d]. This included the *p*-methoxy derivative, which is identical to the natural product piperovatine. A fluorine atom in the meta position gave some improvement in activity against the mustard beetle. Follow-up work provided a series of disubstituted phenyl analogs in which activity was again limited to halogen substitution [101d]. Some improvement was noted with 3,4-dibromo and 3,5-difluoro substitutions, but overall variations in activity were limited to about tenfold. In this work, resort was made to preparing N-2,2-dimethylpropylamides rather than the isobutyl analogs, as this device reduced the propensity of the 2,4-dienamide to rearrange during the final synthesis step.

In continued work, replacement of a phenyl ring in 55 by other aromatic substituents was investigated using 54 variants [112]. Included was the naturally occurring 2-thienyl analog 58 (Figure 11) which proved less active than 55. The most important finding was the enhanced activity, especially against the mustard beetle, for several halo-substituted 2-naphthyl analogs. The best was the 6-chloro-2-napthyl compound 59 (Figure 11) which showed about a 40-fold improvement in activity versus 55 against the mustard beetle and better than half the activity of bioresmethrin against this species.

The Blade group developed a series of target molecules utilizing pipericide and other naturally occurring isobutylamides as models [108]. A part of this effort resulted in the tetraene 60 (Figure 11), which, although only weakly active, showed no cross-resistance to a kdr pyrethroid resistant strain of houseflies. Recognized metabolic instability for 60, combined with a desire for a more easily synthesized candidate, led to the examination of a series of benzyl ethers of structure 61 (Figure 11). Among the best were those where n=4 and x=3-CF$_3$ (housefly) and n=6 and x=3,5-bis-CF$_3$ (mustard beetle) although in each case activity was well below that of the pyrethroid permethrin.

A second type of amide (62, Figure 11), although not as active as the ether types, gave a notably more rapid knockdown. The penetration rate in houseflies was shown to be faster. Neurophysiological studies revealed a response akin to that of S-bioallethrin, a type-1 pyrethroid, whereas the ether's response was more like that of deltamethrin [108].

Several routes to the (2E,4E)-6-arylhexa-2,4-dienamides, e.g., 55, were developed by the Blade group [108]. However, as a strategy for avoiding the reactive methylene group in these amides, schemes were devised for the preparation of a series of benzocycloalkyl types. Of these, the tetrahydronaphthyl series (63, Figure 11) was among the most active, and placement of a halogen or CF$_3$ group on the benzene ring gave further improvement [108].

Table 4 Activity of *N*-isobutyl-12-(phenoxy)-2(2*E*,4*E*)-2,4-dodecadienamides Versus Adzuki Bean Weevil

X	LD$_{50}$ (µg/insect)
3,4-methylenedioxy	2.00
4-Cl	0.15
3-Cl	0.068
3-Br	0.038
3-CF3	0.043
Pyrethrins	0.10
Fenitrothion	0.045

(Reprinted, in part, with permission from Ref. 114. Copyright 1985.)

In synthesis work based on dihydropipericide (*52*, Figure 10) as the model, Miyakado and co-workers showed that saturation of one or both double bonds was detrimental to activity against the adzuki bean weevil [102]. A two-carbon extension, or resort to the more easily synthesized methylenedioxyphenyl ether shown in Table 4, reduced activity only slightly. Addition of a methyl group at C-3 improved activity of the ether. However, potency was raised most strikingly by replacement of methylenedioxy by halogen or CF$_3$. As shown in Table 4, a nearly 50-fold improvement was achieved by the 3-bromo or 3-CF$_3$ derivatives, an activity superior to pyrethrins and equivalent to the organophosphate fenitrothion [102, 114]. Further work with the 3-CF$_3$ analog showed it to be effective against the rice stem borer (*Chilo suppressalis*) and houseflies but inactive against the tobacco cutworm (*Spodoptera litura*).

Although pellitorine (*49*, Figure 10) was shown to have little activity against the housefly, mustard beetle, or adzuki bean weevil in the studies described above, it proved to be quite potent in diet fed to the pink bollworm (*Pectinophora gossypiella*) [115]. By using growth inhibition and toxicity to three lepidopteran species as the guide, pellitorine along with four other previously identified isobutylamides were isolated from the bark of the tree *Fagara macrophylla* (Rutaceae). Pellitorine, with an LD$_{90}$ of 24 ppm was the most toxic and the best growth inhibitor. It did not kill larvae of the tobacco budworm (*Heliothis virescens*), corn earworm (*Heliothis zea*), or fall armyworm (*Spodoptera frugiperda*).

From the species *Piper fadyenii*, C.D.C., Burke and co-workers isolated significant amounts of the butenolides *Z*- and *E*-fadyenolide (*64*, Figure 11) [116]. With amines, the butenolides afforded such unusual enamides as *65* and *66* (Figure 11). These were toxic to flour beetles (*Tribolium confusum*) although 100-fold less than bioresmethrin. They also inhibited ovogenesis in the tick *Boophilus microplus*.

Because of their ability to function like pyrethroids, i.e., to affect the nervous system and produce quick knockdown and kill, the lipophilic unsaturated amides retain much interest. This interest is heightened by the recognition that despite the similarity in action, pyrethroid-resistant insects retain susceptibility to these amides [117, 118]. Indeed, against a resistant (*super kdr*) strain of houseflies, amide *55* (Figure 11) and some analogs were more active than against a pyrethroid-susceptible strain [117].

Additional work is needed to find amides having the necessary combination of potency, breadth of action, and stability to warrant commercialization. Studies with isobutylamide *55* using rat and mouse liver microsomes, rat hepatocytes, and houseflies revealed that persistence and toxicity is limited by ready oxidative metabolism at the benzylic methylene and isobutyl substituents [119]. With *N*-isobutyl-10-phenyl-2*E*,6*E*,8*E*-deca-2,6,8-trienamide, rat microsomes and houseflies generate, through the epoxide, the 6,7-diol [120]. These metabolic reactions as well as the photochemical instability associated with the diene system of *55* [121] reinforce the already perceived need for a less sensitive replacement. A further understanding of degradative pathways for active analogs combined with continued structure/activity studies may yet deliver an agronomically useful product.

VI. CYCLODEPSIPEPTIDES

Cyclodepsipeptides, as described by Russel [122], are cyclic compounds in which the ring is composed of residues of amino and hydroxy acids joined by amide and ester bonds. Many are antibiotics; a number have been recognized for their insecticidal properties.

The destruxins are a series of cyclodepsipeptides consisting of five amino acids and an α-hydroxyacid. As pointed out by Gupta et al. [123], 17 destruxins have been identified from 3 different sources. The presence of what became the first of these materials, destruxins A and B, was reported in 1954 when the blood of silkworm larvae killed by the fungus *Metarhizium anisopliae* proved toxic when injected into healthy worms [124]. Isolation of these two insecticidal compounds was described in 1961 [125] and the structures were reported in 1964 [126]. Their isolation, structural identification, and synthesis (of destruxin B) have been reviewed [127].

From cultured strains of *M. anisopliae*, five destruxins were reported by Tamura et al. [127, 128]. Pais and co-workers reported on an additional nine analogs [129], and two more resulted from work by Gupta and associates [123]. The structures of 16 of these compounds are shown in Figure 13. In the more recent work [123, 129], the isolation procedure involved methylene chloride extraction of the fermentation broth followed by concentration and fractionation first on silica gel and then by reverse-phase HPLC. By using activity against the tobacco budworm (*Heliothis veriscens*) as a guide, the Gupta group was able to isolate 10 of the destruxins from their *M. anisopliae* culture [123].

Early work indicated that destruxins A and B were active against silkworms by injection only. They were inactive topically and larvae refused to eat mulberry leaves sprayed with aqueous solutions [127]. However, feeding to tobacco budworms was effective and Table 5 includes activity for several of the destruxins, including A and B, shown

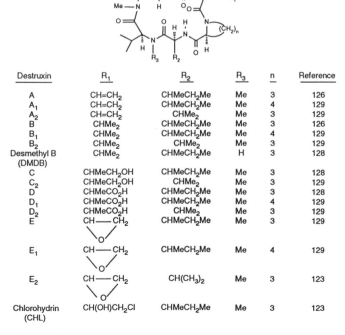

Destruxin	R_1	R_2	R_3	n	Reference
A	CH=CH$_2$	CHMeCH$_2$Me	Me	3	126
A$_1$	CH=CH$_2$	CHMeCH$_2$Me	Me	4	129
A$_2$	CH=CH$_2$	CHMe$_2$	Me	3	129
B	CHMe$_2$	CHMeCH$_2$Me	Me	3	126
B$_1$	CHMe$_2$	CHMeCH$_2$Me	Me	4	129
B$_2$	CHMe$_2$	CHMe$_2$	Me	3	129
Desmethyl B (DMDB)	CHMe$_2$	CHMeCH$_2$Me	H	3	128
C	CHMeCH$_2$OH	CHMeCH$_2$Me	Me	3	128
C$_2$	CHMeCH$_2$OH	CHMe$_2$	Me	3	129
D	CHMeCO$_2$H	CHMeCH$_2$Me	Me	3	128
D$_1$	CHMeCO$_2$H	CHMeCH$_2$Me	Me	4	129
D$_2$	CHMeCO$_2$H	CHMe$_2$	Me	3	129
E	CH——CH$_2$ \\O/	CHMeCH$_2$Me	Me	3	129
E$_1$	CH——CH$_2$ \\O/	CHMeCH$_2$Me	Me	4	129
E$_2$	CH——CH$_2$ \\O/	CH(CH$_3$)$_2$	Me	3	123
Chlorohydrin (CHL)	CH(OH)CH$_2$Cl	CHMeCH$_2$Me	Me	3	123

Figure 13 Destruxins isolated from cultures of *Metarhizium anisopliae*.

Table 5 Insecticidal Activity of Destruxins on Tobacco Budworm
(*Heliothis virescens*)

Destruxin	Concentration (p.p.m.)	% Mortality
A	312	100
	155	100
	78	80
	38	67
B	268	100
	125	40
	63	47
	30	13
E	220	100
	108	100
	93	100
	75	93
A2	250	100
	150	93
	60	73
B2	250	100
	128	73
	75	47
DMDB	225	73
	115	67
	50	40
CHL	265	53
	150	27
	53	33

(Reprinted with permission from Ref. 123. Copyright 1989.)

in Figure 13 [123]. In this case, solutions or suspensions were sprayed
on discs of garbanzo bean leaves that were infested with first instar
larvae. Mortality was assessed after 3 days. Of those tested, A and E
appear the most active, giving 100% kill in the 100-ppm range. These
compounds also showed high toxicity to mice by intraperitoneal injection [127].

A number of other cyclodepsipeptides have been reported to have
insecticidal properties. Valinomycin (Figure 14), originally isolated from
cultures of *Streptomyces fulvissimus* in 1955 [130], was later isolated from
Streptomyces roseochromogenes and shown to be effective against southern armyworm (*Prodenia eridania*) and Mexican bean beetle (*Epilachna
varivestus*) larvae on lima bean leaves treated at 1000 and 100 ppm, re-

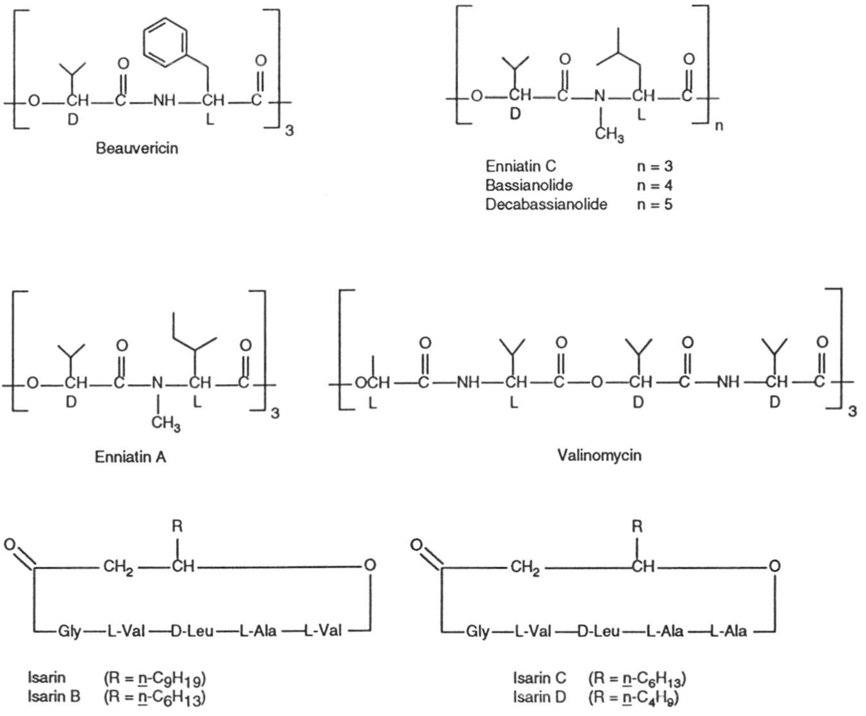

Figure 14 Cyclodepsipeptides.

spectively [131]. Activity against mites (*Tetranychus telarius*) extended down to 10 ppm. Valinomycin has been more recently identified as a mosquito larvicide [132].

Bassianolide (Figure 14) was isolated as the toxic principle from the mycelia of the two entomophagous fungi *Beauveria bassiana* and *Verticillium lecanii* [133]. Bassianolide was lethal to silkworm larvae at 8 ppm in artificial diet. In comparable tests, the earlier reported beauvericin (Figure 14) was inactive at 1000 ppm despite its similarity to bassianolide [134]. Bassianolide was synthesized along with the related cyclodepsipeptides enniatin C and decabassianolide (Figure 14) [135]. Whereas bassianolide killed silkworm larvae when injected above 5 μg/larvae, the others had no activity even at doses tenfold higher. However, enniatin A (Figure 14) in combination with closely related enniatin A1, has been isolated from cultures of *Fusarium avenaceum* present on balsam fir foliage and implicated as the possible causative agent in aiding the collapse of spruce budworm infestations [136].

From the fungus *Isaria felina*, isariins B, C, and D were isolated
and shown [137, 138] to be related to the previously described isariin
(Figure 14) [139]. By topical application at 400 ppm on larvae of the
greater wax moth (*Galleria mellonella*), only isarin D afforded complete
kill. Isariin and isariin B showed no activity.

From the perspective of potential for insecticidal application, perhaps most interesting is the high insecticidal activity (LD_{50} of 4 ppm
versus tobacco budworm larvae) reported for jaspamide (Figure 15)
[140]. This unusual cyclodepsipeptide, which incorporates the two rare
amino acids β-tyrosine and 2-bromoabrine in the unnatural D-config-
uration, was isolated from the bright orange sponge *Jaspis* sp. and structurally defined from NMR and x-ray studies [140, 141]. Total synthesis have been reported by two groups [142, 143]. From the x-ray
structure, Kahn and Su concluded that the two key pharmacophores
(i.e., the bromoindole and the phenol) are contained within a type-II
β-turn [144]. On this basis, the lactam 67 in Figure 15 was synthesized
as a mimetic of jaspamide. Although active by injection to tobacco
budworm and blowfly (*Sarcophaga fulcuta*) larvae, 67 was inactive by
topical application to budworms (144).

VII. LIGHT-ACTIVATED INSECTICIDES

Enhancement by light of the toxic effect of alkaloids on seeds, plants,
and other organisms was reported in 1888 [145]. The modern era of investigation began in 1971 with the study of the photodynamic action
of synthetic dyes, mostly halogenated xanthenes, on houseflies and has

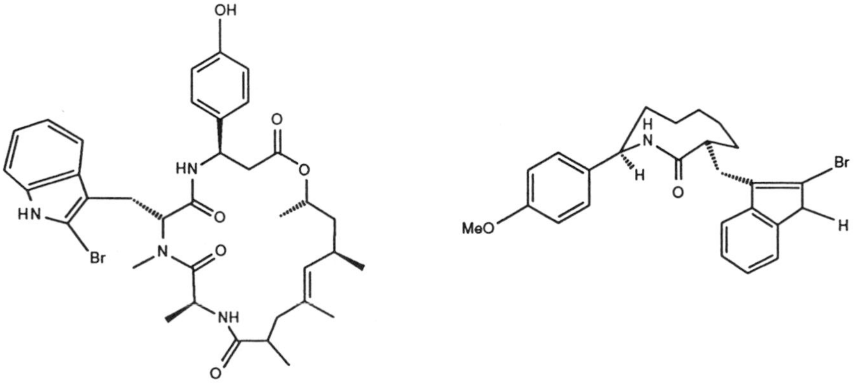

Figure 15 Jaspamide and jaspamide mimetic 67.

been reviewed by Heitz [145]. Efforts to define possible commercial insecticidal utility for erythrosin B (Synerid) has been reviewed by Lemke et al. [146].

Major work demonstrating the importance of light in promoting insecticidal activity for a variety of plant-derived compounds also began in the 1970s. The activity of α-terthienyl (α-T, 68, Figure 16) to nematodes held in an emulsion was shown to be dependent on irradiation with near-ultraviolet light [147]. Berenbaum demonstrated that the toxicity to armyworms (*Spodoptera eridania*) fed the furanocoumarin xanthotoxin (8-methoxypsoralen, 69) was markedly promoted by the presence of light [148]. Camm and co-workers established that plant-

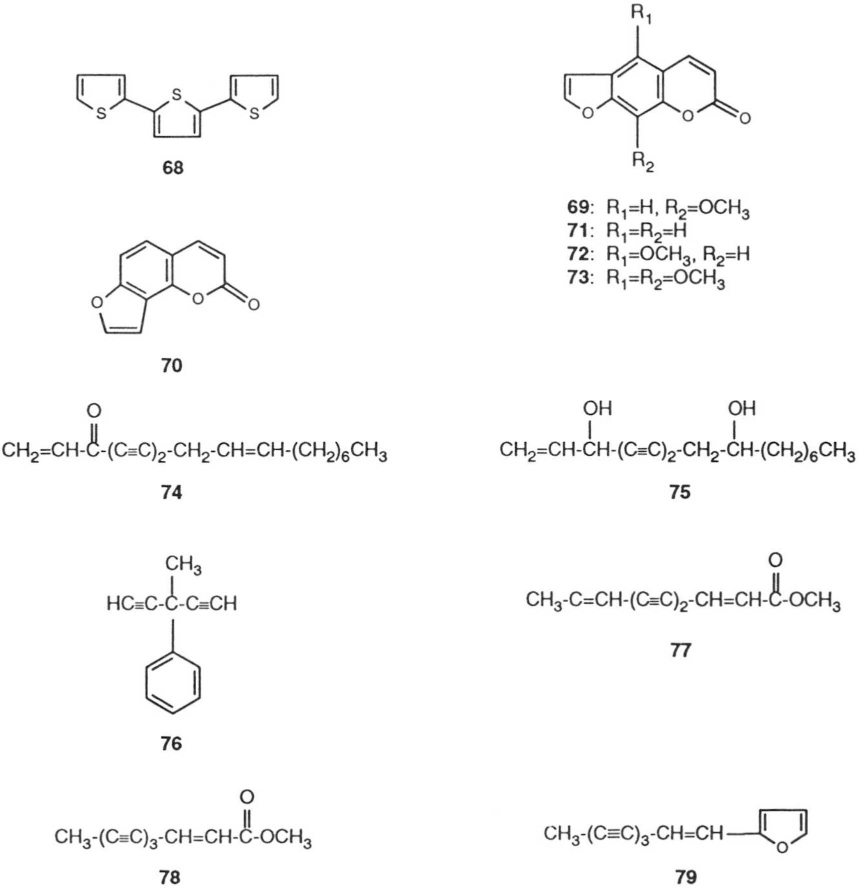

69: R_1=H, R_2=OCH$_3$
71: R_1=R_2=H
72: R_1=OCH$_3$, R_2=H
73: R_1=R_2=OCH$_3$

$CH_2=CH-\overset{O}{\overset{\|}{C}}-(C\equiv C)_2-CH_2-CH=CH-(CH_2)_6CH_3$

74

$CH_2=CH-\overset{OH}{\overset{|}{CH}}-(C\equiv C)_2-CH_2-\overset{OH}{\overset{|}{CH}}-(CH_2)_6CH_3$

75

$HC\equiv C-\overset{CH_3}{\overset{|}{\underset{|}{C}}}-C\equiv CH$

76

$CH_3-C=CH-(C\equiv C)_2-CH=CH-\overset{O}{\overset{\|}{C}}-OCH_3$

77

$CH_3-(C\equiv C)_3-CH=CH-\overset{O}{\overset{\|}{C}}-OCH_3$

78

$CH_3-(C\equiv C)_3-CH=CH-$

79

Figure 16 Light-activated insecticides.

produced polyacetylenes have photosensitizing properties [149]. Since then, a variety of other biologically active phototoxic compounds of differing chemical types have been derived from plant sources [145, 150]. Most prominent as insecticides are the furanocoumarins, the acetylenes, and the thiophenes.

A. Furanocoumarins

The furanocoumarins are typically found in plants of the families Rutaecae and Umbelliferae [148] although their range is much wider [151]. Armyworms are unable or unwilling to survive on wild parsnip, *Daucus carota* L. (Umbelliferae). This fact and the knowledge that wild parsnip contained the furanocoumarin xanthotoxin, known for its ability on light activation to react with pyrimidine bases in RNA and DNA, led to Berenbaum's experiment with armyworms [148]. Unlike the armyworm, swallowtail butterfly larvae (*Papilio polyxenes*) can survive on wild parsnip because they are able to detoxify and excrete xanthotoxin [152]. With this species, however, growth rate and fecundity are reduced by the angular furanocoumarin angelican or isopsoralen *70* (Figure 16) [153]. These toxicity differences are explainable by different rates of metabolism for the two furanocoumarins, and pertinent aspects of these studies have been reviewed [154, 155]. These reviews also describe the broader biological actions and metabolic transformations of the furanocoumarins in humans and animals.

Four linear furanocoumarins isolated from *Thamnosma montana* (Rutaceae) were assayed in diet against tobaccco budworm (*H. virescens*) larvae. Three of these, psoralen (*71*), xanthotoxin (*69*), and bergapten (5-methoxypsoralen, *72*) were phototoxic, whereas isopimpinellin (5,8-dimethoxypsoralen, *73*) was nonphototoxic. All four compounds produced the same level of growth inhibition whether the tests were conducted in a light/dark or a totally dark regimen [7]. The growth inhibition resulted from an antifeedant response that the compounds elicited equally. Dark "toxicity" has been observed for other furanocoumarins [7].

B. Polyacetylenes

Marchant and Cooper have reviewed the relationship between structure and biological activity for a number of naturally occurring polyacetylenes [156]. Most of the biological evaluation for these compounds has been directed at a variety of microorganisms rather than toward insects. Many of the plant-produced biologically active acetylenic compounds do not require photoactivation. Falcarinone (*74*) and falcarindiol (*75*, Figure 16), compounds characteristic of the Apiaceae and Araliaceae, are representative [156]. Of interest, falcarindiol was isolated from the common carrot, *Daucus carota*, and in addition to being toxic to *Daphnia magna* Strauss, it was also quite toxic, by injection, to mice [157].

Although all photoactivated polyacetylenes appear to come from the Asteraceae, this family also produces nonphotoactivated polyacetylenes [156]. Among these, one of the simplest and more interesting is 3-methyl-3-phenyl-1,4-pentadiyne (*76*). Isolated from the desert plant *Artemisia monosperma* Delile (Saleh) as an oil, it is as active as DDT against houseflies and cotton leafworm (*Spodoptera litoralis*) larvae and fivefold more active against rice weevils (*Sitophilus oryzae*) [158]. Matricaria ester (*77*), isolated from the *Solidago altissima* L., functions as an antifeedant to phytophagous insects [156]. On the other hand, the reported [159] ovicidal activity of cis-dehydromatricaria ester (*78*) to eggs of *Drosophila melanogaster* has been shown to be enhanced by long-wavelength UV light [160].

Over two dozen polyacetylenes from the Asteraceae have been extensively tested for photoactivity against a variety of organisms [156]. For aliphatic compounds, photoactivity generally requires three or more conjugated acetylenic bonds [156]. Thus, against mosquito (*Aedes atropalpus*) larvae, 2-(non-1-ene-3,5,7-triynyl)furan (*79*, Figure 16), the only triacetylene of three furanoacetylenes isolated from *Chrysanthemum leucanthemum* L., was the only one which approached α-T in light-promoted activity [161].

Experiments with a variety of microorganisms show that with phenylacetylenes at least two conjugated acetylenic bonds are needed if phototoxicity is to be observed [156]. A good example is phenylheptatryine (*80*, Figure 17), which is widely distributed in Asteraceae species [162] and readily available from the stems and leaves of *Bidens pilosa* [163]. This pantropical weed is noted for its broad antibiotic properties. Phenylheptatryine behaves as a potent *in vitro* photoactivated toxicant to a variety of fungi, yeasts, and bacteria [164]. Its role as an antifeedant in Asteraceae is suggested by demonstration of reduced feeding and weight gain in larvae of the polyphagous insect *Euxoa messoria* when incorporated into artificial diet at 10–300 ppm [162]. In addition, by monitoring for ovicidal activity using *Drosophila melanogaster*, phenylheptatryine was isolated as one of two active components from extracts of *Coleopsis lanceolata* L. [165]. The second toxicant, about half as active, was the thiophene *81*, which is presumably derived from phenylheptatryine. As an ovicide, using eggs of *Drosophila melanogaster*, phenylheptatryine was active in the dark, but 37-fold better on light activation [166]. In the same test, α-T showed a light-induced 4333-fold increase in ovicidal activity, whereas xanthotoxin was active only in the presence of light.

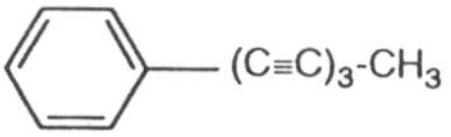

80

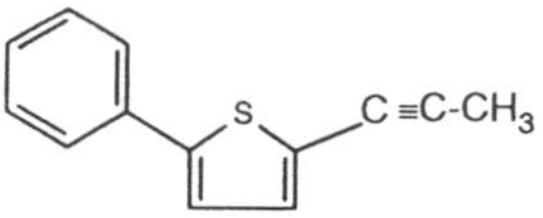

81

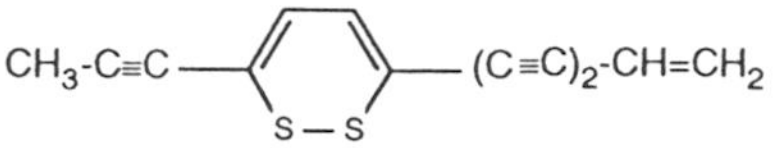

82

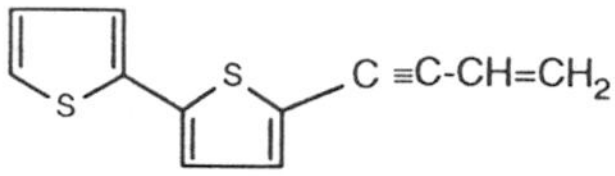

83

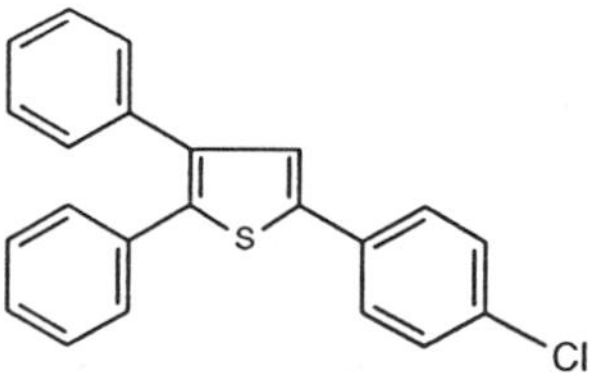

84

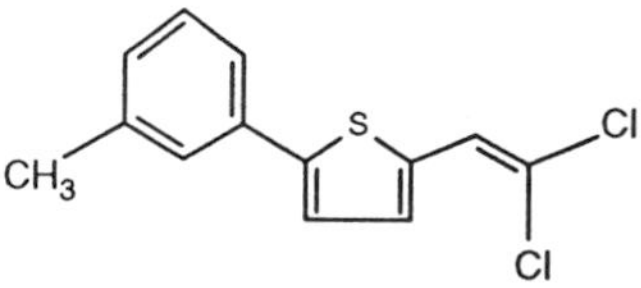

85

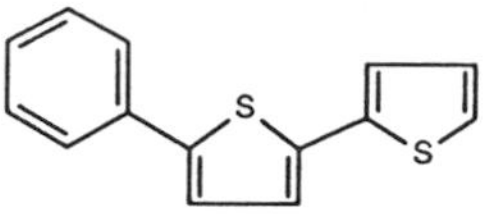

86

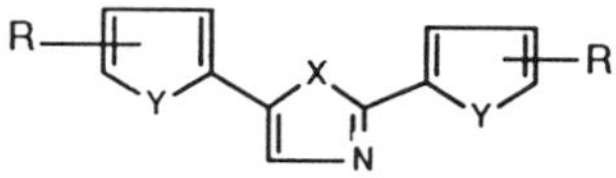

X=O or S
Y=CH=CH or S

87

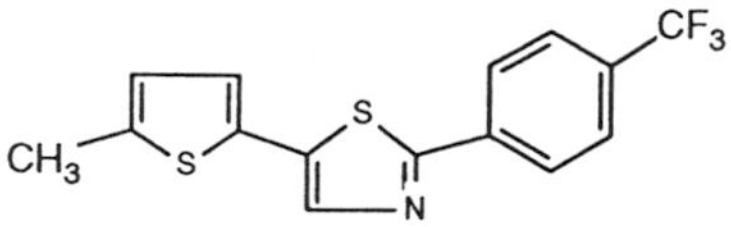

88

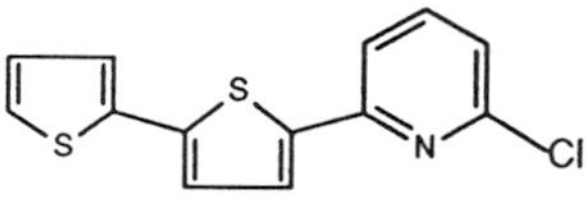

89

Figure 17 Light-activated insecticides.

Another group of acetylenic compounds of great interest for their apparent contribution to the medicinal properties of a variety of plants within the Compositae are the so-called thiarubrines, e.g., 1-(2-methyl-ethynyl)-4-(hexa-1,3-diyn-5-enyl)-2,3-dithiacyclohexa-4,6-diene (thiarubrine A, *82*, Figure 17) [164, 167]. The compound is strongly antibiotic. The soil nematode *C. elegans* was killed by exposure to 0.03 ppm thiarubrine A with irradiation versus 5 ppm in the dark [164]. It also exhibited high activity against the root knot nematode *Meloidogyne incognita*, but in this case light was not a factor [167].

C. Terthienyl

The nematocidal components of the roots of *Tagetes* (African marigold) were shown by Uhlenbroek and Bijloo to be the stable crystalline α-terthienyl (*68*) and the unstable oil, 5-(3-buten-1-ynyl)-2,2′-bithienyl (*83*, Figure 17) [168, 169]. The presence of these compounds has been demonstrated in a large number of species within the genus Compositae. In one study, 16 of 17 species shown to contain these thiophenes suppressed infection by the nematode *P. penetrans* [147]. The ability of α-T to function as a sensitizer and to convert triplet oxygen to the singlet state has been well demonstrated and accounts for its enhanced activity against a variety of organisms in the presence of light [147, 170]. The mystery of how α-T functions within a plant root system, in the dark, to control endoparasitic nemetodes has been explained by a cascade of reactions that begins with peroxidase activation stimulated when roots are wounded [147, 170].

Unlike the photoactivated polyacetylenes, α-T and related thiophenes are relatively stable compounds and were chosen by Arnason and co-workers for extensive investigation. The substance of that work to date is described in two recent reviews [171, 172]. As reported [171], a number of herbivorous insects are sensitive to topically applied α-T in the presence of near-ultraviolet irradiation or sunlight, including the larvae of *Manduca sexta* and *Pieris rapae*. On the other hand, *Heliothis virescens* and some other species are relatively insensitive. A major factor in this selection is accounted for by different rates of excretion. Oxidative processes involving cytochrome P450-based monooxygenases generate polar metabolites and render α-T nonphototoxic [171].

High potency for α-T was reported for light-exposed water-living larvae of the mosquito *Aedes aegypti* and the blackfly *Simulium verecundum*, i.e., an LD_{50} of 19 and 28 ppb, respectively [171]. An improved synthesis of α-T based on a nickel-catalyzed reaction of 2,5-dibromomothiophene with 2-bromothienylmagnesium bromide made α-T available for field scale work [173, 174]. Depending on the formulation, trials in eastern Canada at both boreal and deciduous forest sites resulted in reliable control of *Aedes* spp. larvae at an application rate of 50 or 100 g of α-T per hectare [171]. Good results were also achieved in Tanzania against the malaria mosquito *Anopheles gambiae* [171]. Because of concerns over the effects on nontarget organisms, the suitability of using α-T as a mosquito larvacide has been questioned [175]; the Arnason group presents a more optimistic outlook [172].

D. Terthienyl-based Synthetics

Uhlenbroek and Bijloo reported a number of synthesized dithienyl and terthienyl derivatives, e.g., 5,5"-dimethylterthienyl, to be active *in vitro* to larvae of the nematode *Heterodera rostochiensis* [176]. This was also among a number of synthesized analogs prepared by the Arnason group, but it proved much less active than terthienyl against the mosquito *Aedes atropalpus* and the brine shrimp (*Artemia salina*) [177]. Under light, only the 5-cyano (LD_{50}=79 nM) and the 5-formyl (LC_{50}=133 nM) derivatives of α-T were superior to α-T (LC_{50}=163 nM) against the mosquito. Overall results were reasonably well correlated with a combination of estimated partition coefficients and the measured rates of singlet oxygen production.

In a study of phenylsubstituted thiophenes, Relyea and co-workers found that monophenylthiophenes and tetrasubstituted thiophenes had little miticidal activity [178]. Of the diphenyl homologs, 2,4- and 2,5-diphenylthiophenes were most active; 2,5-diphenylthiophene was phytotoxic. Most of the work centered around 2,3,5-triphenylthiophenes, where light activation was demonstrated and electron-withdrawing groups on the phenyl rings generally enhanced activity. From this work, 5-(4-chlorophenyl)-2,3-diphenylthiophene (*84*, Figure 17), as UBI-T930 or triarathene, was selected as an experimental foliar acaricide for control of citrus rust mite (*Phyllocoptruta oleivora*). In field applications on citrus leaves and fruit, it afforded 9–12 weeks of control at 0.84 kg/ha. Commercial development appears to have been curtailed [179].

Using the terthienyl model, Roush and co-workers synthesized three chemical types with effective miticidal activities [180]. The first was a series of 5-aryl-2-dihalovinylthiophenes. Under light, the better compounds, e.g., 5-(3-methylphenyl)-2-dichlorovinylthiophene (*85*, Figure 17) and 5-(1-naphthyl)-2-dichlorovinylthiophene, were at least as good as α-T (LC_{50}=6 ppm) against the two-spotted spider mite (*Tetranychus urticae*). There was significant cross-resistance to a phosphate-resistant mite strain.

The second set of synthesized compounds, 5-aryl-2,2'-bithiophenes, were generally more active miticides than the 5-aryl-2-dihalovinylthiophenes [180]. 5-Phenyl-2,2'-bithiophene (*86*) itself had been reported earlier to be nematocidal [176]. In the present study, an LC_{50} of 4 ppm was reported against the two-spotted spider mite. A variety of phenyl ring substituents gave analogs with LC_{50}'s in the 0.1–1-ppm range. Small alkyl groups or a chlorine atom at the 5'-position also improved activity in some cases. Activity against phosphate-resistant mites was better than for the dihalovinyl series.

The third class of light-activated miticides incorporated an azole ring as the central component as represented by *87* (Figure 17). Thiazoles were found to be 30–40 times more active than corresponding oxazoles and, in the thiazole series, the 2,5-disubstituted compounds were 10–100 times better than the 2,4-disubstituted analogs. Further work honed in on the 2-aryl-5-thienylthiazoles as the most effective miticides. Based on its high activity to both susceptible and phosphate-resistant mites, i.e., LC_{50}'s of 0.9 and 2.5 ppm, respectively, and its good residual properties relative to other compounds including α-T, 5-(5-methyl-2-thienyl)-2-(4-trifluoromethylphenyl)-thiazole (*88*), as F5183, was selected for field evaluation. In field trials, F5183 gave good control to 23 days against two-spotted spider mites on cotton at 0.2 lb/acre. Against citrus rust mite on orange trees, 33 to 46 day control was reported at 0.025 to 0.1 lb/acre [180].

In a recent study by Sun et al., a variety of compounds modeled after α-T were prepared and evaluated for nematocidal and herbicidal activities [181]. These compounds failed to control the root knot nematode (*Meloidogyne graminicola*) by soil drench or foliar application on sorghum plants even though several, e.g., 2-(6-chloro-2-pyridyl)-5-(2-thienyl)thiophene (*89*), were highly active in a water screen against these nematodes under light.

VIII. CHROMENES

Chromenes (benzopyrans) and benzofurans of diverse structures are known from many species of higher plants, but the majority occur in the Asteraceae. Their distribution and biological activities along with the structures of 167 isolated compounds has been reviewed [182]. Outside of the precocenes, 6,7-dimethoxy-2,2-dimethylchromene and 7-methoxy-2,2-dimethylchromene, noted for their ability to induce precocious metamorphosis in some insects by destroying the gland that secretes juvenile hormone [183], several other chromenes have attracted attention as insecticides.

Species of the desert sunflowers of the genus *Encelia* (Asteraceae) elaborate three major acetylchromenes, i.e., 6-acetyl-7-methoxy-2,2-dimethylchromene (encecalin, *90*), 6-acetyl-7-hydroxy-2,2-dimethylchromene (*91*), and 6-acetyl-2,2-dimethylchromene (*92*, Figure 18) [184]. Encecalin has been shown capable, in artificial diet, of reducing feeding of larvae of bollworms (*Heliothis zea*), loopers (*Plusia gamma*), and cutworms (*Peridroma saucia*) to the point of starvation [184–186]. Even so, in a comparable test with bollworms, encecalin was not as effective as the precocenes in affecting growth rates and survivorship [185].

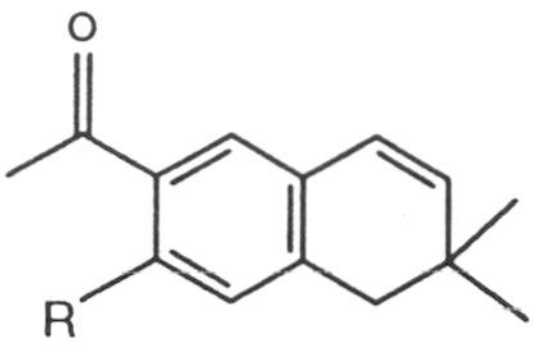

90: R=CH$_3$O
91: R=HO
92: R=H

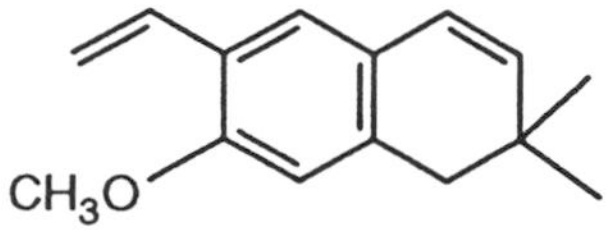

93

Figure 18 Insecticidal chromenes.

Encecalin coated on the inner surfaces of glass vials afforded LC$_{50}$'s by residual contact of 20, 3.0, and 12 µg/cm^2 for neonate milkweed bugs (*Oncopeltus fasciatus*), cutworms, and grasshoppers (*Melanoplus sanguinipes*), respectively [184]. Demethylencecalin (*91*) was significantly less active in this comparison although, by topical application to neonate grasshoppers, the approximate LD$_{50}$ of 38 nmol/insect was about half that of encecalin. In this test, demethoxyencecalin (*92*) was equivalent to encecalin.

A number of additional natural and synthesized chromenes were evaluated in the vial contact test [184]. Analogs having free hydroxyl groups or in which the chromene was reduced to the chromane proved significantly less active. None of the acetylchromenes showed antijuvenile hormone effects characteristic of the precocenes confirming an earlier finding for encecalin [187].

In another study, based on the observation that the tarweed *Hemizonia fitchii* A. Gray (Asteraceae) reduces mosquito populations, the volatile oil from this plant was fractionated to give, among other products, chromenes *90* and *91*, previously known from this plant [188], and 6-vinyl-7-methoxy-2,2-dimethylchromene (*93*) [189]. Against first instar *Culex pipiens*, the LC$_{50}$ was 1.8, 3.0, and 6.6 ppm for compounds *93*, *90*, and *91*, respectively. There was no effect on development noted for survivors through subsequent larval, pupal, and adult stages. Against milkweed bugs, encecalin caused 50% mortality applied topically at 10 and 11 µg to second and third instar nymphs, respectively. Compound *93* was about half as active; *91* showed no effect at 100 µg. Again, no unexpected developmental effects were observed on survivors.

In two articles [184, 189], the authors conclude with thoughts on the suitability of these chromenes as models for future work.

IX. MAMMEINS

The extensive work done on extracts of seeds of the evergreen tree *Mammea americana*, long known to be insecticidal, was recently reviewed by Crombie [190]. Until 1972, a number of closely related coumarins had been isolated and identified, but these were shown to lack the insecticidal activity afforded by the concentrates from which they were isolated [191]. Finally, the active component was isolated as a crystalline material comprised mostly of the coumarins of structure 94 (Figure 19) [191]. These differed from the inactive compounds in the series by having an acetoxy group in the 1' position where the others were unsubstituted. At the time of this discovery, the acetoxy coumarin surangin B (95, Figure 18) was identified as an antibacterial agent from the roots of *Mammea longifola* (192). Surangin B was also shown to be insecticidal and confirmed the importance of the acetoxy group [190, 191]. Coumarins 94 demonstrated uncoupling of oxidative phosphorylation at 0.05 µg/ml [191].

Useful synthesis of these coumarins are outlined in Crombie's review [190].

X. ACETOGENINS

Using lethality to brine shrimp for activity-directed fractionation, asimisin, a trihydroxy-bis-tetrahydrofuran fatty acid γ-lactone (Figure

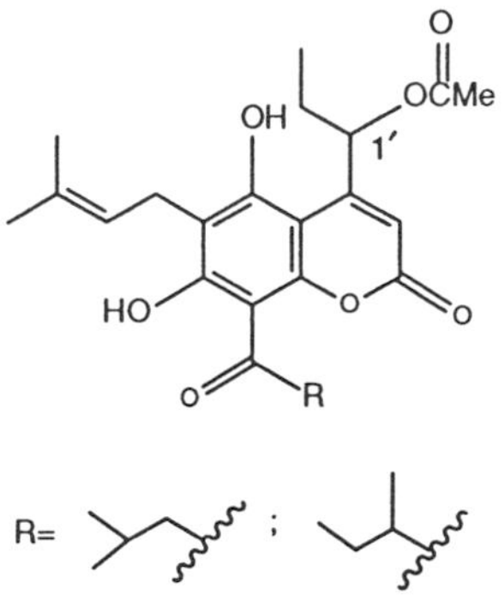

94 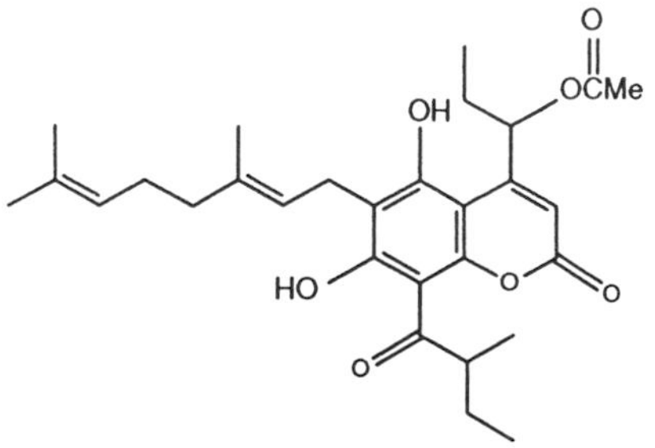   95

Figure 19 Insecticidal coumarins.

20) was isolated from an extract of the bark of the pawpaw tree *Asimina triloba* Dunal. (Annonaceae) [193]. Assimisin is a member of a class of bioactive compounds referred to as Annonaceous acetogenins the first of which, uvaricin, an antileukemic compound, was reported only as recently as 1982 [194]. A review of the status of this group of compounds is provided by Rupprecht and co-workers [195].

As insecticides, asimisin along with annonin I (Figure 20) appear to have been the most thoroughly evaluated and to be the most broadly active of the acetogenins. Several others [193], including bullaticin [196], sylvaticin [197], neoannonin [198], annonin IV [199], goniothalamicin [200], and annonacin [201] are reported to have some insecticidal properties. A patent directed to the control of insects for a broad range of the acetogenins has been issued [202].

Asimisin is active against the Mexican bean beetle (*Epilachna varivestis*) at 10–50 ppm, the cotton aphid (*Aphis gossypii*) at 500 ppm, mosquito larvae (*Aedes aegypti*) at 1 ppm, and the nematode *C. elegans* at 0.1 ppm [203, 204]. It appears to be ineffective against the southern armyworm (*Spodoptera eridania*), southern corn rootworm (*Diabrotica undecimpunctata*), and the mite (*Tetranychus urticae*) [204]. Assimisin proved effective as an antifeedant against the striped cucumber beetle (*Acalymma vittatum*) on melon leaves at 5000 ppm [203].

Recognizing the complexity of the structure of asimisin and the improbability of an economical synthesis, the use of an extract of the bark of Annonaceous trees as an insecticide has been suggested [203]. In one test, such a concentrated extract rich in asimisin but containing at least five other acetogenins was sprayed on rows of bush beans with a natural infestation of bean leaf beetles (*Cerotoma trifurcata*). After three applications over 10 days of 0.5% to 1.0% aqueous formulations,

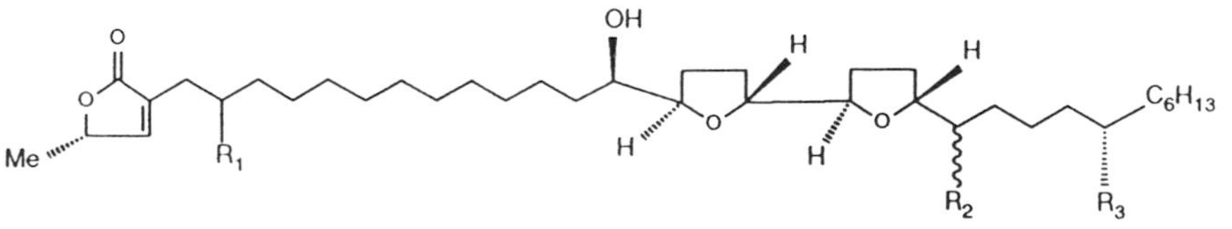

	R_1	R_2	R_3
Asimisin	OH	——— OH	H
Anonin I	H	······ OH	OH

Figure 20 Acetogenin structures. (From Ref. 199.)

there was no visible damage and growth was superior to the untreated row [203].

From the seeds of *Annona squamosa* L., annonin I, annonin IV, asimisin, and several other acetogenins were isolated [205]. From this study, annonin I proved to be identical to squamocin, which had been reported earlier from *A. squamosa* [206]. In a feeding test against the fruit fly (*Drosophila melanogaster*), annonin I showed good ovicidal and larvacidal activity at a rate of 125 µg/2g of diet [198]. Annonin I also gave complete control of mustard beetle (*Phaedon cochleariae*) larvae on cabbage plants, cabbage moth (*Plutella maculipennis*) larvae on cabbage, and aphids (*Myzus persicae*) on beans at 40 ppm [207]. As squamocin, additional activity has been reported by Ohsawa and co-workers [208].

There is some evidence to suggest that the insecticidal activity of the acetogenins is due to their ability to inhibit electron transport at complex 1 of the mitochondrial respiratory chain [209]. The strong cytotoxic action of these compounds raises concerns about their overall toxicity. In the early report on asimisin, the bark extract from which asimisin was isolated proved toxic to mice at 6.25 mg/kg [193].

XI. DOMOIC ACID

Domoic acid (*96*, Figure 21) isolated from the seaweed *Chondria armata*, and, to a lesser extent, α-kainic acid (*97*), from the seaweed *Digenea simplex*, exhibited insecticidal activity when injected into the American cockroach and when applied topically to the housefly and German cockroach [210]. The lethal dose for domoic acid to the American cockroach was 1.3×10^{-9} mol/insect, comparable to that of the pyrethroid phenothrin at 3.2×10^{-10} and that of DDT at 7.4×10^{-9} mol/insect. Kainic acid, at 9×10^{-8} mol/insect, was less effective. By topical application, 50% kill of German cockroaches required 1.9×10^{-9} mol/insect of domoic acid comparable to γ-BHC. The value for the housefly was 3×10^{-10} mol/insect, in the same range as γ-BHC and DDT [210].

Of a number of synthetic analogs, the domoic acid alcohol *98* was about one-tenth as active as domoic acid. Derivatives of α-kainic acid, including the dimethyl ester and N-alkylated analogs were inactive at 100 µg/g by injection in the American cockroach. However, the ketone *99* prepared by ozonolysis of kainic acid was similar to kainic acid in activity [210].

By using insecticidal activity against the American cockroach as the guide, three domoic acid isomers were isolated from *Chondria armata* in follow-up work [211]. They proved to be 12- to 24-fold less active than domoic acid.

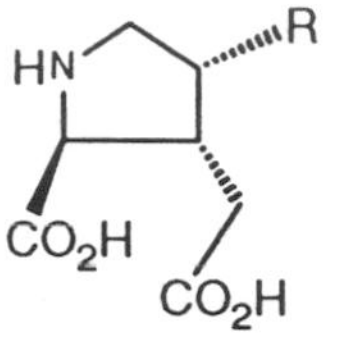

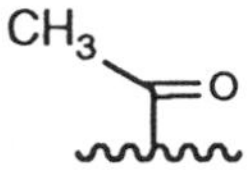

Figure 21 Domoic acid and related compounds.

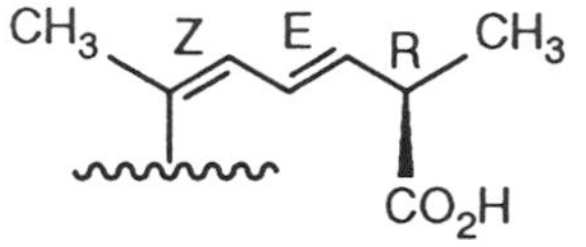

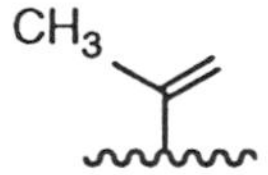

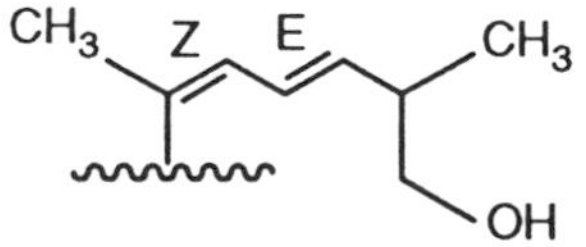

Mode of action studies in American cockroaches have shown that low concentrations of domoic acid enhance sensitivity of insect neuro-muscular junctions to glutamic acid causing contraction of the excised hindgut. The natural neuropeptide proctolin causes the same effect, and evidence is presented suggesting that domoic acid acts by binding to proctolin receptors [212].

XII. OTHER NOVEL STRUCTURES

The organophosphate insecticides, considered as strictly a man-made invention since their discovery by Schrader in 1937, are now known to have their origin in nature. Thus, from cultures of *Streptomyces anti-bioticus* strain DSM 1951 were isolated the cyclic phosphates *100* and *101* (Figure 22) [213]. In agreement with their high activities against *Aph-is craccivora* and *Cydia pomonella*, comparable to those of the synthetic phosphate monocrotophos and the carbamate carbofuran, they were shown to be potent inhibitors ($I_{50} \sim 10^{-7}$ M) of housefly and bovine erythrocyte acetylcholinesterases [213].

The hydantoin phosphonate *102*, isolated from the sponge *Ulosa ruetzleri*, has recently been reported as active (LD_{50} <10 ppm) against the tobacco hornworm (*Manduca sexta*) [35].

From the foliage of the wild tomato *Lycopersicon hirsutum f. glabratum*, known to be resistant to several insect species, was isolated 2-tridecanone *103* [214]. Of a series of methylketones ranging from

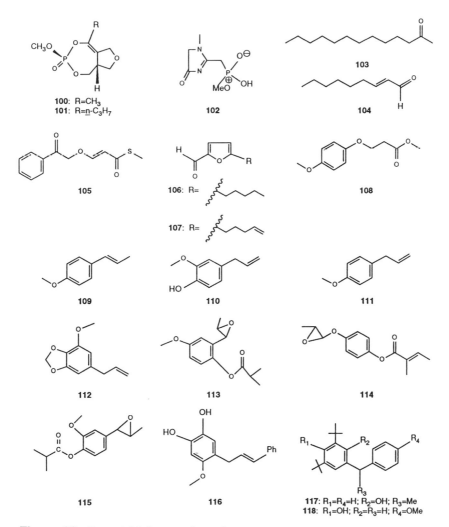

Figure 22 Insecticidal natural products.

1-nonanone to 2-pentadecanone applied to filter paper discs, 2-tridecanone was the most toxic to larvae of the tomato fruitworm, *Heliothis zea* with an $LC_{50}=17$ µg/cm². Changes in the position of the carbonyl group on the carbon chain did not affect toxicity [215]. 2-Tridecanone in wild tomato plants was the major component imparting resistance to the tobacco hornworm and the Colorado potato beetle (*Leptinotarsa decemlineata*) and less so to *H. zea* [216]. Parasitoids and predators of *H. zea* were also adversely affected by the presence of 2-tridecanone and 2-undecanone [216].

Trans-2-nonenal (*104*), a component of the roots of carrot (*Daucus carota* L.), was shown to be toxic as a fumigant to third instar larvae of the carrot fly (*Psila rosae*) with an LD_{50} of 2.17 mg after 24 hr in a 400-cm³ jar [217].

Crude extracts from the South African plant *Aloe pluridens* (Liliacae) exhibited insecticidal activity against a variety of mosquito species and the southern armyworm (*Spodoptera eridania*) [218]. Activity resided mostly in the petroleum ether extracts of dried root material from which the thioester *105*, named pluridone, was isolated. Insecticidal activity for pluridone was not provided; an oxygen for sulfur replacement gave the methyl ester analog, which was inactive.

Of furans *106* and *107* and the phenoxypropionic acid ester *108* (Figure 22), isolated from a culture of the fungus *Irpex lacteus* (IFD 5367) using activity against the rice white-tip nematode (*Aphelenchoides besseyi*) to guide separation, ester *108* proved to have the best activity killing 50% of the nematodes at 25 ppm [219]. As a follow-up, a series of 12 phenoxyacetic and phenoxypropionic acid esters were prepared and tested [220]. Of these, methyl 3-(*p*-chlorophenoxy)-propionate and methyl 3-(3,4-methylenedioxyphenoxy)propionate were equivalent to *108* against the nematode. The ethyl esters were as effective as the methyl esters. The isomers of *108*, in which the methoxy group was moved to the *ortho* or *meta* positions, were much less active. Methyl esters of 2-(*p*-methoxyphenoxy)propionic acid and of *p*-methoxy-phenoxyacetic acid were also weaker in activity [220].

Trans-anethole (*109*), eugenol (*110*), and estragole (*111*) were among several phenols and aromatic ethers isolated from anise plants (*Pimpinella anisum* L.), which showed some topical activity to houseflies [221]. Earlier, myristicin (*112*) had been isolated as the insecticidal component of parsnips (*Pastinaca sativa* L.) and shown to be effective at 0.5% against Mexican bean beetle (*Epilachna varivestis*) and mosquito (*Aedes aegypti*) larvae [222].

Epoxides *113* and *114*, isolated from a species of the genus *Pimpinella*, showed some activity by contact with houseflies and mites

(*Tetranychus telarius*) but were ineffective against several other insects [223]. The synthesized derivative (*115*) of isoeugenol had similar activity.

Synthesis work based on the structures of several phenolic compounds (e.g., *116*, Figure 22) isolated from the Panamanian hardwood *Dolbergia retussa*, noted for its resistance to fungi and marine organisms, afforded several compounds with fly-sterilant and mosquito growth–inhibitory activities [224]. Among these, phenols *117* and *118* were active enough for field investigations as mosquito control agents.

Tenuazonic acid (*119*, Figure 23) was isolated from cultures of the fungus *Alternaria tenuis* using a larvacide test with *Lucilia sericata* as the screen [225]. Follow-up synthesis work in an attempt to improve potency and activity spectrum afforded the bis-aryl analog *120* as among the best, but improvement was only marginal [226].

Using the European corn borer (*Ostrinia nubilalis*) as a bioassay, the cardiotonic glycosides nerifolin (*121*) and 2′-acetylnerifolin (*122*) were isolated from the seeds of the yellow oleander [*Thevetia thevetioides* (HBK) K. Schumm] [227, 228]. Nerifolin, the most active of the two compounds, had an LD_{50} of 30 ppm in diet against corn borers compared to carbofuran at 1–2 ppm. Activity against several other insect species was demonstrated as well as high toxicity to cats (LD_{50}=0.196 mg/kg).

The novel nitro-containing nucleoside clitocine (*123*, Figure 23), isolated from the mushroom *Clitocybe inversa*, was claimed to have strong activity against the pink bollworm (*Pectinophora gossypiella*) [229]. Its synthesis has been reported [230].

From the edible mushroom *Tricholoma muscarium* Kawamura, tricholomic acid (*124*) was isolated as the active compound against flies [231]. Tricholomic acid and its threo isomer were both synthesized [232]. The latter compound showed neither the flyicidal activity nor the good taste attributed to tricholomic acid.

The sesquilignan haedoxan A (*125*, Figure 23), one of number of lignans isolated from the roots of *Phryma leptostachya* L., proved active against several lepidopterous insects by ingestion and to houseflies by topical application (LD_{50}=1.6ng/fly with piperonyl butoxide) [233]. The lack of activity of other natural and synthesized lignans pointed to the importance of the particular 1,4-benzodioxanyl group unique to *125* [234].

The nikkomycins are peptide nucleoside antibiotics isolated from *Streptomyces tendae* Tu 901 cultures which completely inhibit the chitin synthetase of fungi and insects [235, 236]. They show high acaridical activity and low toxicity to rats, but earlier development for agricul-

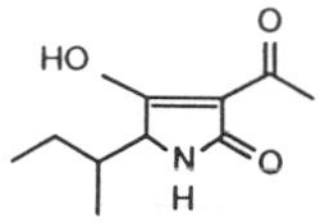

119

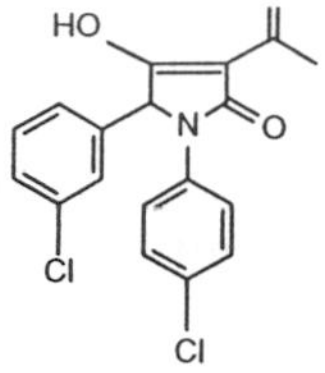

120

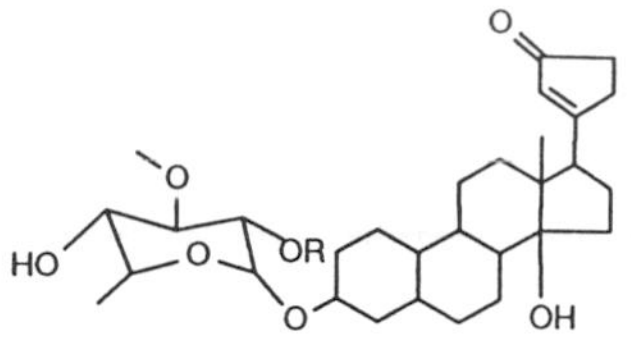

121: R=H
122: R=COMe

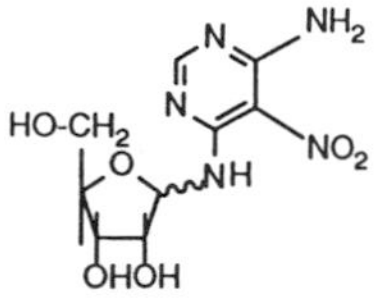

123

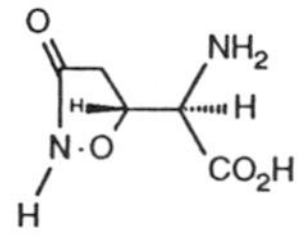

124

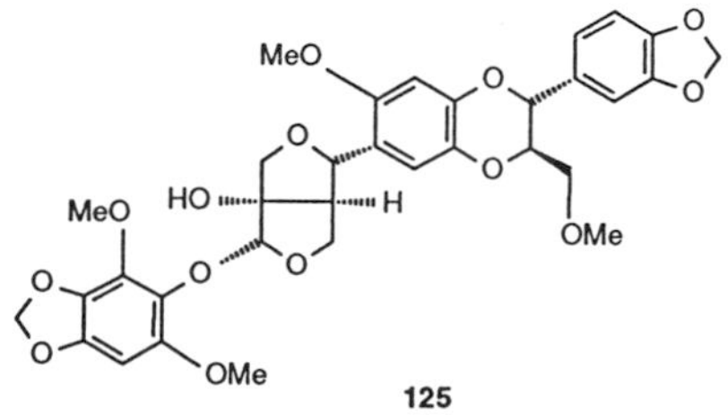

125

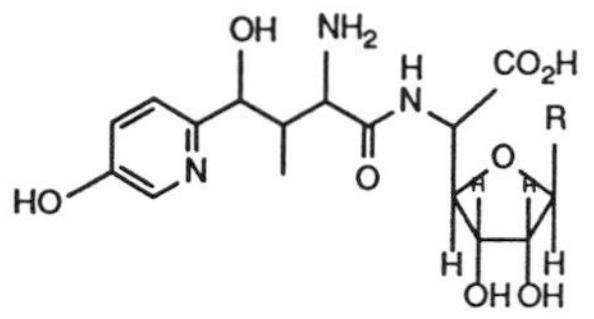

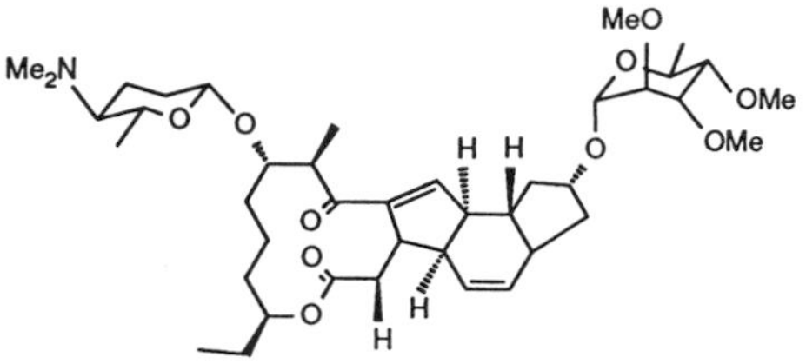

128

126: R= 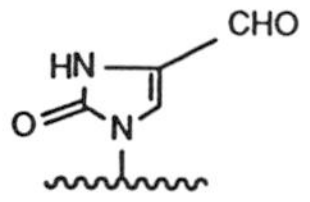

127: R= 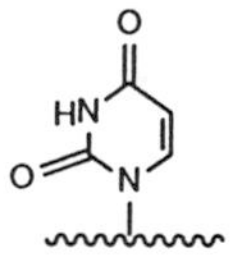

Figure 23 Insecticidal natural products.

tural purposes was discontinued in 1985 [236, and references therein]. Although the nikkomycins encompass a range of structures, nikkomycins X (*126*) and Z (*127*) are the major components.

Using mosquito larvicidal activity as a guide, an active culture was obtained from a soil organism identified as *Saccharopolyspora spinosa* [237]. From this culture, nine structurally related macrolides, of which *128* is representative, were isolated and characterized. Against fourth instar larvae of *Aedes aegypti*, *128* gave a 60% kill at 0.312 ppm after 24 hr. The compounds lacked antibiotic activity.

A series of antileukemic and cytotoxic quasinoids isolated from *Simaba multiflora* A. Juss. and *Soulamea soulameoides* (Gray) Nooteboom were evaluated in diet for their growth-inhibitory and insecticidal activities against the tobacco budworm (*Heliothis virescens*) and for their antifeedant activity against *H. virescens* and the fall armyworm (*Spodoptera frugiperda*) [238]. Insect growth inhibition generally paralleled known potency as antileukemic and cytotoxic agents. The most effective quassinoid was 6α-senecioyloxychaparrinone (*129*, Figure 24), which was equivalent to azadirachtin as a growth inhibitor of newly hatched *H. virescens* but, at an LD_{50} of 7 ppm, 2.5–3.5 times less toxic.

The unusual sulfur-containing alkaloid dithreanitrile (*130*) was isolated from the seeds of *Dithyrea waslizenii* (Cruciferae). As an antifeedant to the fall armyworm and the European corn borer, *130* is notable for its simple structure [239].

A family of compounds, the diabroticins, of which diabroticins A (*131*) and B (*132*) are major components, were isolated from the water-soluble fractions of cultures of *Bacillus subtilis* and *Bacillus cereus* [240]. In diet, diabroticins A and B gave LD_{50}'s of 2–4 and 25–50 ppm, respectively, against the southern corn rootworm (*Diabrotica undecimpunctata*). Extracts containing a mixture of diabroticins were also active against the Colorado potato beetle, the boll weevil, and mosquito larvae.

By screening soil-derived fermentation broths for mite activity, a streptomycete strain afforded the cyclic ether of structure *133* (Figure 24) [241]. Against *Tetranychus urticae* on bean leaves, *133* was more effective than the miticide Dicofol.

Rocaglamide (*134*) and three analogs were isolated from the insecticidal extract of dried twigs of the tropical Asian tree *Aglaia odorata* Lour [242]. Rocaglamide in diet gave a 50% reduction in growth of neonate variegated cutworms (*Peridroma saucia*) at 1.37 mg/kg, about four times greater than that of azadirachtin. Lethal doses causing 50% mortality of fourth instar larvae were 0.32 and 0.34 µg/larva by topical and oral administration, respectively.

Epitaondiol (*135*) was the most insecticidal of several closely related diterpenoids isolated from the alga *Stypopodium flabelliforme* against the fall armyworm *Spodoptera frugiperda* [243].

Aristolochic acid (*136*), isolated from the roots of *Aristolochia albida*, proved only tenfold less effective than azadirachtin as an antifeedant by leaf-disk assay against third instar tobacco cutworm (*Spodoptera litura*, Noctuidae [244]. Less-active derivatives were reported; further synthesis and structure-activity studies are promised.

The antifeedant and insecticidal activities of the alkaloid cocaine, from the leaves of the coca plant (*Erythroaxylum* spp.), have been dem-

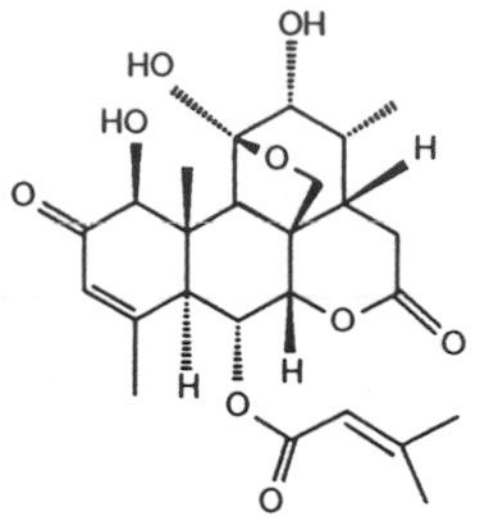

129

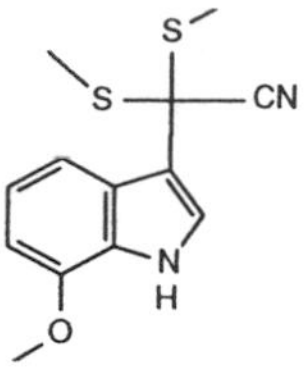

130

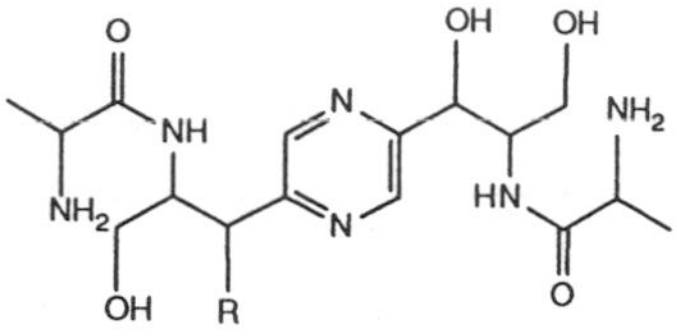

131: R=OH
132: R=H

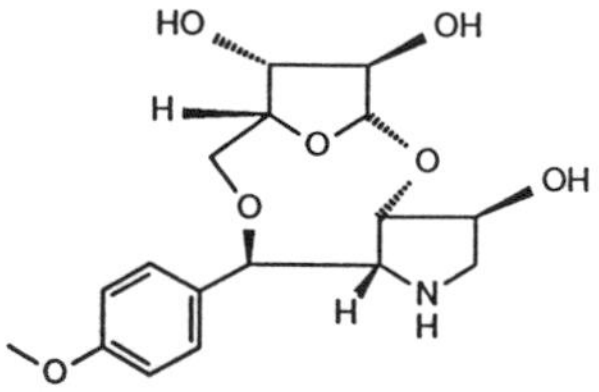

133

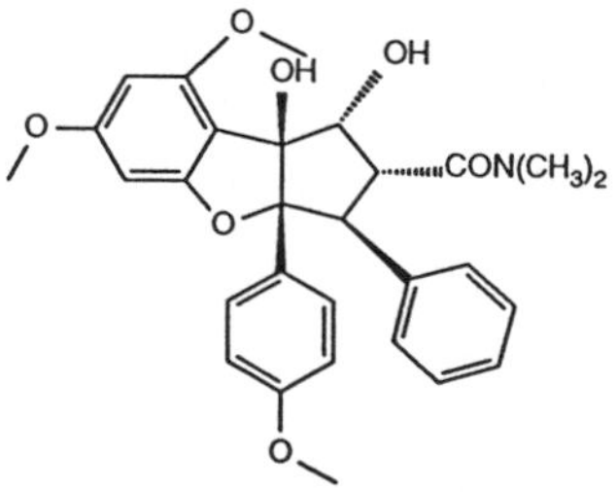

134

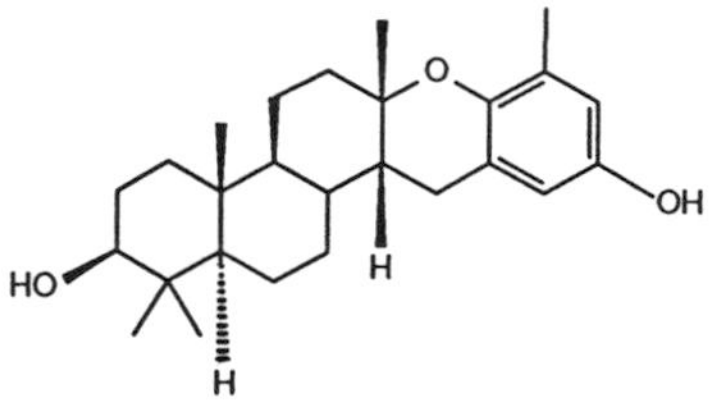

135

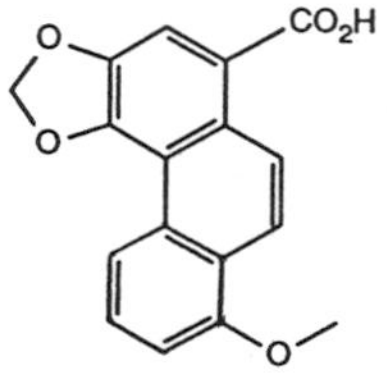

136

Figure 24 Insecticidal natural products.

onstrated using *Manduca sexta* larvae fed treated tomato leaves. Above a cocaine spray concentration of 0.5%, all larvae eventually died and it was estimated that tomato leaves containing an amount of cocaine equivalent to that found in fresh coca leaves would result in 68–88% inhibition of feeding of *Manduca*. Pesticidal activity is attributed to cocaine's ability to block the reuptake of the neurotransmitter octopamine [245].

REFERENCES

1. M. Jacobson and D. G. Crosby (eds.), *Naturally Occurring Insecticides*, Marcel Dekker, New York (1971).
2. H. Fukami and M. Nakajima, *Naturally Occurring Insecticides* (M. Jacobson and D. G. Crosby, eds.), Marcel Dekker, New York, p. 71 (1971).
3. L. Crombie, *Nat. Prod. Rep,* 1:3 (1984).
4. S. Tamura and N. Takahashi, *Naturally Occurring Insecticides* (M. Jacobson and D. G. Crosby, eds.), Marcel Dekker, New York, p. 513 (1971).
5. S. Yoshida and N. Takahashi, *Heterocycles, 10*:425 (1978).
6. I. Kubo and K. Nakanishi, *Pesticide Science and Biotechnology* (R. Greehalgh and T. R. Roberts, eds.), Blackwell Scientific, Boston, p. 248 (1987).
7. J. A. Klocke, M. F. Balandrin, M. A. Barnby, and R. B. Yamasaki, *Insecticides of Plant Origin* (J. T. Arnason, B. J. R. Philogene, and P. Morand, eds.), ACS Symposium Series 387, American Chemical Society, Washington, D.C., p. 136 (1989).
8. a. M. Jacobson, *Insecticides of Plant Origin* (J. T. Arnason, B. J. R. Philogene, and P. Morand, eds.), ACS Symposium Series 387, American Chemical Society, Washington, D.C., p. 1 (1989); b. C. A. Elliger and A. C. Waiss, Jr., *Naturally Occurring Insect Bioregualators* (P. A. Hedin, ed.), ACS Symposium Series 449, American Chemical Society, Washington, D.C., p. 210 (1991).
9. A. Marston and K. Hostettmann, *Phytochemistry, 24*:639 (1985).
10. J. E. Casida, L. M. Cole, J. E. Hawkinson, and C. J. Palmer, *Recent Advances in the Chemistry of Insect Control II* (L. Crombie, ed.), Spec. Pub. 79, The Royal Society of Chemistry, Cambridge, p. 212 (1990).
11. C. J. Palmer, L. M. Cole, J. P. Larkin, I. H. Smith, and J. E. Casida, *J. Agric. Food Chem,* 39:1329 (1991).
12. C. J. Palmer, L. M. Cole, I. H. Smith, M. D. V. Moss, and J. E. Casida, *J. Agric. Food Chem,* 39:1335 (1991).
13. M. Elliott, D. A. Pulman, J. P. Larkin, and J. E. Casida, *J. Agric. Food Chem,* 40:147 (1992).
14. C. J. Palmer and J. E. Casida, *J. Agric. Food Chem,* 40:492 (1992).
15. T. J. Mabry and J. E. Gill, *Herbivores* (G. A. Rosenthal and D. H. Janzen, eds.), Academic Press, New York, p. 501 (1979).
16. J. R. Coats, L. L. Karr, and C. D. Drewes, *Naturally Occurring Pest Bioregulators, Potential Use in Agriculture* (P. H. Hedin, ed.), ACS Symposium Series 449, American Chemical Society, Washington, D.C., p. 305 (1991).
17. A. A. Bell, R. D. Stipanovic, G. W. Elzen, and H. J. Williams, Jr., *Allelochemicals: Role in Agriculture and Forestry* (G. R. Walter, ed.), ACS Symposium Series 330, American Chemical Society, Washington, D.C., p. 477 (1987).

18. A. San-Martin, R. Negrete and J. Rovirosa, *Phytochemistry, 30*:2165 (1991).

19. D. L. Grundy and C. C. Still, *Pest. Biochem. Physiol. 23*:378 (1985).

20. a. D. L. Grundy and C. C. Still, *Pest. Biochem. Physiol. 23*:383 (1985); b. I. Konstantopoulou, L. Vassilopopoulou, P. Mavragani-Tsipidou, and Z. G. Scouras, *Experientia, 48*:616 (1992).

21. A. K. Picman, *Biochem. System. Ecol. 14*:255 (1986).

22. J. T. Arnason, M. B. Isman, B. J. R. Philogene, and T. G. Waddell, *J. Nat. Prod., 50*:690 (1987).

23. W. Maotian, Q. Hailin, K. Man, and L. Yanzi, *Phytochemistry, 30*:3931 (1991).

24. N. Wakabayashi, W. J. Wu, R. M. Waters, R. E. Redfern, G. D. Mills, Jr., A. B. DeMilo, W. R. Lusby, and D. Andrejewski, *J. Nat. Prod., 51*:537 (1988).

25. J. K. Liu, Z. J. Jia, D. G. Wu, J. Zuou, and Q. G. Wang, *Phytochemistry, 29*:2503 (1990).

26. G. A. Cooper-Driver and P. W. Le Quesne, *Allelochemicals: Role in Agriculture and Forestry* (G. R. Walter, ed.), ACS Symposium Series 330, American Chemical Society, Washington, D.C., p. 534 (1987).

27. S. O. Duke, R. N. Paul, Jr., and S. M. Lee, *Biologically Active Natural Products, Potential Use in Agriculture* (H. G. Cutler, ed.), ACS Symposium Series 380, American Chemical Society, Washington, D.C., p. 318 (1988).

28. E. Rodriguez, *Bioregulators for Pest Control* (P. A. Hedin, ed.), ACS Symposium Series 276, American Chemical Society, Washington, D.C., p. 447 (1985).

29. G. T. Marshall, J. A. Klocke, L. J. Lin, and A. D. Kinghorn, *J. Chem. Ecol., 11*:191 (1985).

30. D. G. Crosby, *Naturally Occurring Insecticides* (M. Jacobson and D. G. Crosby, ed.), Marcel Dekker, New York, p. 224 (1971).

31. I. C. Alexander, K. O. Pascoe, P. Manchard, and L. A. D. Williams, *Phytochemistry, 30*:1801 (1991).

32. J. A. Klocke, M. Y. Hu, S. F. Chiu, and I. Kubo, *Phytochemistry, 30*:1797 (1991).

33. I. Kubo, T. Matsumoto, and J. A. Klocke, *J. Chem. Ecol., 10*:547 (1984).

34. M. Zhang, B. P. Ying, and I. Kubo, *J. Nat. Prod., 55*:1057 (1992).

35. J. H. Cardellina II, *Biologically Active Natural Products, Potential Use in Agriculture* (H. G. Cutler, ed.), ACS Symposium Series 380, American Chemical Society, Washington, D.C., p. 305 (1988).

36. J. H. Carellina II, *Pure Appl. Chem., 58*, 365 (1986).

37. P. S. Jones, S. V. Ley, E. D. Morgan, and D. Santafianos, *Phytochemical Pesticides, Volume 1, The Neem Tree* (M. Jacobson, ed.), CRC Press, Boca Raton, FL, p. 19 (1989).

38. J. H. Butterworth and E. D. Morgan, *Chem. Commun.*, 23 (1986).

39. H. Rembold, *Focus On Phytochemical Pesticides, Volume 1, The Neem Tree* (M. Jacobson, ed.), CRC Press, Boca Raton, FL, p. 47 (1989).

40. H. Rembold, *Insecticides of Plant Origin* (J. T. Arnason, B. J. Philogene,

and P. Morand, eds.), ACS Symposium Series 387, American Chemical Society, Washington, D.C., p. 150 (1989).

41. S. M. Lee, J. A. Klocke, M. A. Barnby, R. B. Yamasaki, and M. F. Balandrin, *Naturally Occurring Pest Bioregulators* (P. H. Hedin, ed.), ACS Symposium Series 449, American Chemical Society, Washington, D.C., p. 293 (1991).

42. a. J. C. Anderson, S. V. Ley, D. Santafianos, and R. N. Sheppard, *Tetrahedron, 47*:6813 (1991); b. H. C. Kolb, S. V. Ley, A. M. Z. Slawin, and D. J. Williams, *J. Chem. Soc. Perkin Trans.* 1, 2735 (1992); c. S. V. Ley, *Pesticidal Chemistry, Advances in International Research, Development, and Legislation* (H. Frehse, ed.), VCH Publishers, New York, p. 97 (1991); d. P. Aldhous, *Science, 258*: 893 (1992); e. S. V. Ley, H. Lovell, and D. J. Williams, *Chem. Commun.*, 1304 (1992).

43. S. M. Lee, J. A. Klocke, and M. F. Balandrin, *Tetrahedron Lett. 28*:3543 (1987).

44. R. O. Larson, *Focus on Phytochemical Pesticides, Volume 1, The Neem Tree* (M. Jacobson, ed.), CRC Press, Boca Raton, FL, p.155 (1989).

45. R. Stone, *Science, 255*:1070 (1992).

46. H. Schumutterer and C. Hellpap, *Focus on Phytochemical Pesticides, Volume 1, The Neem Tree* (M. Jacobson, ed.), CRC Press, Boca Raton, FL, p. 69 (1989).

47. D. E. Champagne, M. B. Isman, and G. H. N. Towers, *Insecticides of Plant Origin* (J. T. Arnason, B. J. Philogene, and P. Morand, eds.), ACS Symposium Series 387, American Chemical Society, Washington, D.C., p. 95 (1989).

48. J. A. Klocke, *Allelochemicals: Role in Agriculture and Forestry* (G. R. Waller, ed.), ACS Symposium Series 300, American Chemical Society, Washington, D.C., p. 396 (1987).

49. T. Robinson, *Herbivores* (G. A. Rosenthal and D. H. Janzen, eds.), Academic Press, New York, p. 413 (1979).

50. a. H. Z. Levinson, *Experientia, 32*:408 (1976); b. J. C. Cavin and E. Rodriguez, *J. Chem. Ecol. 14*:475 (1988).

51. I. Schmeltz, *Naturally Occurring Insecticides* (M. Jacobson and D. G. Crosby, eds.), Marcel Dekker, New York, p. 99 (1971).

52. R. F. Severson, R. F. Arrendale, H. G. Cutler, D. Jones, V. A. Sisson, and M. G. Stephenson, *Biologically Active Natural Products, Potential Use in Agriculture* (H. G. Cutler, ed.), ACS Symposium Series 380, American Chemical Society, Washington, D.C., p. 335 (1988).

53. R. L. Metcalf, *Organic Insecticides*, Interscience Publishers, New York, p. 1 (1955).

54. I. Yamamoto, *Advances in Pest Control Research* (R. L. Metcalf, ed.), Vol. 6, Interscience Publishers, New York, p. 231 (1965).

55. S. B. Soloway, A. C. Henry, W. D. Kollmeyer, W. M. Padgett, J. E. Powell, S. A. Roman, C. H. Tieman, R. A. Corey, and C. A. Horne, *Advances in Pesticide Science* (H. Geissbuhler, ed.), Part 2, Pergamon Press, New York, p. 206 (1979).

56. D. B. Sattelle, S. D. Buckingham, K. A. Wafford, S. M. Sherby, N. M. Bakry, A. T. Eldefrawi, M. E. Eldefrawi, and T. E. May, *Vertebrate Nicotinic Acetylcholine Receptors Proc. Roy. Soc. (Lond.) B*, *237* (1289):501 (1989).

57. M. E. Schroeder and R. F. Flattum, *Pestic. Biochem. Physiol.*, 22:148 (1984).

58. a. A. Elbert, H. Overbeck, K. Iwaya, and S. Tsuboi, *Brighton Crop Protection Conference—Pests and Diseases*, Vol. 1, The British Crop Protection Council, Surrey, p. 21 (1990); b. H. Cheung, B. S. Clarke, and D. J. Beadle, *Pestic. Sci.*, 34:187 (1992); c. K. Shiokawa, K. Moriya, K. Shibuya, Y. Hattori, S. Tsuboi, and S. Kagabu, *Biosci. Biotech. Biochem.*, 56:1364 (1992).

59. K. D. Jennings, D. G. Brown, and D. P. Wright, Jr., *Experientia*, 42:611 (1986).

60. K. D. Jennings, D. G. Brown, D. P. Wright, Jr. and A. E. Chalmers, *Site of Action for Neurotoxic Pesticides* (R. M. Hollingworth and M. B. Green, eds.), ACS Symposium Series 356, American Chemical Society, Washington, D.C., p. 274 (1987).

61. S. A. Ross and S. W. Pelletier, *Heterocycles*, 27:1381 (1988).

62. S. W. Pelletier and S. A. Ross, *Heterocycles*, 31:671 (1990).

63. S. A. Ross and S. W. Pelletier, *Heterocycles*, 32:1307 (1991).

64. J. A. Grina, D. R. Schroeder, E. T. Wydallis, and F. R. Stermitz, *J. Organ. Chem.*, 51:390 (1986).

65. S. Nitta, *Yakagaku Zasshi*, 54:648 (1934).

66. K. Konishi, Insecticides, *Proceedings of the Second International IUPAC Congress of Pesticide Chemistry* (A. S. Tahori, ed.), Vol I, Gordon and Breach, New York, p. 178 (1972).

67. M. Sakai, *Rev. Plant Protect. Res.*, 2:17 (1969).

68. S. M. Sherby, A. T. Eldefrawi, J. A. David, D. B. Sattelle, and M. E. Eldefrawi, *Arch. Insect Biochem. Physiol.*, 3:431 (1986).

69. C. R. Worthing, (ed.), *The Pesticide Manual*, 9th ed., British Crop Protection Council, Surrey, p. 132 (1991).

70. C. R. Worthing, (ed.), *The Pesticide Manual*, 9th ed., British Crop Protection Council, Surrey, p. 64 (1991).

71. W. Berg and H. J. Knutti, Proceedings of the 8th British Insecticide and Fungicide Conference, Brighton, 683–691 (1975).

72. H. Mitsudera, K. Konishi, and Y. Sato, U.S. Patent 4,640,929 (1987).

73. A. Anthoni, C. Christophersen, J. O. Madsen, S. Wium-Andersen, and N. Jacobsen, *Phytochemistry*, 19:1228 (1980).

74. N. Jacobsen and L. E. K. Pedersen, *Pestic. Sci.*, 14:90 (1983).

75. H. Mitsudera, T. Kamikado, H. Uneme, and Y. Kono, *Agric. Biol. Chem.*, 54:1719 (1990).

76. H. Mitsudera, T. Kamikado, H. Uneme, and Y. Manabe, *Agric. Biol. Chem.*, 54:1723 (1990).

77. M. Takasugi, Y. Yachida, M. Anetai, T. Masamune, and K. Kegasawa, *Chem. Lett. 43* (1975).

78. B. V. McInerney, R. P. Gregson, M. J. Lacey, R. J. Akhurst, G. R. Lyons, S. H. Rhodes, D. R. J. Smith, L. M. Engelhardt, and A. H. White, *J. Nat. Prod.*, 54:774 (1991).

79. D. G. Crosby, *Naturally Occurring Insecticides* (M. Jacobson and D. G. Crosby, eds.), Marcel Dekker, New York, p. 186 (1971).

80. I. Ujváry, B. K. Eya, R. L. Grendell, R. F. Toia, and J. E. Casida, *J. Agric. Food Chem.,* 39:1875 (1991).

81. D. G. Crosby, *Naturally Occurring Insecticides* (M. Jacobson and D. G. Crosby, eds.), Marcel Dekker, New York, p. 198 (1971).

82. P. R. Jeffries, R. F. Toia, B. Brannigan, I. Pessah, and J. E. Casida, *J. Agric. Food Chem.,* 40:142 (1992).

83. I. N. Pessah, *Recent Advances in the Chemistry of Insect Control II* (L. Crombie, ed.), Spec. Pub. 79, Royal Society of Chemistry, Cambridge, p. 278 (1990).

84. J. E. Casida, I. N. Pessah, J. Seifert, and A. L. Waterhouse, *Pesticide Science and Biotechnology* (R. Greenhalgh and T. R. Roberts, eds.), Blackwell Scientific, Boston, p. 177 (1987).

85. A. L. Waterhouse, I. N. Pessah, A. O. Francini, and J. E. Casida, *J. Med. Chem.,* 30:710 (1987).

86. P. Deslongchamps, A. Belanger, D. J. F. Berney, H. J. Borschberg, R. Brousseau, A. Doutheau, R. Durand, H. Katayama, R. Lapalme, D. Leturc, C.-C. Liao, F. N. MacLachlan, J.-P. Maffrand, F. Marazza, R. Martino, C. Moreau, L. Ruest, L. Saint-Laurent, R. Saintonge, and P. Soucy, *Can. J. Chem,* 68:186 (1990).

87. H. Nakamura, K. Shiomi, H. Iinuma, H. Naganawa, T. Obata, T. Takeuchi, and H. Umezawa, *J. Antibiotics,* 40:899 (1987).

88. K. Yano, J. Oono, K. Mogi, T. Asaoka, and T. Nakashima, *J. Antibiotics,* 40:961 (1987).

89. M. Koyama, Y. Kodama, T. Tsurouoka, N. Ezaki, T. Niwa, and S. Inouye, *J. Antibiotics,* 34:1569 (1981); N. Ezaki, M. Koyama, T. Shomura, T. Tsuruoka, and S. Inouye, *J. Antibiotics,* 36:1263 (1983).

90. G. T. Carter, J. N. Nietsche, J. J. Goodman, M. J. Torrey, T. S. Dunne, D. B. Borders, and R. T. Testa, *J. Antibiotics,* 40:233 (1987).

91. R. W. Addor, T. J. Babcock, B. C. Black, D. G. Brown, R. E. Diehl, J. A. Furch, V. Kameswaran, V. M. Kamhi, K. A. Kremer, D. G. Kuhn, J. B. Lovell, G. T. Lowen, T. P. Miller, R. M. Peevey, J. K. Siddens, M. F. Treacy, S. H. Trotto, and D. P. Wright, Jr., *Synthesis and Chemistry of Agrochemicals III* (D. R. Baker, J. G. Fenyes, and J. J. Steffens, eds.), ACS Symposium Series 504, American Chemical Society, Washington, D.C., p. 283 (1992).

92. D. T. Saggers and M. L. Clark, *Nature,* 215:275 (1967).

93. J. B. Lovell, D. P. Wright, Jr., I. E. Gard, T. P. Miller, M. F. Treacy, R. W. Addor, and V. M. Kamhi, *Brighton Crop Protection Conference—Pests and Diseases—1990,* Vol. 1, The British Crop Protection Council, Surrey, p.37 (1991).

94. a. AC 303,630, *Experimental Insecticide-Miticide,* American Cyanamid Agricultural Research Division, Princeton, NJ (1992); b. M. Treacy, T. Miller, B. Black, I. Gard, and D. Hunt, *J. Biochem. Trans.,* 22:244 (1994).

95. M. F. Treacy, T. P. Miller, I. E. Gard, J. B. Lovell, and D. P. Wright, Jr., "Proceedings of the Beltwide Cotton Conferences," Memphis, TN, Vol. 2, 738–741 (1991).

96. T. P. Miller, M. F. Treacy, I. E. Gard, J. B. Lovell, and D. P. Wright, Jr., R. W. Addor, and V. M. Kamhi, *Brighton Crop Protection Conference—Pests and Diseases—1990*, Vol. 1, British Crop Protection Council, Surrey, p. 43 (1991).

97. D. G. Kuhn, V. M. Kamhi, J. A. Furch, R. E. Diehl, S. H. Trotto, G. T. Lowen, and T. J. Babcock, *Synthesis and Chemistry of Agrochemicals III* (D. R. Baker, J. G. Fenyes, and J. J. Steffens, eds.), ACS Symposium Series 504, American Chemical Society, Washington, D.C., p. 298 (1992).

98. H. Irie, N. Masaki, K. Ohno, K. Osaki, T. Taga, and S. Uyeo, *Chem. Commun.*, 1066 (1970).

99. K. Sakata, K. Aoki, C. Chang, A. Sukurai, S. Tamura, and S. Murakoshi, *Agri. Biol. Chem.*, 42:457 (1978).

100. M. Jacobson, *Naturally Occurring Insecticides* (M. Jacobson and D. G. Crosby, eds.), Marcel Dekker, New York, p. 137 (1971).

101. M. Elliott, A. W. Farnham, N. F. Janes, D. M. Johnson, and D. A. Pulman, *Pestic. Sci.*, 18:a, 191 (part 1); b, 203 (part 2); c, 211 (part 3); d, 223 (part 4); e, 229 (part 5); f, 239 (part 6) (1987).

102. M. Miyakado, I. Nakayama, and N. Ohno, *Insecticides of Plant Origin* (J. T. Arnason, B. J. Philogene, and P. Morand, eds.), ACS Symposium Series 387, American Chemical Society, Washington, D.C., p. 173 (1989).

103. M. Miyakado, I. Nakayama, N. Ohno, and H. Yoshioka *Natural Products for Innovative Pest Management* (D. L. Whitehead and W. S. Bowers, eds.), Pergamon Press, Oxford, p. 369 (1983).

104. E. K. Harvil, A. Hartzell, and J. M. Arthur, *Contrib. Boyce Thompson Inst.*, 13:87 (1943).

105. Y. Hwang and M. S. Mulla, *J. Agric. Food Chem.*, 28:1118 (1980); Y. Hwang, M. S. Mulla, M. S. Pope, and C. Rodriguez, *Pestic. Sci.*, 13:517 (1982).

106. M. Elliott, *Recent Advances in the Chemistry of Insect Control* (N. F. Janes, ed.), Spec. Public. 53, The Royal Society of Chemistry, London, p. 89 (1985).

107. H. Greger, *Chemistry and Biology of Naturally-Occurring Acetylenes and Related Compounds: Bioactive Molecules* (J. Lam, H. Breteler, T. Arnason, and L. Hansen, eds.), Vol 7, Elsevier, New York, p. 159 (1988).

108. R. J. Blade, *Recent Advances in the Chemistry of Insect Control* (L. Crombie, ed.), Spec. Publ. 79, The Royal Society of Chemistry, Cambridge, p. 151.(1990).

109. L. Crombie and D. Fisher, *Tetrahedron Lett.*, 26:2477 (1985).

110. L. Crombie, M. A. Horsham, and R. J. Blade, *Tetrahedron Lett.*, 28:4879 (1987).

111. E. Winterfeldt, *Chem. Ber.*, 96:3349 (1963).

112. M. Elliott, A. W. Farnham, N. F. Janes, D. M. Johnson, and D. A. Pulman, *Pestic. Sci.*, 26:199 (1989).

113. A. Isogai, T. Horii, A. Suzuki, S. Murakoshi, K. Ikeda, S. Sata, and S. Tamura, *Agric. Biol. Chem.*, 39:739 (1975).

114. M. Miyakado, I. Nakayama, A. Inoue, M. Hatakoshi, and N. Ohno, *J. Pestic. Sci.*, 10:25 (1985).

115. I. Kubo, T. Matsumoto, J. A. Klocke, and T. Kamikawa, *Experientia,* 40:340 (1984).
116. M. G. Nair, A. P. Mansingh, and B. A. Burke, *Agric. Biol. Chem.,* 50:3053 (1986).
117. M. Elliott, A. W. Farnham, N. F. Janes, D. M. Johnson, D. A. Pulman, and R. M. Sawicki, *Agric. Biol. Chem.,* 50:1347 (1986).
118. M. Hatakoshi, M. Miyakado, N. Ohno, and I. Nakayama, *Appl. Entomol. Zool.,* 19:288 (1984).
119. J. J. Johnston, M. A. Horsham, T. J. Class, and J. E. Casida, *J. Agric. Food Chem.,* 37:781 (1989).
120. C. J. Brealey, *Biochem. Soc. Trans.,* 15:1102 (1987).
121. L. O. Ruzo, S. J. Holloway, J. E. Casida, and V. V. Krishnamurthy, *J. Agric. Food Chem.,* 36:841 (1988).
122. D. W. Russell, *Quart. Rev.* 20:559 (1966).
123. S. Gupta, D. W. Roberts, and J. A. Renwick, *J. Chem. Soc. Perkin Trans. 1,* 2347 (1989).
124. Y. Kodaira, *Res. Repts. Fac. Textile Sericult. Shinshy University,* No. 4, 1 (1954).
125. Y. Kokaira, *Agric. Biol. Chem.,* 25:261 (1961).
126. S. Tamura, S. Kuyma, Y. Kodaira, and S. Higashikawa, *Agric. Biol. Chem.* 28:137 (1964).
127. S. Tamura and N. Takahashi, *Naturally Occurring Insecticides* (M. Jacobson and D. G. Crosby, eds.), Marcel Dekker, New York, p. 499 (1971).
128. A. Suzuki, H. Tagushi, and S. Tamura, *Agric. Biol. Chem.,* 34:813 (1970).
129. M. Pais, B. C. Das, and P. Ferron, *Phytochemistry,* 20:715 (1981).
130. H. Brockmann and G. Schmidt-Kastner, *Chem. Ber.,* 88:57 (1955).
131. E. L. Patterson and D. P. Wright, Jr., U.S. Patent 3,520,973 (1970).
132. R. M. Heisey, S. K. Mishra, A. R. Putnam, J. R. Miller, C. J. Whitenack, J. E. Keller, and J. Huang, *Biologically Active Natural Products, Potential Use in Agriculture* (H. G. Cutler, ed.), ACS Symposium Series 380, American Chemical Society, Washington, D.C., p. 74 (1988).
133. M. Kanaoka, A. Isogai, S. Murakoshi, M. Ichinoe, A. Suzuki, and S. Tamura, *Agric. Biol. Chem.,* 42:629 (1978).
134. R. L. Hamill, C. E. Higgins, M. E. Boaz and M. Gorman, *Tetrahedron Lett.,* 4255 (1969).
135. M. Kanaoka, A. Isogai, and A. Suzuki, *Agric. Biol. Chem.* 43:1079 (1979).
136. D. B. Strongman, G. M. Strunz, P. Giguere, C. Yu, and L. Calhoun, *J. Chem. Ecol.,* 14:753 (1988).
137. R. Baute, G. Deffieux, D. Merlet, M. Baute, and A. Neveu, *J. Antibiotics,* 34:1261 (1981).
138. G. Deffieux, D. Merlet, R. Baute, G. Bourgeois, and M. Baute, A Neveu, *J. Antibiotics,* 34:1266 (1981).
139. L. C. Vining and W. A. Taber, *Canad. J. Chem.,* 40:1579 (1962).
140. T. M. Zabriskie, J. A. Klocke, C. M. Ireland, A. H. Marcus, T. F. Molinski, D. J. Faulkner, C. Xu, and J. C. Clardy, *J. Am. Chem. Soc.,* 108:3123 (1986).

141. P. Crews, L. V. Manes, and M. Boehler *Tetrahedron Lett.,* 27:2797 (1986).
142. P. A. Grieco, Y. S. Hon, and A. Perez-Medrano, *J. Am. Chem. Soc.,* 110:1630 (1988).
143. K. S. Chu, G. R. Negrete, and J. P. Konopelski, *J. Organ. Chem.,* 56:5196 (1991).
144. M. Kahn, H. Nakanishi, T. Su, J. Y. H. Lee, and M. E. Johnson, *Int. J. Peptide Res.,* 38:324 (1991).
145. J. R. Heitz, *Light-Activated Pesticides* (J. R. Heitz and K. R. Downum, eds.), ACS Symposium Series 339, American Chemical Society, Washington, D.C., p. 1 (1987).
146. L. A. Lemke, P. G. Koehler, R. S. Patterson, M. B. Feger, and T. Eickoff, *Light-Activated Pesticides* (J. R. Heitz and K. R. Downum, eds.), ACS Symposium Series 339, American Chemical Society, Washington, D.C., p. 156 (1987).
147. F. J. Gommers and J. Bakker, *Chemistry and Biology of Naturally-Occurring Acetylenes and Related Compounds; Bioactive Molecules* (J. Lam, H. Breteler, T. Arnason, and L. Hansen, eds.), Vol 7, Elsevier, New York, p. 61 (1988).
148. M. Berenbaum, *Science,* 201:532 (1978).
149. E. L. Camm, G. H. N. Towers, and J. C. Mitchell, *Phytochemistry,* 14:2007 (1975).
150. M. R. Berenbaum, *Light-Activated Pesticides* (J. R. Heitz and K. R. Downum, eds.), ACS Symposium Series 339, American Chemical Society, Washington, D.C., p. 206 (1987).
151. L. A. Swain and K. R. Downum, *Naturally Occurring Pest Bioregulators* (P. H. Hedin, ed.), ACS Symposium Series 449, American Chemical Society, Washington, D.C., p. 361 (1991).
152. G. W. Ivie, D. L. Bull, R. C. Beier, N. W. Pryor, and E. H. Oertli, *Science,* 221:374 (1983).
153. M. Berenbaum and P. Feeny, *Science,* 212:927 (1981).
154. G. W. Ivie, D. L. Bull, R. C. Beier, and N. Waller, *Allelochemicals: Role in Agriculture and Forestry* (G. R. Waller, ed.), ACS Symposium Series 330, American Chemical Society, Washington, D.C., p. 455 (1987).
155. G. W. Ivie, *Light-Activated Pesticides* (J. R. Heitz and K. R. Downum, eds.), ACS Symposium Series 339, American Chemical Society, Washington, D.C., p. 217 (1987).
156. Y. Y. Marchant and G. K. Cooper, *Light-Activated Pesticides* (J. R. Heitz and K. R. Downum, eds.), ACS Symposium Series 339, American Chemical Society, Washington, D.C., p. 241 (1987).
157. D. G. Crosby and N. Aharonson, *Tetrahedron,* 23:465 (1967).
158. M. A. Saleh, *Phytochemistry,* 23:2497 (1984).
159. K. Kawazu, M. Ariwa, and Y. Kii, *Agric. Biol. Chem.,* 41:223 (1977).
160. J. Kagan, C. P. Kolyvas, and J. Lam, *Experientia,* 40:1396 (1984).
161. J. T. Arnason, B. J. R. Philogene, C. Berg, A. MacEachern, J. Kaminski, L. C. Leitch, P. Morand, and J. Lam, *Phytochemistry,* 25:1609 (1986).
162. D. McLachlan, J. T. Arnason, B. J. R. Philogene, and D. Champagne, *Experientia,* 38:1061 (1982).

163. C. K. Wat, R. K. Biswas, E. A. Graham, L. Bohm, G. H. N. Towers, and E. R. Waygood, *J. Nat. Prod.,* 42:103 (1979).

164. G. H. N. Towers and D. E. Champagne, *Chemistry and Biology of Naturally-Occurring Acetylenes and Related Compounds; Bioactive Molecules* (J. Lam, H. Breteler, T. Arnason, and L. Hansen, eds.), Vol 7, Elsevier, New York, p. 139 (1988).

165. S. Nakajima and K. Kawazu, *Agric. Biol. Chem.,* 44:1529 (1980).

166. J. Kagan and G. Chan, *Experientia,* 39:402 (1983).

167. E. Rodriguez, *Biologically Active Natural Products, Potential Use in Agriculture* (H. G. Cutler, ed.), ACS Symposium Series 380, American Chemical Society, Washington, D.C., p. 432 (1988).

168. J. H. Uhlenbroek and J. D. Bijloo, *Rec. Trav. Chim.,* 77:1004 (1958).

169. J. H. Uhlenbroek and J. D. Bijloo, *Rec. Trav. Chim.,* 78:382 (1959).

170. H. Wijnberg, *Chemistry and Biology of Naturally-Occurring Acetylenes and Related Compounds; Bioactive Molecules* (J. Lam, H. Breteler, T. Arnason, and L. Hansen, eds.), Vol 7, Elsevier, New York, p. 21 (1988).

171. J. T. Arnason, B. J. R. Philogene, P. Morand, K. Imrie, S. Iyengar, F. Duval, C. Soucy-Breau, J. C. Scaiano, N. H. Werstiuk, B. Hasspieler, and A. E. R. Downe, *Insecticides of Plant Origin* (J. T. Arnason, B. J. R. Philogene, and P. Morand, eds.), ACS Symposium Series 387, American Chemical Society, Washington, D.C., p. 164 (1988).

172. J. T. Arnason, B. J. R. Philogene, P. Morand, J. C. Scaiano, N. Werstiuk, and J. Lam, *Light-Activated Pesticides* (J. R. Heitz and K. R. Downum, eds.), ACS Symposium Series 339, American Chemical Society, Washington, D.C., p. 255 (1987).

173. K. Tamao, S. Kodama, I. Nakajima, M. Kumada, A. Minato, and K. Suzuki, *Tetrahedron,* 38:3347 (1982).

174. B. J. R. Philogene, J. T. Arnason, C. W. Berg, F. Duval, D. Champagne, R. G. Taylor, L. C. Leitch, and P. Morand, *J. Econ. Entomol.,* 78:121 (1985).

175. J. Kagan, W. J. Bennett, E. D. Kagan, J. L. Maas, S. A. Sweeney, I. A. Kagan, E. Seigneruie, and V. Bindokas, *Light-Activated Pesticides* (J. R. Heitz and K. R. Downum, eds.), ACS Symposium Series 339, American Chemical Society, Washington, D.C., p. 176 (1987).

176. J. H. Uhlenbroek and J. D. Bijloo, *Rec. Trav. Chim.,* 79:1181 (1960).

177. R. J. Marles, R. L. Compadre, C. M. Compadre, C. Soucy-Breau, R. W. Redmond, F. Duval, B. Mehta, P. Morand, J. C. Scaiano, and J. T. Arnason, *Pest. Biochem. Physiol.* 41:89 (1991).

178. D. I. Relyea, R. C. Moore, W. L. Hubbard, and P. A. King, *10th International Congress of Plant Protection 1983,* The British Crop Protection Council, Vol. 1, p/2C-S9.

179. C. R. Worthington and R. J. Hance (eds.). *The Pesticide Manual,* 9th Ed., British Crop Protection Council, Surrey, p. 931 (1991).

180. D. M. Roush, K. A. Lutomski, R. B. Phillips, and S. E. Burkart, *Synthesis and Chemistry of Agrochemicals II* (D. R. Baker, J. G. Fenyes, and W. K. Moberg, eds.), ACS Symposium Series 443, American Chemical Society, Washington, D.C., p. 352 (1991).

181. K. Sun, K. H. Pilgrim, D. A. Kleier, M. E. Schroeder, and A. Y. S. Yang, *Synthesis and Chemistry of Agrochemicals II* (D. R. Baker, J. G. Fenyes, and W. K. Moberg, eds.), ACS Symposium Series 443, American Chemical Society, Washington, D.C., p. 371 (1991).

182. P. Proksch and E. Rodriguez, *Phytochemistry, 22*:2335 (1983).

183. W. S. Bowers, *Insecticide Mode of Action* (J. R. Coates, ed.), Academic Press, New York, p. 403 (1982).

184. M. B. Isman, *Insecticides of Plant Origin* (J. T. Arnason, B. J. R. Philogene, and P. Morand, eds.), ACS Symposium Series 387, American Chemical Society, Washington, D.C., p. 44 (1989).

185. C. S. Wilson, J. T. Smiley, and E. Rodriguez, *J. Econ. Entomol., 76*:993 (1983).

186. M. B. Isman and P. Proksch, *Phytochemistry, 24*, 1949 (1985).

187. P. Proksch, M. Proksch, G. H. N. Towers, and E. Rodriguez, *J. Nat. Prod., 46*:331 (1983).

188. F. Bohlmann, J. Jakupovic, M. Ahmed, M. Wallmeyer, H. Robinson, and R. M. King, *Phytochemistry, 20*:2383 (1981).

189. J. A. Klocke, M. F. Balandrin, R. P. Adams, and E. Kingsford, *J. Chem. Ecol., 11*:701 (1985).

190. L. Crombie, *Recent Advances in the Chemistry of Insect Control II* (L. Crombie, ed.), Spec. Pub. 79, The Royal Society of Chemistry, Cambridge, p. 23 (1990).

191. L. Crombie, D. E. Games, N. J. Haskins, and G. F. Reed, *J. Chem. Soc., Perkin Trans, I*,2255 (1972).

192. B. S. Joshi, Y. N. Karnat, T. R. Govlindachari, and A. K. Ganguly, *Tetrahedron, 25*:1453 (1969).

193. J. K. Rupprecht, C.-J. Chang, J. M. Cassady, and J. L. McLaughlin, *Heterocycles, 24*:1197 (1986).

194. S. D. Jolad, J. J. Hoffman, K. H. Schram, and J. R. Cole, *J. Organ. Chem., 47*:3151 (1982).

195. J. K. Rupprecht, Y.-H. Hui, and J. L. McLaughlin, *J. Nat. Prod., 53*:237 (1990).

196. Y.-H. Hui, J. K. Rupprecht, Y. M. Liu, J. E. Anderson, D. L. Smith, C.-J. Chang, and J. L. McLaughlin, *J. Nat. Prod., 52*:463 (1989).

197. K. J. Mikolajczak, R. V. Madrigal, J. K. Rupprecht, Y.-H. Hui, Y.-M. Liu, D. L. Smith, and J. L. McLaughlin, *Experientia, 46*:324 (1990).

198. K. Kawazu, J. P. Alcantara, and A. Kobayashi, *Agric. Biol. Chem., 53*:2719 (1989).

199. M. Nonfon, F. Lieb, H. Moeschler, and D. Wendisch, *Phytochemistry, 29*:1951 (1990).

200. A. Alkofahi, J. K. Rupprecht, D. L. Smith, C.-J. Chang, and J. L. McLaughlin, *Experientia, 44*:83 (1988).

201. T. G. McCloud, D. L. Smith, C.-J. Chang, and J. M. Cassady, *Experientia, 43*:947 (1987).

202. K. L. Mikolajczak, J. L. McLaughlin, and J. K. Rupprecht, U.S. Patent 4,721,727 (1988).

203. A. Alkofahl, J. K. Rupprecht, J. E. Anderson, J. L. McLaughlin, K. L. Mikolajczak, and B. A. Scott, *Insecticides of Plant Origin* (J. T. Arnason, B. J. R. Philogene, and P. Morand, eds.), ACS Symposium Series 387, American Chemical Society, Washington, D.C., p. 25 (1989).

204. K. L. Mikolajczak, J. L. McLaughlin, and J. K. Rupprecht, U.S. Patent 4,855,319 (1989).

205. L. Born, F. Lieb, J. P. Lorentzen, H. Moeschler, M. Nonfon, R. Soellner, and D. Wendisch, *Planta Med.*, 56:312 (1990).

206. Y. Fujimoto, T. Eguchi, K. Kakinuma, N. Ikekawa, M. Sahai, and Y. K. Gupta, *Chem. Pharm. Bull.*, 36:4802 (1988).

207. H. F. Moeschler, W. Pfluger, and D. Wendisch, U.S. Patent 4,689,232 (1987).

208. K. Ohsawa, S. Atsuzawa, T. Mitsui, and I. Yamamoto, *J. Pestic. Sci.*, 16:93 (1991).

209. R. M. Hollingworth, K. I. Ahammadsahib, G. G. Gadelhak, and J. L. McLaughlin, 203rd American Chemical Society National Meeting, San Francisco, CA., Abstract 156, (1992).

210. M. Maeda, T. Kodama, T. Tanaka, Y. Ohfune, K. Nomoto, K. Nishimura, and T. Fujita, *J. Pestic. Sci.*, 9:27 (1984).

211. M. Maeda, T. Kodama, T. Tanaka, H. Yoshizumi, T. Takemoto, K. Nomoto, and T. Fujita, *Chem. Pharm. Bull.*, 34:4892 (1986).

212. M. Maeda, T. Kodama, M. Saito, T. Tanaka, H. Yoshizumi, K. Nomoto, and T. Fujita, *Pestic. Biochem. Physiol.*, 28:85 (1987).

213. R. Neumann and H. H. Peter, *Experientia*, 43:1235 (1987).

214. W. G. Williams, G. G. Kennedy, R. T. Yamamoto, J. D. Thacker, and J. Bordner, *Science*, 207:888 (1980).

215. G. G. Kennedy and M. B. Dimock, *Pesticide Chemistry: Human Welfare and the Environment* (J. Miyamoto and P. Kearney, eds.), Pergamon Press, New York, Vol. 2, p. 123 (1983).

216. G. G. Kennedy, R. R. Farrar, Jr., and R. K. Kashyap, *Naturally Occurring Pest Bioregulators* (P. H. Hedrin, ed.), ACS Symposium Series 449, American Chemical Society, Washington, D.C., p. 150 (1991).

217. P. M. Guerin and M. F. Ryan, *Experientia*, 36:1387 (1980).

218. P. N. Confalone, E. M. Huie, and N. G. Patel, *Tetrahedron Lett.*, 24:5563 (1983).

219. M. Hayashi, K. Wada, and K. Munakata, *Agric. Biol. Chem.*, 45:1527 (1981).

220. M. Hayashi, K. Wada, and K. Munakata, *Agric. Biol. Chem.*, 47:2653 (1983).

221. C. Marcus and E. P. Lichtenstein, *J. Agric. Food Chem.*, 27:1217 (1979).

222. E. P. Lichtenstein and J. E. Casida, *J. Agric. Food Chem.*, 11:410 (1963).

223. J. Reichling, B. Merkel, and P. Hofmeister, *J. Nat. Prod.*, 54:1416 (1991).

224. L. Jurd and G. D. Manners, *J. Agric. Food Chem.*, 28:183 (1980).

225. M. Cole and G. N. Rolinson, *Appl. Microbiol.*, 24:660 (1972).

226. E. A. S. la Croix, S. E. Mhasalkar, P. Mamalis, and F. P. Harrington, *Pestic. Sci.*, 6:491 (1975).

227. J. L. McLaughlin, B. Freedman, R. G. Powell, and C. R. Smith Jr., *J. Econ. Entomol.*, 73:398 (1980).
228. A. Alkofahi, J. K. Rupprecht, J. E. Anderson, J. L. McLaughlin, K. L. Mikolajzak, and B. A. Scott, *Insecticides of Plant Origin* (J. T. Arnason, B. J. R. Philogene, and P. Morand, eds.), ACS Symposium Series 387, American Chemical Society, Washington, D.C., p. 25 (1989).
229. I. Kubo, M. Kim, W. F. Wood, and H. Naoki, *Tetrahedron, Lett.*, 27:4277 (1986).
230. T. Kamikawa, S. Fujie, Y. Yamagiwa, M. Kim, and H. Kawaguchi, *Chem. Commun.*, 195 (1988).
231. T. Takemoto and T. Nakajima, *J. Pharm. Soc. Japan*, 84:1183 (1964).
232. H. Iwasaki, T. Kamiya, O. Oka, and J. Ueyanagi, *Chem. Pharm. Bull.*, 13:753 (1965).
233. E. Taniguchi, K. Imamura, F. Ishibashi, T. Matsui, and A. Nishio, *Agric. Biol. Chem.*, 53:631 (1989).
234. S. Yamauchi and E. Taniguchi, *Biosci. Biotech. Biochem.*, 56:412 (1992).
235. U. Mothes and K. A. Seitz, *Pestic. Sci.*, 13:426 (1982).
236. H. Decker, C. Bormann, H. P. Fiedler and H. Zahner, *J. Antibiotics*, 42:230 (1989).
237. H. A. Kirst, K. H. Michel, J. S. Mynderase, E. H. Chio, R. C. Yao, W. M. Nakasukasa, L. D. Boeck, J. L. Occlowitz, J. W. Paschal, J. B. Deeter, and G. D. Thompson (D. R. Baker, J. G. Fenyes, and J. J. Steffens, eds.), ACS Symposium Series 504, American Chemical Society, Washington, D.C., p. 214 (1992).
238. J. A. Klocke, M. Arisawa, S. S. Handa, A. D. Kinghorn, G. A. Cordell, and N. R. Farnsworth, *Experientia*, 41:379 (1985).
239. R. G. Powell, K. L. Mikolajczak, B. W. Zilkowski, H. S. M. Lu, E. K. Mantus, and J. Clardy, *Experientia*, 47:304 (1991).
240. R. J. Stonard, B. Isaac, J. T. Letendre, T. E. Nickson, P. B. Lavrik, N. LeVan, and S. W. Ayer, Special Conference V, American Chemical Society, (1992).
241. K. Kanbe, Y. Mimura, T. Tamamura, S. Yatagai, H. Naganawa, H. Nakamura, T. Takeuchi, and Y. Iitaka, *J. Nat. Prod.*, 45:458 (1992).
242. C. Satasook, M. B. Isman, and P. Wiriyachitra, *Pestic. Sci.*, 36 (1992).
243. J. Rovirosa, M. Sepulveda, E. Quezada, and A. San-Martin, *Phytochemistry*, 31:2679 (1992).
244. L. Lajide, P. Escoubas, and J. Mizutani, *J. Agric. Food Chem.*, 41:669 (1993).
245. J. A. Nathanson, E. J. Hunnicutt, L. Kantham, and C. Scavone, *Proc. Nat. Acad. Sci. USA*, 90:9645 (1993).

2

Pyrethroids

Clive A. Henrick

Sandoz Agro, Inc., Palo Alto, California

I. INTRODUCTION

The commercial development of pyrethroids (synthetic analogs of the natural pyrethrins) is one of the major success stories in the use of natural products as a source of leads for novel compounds possessing useful insecticidal activity. What began in the early years of this century as attempts to elucidate the chemical structures of the insecticidally active constituents of "pyrethrum extract" led, in the period from the late 1940s to the late 1960s, to the synthesis of several very effective household insecticides that were analogs of the natural pyrethrins. The breakthroughs that occurred in the early 1970s, with the discovery of a number of light-stable analogs suitable for field applications, catapulted the pyrethroids into a major area of industrial and academic research. In less than a decade this intense effort led to extensive commercial applications of pyrethroids, so that today they are the second largest class of insecticides.

In general, natural products have not been particularly useful as leads for the development of commercial insecticides. Many biologically active natural products are highly toxic to mammals while having only moderate activity against insect pests. The pyrethroids are an exception, having generally favorable toxicological and environmental properties. They are very active broad-spectrum insecticides and, although not

63

harmless to mammals, are generally not hazardous under normal circumstances due to their rapid metabolism and excretion in mammals.

Pyrethroids (and the pyrethrins) can be grouped together on the basis of their similar overall shape, physical properties (especially polarity), and experimentally observed action at nerve membranes. All of the compounds discussed in this chapter can be assigned reasonably to this class of insecticide. Pyrethroid research is a mature field and many aspects of this area have been extensively reviewed [1–6]. Synthetic aspects have been very well covered by Naumann [2]. This chapter will present a general overview of the field with emphasis on structure-activity relationships.

II. PYRETHRUM AND PYRETHRINS

A. Pyrethrum

Pyrethrum is the common name for the dried flowers, and the active insecticidal ingredients present in pyrethrum are known as pyrethrins. There is considerable confusion in the literature about the taxonomy of the plant species from which pyrethrum has been obtained. Recent revisions transfer these species from the genus *Chrysanthemum* to the genus *Tanacetum* [7]. There appear to be only two species of plants with historical and commercial importance, the painted daisy or Persian insect flower (red flower), *Tanacetum coccineum* (Willd.) Grierson, and the Dalmatian insect flower (white flower), *Tanacetum cinerariifolium* (Trevir.) Schultz-Bip. These are both daisylike herbaceous perennials of the sunflower family, Compositae. Although a few other closely related plant species contain significant amounts of pyrethrins, most related species, including the common field daisy, contain negligible amounts of pyrethrins [1, 8, 9].

The only commercially important pyrethrum plant species today is *T. cinerariifolium* (identical to *Chrysanthemum cinerariifolium* and *Pyrethrum cinerariifolium*). This species is a native plant in the Adriatic coastal regions of Yugoslavia (Dalmatia) and some parts of Caucasus (Caucasia). The painted daisy, *T. coccineum* (apparently identical to *Chrysanthemum coccineum, C. roseum, C. carneum, Pyrethrum coccineum, P. roseum, and P. carneum*) appears to be the oldest known source of pyrethrum. Although it has often been referred to in the early literature under these various names, designating different species, they all appear to be the same species. *T. coccineum* is a native plant of the plateau region of the Caucasus mountains and of Iran (Persia). The use of the dried and powdered flower heads of *T. coccineum* as an insecticide probably dates back many centuries. Although the discovery of the insecticidal activity of this "Persian dust" or "Persian insect powder" (pyrethrum) is lost in the distant past, its activity was certainly well known to the people of eastern Europe and territories further to the east prior to the year 1800. However, they kept it a secret from western Europe until early in the nineteenth century. A fascinating summary of the early literature and the conflicting reports on the introduction of pyrethrum into western Europe is given by McDonnell and his co-workers [8, 9]. By the 1820s, pyrethrum (prepared from *T. coccineum*) was being exported in large quantities by Armenian merchants from Caucasia to Russia and Europe. Russians had become aware of the material at an earlier date, presumably because of their military incursions into the region between the Black and Caspian seas.

The Dalmatian insect powder, prepared from *Tanacetum cinerariifolium*, had also been used for many years by the local population in this part of Yugoslavia. A German woman, Anna Rosauer living in Dubrovnik, is often credited with the rediscovery of the insecticidal activity of the flowers of this species [8–10]. She reportedly observed dead insects surrounding some discarded withered flowers, but this story seems somewhat oversimplified, as the flowers would have had to be crushed and powdered to be effective. Nevertheless, she was apparently involved in exporting this pyrethrum to Europe. By 1860, "insect powders" were being produced and exported to Europe and the United States from Caucasia, Persia, and Dalmatia. The subsequent production of pyrethrum around the world was heavily influenced by the two world wars. The two plant species (primarily *T. cinerariifolium*) were introduced into Japan in 1885. By the end of the First World War, Japan had become the principal supplier to the United States because the supply from Europe had been cut off. *T. cinerariifolium* was introduced into Kenya in 1928, and by the Second World War, this country had replaced Japan as the main supplier to the United States. Attempts to grow pyrethrum have been made in many countries around the world but usually without economic success [11]. Pyrethrum is still grown today in a number of countries for local use, but the principal exporters are East African countries; especially Kenya and to a lesser extent Tanzania, Zimbabwe, and Uganda. East Africa produces about 85% of the pyrethrum for the world market and the rest comes from South America, New Guinea, and even Australia (Tasmania) to a limited extent.

The flowers of *Tanacetum cinerariifolium* are picked by hand shortly after flowering and contain about 80% water. They are dried, either in the sun or using mechanical dryers, and ground. The finely ground material, containing about 1.3% pyrethrins by weight, is still used in the manufacture of mosquito coils, but since 1919, most of the pyrethrum has been refined by solvent extraction. The powder is usually extracted with a solvent such as hexane or kerosene to yield a dark viscous oleoresin concentrate containing about 30% pyrethins ("pyrethrum extract"). Various additional refining processes (mostly solvent based) are used to prepare a decolorized and dewaxed material suitable for use in aerosols. This further purification is necessary to remove components of the flowers that could cause allergic responses in some humans.

Although pyrethrins are present in small amounts throughout all parts of the plant, the highest concentration is in the flower head. The pyrethrin content in the flowers is at an optimum when they are almost fully open [12]. More than 90% of the pyrethrins in the flower head are localized in the secretory ducts of the achenes [12]. Here the material is protected from photodecomposition and isolated so that it is not readily available to insects feeding on or visiting the pyrethrum flowers. Today it is a widely published view that many secondary metabolites in plants are present in order to protect the plant against potential insect predators, thus conferring a selective advantage. Although this may be true in some cases, this argument is applied too widely without much thought or careful analysis. For example, the pyrethrins are generally much more active against insects by contact than by ingestion. In the case of *T. cinerariifolium*, the flowers are not an effective insecticide until they are crushed and powdered, nor is the plant free of insect and spider mite pests when in field cultivation (the major insect pests are various thrips species [8, 11]).

B. Pyrethrins

A total of six closely related insecticidally active esters (pyrethrins) have been isolated from the pyrethrum extract. Their structures are given in Figure 1. They are esters of two carboxylic acids, chrysanthemic acid ($R^1 = CH_3-$) and pyrethric acid ($R^1 = CH_3O_2C-$), with three alcohols (rethrolones), pyrethrolone, cinerolone, and jasmolone. The esters differ only in the terminal substituents on the side chains of the acid and alcohol moieties. The pyrethrum extract usually contains about equal quantities of chrysanthemic and pyrethric acid esters, with pyrethrin I (*1*) and pyrethrin II (*2*) accounting for 65–75% of the mixture of esters [12]. The absolute stereochemistry in the cyclopropane ring of the acid portion is 1R, 3R and the α-carbon of the alcohol position has the (S) configuration. However, for convenience, the stereochemistry of the substituent at the C-3 carbon atom of the cyclopropane ring of pyre-

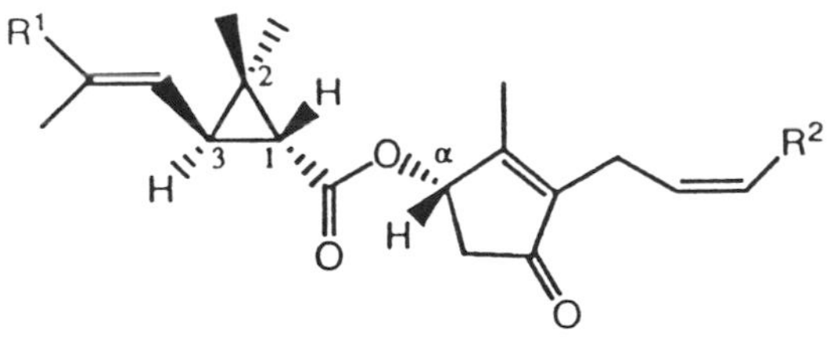

$$E; (1R)\text{-}trans; \alpha S; Z'$$

pyrethrin I [1]	$R^1 = CH_3 \cdot$	$R^2 = \cdot CH=CH_2$
jasmolin I	$R^1 = CH_3 \cdot$	$R^2 = \cdot CH_2CH_3$
cinerin I	$R^1 = CH_3 \cdot$	$R^2 = \cdot CH_3$

pyrethrin II [2]	$R^1 = CH_3O_2C \cdot$	$R^2 = \cdot CH=CH_2$
jasmolin II	$R^1 = CH_3O_2C \cdot$	$R^2 = \cdot CH_2CH_3$
cinerin II	$R^1 = CH_3O_2C \cdot$	$R^2 = \cdot CH_3$

Figure 1 Active components of pyrethrum.

thrins is generally given as cis or trans relative to the ester group with respect to the plane of the cyclopropane ring [13]. This nomenclature is used to avoid the confusion that may arise, using the R and S designation, as the C-3 substituent is varied. Thus, the pyrethrins are designated as having the (1R)-trans stereochemistry in the acid moiety.

Complete structural determination of the active constituents of pyrethrum took many years. Probably the earliest published study on the active insecticidal components was that of Ragazzini, as cited by De Visiani in 1854 [14]. Along with a description of the early uses and trade of insect powder from *Tanacetum coccineum* and *T. cinerariifolium*, this article summarized attempts by Ragazzini to isolate the active materials by alcohol extraction and distillation. He concluded that the insecticidal action of the smoke obtained by burning the powder on activated charcoal was not due to the individual decomposition products. All of the early investigations have been reviewed in detail [8, 11]. The important work of Staudinger and Ruzicka, which elucidated the main structural features, was carried out from 1910 to 1916 (but not published until 1924). This research, and that of LaForge, Schechter, Barthel, and their co-workers in the USDA carried out from 1935 to 1945, have been well summarized [11]. The overall structural elucidation of the pyrethrins, including establishment of their complete absolute stereochemistry in the early 1970s, has been reviewed by a number of authors [4, 5, 15].

As all of the natural pyrethrins are unstable in air and light, their principal use has been in the household sector. All six of the esters (Figure 1) show both kill and knockdown activities against the housefly, but 1 is the most active for kill and 2 is the most effective as a knockdown agent for flying insects. The pyrethrins are lethal against a wide range of insect species, but their relative toxicities vary with the insect species and the conditions of the treatment [15].

III. EARLY ANALOGS AND DISCOVERY OF PHOTOSTABLE PYRETHROIDS

A. Alcohol Variations

As soon as the general structural features of the pyrethrins had been elucidated, synthesis work was begun with the aim of developing analogs with improved physical and biological properties. Between 1910 and 1916, Staudinger and Ruzicka synthesized acids structurally related to chrysanthemic acid and esterified them with (S)-pyrethrolone. They also prepared various esters of chrysanthemic acid. Although none of their analogs were particularly active as insecticides, they did make the important discovery that some substituted benzyl esters [e.g., piperonyl (3)] of chrysanthemic acid had measurable insecticidal activity [16]. They also prepared inactive analogs, such as 4, without the cyclopropane ring

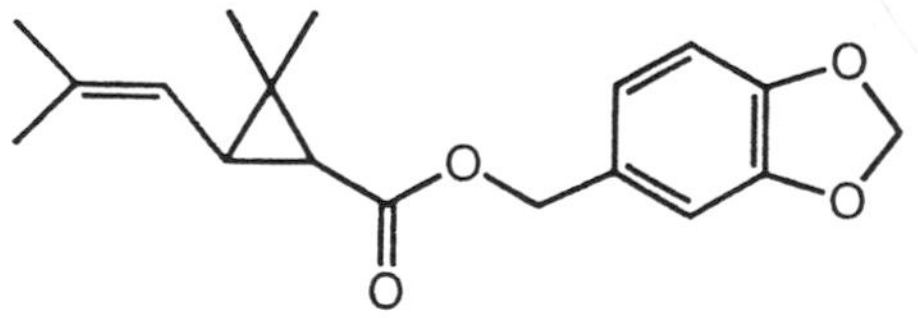

(3)

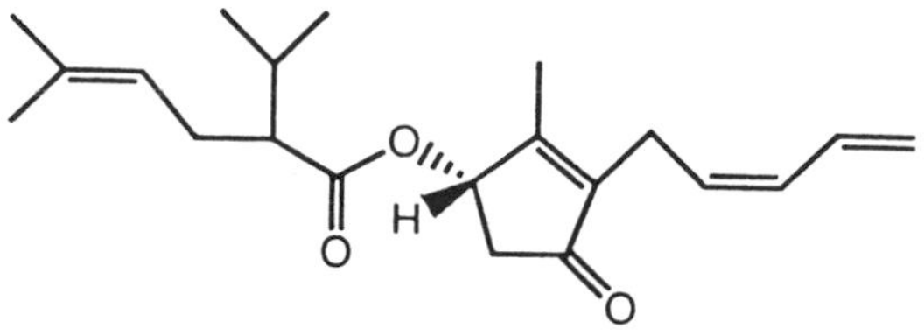

(4)

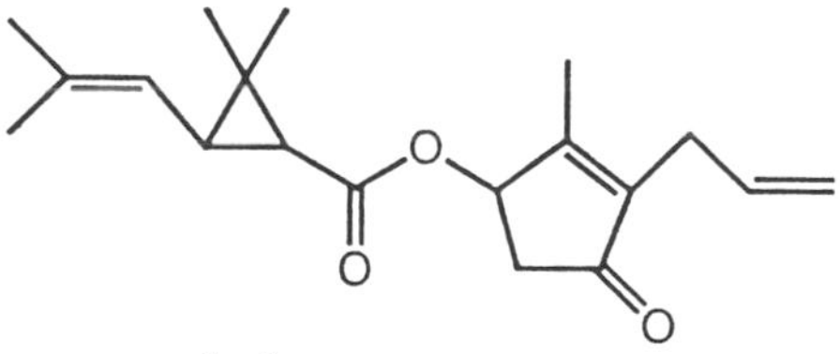

allethrin
(1RS)-*cis-trans;* αRS

(5)

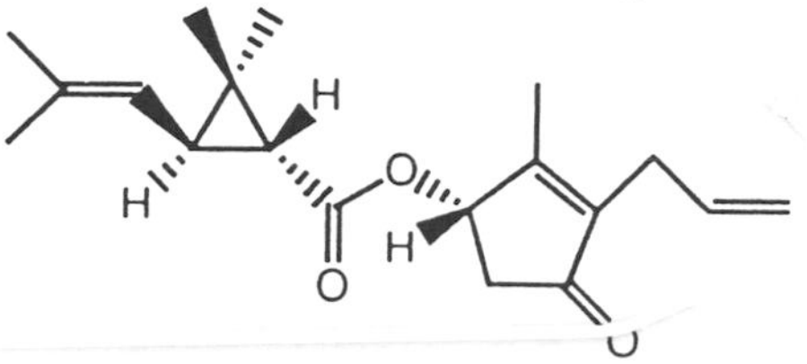

S-bioallethrin

(6)

Analog synthesis in the period from the 1940s until around 1970 produced a number of commercial chrysanthemates that were effective substitutes for pyrethrum. Their lack of stability in sunlight essentially limited their application to indoor household markets. These pyrethroids were usually introduced commercially as mixtures of stereoisomers and, subsequently, as their partially or completely resolved forms. LaForge et al. corrected some details in the pyrethrin structure and synthesized allethrin (5) [17]. This compound was the first commercial pyrethroid and the first analog to show useful insecticidal activity. It was initially introduced in 1952 as a mixture of eight stereoisomers (with a typical cis:trans ratio of 1:3). It is more effective than the pyrethrins in mosquito coils and mats (thermal fumigation) due to its increased volatility and thermostability. Early commercial analog synthesis emphasized "knockdown" activity against flying insects (rather than "kill" activity). For example, the invention of the mosquito coil was the driving force for early pyrethroid research in Japan [18]. Subsequently, several partially resolved mixtures of isomers of 5 were commercialized, including bioallethrin,* which contains the esters from the (1R)-trans acid and the racemic alcohol. This was eventually followed by the introduction of S-bioallethrin (6), which is the ester of the (1R)-trans acid and (S)-allethrolone. Although the spectrum of insecticidal activity of 5 is narrow, its good activity against insects such as the housefly, *Musca domestica* (Table 1), has enabled it and the more active resolved products bioallethrin and S-bioallethrin to be used for 40 years in thermal fumigation and as knockdown agents in household sprays. Mixtures of the isomers of 5 show considerable antagonism in

---

*The prefix bio usually designates that the ester is prepared from the natural (1R)-trans isomer.

Table 1 Toxicity of Pyrethroids at 25°C

Compound	Heliothis virescens (Topical) (μg/third instar larva)[a] LD50	Spodoptera exigua (Topical) (μg/third instar larva)[a] LD50	Musca domestica (Topical) (μg/adult)[b] LD50	Aphis fabae Foliar residue)[c] Adults LC50 (ppm)	Tetranychus urticae (Contact and Repellency)[c] Adults LC50 (ppm)
Pyrethrins[d]	0.48	—	1.6	>1000	>1000
Allethrin (5)	10	—	1.0	310	>1000
Bioallethrin	2.9	—	0.62	430	>1000
Tetramethrin (7)	0.47	—	0.23	>1000	420
Resmethrin (8)	0.080	0.022	0.030	250	>1000
Bioresmethrin	0.034	—	0.018	—	—
Dimethrin	22	—	1.6	>1000	—
(1R)-Phenothrin (10)[e]	0.22	—	0.095	63	>1000
Fenpropathrin (14)	0.027	0.0046	0.053	—	>1000
Kadethrin (17)	0.035	—	0.15	—	0.43
Permethrin (18)[f]	0.018	0.0070	0.031	18	16
Cypermethrin (19)[f]	0.0069	0.005	0.0065	—	4.5
Deltamethrin (20)	0.00084	0.003	0.0019	—	2.3
Fenvalerate (22)	0.012	0.018	0.042	8.5	4.6

[a] Mortality scored after 72 hr.
[b] Mortality scored after 24 hr.
[c] Mortality scored after 48 hr.
[d] Pyrethrum containing 20% pyrethrins; bioassay results are calculated for active ingredient.
[e] Isomer ratio of cis : trans is 1 : 4.
[f] Isomer ratio of cis : trans is 3 : 7.

their action against the housefly. This antagonism is correlated with the content of (R)-allethrolone esters present. Thus, to observe the full insecticidal potential, the pure isomer 6 is needed [19].

The discovery that imidomethyl chrysanthemates are also active against the housefly [4] led to the commercial development of tetramethrin (7), another effective knockdown agent against public health pests [20]. It also shows good toxic activity against the housefly (Table 1). Tetramethrin is usually used in mixtures with other pyrethroids along with synergists to improve its spectrum of insect toxicity.

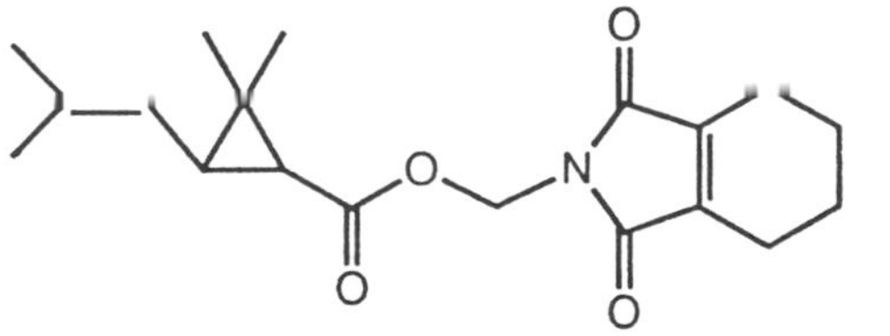

tetramethrin
(1RS)-cis-trans
(7)

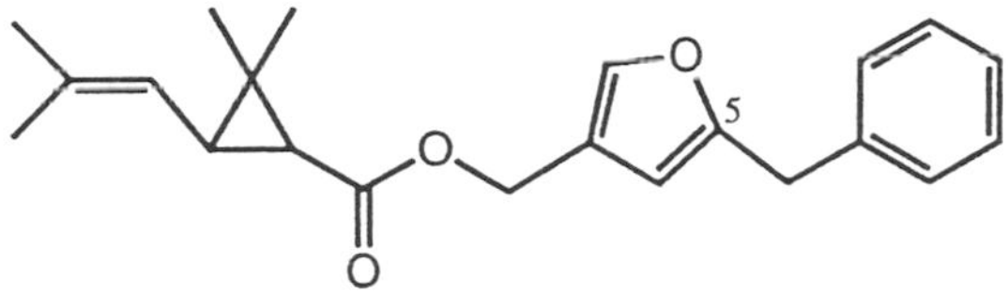

resmethrin
(1RS)-cis-trans
(8)

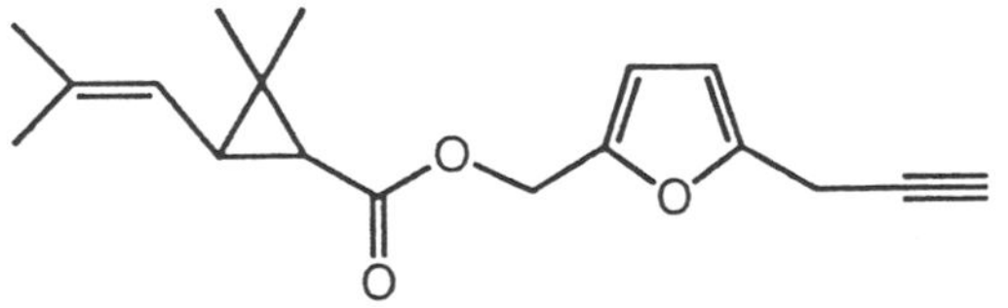

prothrin
(1RS)-cis-trans
(9)

A major advance in pyrethroid research was the discovery of resmethrin (8) by Elliott and co-workers at the Rothamsted Experimental Station in the United Kingdom [21, 22]. Although it has lower knockdown activity, this 5-benzyl-3-furylmethyl ester was the first analog to show much higher lethal activity than pyrethrum against many insect species (Table 1). Bioresmethrin is the corresponding ester prepared from (1R)-trans-chrysanthemic acid. Both resmethrin and bioresmethrin have remarkably low mammalian toxicity even by intravenous dosing. They are unstable in air and sunlight [23] and, thus, are used as nonresidual contact sprays in consumer, public health, and greenhouse applications. There is an advantage to using pure bioresmethrin, because as with allethrin, there is antagonism among some of the isomers in resmethrin that results in lower than expected toxicity against the housefly [19]. There are a number of other substituted furylmethyl esters that show insecticidal activity [4, 22]. For example, prothrin (9) is more volatile than the pyrethrins and is an effective knockdown agent for flying insect control, especially in mosquito coils [24, 25]. However, it is much less toxic than resmethrin in topical applications against the housefly and the German cockroach [22, 25].

Many substituted benzyl esters of chrysanthemic acid have been prepared since the pioneering work of Staudinger and Ruzicka. In the early 1960s, some simple esters such as the 2,4-dimethylbenzyl ester (dimethrin, Table 1) and the 6-chloro-3,4-methylenedioxybenzyl ester (barthrin) were shown to have low activity [4]. Elliott and co-workers then observed that the 4-allylbenzyl and 4-allyl-2,6-dimethylbenzyl esters showed better insecticidal activity [26]. As part of their extensive investigations into the structure-activity relationships of aromatic and olefinic esters of chrysanthemic acid, they also substituted an aromatic ring for the (Z)-pentadienyl side chain in the alcohol moiety of pyrethrin I and prepared the 4-benzylbenzyl and related esters. This work led to the discovery of the 5-benzyl-3-furylmethyl esters such as (8) described earlier [27–30].

(1R)-phenothrin
(1R)-cis-trans   (10)

(1R)-cyphenothrin
(1R)-trans-isomer-rich   (11)

A very important advance, made independently in England and Japan at about the same time, was the idea of using the known 3-phenoxybenzyl alcohol to esterify chrysanthemic acid. In Japan, studies on 3-benzylbenzyl esters led to the synthesis of the 3-phenoxybenzyl ester (phenothrin) in the late 1960s [18, 31]. This pyrethroid was introduced commercially as the partly resolved (1R)-cis-trans isomeric mixture, (1R)-phenothrin (10), for household and public health use. It is more stable than resmethrin and shows good toxicity against many insect species although it is, in general, not as active as resmethrin (Table 1). It also shows much lower knockdown activity against the housefly than pyrethrum or tetramethrin. (1R)-Phenothrin is one of the safest pyrethroids, having very low mammalian toxicity even by the intravenous route [31, 32]. In England, Elliott and co-workers had postulated that the activity of 8 was associated with the suitable spacing of the unsaturated rings. Extending their work on 4-benzylbenzyl esters, they discovered that 3-phenoxybenzyl esters were more active than the corresponding 4-benzylbenzyl and 3-benzylbenzyl esters [27].

In Japan, chemists at Sumitomo Chemical Company in the early 1970s made the very important discovery that an α-cyano substituent on the benzylic carbon atom of the alcohol moiety increased the activity of 10 and related esters [33]. Thus, (1R)-cyphenothrin (11) is two to four times more toxic than 10 against a variety of insect pests [34]. These discoveries were a significant step toward the development of photostable pyrethroids.

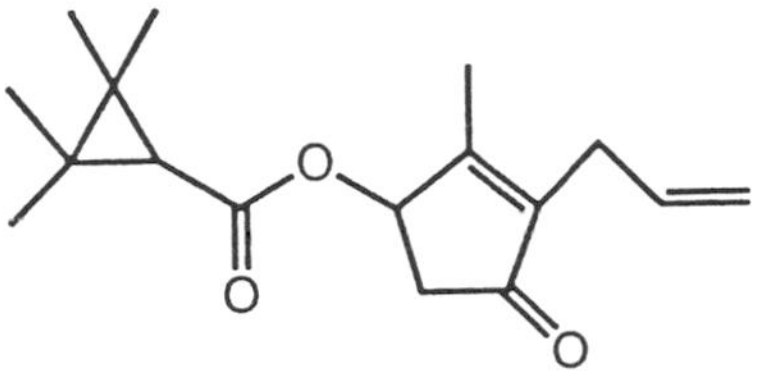

terallethrin   (12)

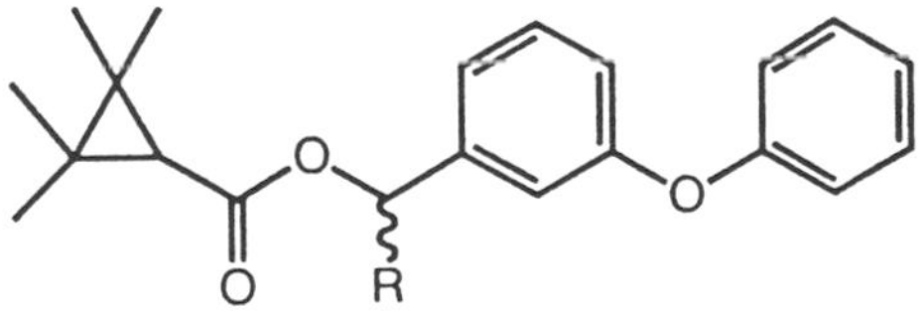

R=H   (13)

R=CN, fenpropathrin   (14)

B. Acid Variations

Early investigators had carried out variations in the acid moiety of the pyrethrins without significant success, although a number of substituted 3-phenyl-2,2-dimethylcyclopropanecarboxylic esters showed moderate activity [4]. However, esters of 2,2,3,3-tetramethylcyclopropanecarboxylic acid were shown to have insecticidal activity comparable to that of the chrysanthemates [35]. Terallethrin (12) was found to be a more volatile knockdown agent than allethrin. Its discovery was followed by the synthesis of 13 and 14 in the early 1970s [33]. The esters 13 and 14 (fenpropathrin) are much more photostable than any previously known analog and were undoubtably the first pyrethroids to be discovered that are stable enough to be used in crop applications. However, even though fenpropathrin shows high activity against insect and mite pest species (Table 1), it was not introduced commercially into agriculture until 1980, possibly due to its unusually high mammalian toxicity [LD_{50} (rats, oral) of ~ 25–70 mg/kg].

(1RS)-trans; αRS   (15)

bioethanomethrin
(1R)-trans   (16)

Kadethrin (RU 15525)
(E);(1R)-cis   (17)

Another very important discovery was the synthesis of both the cis and trans isomers of the dichlorovinyl analog of chrysanthemic acid by Farkaš and co-workers in the late 1950s [36]. They prepared the allethrolone ester (15) of the trans acid and showed that it had activity similar to that of allethrin against the housefly; however, they did not follow up this lead and the significance of their discovery was not recognized at that time. Other variations in the acid moiety by chemists at Roussel-Uclaf resulted in the discovery of bioethanomethrin (16) [37], which is a little more toxic than bioresmethrin against some insect species [22], and of Kadethrin™ (17) [38]. The latter analog is very active as a knockdown agent against the housefly and the yellowfever mosquito, *Aedes aegypti* [38] and shows considerable lethal activity against many insect species (Table 1). However, it does not show good knockdown activity against the mosquito *Culex pipiens pallens* [39].

permethrin (R=H)
(1RS)-cis-trans   (18)

cypermethrin (R=CN)
(1RS)-cis-trans; αRS   (19)

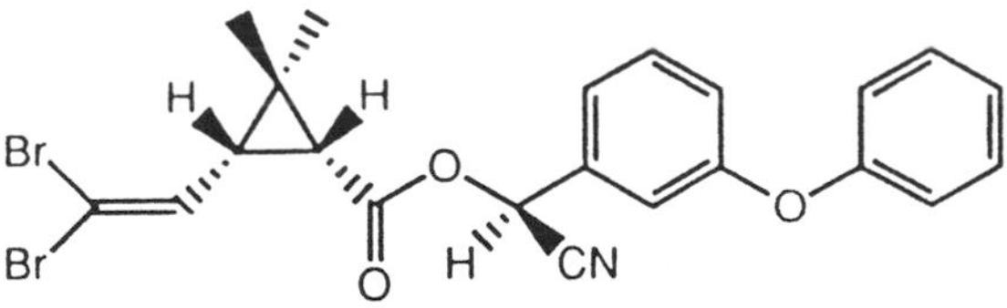

deltamethrin
(1R)-*cis*; αS

(20)

Two major breakthroughs occurred in the early 1970s. First, Elliott and co-workers, while continuing their systematic structure-activity studies on the acid moiety, studied variations in the 3-substituent on the cyclopropane ring. They prepared active derivatives in which the 3-dimethylvinyl group was replaced by (Z)-butadienyl, ethynyl, monochlorovinyl, and, finally, dichlorovinyl, Thus, they discovered permethrin (*18*) [40], cypermethrin (*19*), and deltamethrin (*20*) [41]. Permethrin is at least 20 times more stable in air and sunlight than chrysanthemates such as resmethrin (*8*) and is more toxic to insects than was predicted by structure-activity studies [30]. Second, a significant breakthrough was achieved by Ohno and co-workers at Sumitomo Chemical Company in Japan. They had been studying α-substituted phenylacetates and observed toxicity against the housefly with the 5-benzyl-3-furylmethyl ester (*21*) prepared from 2-ethylphenylacetic acid [42]. This work led to the synthesis of fenvalerate (*22*) [43, 44], which was introduced commercially in 1976 as the racemate. This was the first photostable pyrethroid to be introduced into commerce. The fully resolved (*S,S*) isomer, esfenvalerate (*23*), was eventually commercialized in 1986.

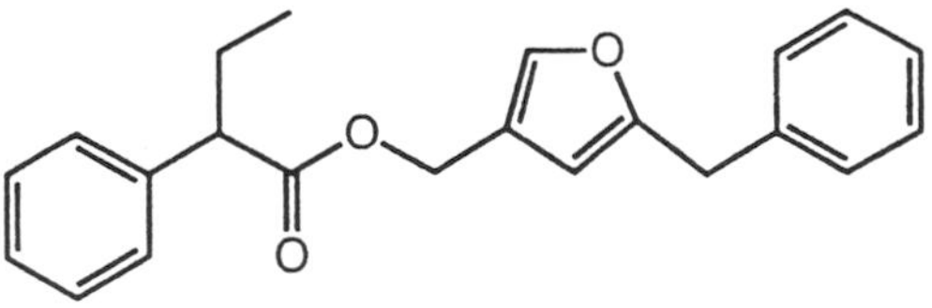

(21)

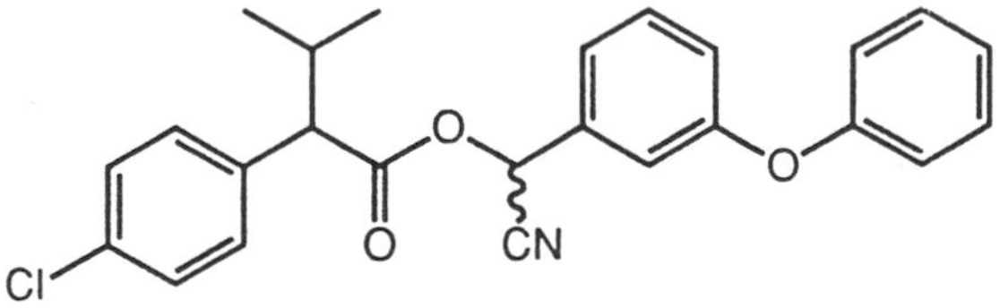

fenvalerate
RS; αRS

(22)

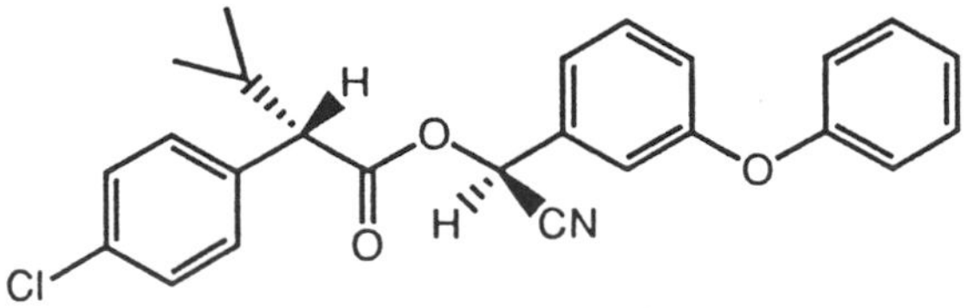

esfenvalerate
S;S

(23)

Thus, it is possible to replace the photolabile centers in both the acid and alcohol parts of pyrethrin I (*1*) with alternative groups and obtain analogs that have both increased insecticidal activity and much greater stability in air and light. Both types of structures, represented by *18* and *22*, show much higher insecticidal activity than the pyrethrins (Table 1). Their photostability expands the range of utility, especially for applications in agriculture. Following these discoveries, essentially every major agrochemical company as well as many university groups initiated synthetic programs on pyrethroids that resulted in the rapid expansion of this field, the discovery of many insecticidally active structures, and the commercialization of a number of additional pyrethroids [1, 2]. The pyrethroids *18*, *19*, *20*, and *22* are broad-spectrum insecticides with high activity, especially against lepidopterous species. The fully resolved analog deltamethrin (*20*) is the most active of this group (Table 1), although the corresponding pure (1R)-*cis;αS* isomer of cypermethrin (*19*) has similar activity to *20* [45].

The toxicity at 25°C of most of the pyrethroids discussed in this section against the tobacco budworm, *Heliothis virescens*, the beet armyworm, *Spodoptera exigua*, the housefly, *Musca domestica*, the bean aphid, *Aphis fabae*, and the two-spotted spider mite, *Tetranychus urticae*, are given in Table 1. All of the bioassay results presented in the tables in this chapter were carried out in Sandoz Agro, Inc. laboratories using published bioassay methods [46, 47]. Elliott and co-workers have also listed the relative toxicities for most of these analogs [48] and Naumann has extensively summarized the bioassay results of various research groups [1]. In general, insects do not die rapidly on treatment with pyrethroids, and insects that have been knocked down can recover. For example, it was observed that with the lepidopterous species *H. virescens, Spodoptera frugiperda*, and *Trichoplusia ni*, mortality in many cases continued to increase until at least 72 hr after treatment, so that LD_{50} values based on mortality after less than 48 hr did not accurately reflect the toxicity of a pyrethroid [49]. Similar observations have been made for the armyworm *Pseudaletia separata* [50] and the migratory grasshopper, *Melanoplus sanguinipes* [51]. Thus, it is important with many insect bioassays, particularly against lepidopterous species, to use 72 hr mortality counts for comparative purposes (see Table 1). For the housefly, *M. domestica*, mortality scored after 24 hr is satisfactory.

IV. STRUCTURE-ACTIVITY RELATIONSHIPS

The high insecticidal activity of pyrethroids depends on the overall shape and asymmetry of the molecule and on its lipophilicity [1, 3, 28,

29, 52]. There are large differences in activity among the various stereoisomers of individual pyrethroids; hence, the appropriate chirality of the molecule is particularly important. Other properties such as electron density and polarizability are of secondary importance. Pyrethroids are flexible molecules in solution, with the various segments able to rotate readily about the single bonds, so that the relevant conformation for biological activity may not be evident from standard ground-state molecular modeling studies. Interestingly, almost every structural element of pyrethrin I (1) (Figure 1) can be replaced by an analogous suitable group to give analogs with high insecticidal activity. Thus, the structural requirements for pyrethroid activity are much broader than originally thought [27]. In this section, a generally accepted definition of a pyrethroid has been adopted that includes all structural types that have the overall shape, stereochemical features, and biological properties typically assigned to this class of insecticide.

The common structural features of most of the more active pyrethroids are the presence of two suitably spaced centers of unsaturation (aromatic rings or olefinic bonds) at the extremities of the molecule and a geminal dimethyl group, or its steric equivalent, beta to the ester or other linking group. A large number of variations have been studied in both the acid and alcohol moieties of 1 and in the linking group. However, the effect on activity of changing one part of the molecule depends to some extent on the nature of the other groups present. Therefore, one cannot assume that a change in the "acid" moiety will produce changes in activity that can be generalized for esters (and other linkages) prepared from all of the various types of "alcohols" and vice versa. In addition, different structural modifications can produce considerable variations in activity with different insect species, thus making overall structure-activity generalizations difficult. Within a particular series of analogs, good structure-activity correlations can be obtained for a specific insect bioassay, but small changes in the structure, such as minor substituent variations, can dramatically alter the biological activity (e.g., [46, 52]). This could be due, at least in part, to changes in the preferred conformation of the molecule.

The lipophilicity of the pyrethroid must be within an optimum range in order for it to show high insecticidal activity. In structure-activity work, extensive use has been made of octanol-water partition coefficients, which are often estimated indirectly by reverse-phase chromatography [53, 54]. With the housefly, there appear to be different optimum partition values for lethality and for knockdown activity. Rapid penetration of the pyrethroid into the insect, observed as fast knockdown, does not correlate well with the highest lethality [55]. In this chapter, the emphasis is on the insect lethality of the pyrethroids.

A. Chrysanthemates

A number of structure-activity generalizations can be made for chrysanthemic acid derivatives [1, 3, 28, 29]. Insecticidally active esters of 3-substituted acids all have the natural (1R) configuration (as in Figure 1), whereas the corresponding (1S) isomers are inactive. The stereochemistry at C-3 is generally not critical; both the (1R)-*cis* and (1R)-*trans* isomers can show high activity. Which of the two is more active depends on the overall structure and the insect test species (see below). The cyclopropane ring requires a geminal dimethyl group at C-2 for the highest activity, although the 2,2-dichloro analogs and some spiroalkyl derivatives show good activity. The function of these substituents is probably mainly steric. There is considerable variation in substituents at C-3 on the cyclopropane ring that can give analogs with high activity, although there are steric limitations to the bulk of many of the C-3 groups for most insect species. The 2,2,3,3-tetramethylcyclopropanecarboxylates, such as *12–14*, show excellent activity [35, 56, 57], but this is an exceptional case. With the 2,2-dimethylcyclopropanecarboxylates only a limited number of other 3,3-disubstitutions, such as dichloro and certain spiro derivatives, give analogs with any reasonable activity [57]. With most other 3-substituted-2,2-dimethylcyclopropanecarboxylates, the introduction of an additional methyl group at C-3 dramatically lowers the activity [58]. Therefore, except for certain spiro derivatives (see below), steric limitations severely restrict the size of the substituents in 3,3-disubstituted analogs. Introduction of an additional methyl group at C-1 also considerably reduces activity and the pentamethylcyclopropane derivative (*24*) is inactive [59].

(24)

R= (Z)CH$_3$CH$_2$CH=CH·   (25)

R= (E,E)CH$_3$CH=CHCH=CH·   (26)

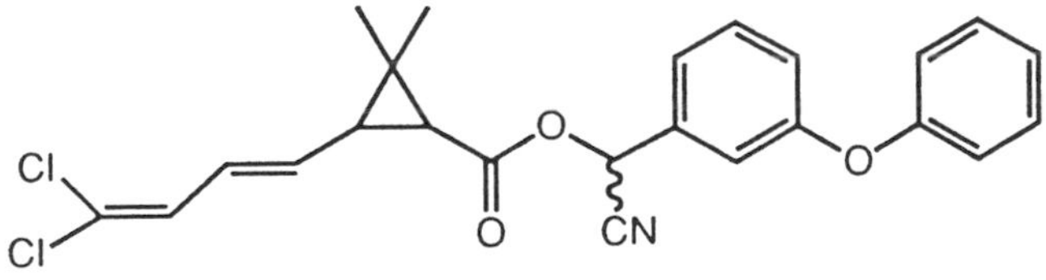

(1RS)-cis; αRS (27)

Unsaturated substituents (olefinic or aromatic) at C-3 usually have much higher activity than saturated substituents [60, 61]. The (Z)-1-butenyl analog (25) and the ethano-bridged analog (16) both have a little higher activity than the corresponding 3-isobutenyl analog, bioresmethrin. In this (1R)-trans series, dienyl chains at C-3 such as (E)- and (Z)-1,3-butadienyl and 1,3-pentadienyl give analogs, e.g., 26, which show even higher activity [60]. Of the other groups at C-3 in this series studied by Elliott and co-workers, the 3-methoxyiminomethyl group gives quite active analogs [61].

An extensive investigation has been carried out on halovinyl substituents at C-3. As discussed in Section IIB, the 2,2-dichlorovinyl group gives analogs (e.g., 18 and 19) that are somewhat more active than the corresponding chrysanthemic acid derivatives [40, 62, 63]. The 2,2-difluorovinyl and especially the 2,2-dibromovinyl derivatives (e.g., 20) are also very active pyrethroids. In contrast to the chrysanthemates, in the 3-(dihalovinyl) series the (1R)-cis isomer often shows higher activity than the corresponding (1R)-trans isomer, although which isomer is the more active depends on the halo group, the type of ester, and the insect species (see below). The (E)- and (Z)-3-(monohalovinyl) analogs in 3-phenoxybenzyl and α-cyano-3-phenoxybenzyl esters are generally less active than the corresponding 3-(dihalovinyl) side chain analogs [64]. A number of halogenated 1,3-dienyl substituents at C-3 have also been investigated [65]. High activity in these halogenated dienes appears to require coplanarity of the conjugated double bonds of the chain. The analog 27 shows similar activity to cypermethrin (19).

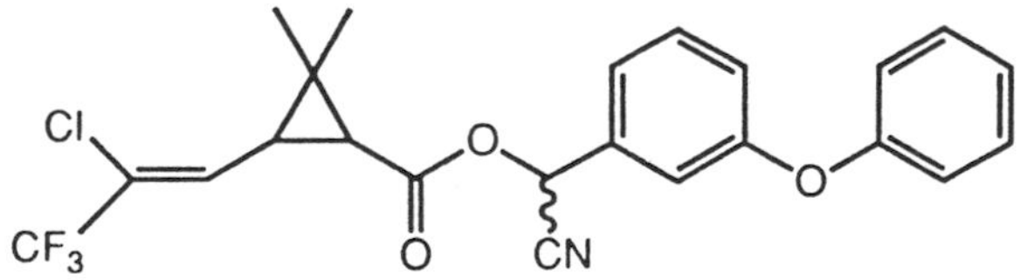

cyhalothrin
Z; (1RS)-cis; αRS (28)

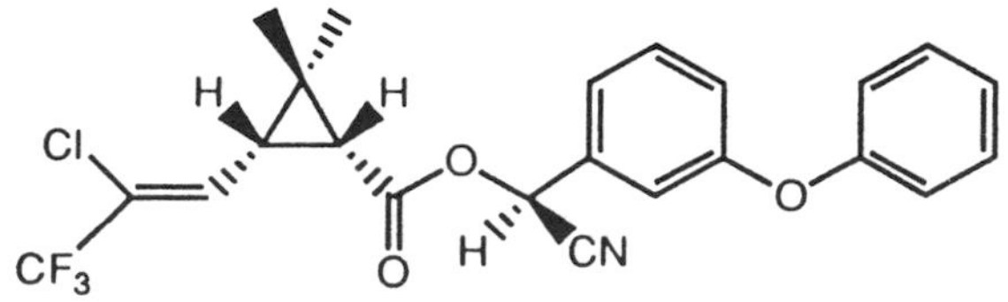

Z; (1R)-cis; αS (29)

Several companies have investigated the substitution of one or both of the halo groups in *19* and *20* with various haloalkyl groups, especially fluorinated alkyl. A number of very active analogs were discovered and the overall best 3-substituents appear to be the *cis*-[(Z)-2-chloro-3,3,3-trifluoro-1-propenyl] group and its corresponding *E* isomer [66]. This work by ICI led to the development of cyhalothrin (*28*), which has excellent insecticidal activity (Table 2) as well as high activity against ticks, such as the southern cattle tick, *Boophilus microplus*, infesting sheep and cattle [67]. Subsequently, ICI introduced a partially resolved mixture, λ-cyhalothrin, which contains equal quantities of the Z; (1R)-*cis*;αS isomer (*29*) and its inactive enantiomer [68]. The pure isomer (*29*) is more active (2.5-fold) than deltamethrin (*20*) against the housefly, making it one of the most active insecticides known [66]. λ-Cyhalothrin controls a wide range of foliar, soil-surface, and public health pests at rates of 5–30 g of active ingredient per hectare.

flumethrin
EZ; (1RS)-*cis-trans;* αRS

(*30*)

Table 2 Toxicity of Some Chrysanthemate Analogs at 25°C

Compound	*Heliothis virescens* (Topical) (μg/third instar larva) LD_{50}	*Spodoptera exigua* (Topical) (μg/third instar larva) LD_{50}	*Musca domestica* (Topical) (μg/adult) LD_{50}
Cyhalothrin (*28*)	0.0037	0.014	0.0017
Tralomethrin (*33*)[a]			
Diastereomer A	0.002	0.003	0.0010
Diastereomer B	0.003	0.004	0.0009
CGA 74055 (*35*)[b]	0.029	0.018	0.0069
38[c]	0.17	—	0.16
41[c]	0.46	—	0.17
42[c]	0.061	—	0.24
Cypothrin (*43*)	0.035	0.050	0.053

[a] Tralomethrin was chromatographically separated into its two diastereomers.
[b] Mixture of 16 isomers.
[c] Mostly trans isomer.

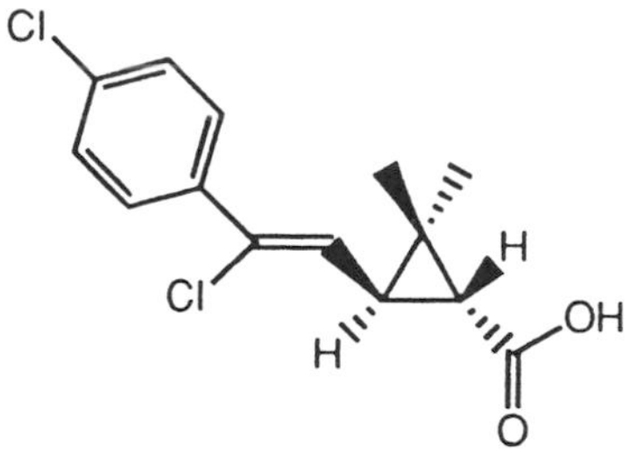

Z; (1R)-*trans* (31)

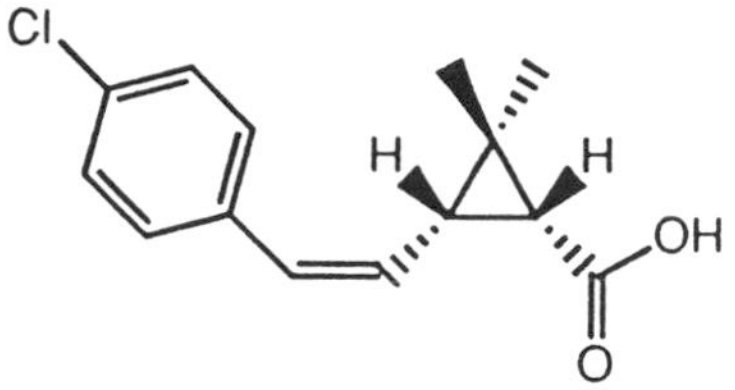

Z; (1R)-*cis* (32)

Another interesting C-3 variation is the replacement of one of the chlorine atoms in the dichlorovinyl group of *19* and related esters with a 4-chlorophenyl group. This led to the discovery of flumethrin (*30*) by Bayer AG, which evaluated it extensively for the control of ticks. This analog is highly active against susceptible and resistant strains of many one-host and multihost tick species and is very safe for livestock [69]. The 4-fluoro substituent in the benzylic aromatic ring of the alcohol moiety gives a large increase in activity against ticks [1]. One of the isomers of flumethrin, presumably the Z;(1R)-*trans*;αS isomer, is extraordinarily toxic to the southern cattle tick, *Boophilus microplus*, being 50 times more toxic than *cis*-permethrin and deltamethrin to the larvae [70]. All four (1R) isomers of the acid moiety of flumethrin have been prepared and esterified with (S)-α-cyano-3-phenoxybenzyl alcohol and (S)-α-cyano-4-fluoro-3-phenoxybenzyl alcohol [71]. The (1R)-*trans* isomer (*31*), with the (Z) geometry of the double bond, gives the most insecticidally active analogs against the housefly, followed by the E;(1R)-*cis* isomer. The other two isomers give analogs with very low activity. When the vinylic chlorine group is absent, the Z;(1R)-*cis* isomer (*32*) (with the stereochemistry of the double bond inverted) now gives more active analogs than the corresponding E;(1R)-*trans* isomer. The other two isomers show very low activity [71].

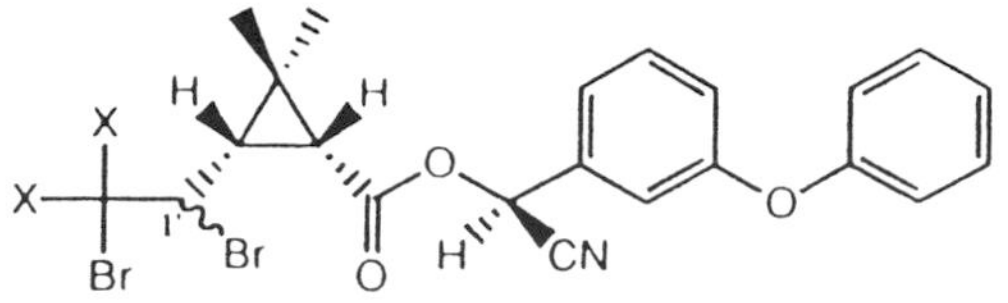

tralomethrin (X=Br)
1'RS; (1R)-*cis*; αS (33)

tralocythrin (X=Cl) (34)

CGA 74055 (16 stereoisomers) (35)

Bromination of the double bond in the 3-dihalovinyl group also gives active analogs. For example, bromination of the single isomer deltamethrin (20) affords tralomethrin (33), which is a mixture of two diastereoisomers. Although an additional asymmetric center has been introduced in the tetrabromoethyl group of 33, both of the diastereoisomers have essentially the same biological activity (Table 2). In an analogous fashion, the dichlorovinyl group of cypermethrin (19) has been brominated [45]. The resulting mixture of 16 stereoisomers, CGA 74055 (35), is an active insecticide (Table 2). All of the isomers of 35 have been separated and bioassayed, and the two (1R)-cis;αS diastereoisomers have by far the highest insecticidal activity, similar to that of the resolved (1R)-cis;αS isomer of cypermethrin [45]. Again both the 1' R and 1' S isomers have similar activity, indicating that the configuration at this asymmetric center is not important for activity. The 3-tetrahaloethyl substituents in 33 and 34 readily undergo debromination to regenerate deltamethrin (20) and (1R,αS)-cis-cypermethrin, respectively, in insects [72] and rats [73], and upon photolysis [74]. This debromination occurs rapidly following topical application to the housefly or feeding to cabbage looper larvae, thus the insecticidal activity of 33 and 34 is probably due, at least in part, to the formation of deltamethrin (20) and (1R, αS)-cis-cypermethrin within the insect [72]. There is some dispute over whether or not these tetrahaloethyl analogs are themselves intrinsically insecticidal [75], but they do penetrate unchanged into housefly adults in significant amounts, so that the intact molecule probably contributes to the observed insecticidal activity.

R=CH₃   (36)

acrinathrin [R=(CF₃)₂CH·]   (37)
Z; (1R)-cis; αS

The natural pyrethric acid moiety in pyrethrin II (2) has the E; (1R)-trans geometry and some very interesting pyrethroids have been discovered by varying the methoxycarbonyl-substituted side chain of 2 [71]. The ester prepared from pyrethric acid and (S)-α-cyano-3-phenoxybenzyl alcohol has only 7% of the toxicity of deltamethrin (20) against the housefly, and the other three 1R isomers of this ester show even lower activity. However, replacing the vinylic methyl group in the four isomers with a hydrogen atom gives a series of nor-pyrethric diesters in which the Z;(1R)-*cis* isomer (*36*) is now the most toxic of the isomers to the housefly, and in addition, *36* shows a considerable increase in activity over the pyrethric diesters. Variations in the methoxycarbonyl group in *36* led to the discovery of a series of analogs with high insecticidal activity. Several of these analogs having bulky ester groups in this side chain [e.g., R = $(CH_3)_3C$-] show high acaricidal activity against ticks, such as the southern cattle tick, *Boophilus microplus* [71]. This research led to the commercial introduction of acrinathrin (*37*) by Roussel Uclaf in 1991. This pyrethroid has about half of the insecticidal activity of deltamethrin (*20*) against the housefly and is also active as an acaracide against a wide range of phytophagous mite species.

(1RS)-*trans*; αRS (*39*)

R= 4-Cl-C_6H_4· (*38*)

R= $CH_3(CH_2)_4O$· (*40*)

R= $CH_3(CH_2)_3O$· (*41*)

R= Cl-C_6H_4-O· (*42*)

The isobutenyl group at the 3-position on the cyclopropane ring of chrysanthemic acid has also been replaced with a number of alkoxy, alkylthio, aryloxy, phenylthio, aryl, and analogous substituents [4, 52, 58, 76]. Analogs with substituted phenyl groups at the 3-position were first prepared in the late 1950s but, in general, they show only moderate activity (e.g., *38*, Table 2) [4, 52]. However, the 3-(4-*tert*-butylphenyl)-*trans* analog (*39*) is reported to show excellent insecticidal and miticidal activity [77]. Several of the 3-alkoxy and 3-aryloxy-substituted analogs show interesting activity. For example, the 3-*n*-pentoxy analog (*40*) (cf. *41*, Table 2) is reported to have toxicity similar to phenothrin and fenvalerate (*22*) against the housefly [76]. The 3-(4-chlorophenoxy) analog (*42*) also shows good activity (Table 2) [52, 58].

Several aromatic spiro-substituted cyclopropanecarboxylates have good activity. For example, the analog cypothrin (*43*) is active against insect and tick species but is not stable enough for field use [78] (see

cypothrin
(1RS)-cis-trans; αRS   (43)

(1RS); αRS   (44)

Table 2). The spiro derivative (44) is more active insecticidally than 43, being as good as cypermethrin (19) against the housefly but is less active than 19 against several other insect species [79].

a) R^2= 5-benzyl-3-furylmethyl
b) R^2= 3-phenoxybenzyl   (45)
c) R^2= α-cyano-3-phenoxybenzyl

The cyclopropanecarboxylic acid moiety of pyrethrin I (1) has the (1R)-trans configuration and, although the natural (1R) configuration is essential for insecticidal activity, the superiority of the (1R)-trans isomer over the (1R)-cis isomer is not a general rule. Which of the two stereoisomers gives analogs with the higher activity depends on the structure of the alcohol moiety, the nature of the substituent C-3, and the insect species being treated [63, 71]. For example, against the housefly, in the case of the chrysanthemic acid esters (45; R^1 = CH$_3$), the (1R)-trans geometry generally produces more active esters than the (1R)-cis geometry for alcohol types a, b, and c and the natural rethrolones (especially for esters of type a). For the dihalovinyl analogs (45; R^1 = halo) results are variable and depend both on the alcohol and the specific halo group. For 5-benzyl-3-furylmethyl esters (type a) and R^1 = chloro, the cis and trans isomers show essentially identical activity against the housefly, whereas for R^1 = fluoro, the trans isomer is the more active of the two. For R^1 = bromo, the cis isomer is the more active. However, for esters of types b and c where R^1 = halo, the (1R)-cis isomers are almost always more insecticidally active than the corresponding (1R)-trans isomers. For example, the fully resolved (1R,αS)-cis-cypermethrin is approximately 23 times more active topically against larvae of *Heliothis virescens* than the corresponding trans isomer. This cis isomer is also more active topically than the trans isomer against adults of *Blattella germanica*, but against adults of *Calliphora erythrocephala*, the trans isomer is as active as the cis isomer [45]. In the corresponding (*E*)- and (*Z*)-3-(monohalovinyl) analogs, with alcohol moieties of types b and c, there is little consistent difference in activity against the housefly and mustard beetle between the cis and trans series, or between the *E* and *Z* isomeric forms of the side chain [64]. In the Kadethrin type of structure (*17*), the cis isomer is superior to the trans, whereas for cycloalkylidenemethyl side chains, such as in bioethanomethrin (*16*), the (1*R*)-trans isomer is the more active of the two [71].

B.   α-Substituted Phenylacetates

A large number of analogs have been prepared in the α-substituted phenylacetate series based on fenvalerate (*22*). Active analogs in this series have the (2*S*) configuration in the acid moiety, which corresponds structurally to the (1*R*) configuration in the chrysanthemates [44, 80]. Phenylacetate analogs where the α-substituent is ethyl, isopropyl, isopropenyl, cyclopropyl, or *tert*-butyl have a good activity, whereas analogs with α substituents such as methyl and *n*-butyl have low activity [42–44, 52, 81]. The best reported α-substituents are isopropyl and cyclopropyl with isopropyl often giving the more active analog (Table 3) [52, 81]. Various substituents in the meta and para positions of the acid phenyl ring improve the activity of *46* and *47*, Table 3, especially the 4-chloro, 4-methoxy, 4-difluoromethoxy, and 3,4-methylenedioxy groups. A 4-*tert*-butyl substituent in this phenyl ring gives analogs, such as *53*, that show low activity against insects [39, 42] but have very high acaricidal activity [39]. Introduction of substituents into the ortho position of the acid phenyl group of fenvalerate analogs substantially decreases the insecticidal activity [42–44]. Even the introduction of an *ortho*-fluoro group into the 4-chlorophenyl ring of *22* dramatically lowers the activity (Table 3) [52].

(*53*)

flucythrinate
2S; αRS   (54)

The fully resolved (S,S) isomer of fenvalerate [esfenvalerate (23)] is 3.5 to 4 times more toxic than the racemate (22) against insect species in laboratory assays [42] and is also 4 times as active as 22 under actual field conditions [39, 42]. It is also 4–5 times more toxic than the racemate to mammals and to aquatic animals such as fish. Thus, the insecticidal potency appears to parallel the mammalian and fish toxicities.

Flucythrinate (54) is another analog in this series which was introduced commercially [82]. This pyrethroid has the (2S) configuration in the acid moiety, but the asymmetric center in the alcohol portion is not resolved [the sample (52) in Table 3 is racemic].

(55)

tau-fluvalinate
2R; αRS   (66)

Analogs of the general structural type 55 show very low activity when X is methylene, $-NHCH_2-$, $-CH_2NH-$, $-CONH-$, or sulfur [28, 46]. When X is oxygen the analogs have moderate activity [46, 52, 83] but substituted 2-anilino-3-methylbutanoates, where X is NH, show considerably higher activity [46, 52]. In this anilino series, the introduction of a methyl, chloro, methoxy, or trifluoromethyl substituent in the 4-position of the phenyl group gives compounds with high insecticidal activity [46]. In contrast to the fenvalerate series (Table 3), the introduction of one *ortho*-fluoro group into the phenyl ring produces a considerable increase in activity against most pest species (Table 4). When the 4-substituent is a trifluormethyl group, the introduction of one *ortho*-chloro group into the phenyl ring also gives analogs that show,

Table 3 Toxicity of Some α-Substituted Phenylacetates

No.	R^1	R^2	R^3	Heliothis virescens (Topical) (µg/third instar larva) LD$_{50}$	Spodoptera exigua (Topical) (µg/third instar larva) LD$_{50}$	Musca domestica (Topical) (µg/adult) LD$_{50}$
46	H	Me$_2$CH-	H	1.0	—	0.20
47	H	Me$_2$CH-	CN	0.086	—	0.20
48	4-Cl	Me$_2$CH-	H	0.090	6.0	0.091
22[b]	4-Cl	Me$_2$CH	CN	0.012	0.018	0.042
49	4-Cl	Cyclopropyl	CN	0.028	0.021	0.083
50	4-Cl; 2-F	Me$_2$CH-	H	10	—	5.1
51	4-Cl; 2-F	Me$_2$CH-	CN	0.32	—	0.52
52[c]	4-OCHF$_2$	Me$_2$CH-	CN	0.017	0.036	0.059

[a] All compounds are racemic.
[b] Fenvalerate.
[c] Racemic flucythrinate.

Table 4 Toxicity of Some 2-Anilino-3-methylbutanoates

No.	Compound[a] R^1	R^2	Heliothis virescens (Topical) (μg/third instar larva) LD$_{50}$	Spodoptera exigua (Topical) (μg/third instar larva) LD$_{50}$	Musca domestica (Topical) (μg/adult) LD$_{50}$	Tetranychus urticae (Contact and Repellency) Adults LC$_{50}$
56	4-Cl	H	0.11	0.010	0.21	130
57	4-CF$_3$	H	0.12	0.10	0.35	2.6
58	2-F; 4-Cl	H	0.035	0.041	0.081	4.2
59	2-F; 4-CF$_3$	H	0.031	0.038	0.11	0.24
60	2-Cl; 4-CF$_3$	H	0.055	0.080	0.23	0.77
61	4-Cl	CN	0.051	0.027	0.20	44
62	4-CF$_3$	CN	0.059	0.58	0.084	9.2
63	2-F; 4-Cl	CN	0.033	0.15	0.11	5.2
64	2-F; 4CF$_3$	CN	0.018	0.36	0.052	0.65
65[b]	2-Cl; 4CF$_3$	CN	0.080	0.19	0.16	2.3

[a] All esters are racemic.
[b] Racemic fluvalinate.

in general, increased activity (Table 4). Both *64* and *65* show good activity against a wide range of insect and mite species, although the activity against *Spodoptera exigua* is only moderate. The 2-chloro-4-trifluoromethylanilino analogs (*60* and *65*) also show enhanced foliar stability, compared with the analogs *57* and *62* which lack the *ortho*-chloro substituent [46]. These observations led to the commercial introduction of τ-fluvalinate (*66*) which has the (*R*) configuration at C-2 in the acid moiety but is unresolved at the α-carbon in the alcohol portion. The active (2*R*) configuration is structurally analogous to the biologically active absolute configuration at the chiral carbon atom α to the carboxylate group in both the (1*R*)-chrysanthemates and the (*S*)-2-aryl-3-methylbutanoates [46]. τ-Fluvalinate (*66*) shows broad-spectrum insecticidal and miticidal activity without the resurgence of mite populations commonly observed with the field use of many other pyrethroids. It is also essentially nontoxic to the honey bee, *Apis mellifera*, being 200 to 400 times less toxic topically than permethrin (*18*), cypermethrin (*19*), or fenvalerate (*22*) [84]. All four of the stereoisomers of fluvalinate (*65*) have been prepared in high optical purity and, as expected, the isomer prepared from the *R* acid and *S* alcohol is far more active insecticidally than the other three isomers. The active isomer is approximately four times more active than the racemate (*65*), indicating that there is little or no antagonism by the other isomers [85]. Examination of molecular models demonstrated that *61* (Table 4) and fenvalerate (*22*) can readily assume similar conformations [46]. However, in contrast to the fenvalerate series, much less variation in the C-2 alkyl group is tolerated in analogs of fluvalinate. Replacement of the α-isopropyl group at C-2 with an ethyl group dramatically lowers the activity and the corresponding α-*tert*-butyl analog is completely inactive [46].

(*68*)

Several research groups have prepared analogs of fenvalerate (*22*) in which non-aryl groups replace the 4-chlorophenyl group. Simple alkyl, alkenyl, and cycloalkyl analogs show very low activity [52] although the 5,5-dichloro-2-isopropyl-4-pentenoate (*67*) does show some activity (Table 5). The related α-cyano-3-phenoxybenzyl analog (*68*), with an additional methyl substituent at C-3 in the acid moiety, shows moderately good broad-spectrum insecticidal activity [86]. The introduc-

Table 5   Toxicity of 3-Phenoxylbenzyl 2-Substituted-3-methylbutanoates

No.	R^1	R^2	Heliothis virescens (Topical) (μg/third instar larva) LD$_{50}$	Musca domestica (Topical) (μg/adult) LD$_{50}$
67	$(Cl)_2C=CHCH_2-$	H	2.4	0.79
69	$(Cl)_2C=CHCH_2CH_2-$	H	5.1	2.4
70	(E) $CH_2=CHCH=CH-$	CN	0.70	0.44
71	(E, E) $CH_3CH=CHCH=CH-$	CN	0.29	0.23
72	$(Cl)_2C=CHCH_2NH-$	H	5.0	1.1
73	$CH_3C(Cl)=CHCH_2NH-$	H	2.2	2.8
74	(E) $4\text{-}F\text{-}C_6H_4\text{-}CH=CH-$	CN	0.044	0.13
75	2-naphthyl	H	0.54	0.74
76	2-naphthyl	CN	0.094	0.25
77	2-indenyl	H	0.35	0.25
78	2-benzo [b] thienyl	H	0.20	0.25
79	2-benzo [b] thienyl	CN	0.071	0.12
80	2-benzo [b] furanyl	H	0.83	0.41
81	3-benzo [b] thienyl	CN	0.78	0.40

[a] All esters are racemic.

tion of this methyl group at C-3 enhances the activity, in general, but the corresponding 3,3-dimethyl analog is inactive. Analogs such as *70* and *71* (Table 5), with a conjugated diene function replacing the 4-chlorophenyl group of *22*, show improved activity over monoolefinic analogs [52]. In these diene structures, the *E* isomers are much more active than the corresponding *Z* isomers. Some substituted amino analogs such as *72* and *73* (Table 5) also show modest activity [46, 52].

(74)

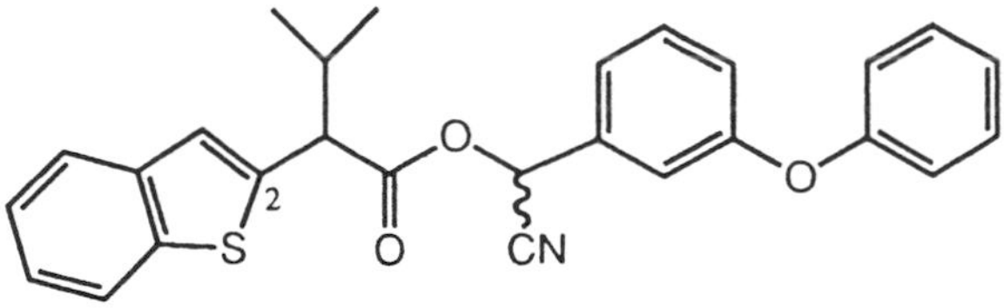

(79)

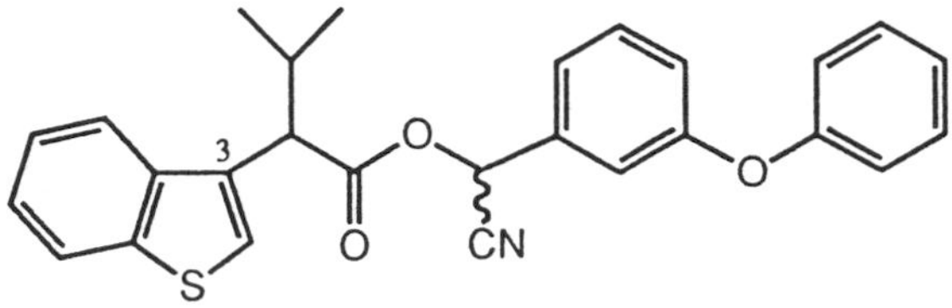

(81)

Substantial activity is retained upon extending the distance between the aromatic ring and the C-2 carbon atom of fenvalerate (22)
with an (E)-olefinic linkage. Thus, vinylogous analogs such as the 4-
(4-fluorophenyl)-2-isopropyl-3-butenoate (74) show considerable insecticidal activity (Table 5) although they are not quite as active as
fenvalerate [47, 52]. Several other aromatic heterocyclic, and related
groups can be substituted for the phenyl ring at C-2 in fenvalerate.
Thus, the 2-(2-naphthyl)- (75 and 76), 2-(2-indenyl)- (77), 2-(2-
benzo[b]thienyl)- (78 and 79), and 2-(2-benzo[b]-furanyl)- (80) substituted analogs all show good insecticidal activity [47, 52, 87]. The corresponding 2-(1-naphthyl)-, 2-(3-indenyl)-, and 2-(3-benzo[b]thienyl)
(81) derivatives are less active (cf. 79 versus 81 in Table 5).

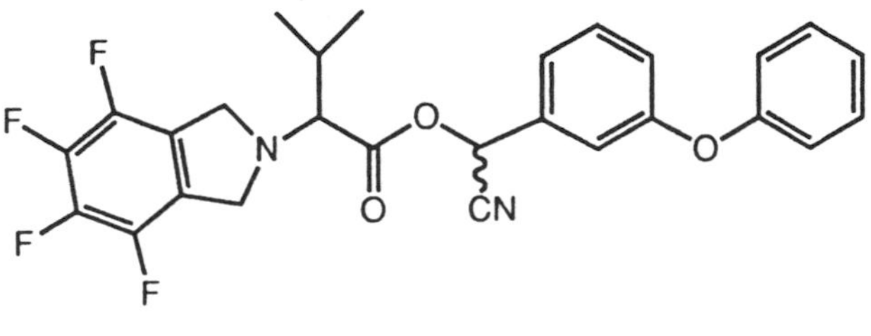

(86)

Another group of compounds investigated were those in which
the phenyl ring of the acid moiety of 22 was replaced with a heterocyclic ring containing a nitrogen atom attached at C-2. A number of
substituted 2-(3-pyrrolin-1-yl), 2-(1-pyrrolyl), 2-(1-indolinyl), 2-(1H-
indol-1-yl), and related structures show moderate activity (Table 6). The
most active analog in this series is the 2-(4,5,6,7-tetrafluoro-2-
isoindolinyl) derivative (86) [46, 52]. However, these 2-isoindolinyl
analogs are not stable in sunlight [46].

Table 6 Toxicity of 2-(2-Isoindolinyl)-3-methyl-butanoates and Analogs

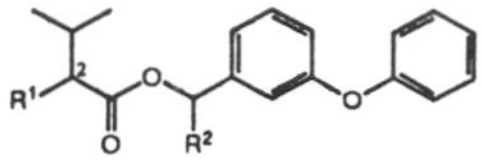

Compound[a] No.	R^1	R^2	*Heliothis virescens* (Topical) (μg/third instar larva) LD_{50}	*Musca domestica* (Topical) (μg/adult LD_{50}	*Tetranychus urticae* (Contact and Repellency) Adults LC_{50} (ppm)
82	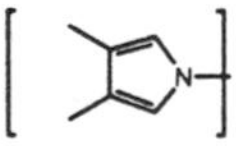	H	0.26	0.096	90
83	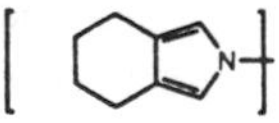	H	0.15	0.13	6.5
84	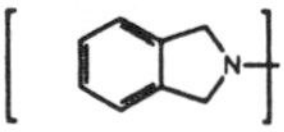	H	0.15	0.18	38
85	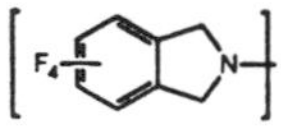	H	0.14	0.049	7
86	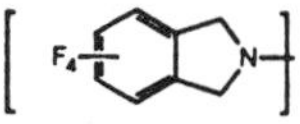	CN	0.028	0.063	>100
87	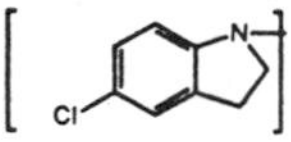	H	1.0	0.40	60
88	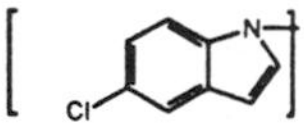	H	0.56	0.56	240
89	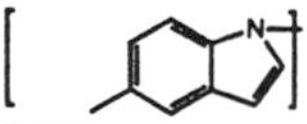	H	0.21	0.26	34

[a] All esters are racemic.

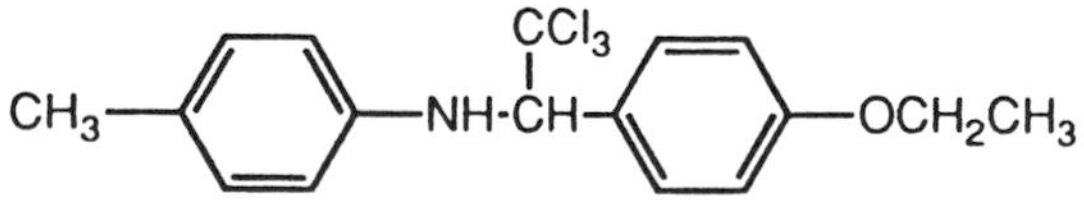

(*90*)

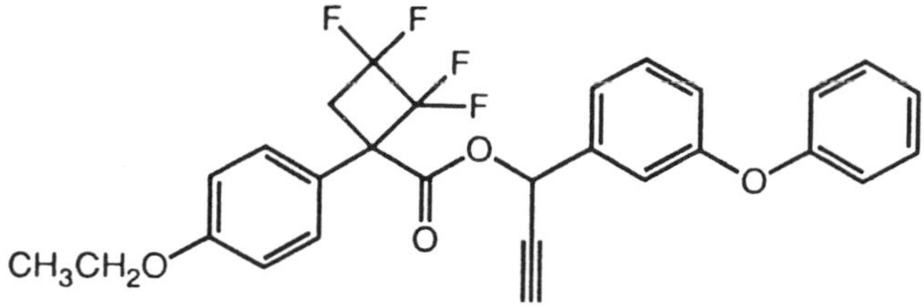

GH 601 (*94*)

There are some similarities between the structure-activity relationships of pyrethroids and DDT. In reference to the anilino analogs discussed earlier (Table 4), it is interesting that insecticidal activity is not appreciably reduced by the insertion of a nitrogen atom between one of the aryl rings and the -$CHCCl_3$ group of DDT. Thus, the anilino analog (90) has similar activity to DDT against the housefly and mosquito larvae [88]. Similarly, as for analogs of 55, the corresponding ether analogs of DDT are much less active than the anilino derivatives and the related sulphur analogs are much less active than the ethers (cf. [46]). Holan and co-workers at CSIRO, Australia have extensively investigated analogs that combine some of the structural features of DDT (and DDT-type insecticides) and the pyrethroids. This work led to the discovery and commercial development of cycloprothrin (93). Compounds such as 91, 92, and 93 show only moderate activity against *Heliothis virescens* and the housefly (Table 7), although 93 has good activity against the rice water weevil, *Lissorhoptrus oryzophilus*, and the Australian sheep blowfly, *Lucilia cuprina* [89, 90], with very low toxicity to both fish and mammals. For the compound 92, the (S)-acid gives much more active esters against the housefly than does the (R)-acid [91]. (The configurations assigned in Ref. 89 are incorrect.) Some aryltetrafluoro-cyclobutane esters such as 94 show excellent topical activity against *L. cuprina*, good activity against the German cockroach, *Blattella germanica*, and moderate activity against the budworm, *Heliothis punctigera* [91, 92]. The pyrethroid 94 also has very low acute mammalian toxicity.

Table 7 Toxicity of DDT-Type Pyrethroids

No.	R¹	R²	*Heliothis virescens* (Topical) (μg/third instar larva) LD$_{50}$	*Musca domestica* (Topical) (μg/adult) LD$_{50}$	*Tetranychus urticae* (Contact and Repellency) Adults LC$_{50}$ (ppm)
				Compound[a]	
91	Cl-	H	1.20	0.62	57
92	EtO-	H	0.43	0.66	>100
93[b]	EtO-	CN	0.14	0.31	46

[a] All esters are racemic.
[b] Cycloprothrin.

C. Alcohol Variations

There are a number of variations in the alcohol moiety of the pyrethroids that can give very active esters. However, the effect on activity of varying the alcohol depends on which active acid moiety is involved and which insect species is being studied. The alcohols that give
esters with the highest toxicity to insects contain two centers of
unsaturation appropriately spaced and share steric characteristics that
allow them to adopt similar conformations in which the two unsaturated groups are not coplanar. Many of the alcohols that give active
pyrethroids are benzylic or allylic, with at least one hydrogen atom remaining on the tetrahedral carbon atom attached to the oxygen atom
of the alcohol [28, 29, 52]. The corresponding phenyl esters are inactive
[26].

Some simple benzyl esters such as the 2,4,6- and 2,3,6-trimethyl-
benzyl chrysanthemates show activity against the housefly and mustard beetle [93], but various 3- and 4-alkenyl-, alkynyl-, and dienyl-sub-
stituted benzyl esters are somewhat more active [94, 95]. With simple
allylbenzyl chrysanthemates, the 4-allylbenzyl analog is more active
against the housefly than the 3-allylbenzyl analog, but with benzylbenzyl esters, the 3-benzylbenzyl derivative is considerably more active, against both the housefly and the mustard beetle, than the 4-
benzylbenzyl ester [27]. The corresponding 4-phenoxybenzyl ester is
even less active than the 4-benzylbenzyl ester ([27];cf. fluvalinate series, [46]). For the 3-substituted benzyl esters, benzoyl, benzyl,
phenoxy, phenyl, phenylthio, and anilino substituents all give active
esters. Of these, the 3-phenoxybenzyl group usually gives the most
active pyrethroid [27, 31, 96, 97].

(1R)-trans   (95)

bifenthrin   (96)
Z; (1RS)-cis

(*97*)

The 3-xanthenylmethyl chrysanthemate (*95*) is inactive and this observation led Elliott to postulate that the bridging atoms in both 5-benzyl-3-furylmethyl and 3-phenoxybenzyl esters allow the molecules to adopt an optimum active conformation in which the two centers of unsaturation are not coplanar [27, 98]. More recently, various dibenzofuranmethyl esters, in which the aromatic centers are held coplanar, have also been shown to be essentially inactive [99]. All 3-substituted benzyl esters appear to belong to a congeneric series in which the lipophilicity of the 3-substituent is an important factor [100]; the activity of biphenyl-3-ylmethyl esters, where the 3-substituent is a phenyl group, demonstrates that a bridging atom between the two centers of unsaturation is not a requirement for insecticidal activity [96]. These unsubstituted ([1,1'-biphenyl]-3-yl)methyl esters are, in general, less active than the corresponding 3-phenoxybenzyl analogs [96]. However, a careful study of 2-substituted biphenyl alcohols demonstrated that (2-fluoro- and 2-methyl-[1,1'-biphenyl]-3-yl)methyl alcohols give esters with increased activity [100, 101]. Biphenyl itself is essentially planar in solution, but a substituent in the 2-position disrupts the coplanarity of the two aromatic rings. Thus, the improved activity of the 2-fluoro and 2-methyl derivatives appears to be related to the conformational preference for a twist angle of about 50° between the two aromatic rings of the alcohol moiety, which allows a better fit at the appropriate binding site [100]. This work led to the commercialization of bifenthrin (*96*) which has the same active $Z;(1\text{-}RS)\text{-}cis$ acid moiety as cyhalothrin (*28*). This pyrethroid has excellent broad-spectrum insecticidal properties (Table 8) and also controls some species of spider mites. However, it is also one of the most toxic commercial pyrethroids to fish and mammals (acute oral LD_{50} for rats is ~55 mg/kg).

$Z; (1R)\text{-}cis; \alpha S$ (*98*)

(1RS)-cis-trans   (99)

(1RS)-cis-trans   (100)

Various heterocyclic rings can be substituted for the phenyl group in structures such as *96*, with a 1-pyrrole ring being the most active of the heterocyclic substituents, but none of these compounds are as insecticidally active as the corresponding phenyl analog [102]. Bridged biphenyl and heteroaromatic alcohols have also been prepared in order to fix the conformation of the alcohol moiety. Esters such as *97* show high acaricidal and insecticidal activity against beetles and aphids but are less active than *96* against lepidopterous species [100]. The structure *97* has a twist angle of about 50° between the two aromatic rings, which is similar to that of *96*. In studying other changes in the biphenyl type of alcohol, it was discovered that 4-phenyl-2-indanyl esters have high activity, whereas the corresponding 1-indanyl esters are much less active [103, 104]. For example, the ester *98* shows excellent topical and foliar insecticidal and acaricidal activity. The chiral carbon atom of the alcohol moiety in the biologically active isomer has the (*S*) configuration. The activity of *98* was unexpected because the carbon atom α to the oxygen of the alcohol is not benzylic. However, examination of the spatial overlap of *98* and *96* demonstrated that the two alcohol moieties could fit the same receptor site with little conformational distortion [103]. Some cyclic derivatives of benzyl esters, such as 4-allyl-1-indanyl chrysanthemate (*99*) and the analogous 7-substituted-2,3-dihydrobenzo-[b]furan-3-yl chrysanthemates (e.g., *100*), have been prepared and shown to be more potent than allethrin against the American cockroach [104–106].

Substitution on the carbon atom α to the alcohol oxygen in these various esters causes large variations in their biological activity. The insect activity depends on the structure of the alcohol, the nature of the α-substituent, and the type of acid moiety present in the ester. Most α-substituents lower the activity considerably and in only a limited

Table 8 Toxicity of Some Alcohol Variation in Pyrethroids

No.	Compound[a]	Heliothis virescens (Topical) (μg/third instar larva) LD_{50}	Spodoptera exigua (Topical) (μg/third instar larva) LD_{50}	Musca domestica (Topical) (μg/adult) LD_{50}	Tetranychus urticae (Contact and Repellency) Adults LC_{50} (ppm)
96	Bifenthrin	0.005	0.0006	0.024	1.5
101	Cyfluthrin	0.0022	0.006	0.0035	5.4
102[b]	Trans-fenfluthrin	0.32	0.039	0.016	0.22
103	Cis-fenfluthrin	0.29	0.028	0.061	—
105	Tefluthrin	0.022	0.0046	0.059	470
106[c]	Fenpirithrin	0.025	0.039	0.027	—
110		0.020	0.061	0.15	0.18
111		0.083	0.048	0.30	8.8
112		0.066	—	0.89	—
113		0.025	0.066	0.088	0.39
114		0.031	0.032	0.17	6.4
115		0.037	0.038	0.19	0.11
116		0.029	0.036	0.12	—
117	Cis isomer	0.0042	0.010	0.0070	—
118	Trans isomer	0.012	—	0.0074	—
122	Empenthrin	0.36	—	0.24	190

[a] All esters are racemic.
[b] Racemic fenfluthrin.
[c] Cis : trans isomer ratio = 3 : 7 (cf. Dowco 417 has a cis : trans isomer ratio of 4 : 6).

number of 3-substituted benzyl esters does an α-substituent increase the
activity. For example, in the α-substituted 3-phenoxybenzyl esters of
(1R)-*trans*-chrysanthemic acid, (1R)-*cis*-3-(2,2-dibromovinyl)-2,2-
dimethylcyclopropanecarboxylic acid, and some related pyrethroids, an
α-cyano substituent gives the most active analogs and they show a general increase in activity over the unsubstituted derivatives [compare permethrin (*18*) with cypermethrin (*19*) in Table 1]. An α-ethynyl group
in these series can give analogs with increased, similar, or somewhat
reduced activity depending on the particular analog and the insect species [34, 107–109]. Elliott and co-workers showed that for such α-cyano
substitution, only the (S)-α-cyano group enhances activity and (RS)-
α-cyano mixtures are generally less active than expected, implying that
antagonism can occur between the (R) and (S) isomers [110]. The (R)
isomers of the α-ethynyl derivatives, which are structurally analogous
to the (S)-α-cyano isomers, are the active isomers in this series (the
Cahn-Ingold-Prelog sequence rules change the specification from S to
R for the α-ethynyl analogs).

The introduction of an α-cyano group into 3-phenoxybenzyl esters of (*E*)- and (*Z*)-3-monohalovinyl-2,2-dimethylcyclopropane carbox-
ylates gives a variable effect on the activity against the housefly with
a decrease in activity in many cases. However, against the mustard
beetle there is usually an increase in activity in this series [64]. In the
fenvalerate series, the introduction of an α-cyano group into the 3-
phenoxybenzyl ester produces, in general, a significant increase in activity (compare *22* with *48* in Table 3) [44, 52]. However, the
unsubstituted analog *48* is reported to be much more effective than *22*
against the armyworm, *Pseudaletia separata* [50]. The α-ethynyl derivative in this series is three to five times less active than the α-cyano
analog (*22*) against the housefly and *Spodoptera litura* [39]. In contrast,
with the anilino esters of the fluvalinate type (*65*, Table 4), the introduction of a racemic α-cyano group into the 3-phenoxybenzyl esters
produces at most a small increase and sometimes a decrease in activity depending on the substituents in the anilino ring [46, 52]. The introduction of an α-ethynyl group in this anilino series produces analogs with similar activity against *H. virescens* and the housefly to that
shown by the unsubstituted 3-phenoxybenzyl esters [52]. In the
isoindolinyl derivative (*86*) there is fivefold increase in activity against
H. virescens on the introduction of an α-cyano group (cf. *86* versus *85*,
Table 6), but there is a slight decrease in activity against the housefly
[46]. The introduction of an α-cyano group into the alcohol moiety of
the anilino analogs *56*, *57*, *59*, and *60* (to give *61*, *62*, *64*, and *65*, respec-

tively, Table 4) reduced stability in sunlight [46]. A similar observation was noted in the permethrin series [111].

The introduction of a racemic α-methyl group into 3-phenoxybenzyl esters of (1R)-*trans*-chrysanthemic acid and (1R)-*cis*-3-(2,2-dibromovinyl)-2,2-dimethylcyclopropanecarboxylic acid as well as related analogs has been reported to result in a considerable reduction in insecticidal activity [28, 34, 98, 107, 108]. In contrast, our bioassays indicate that the insertion of an α-methyl group into permethrin only slightly decreases activity against *Heliothis virescens* or *Musca domestica* [52]. In the fenvalerate series, however the α-methyl analog shows considerably lower activity than the corresponding unsubstituted analog *48* [52]. While in the fluvalinate series, the substitution of an α-methyl group in the 4-chlorophenyl derivative (*56*) (Table 4) produces a 2.5-fold loss in activity against *H. virescens* but little change in activity against the housefly [52]. As expected, the (*R*)-α-methyl analog of *56* is much more active than the corresponding (*S*) isomer [112]. As esters prepared from 3-phenoxybenzyl alcohols with two α-substituents are inactive [71, 107], at least one hydrogen atom at the α-position is necessary for insecticidal activity with this type of ester.

Although an α-cyano substituent may enhance the activity of chrysanthemate-type esters of some 3-substituted benzyl alcohols, it can dramatically lower the activity of esters of the corresponding 4-substituted benzyl alcohol [94, 95, 107, 113]. For example, introduction of an α-cyano group into 4-(haloallyl)benzyl esters of this type drastically lowers the insecticidal activity [94, 95]. With 5-benzyl-3-furylmethyl esters, the introduction of an α-cyano, α-ethynyl, or α-methyl group produces a dramatic loss in activity [28, 34, 108]. In addition, the introduction of an α-cyano group into most other allylic or benzylic esters, including the biphenyl type (*96*), gives rise to a considerable decrease in insecticidal activity [1, 34, 96, 108].

cyfluthrin
(1RS)-*cis-trans*; αRS   (*101*)

Introduction of substituents into the phenoxy ring of the alcohol moiety of phenothrin (*10*), fenvalerate (*22*), or fluvalinate (*65*) generally diminishes activity, although the effect on activity is negligible with a 4-fluoro substituent [34, 39, 52]. Substitution in the benzyl ring of these pyrethroids also tends to reduce the activity [34], except in the case of a fluoro atom in the 4-position, which usually gives analogs with a twofold to fivefold increase in activity. However, the effect on activity varies with the acid moiety and the pest species [1]. Such substitution in cypermethrin (19) led to commercial development of cyfluthrin (101) [114, 115]. This pyrethroid has enhanced activity over 19 against *Heliothis virescens* and the housefly but not against *Spodoptera exigua* (Table 8). Subsequently, a partially resolved product β-cyfluthrin was introduced. This product is a mixture of the (1R)-*cis*;α(S) and the (1R)-*trans*;α(S) isomers plus their inactive enantiomers. Interestingly, this 4-position is not a general site for metabolic hydroxylation and introduction of a fluoro atom into other positions in the benzyl ring either has little effect or produces a substantial loss in activity [1]. The introduction of a 4-fluoro group into the benzyl ring of the ester moiety of fenvalerate and fluvalinate also gives an increase in activity against *H. virescens* and the housefly of twofold to fourfold [52, 116].

fenfluthrin (R=F); (1R)-*trans*   (102)

benfluthrin (R=H); (1R)-*trans*   (104)

tefluthrin
Z; (1RS)-*cis*;   (105)

Esterification of some simple tetrafluorobenzyl and pentafluorobenzyl alcohols with various cyclopropanecarboxylic acids can give pyrethroids with relatively high volatility and interesting activity against some insect species. Thus, the pentafluorobenzyl ester (102), as the (1R)-*trans* isomer fenfluthrin, was evaluated by Bayer AG for the control of household and public health insect pests [117]. The corresponding 2,3,5,6-tetrafluorobenzyl ester, benfluthrin (104), also shows rapid knockdown activity against flies and mosquitos [118]. The ester, tefluthrin (105), prepared from the acid moiety of cyhalothrin (28) and 2,3,5,6-tetrafluoro-4-methylbenzyl alcohol, shows good insecticidal activity (Table 8). Unlike most other pyrethroids, it is active against a wide range of soil insect pests, including various corn rootworm species (*Diabrotica* spp.) [119]. Tefluthrin (105) has a relatively high vapor pressure compared to most other pyrethroids, which assists its mobility in the soil. Relatively nonvolatile pyrethroids such as cypermethrin (*19*) and cyhalothrin (*28*) are active against *Diabrotica* spp. in laboratory tests, but are not very active when incorporated into soil, presumably due to their being bound to organic matter [119]. Various formulations of tefluthrin have been developed to control a variety of soil insect pests [120].

The corresponding 2,3,5,6-tetrafluoro-4-methylbenzyl ester of the τ-fluvalinate (*66*) acid moiety does not show useful activity against *Diabrotica* spp. [52, 116]. The introduction of an α-cyano, α-ethynyl, or α-methyl group into benzyl esters such as *102* and *105* produces a significant decrease in insecticidal activity [1, 119]. Fenfluthrin (*102*) and especially tefluthrin (*105*) are quite toxic to mammals when compared with many other pyrethroids (LD_{50} acute oral for rats of 85–120 mg/kg and 22–35 mg/kg, respectively).

fenpirithrin (R=CN)   (*106*)

R=H
(1RS)-*cis-trans*; αRS   (*107*)

Another allowable variation in the alcohol moiety is demonstrated by the activity of esters of (6-phenoxy-2-pyridyl)methyl alcohol. The α-cyano ester fenpirithrin (*106*) has lower activity than cypermethrin against the insect species in Table 8, but shows higher activity against some sucking insect pests [121]. Both the cis and trans isomers of *107* show activity similar to *106* against *Heliothis virescens* and, thus, in this series, the presence of an α-cyano group does not, in general, result in increased activity over the unsubstituted derivative [52]. However, the introduction of an α-methyl group into the cis and trans isomers of *107* gives analogs with very similar activity to *107* against *H. virescens* and the housefly. The corresponding *S*-alkyl thioester (*108*) with an α-methyl group shows excellent activity against these two insect species [52]. Interestingly, the (6-phenoxy-2-pyridyl)methyl ester of the fenvalerate acid moiety shows only moderate activity, as does the corresponding α-methyl analog. However, the thioester *109* shows surprisingly good activity against *H. virescens* [52].

(*108*)

(109)

R¹	X	R²	
Cl	O	H	(110)
Cl	O	CN	(111)
Cl	O	CH₃	(112)
F	O	H	(113)
F	O	CN	(114)
F	S	H	(115)
F	S	CH₃	(116)

Corresponding analogs have been studied in the fluvalinate series. Of the various possible isomeric (phenoxy-2-pyridyl)methyl esters, the 6-phenoxy compounds show the highest activity presumably due to steric shielding of the basic nitrogen atom in the pyridine ring in this isomer [52]. Introduction of a racemic α-cyano group produces either a decrease or little change in activity against *Heliothis virescens*, and a large decrease in activity against *Tetranychus urticae* (cf. *110* versus *111* and *113* versus *114*, Table 8). Analogs of this type, with an α-methyl group (e.g., *112*), have activity similar to that of the corresponding α-cyano derivatives [52]. Thioesters in this series (e.g., *115* and *116*) also show interesting activity. Esters without an α-substituent, such as *110*, *113*, and *115*, show good activity against all of the species in Table 8, especially against *Tetranychus urticae* and have good stability under field conditions [52]. Replacement of the phenoxy ring in the 3-phenoxybenzyl esters with a 2- or 3-pyridyl ring gives analogs with much lower activity [52].

In esters of the α-cyano(6-phenoxy-2-pyridyl)methyl type, both the (*R*) and (*S*) isomers are reported to have *identical* insecticidal activity, in contrast to the corresponding α-cyano-3-phenoxybenzyl esters. This is presumably due to rapid racemization at the α-carbon atom in both of the isomers within the insect. When the α-cyano group is replaced by an α-methyl group, where rapid epimerization is less likely, only the (*R*) isomer, with the same configuration as the (*S*)-α-cyano isomer, shows activity [71, 122].

Analogs have been prepared in which the furyl ring of resmethrin (*8*) and related cyclopropanecarboxylates has been replaced with other heterocyclic rings. For example, the (3-benzyl-1-pyrrolyl)methyl esters *117* and *118* are more active than resmethrin (*8*) and permethrin (*18*) against *Heliothis virescens* and the housefly (Table 8) [52, 123], but in the fenvalerate and fluvalinate series this type of ester shows only moderate activity [52, 123]. Interestingly, the 5-benzyl-3-furylmethyl esters in the fluvalinate series also show lower insecticidal activity than the corresponding 3-phenoxybenzyl esters, in contrast to the results obtained in the permethrin and fenvalerate series [28, 29, 44, 46]. The esters *117* and *118* are unstable in air and light. Other related heterocyclic derivatives were synthesized and found to be much less active [123, 124]. The alcohol moieties in 5-benzyl-3-furylmethyl and (3-benzyl-1-pyrrolyl)methyl esters are similar to each other in hydrophobicity and in molecular shape in their optimized stable conformations [124].

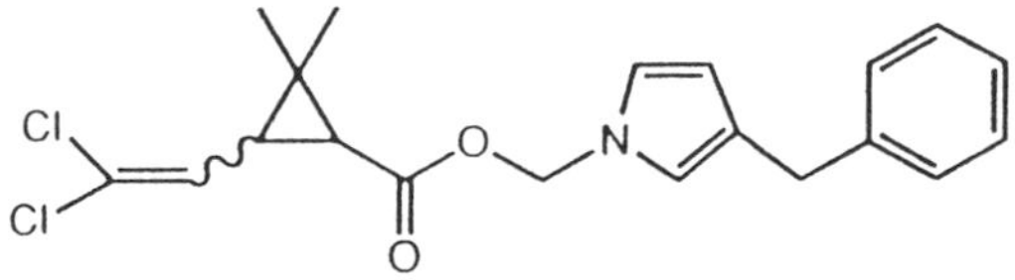

(1RS)-*cis*   (*117*)

(1RS)-*trans*   (*118*)

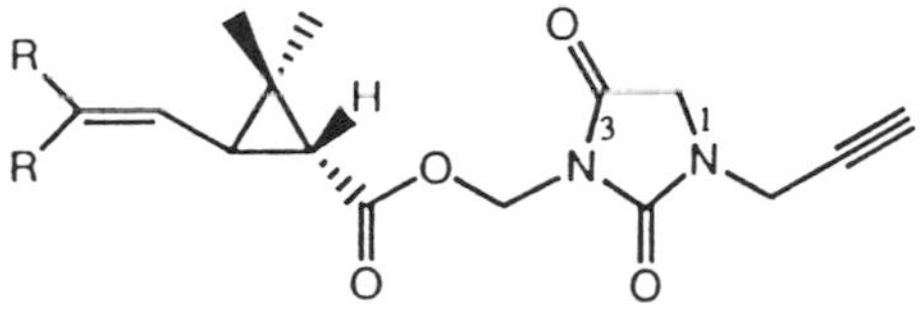

R=Cl   (*119*)

imiprothrin (R=·CH₃)   (*120*)
(1R)-*cis-trans*

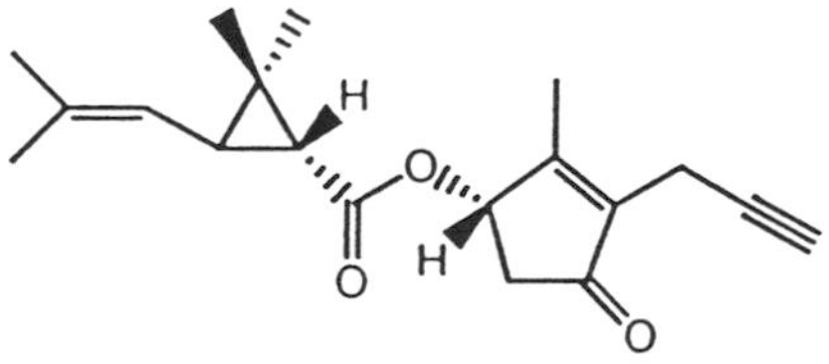

prallethrin   (*121*)
(1R)-*cis-trans*; αS

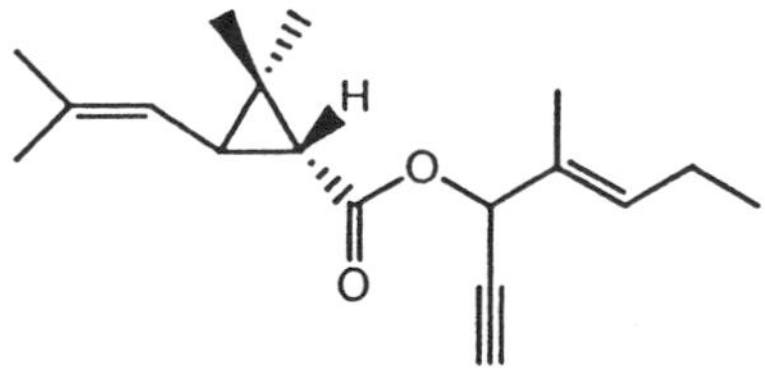

empenthrin   (*122*)
(1R)-*cis-trans*; αRS

A number of relatively simple alcohol moieties have been investigated in attempts to find pyrethroids with high volatility and/or high knockdown activity against mosquitos and flies for use in household markets. For example, (2,4-dioxo-1-prop-2-ynylimidazolidin-3-yl)methyl esters such as *119*, as the (1R)-*trans* isomer, show better knockdown activity than Kadethrin (*17*) against flying insects and the German cockroach in laboratory tests when formulated in oil solution. However, when formulated in aerosol pressure packs, these esters did not show good knockdown activity [125]. Sumitomo Chemical Co. has evaluated the closely related pyrethroid imiprothrin (*120*) as a mixture of the (1*R*)-*cis* and (1*R*)-*trans* isomers. Other analogs from Sumitomo Chemical Company include prallethrin (*121*) and empenthrin (*122*). Prallethrin (*121*), the propargyl analog of allethrin (*5*), has been introduced as a household and public health insecticide. It is more effective than bioallethrin against the housefly and mosquitoes and has outstanding flushing activity against cockroaches [126]. Empenthrin (*122*) has superior volatility to most other pyrethroids and is moderately toxic against insects (Table 8). It has low toxicity to mammals and good fumigant action, due to its high vapor pressure, allowing it to be useful in controlling insect pests attacking fabrics [39].

The active isomer in acyclic secondary alcohol moieties, such as that in empenthrin (*122*), has the (*S*) configuration. This is consistent with the configuration of (*S*)-allethrolone, but opposite in stereochemistry with that of (*S*)-α-cyano-3-phenoxybenzyl alcohol (*125*), when considering the hydrogen atom and hydroxyl group attached to the asymmetric α carbon atom (Figure 2) [127]. A study of the *R* and *S* isomers of various α-ethynylbenzyl esters of (*RS*)-*cis-trans* permethrin acid demonstrated that the insecticidal activity is dependent on both the position and nature of the substituents on the benzene ring. The toxicity ratio against the housefly of the *R* and *S* isomers varies considerably with different alcohol moieties. The most active configuration of the secondary alcohol portion can be reversed merely by changing the position of the substituent in the benzyl ring. For example, the ester of the (*R*)-3-allybenzyl alcohol (*126*) is more active than the corresponding (*S*) isomer, whereas the ester of the (*S*)-4-allybenzyl alcohol (127) is much more active than the corresponding (*R*) isomer [127]. However, molecular modeling studies have shown that the (*R*)-3-allybenzyl moiety is reasonably superimposable with the (*S*)-4-allybenzyl group in certain conformations. The *S* and *R* isomers of unsubstituted α-ethynyl- and α-cyanobenzyl esters have *R*:*S* toxicity ratios close to 1:1. The α-ethynyl- and α-cyano-3-phenoxybenzyl esters, prepared from *124* and *125*, respectively, have the highest toxicity ratios, favoring the configuration shown in Figure 2, whereas the α-ethynyl-4-allybenzyl esters have the highest opposite toxicity ratios [127]. In expanding this investigation to cinnamyl esters, it was discovered that the esters of the (*S*)-alcohol (*128*) are more active than the corresponding (*R*) isomers, as observed in the case of *123*. However, in the corresponding phenoxyphenyl derivatives, the (*R*)-α-ethynyl isomer (*129*) and the

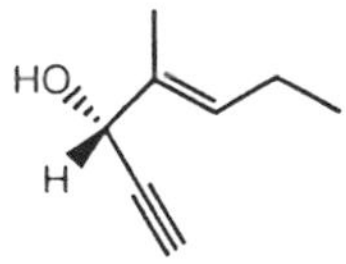

123, (*S*)

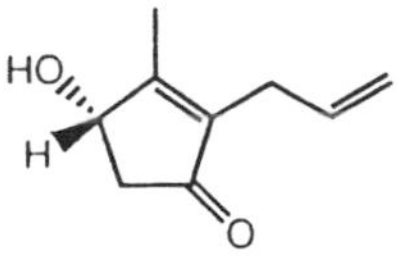

(*S*)-allethrolone

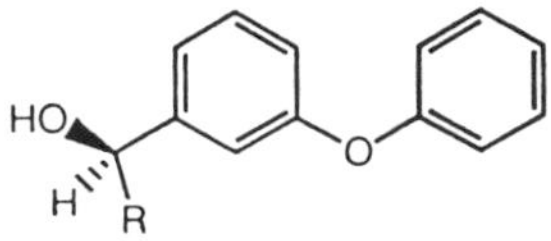

124, (*R*); R= ·C≡CH

125, (*S*); R= ·CN

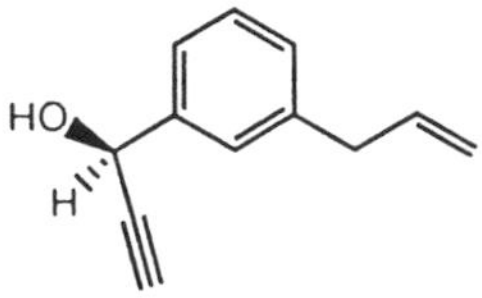

126, (*R*)

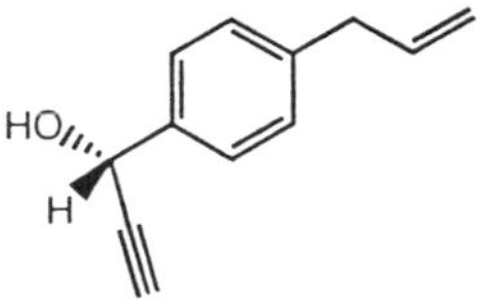

127, (*S*)

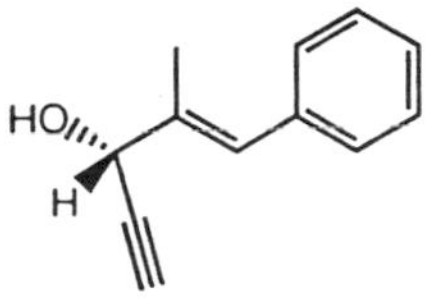

128, (*S*)

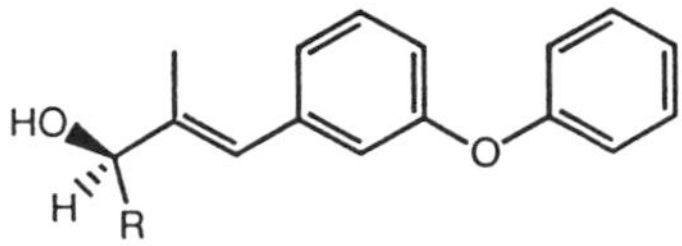

129, (*R*); R= ·C≡CH

130, (*S*); R= ·CN

Figure 2 Absolute configuration of alcohol moiety required for highest insecticidal activity.

stereochemically equivalent (S)-α-cyano isomer *130*, are now twice as active as the corresponding racemates against the housefly [128]. The introduction of a phenoxy group into the 3-position of the benzene ring in *128* reverses the absolute configuration required for the highest insecticidal activity. Therefore, the insecticidally dominant configuration in the alcohol moiety of pyrethroids containing a 3-phenoxyphenyl group, such as *124*, *125*, *129*, and *130*, is opposite from that of allethrolone and α-ethynyl alcohols like *123* and *128*.

D.   Nonester Pyrethroids

The central ester linkage in the pyrethroids discussed above is not essential for insecticidal activity and can be replaced by other suitable groups. Early efforts in changing the ester linkage of chrysanthemates gave compounds with low activity [129], although this work did demonstrate that the ester linkage is not required. The first nonester analogs to be discovered with good insecticidal activity were some alkyl aryl ketone oxime *O* ethers such as *131* and *132* [130, 131]. Only the *E* isomers in this series show good activity (Table 9) [52, 130]; cf. also [132]. They are indistinguishable from fenvalerate analogs in their symptoms of poisoning. However, these analogs, using an oxime linkage instead of an ester, are unstable in air and light. In this series, the replacement of the 3-phenoxybenzyl group in *132* with a (2-methyl[1,1'-biphenyl]-3-yl)methyl group (cf. *96*) gives an increase in activity against the southern armyworm, *Spodoptera eridania* [133], whereas replacement with a (6-phenoxy-2-pyridyl)methyl group (cf. *107*) gives an analog with lower activity against *Heliothis virescens* and *Musca domestica* [52].

R= ·CH(CH$_3$)$_2$ (*131*)

R= cyclopropyl   (*132*)

Most examples of nonester pyrethroids with high insecticidal activity have a structural resemblance to fenvalerate (*22*), but there are considerable differences in which groups can be placed at the isopropyl position in *22*. The different linkage groups have different structural requirements for the "acid" and "alcohol" moieties, and the effects of changes in the "acid," "alcohol," and linking moieties are nonadditive. There is also much less variation allowable in the "alcohol" moiety compared to pyrethroid esters.

Table 9 Toxicity of Some Nonester Pyrethroids

No.	Compound[a]	Heliothis virescens (Topical) (µg/third instar larva) LD50	Spodoptera exigua (Topical) (µg/third instar larva) LD50	Musca domestica (Topical) (µg/adult) LD50	Tetranychus urticae (Contact and Repellency) Adults LC50 (ppm)
131	E isomer	0.17	0.027	0.20	—
	Z isomer	13		7.9	—
132	E isomer	0.083	0.013	0.072	—
	Z isomer	1.1		0.64	—
133[b]	Etofenprox	0.18	0.0035	0.047	360
134	MTI-790	0.049	0.018	0.037	27
135	MTI-800	0.018	0.008	0.036	68
137		2.0	—	1.3	—
147		0.097	0.009	0.072	—
148		0.076	0.0030	0.036	—
149		0.061	0.033	0.12	—
150		0.017	0.0029	0.034	—
155		0.22	0.049	—	110
157		0.74	0.13	0.63	540

[a] All compounds are racemic.
[b] MTI-500.

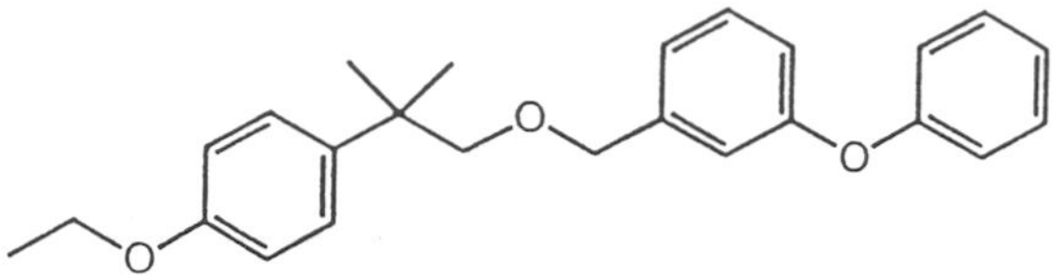

etofenprox (MTI-500)   (133)

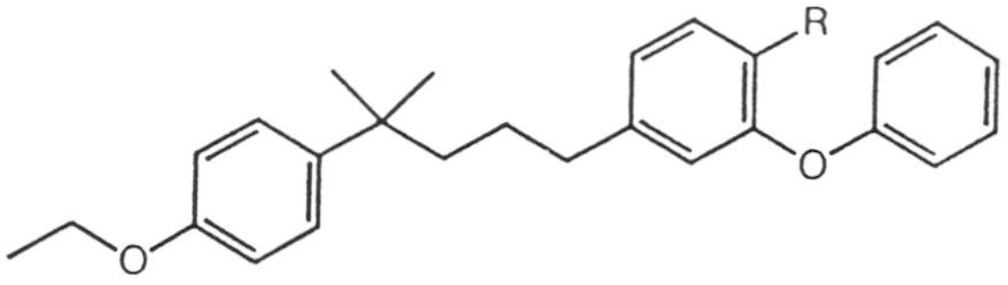

R=H; MTI-790   (134)

R=F; MTI-800   (135)

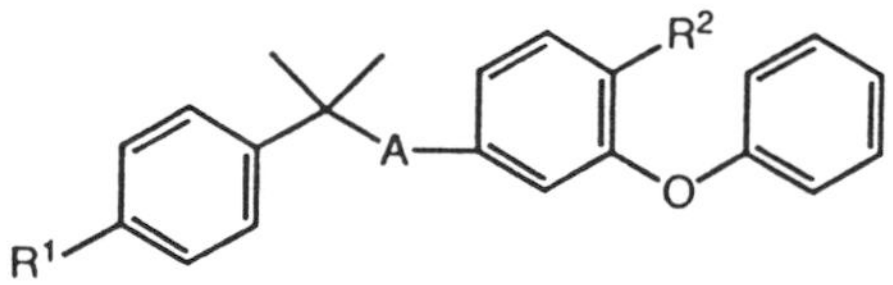

a) A= ·CH$_2$CH$_2$CH$_2$·
b) A= ·CH$_2$OCH$_2$·   (136)
c) A= ·CH=CHCH$_2$·

An important breakthrough in this area was the discovery of the ether etofenprox (MTI-500) (133) and related alkane-linked analogs such as MTI-800 (135) by chemists at Mitsui Toatsu Chemicals [134]. Etofenprox shows good insecticidal activity (Table 9) and low toxicity to fish and mammals. The alkane-linked analog MTI-800 (135) is, in general, more active than 133, except against *Spodoptera* spp., and shows similar activity to permethrin (18) against the pest species in Table 9 (cf. Table 1). MTI-800 also has very low toxicity to fish. For structures of the general type 136, the analogs 136a where A is -CH$_2$CH$_2$CH$_2$- are reported to be equal or superior in activity to the corresponding ethers (136b), and the ethers are slightly more active than the alkene analogs (136c). The introduction of a 4-fluoro substituent (R^2 = F) generally increases activity, and good activity is obtained when the substituent R^1 is chloro, ethoxy, and difluoromethoxy [134]. Although these analogs are not as active as some of the ester pyrethroids against many lepidopterous species, etofenprox (133) and, especially, 135 are very effective against the brown planthopper, *Nilaparvata lugens*, without causing a resurgence of the pest in rice paddy fields. They are active against a number of other rice pests, including various plant and leaf hoppers, stemborers, and the rice water weevil. This activity, along with their low fish toxicity, allows them to be targeted at the control of insect pests in the rice market [134, 135]. These pyrethroids are much more active than cycloprothrin (93), Table 7, against rice pests such as *N. lugens* [135]. The discovery of etofenprox prompted many research groups to investigate analogs of this type in order to extend the utility of pyrethroids into irrigated crop situations such as rice.

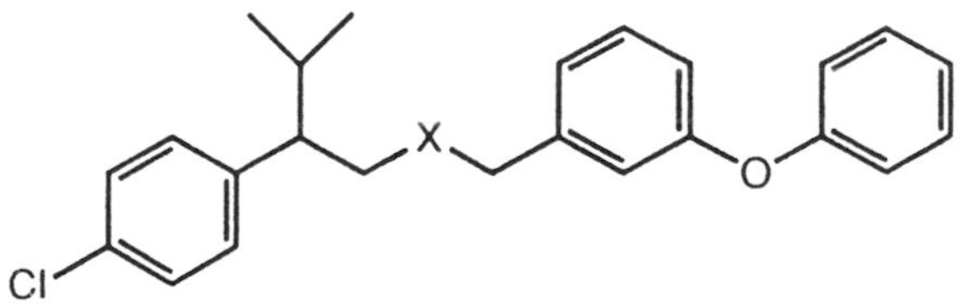

X=O (137)

X=CH$_2$ (138)

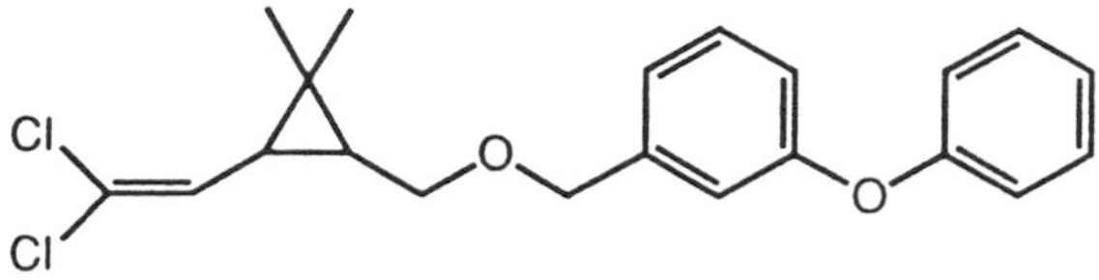

(1RS)-cis (139)

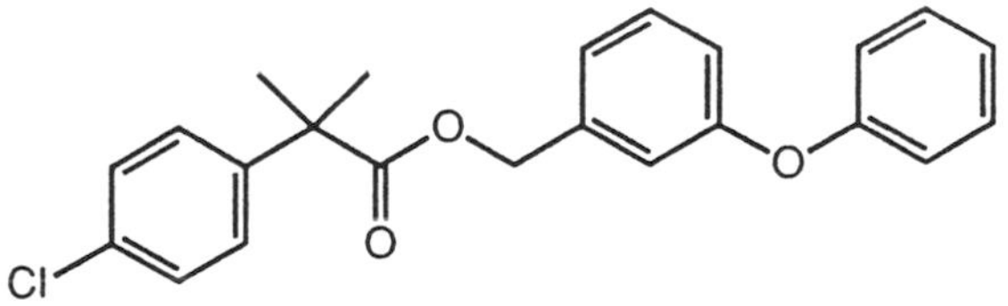

(140)

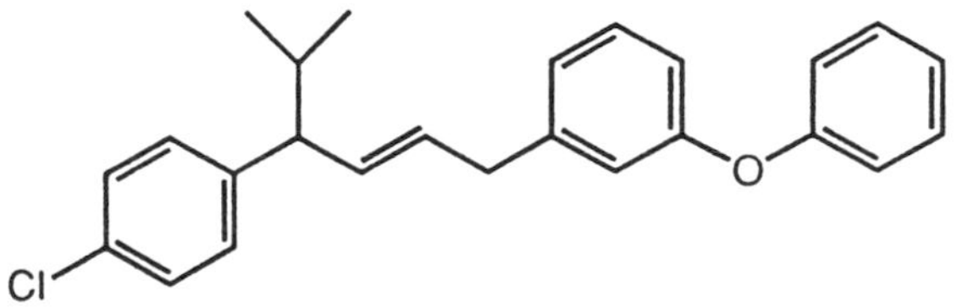

(141)

Interestingly, closely related ethers and alkane-linked analogs such as 137 (Table 9), 138, 139, and esters such as 140 have much lower insecticidal activity than 133 [52, 136, 137]. The replacement of the central ester linkage in fenvalerate and permethrin with other groups is, in general, not successful. Although a few selected derivatives with alkyl ketone and (E)-alkene linkages (e.g., 141) show modest activity, they are much less active than etofenprox (133) [136, 137; cf 138]. However, further investigations of the nonester fenvalerate analogs 137, 138, and 141 demonstrated that very active pyrethroids can be obtained with suitable structural variations. Thus, ethers such as 142, with a cyclopropyl group replacing the isopropyl group, are very active insecticides and miticides, being much more active than etofenprox (133) in foliar tests against a number of pest species [139]. The (−)-enantiomer of an analog of 142 (4-trifluoromethoxy replaced by 4-chloro) is more active than the corresponding (+)-enantiomer. The fish toxicity of the racemic 4-chloro analog is similar to that of etofenprox. In this series the 5-benzyl-3-furylmethyl and (2-methyl[1,1'-biphenyl]-3-yl)methyl ethers are considerably less active than the corresponding 4-fluoro-3-phenoxybenzyl ethers [139]. Thus, the range of "alcohol" moieties that is effective is limited. Similar results have been obtained in other ether series of analogs [137]. The corresponding (E)-2-alkenes, such as *143*, also show good foliar insecticidal activity, but have only moderate activity against plant hoppers and spider mites. Their fish toxicity is similar to that of etofenprox [140]. The related alkanes (*144* R = Cl or CF_3) show good activity against lepidopterous species and plant hoppers, comparable to that of MTI-800, but lower than that of the corresponding ethers [141]. Interestingly, the analog obtained by replacing the cyclopropyl group in *144* (R = Cl) with an isopropyl group (cf. *137*) is essentially inactive! The analogs of the type *144*;(R = Cl) also have relatively low fish toxicity, but they are much more toxic to mammals than etofenprox. Generally, in these three series, the ethers of type *142* are more active than the analogous 2-butenes (*143*), which are slightly more active than the corresponding alkanes (*144*), against insect species. In all three groups, the introduction of the 4-fluoro substituent in the benzyl ring gives a considerable increase in activity [140]. The introduction of a cyano group at C-4, α to the phenyl ring, in analogs such as *144* (R = Cl) generally gives an overall decrease in activity [142].

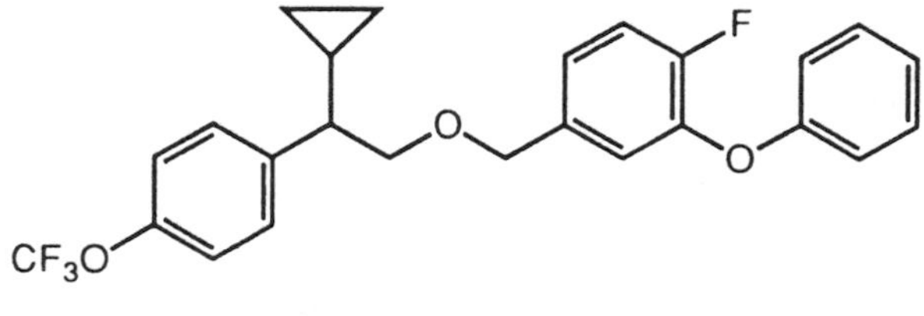

(*142*)

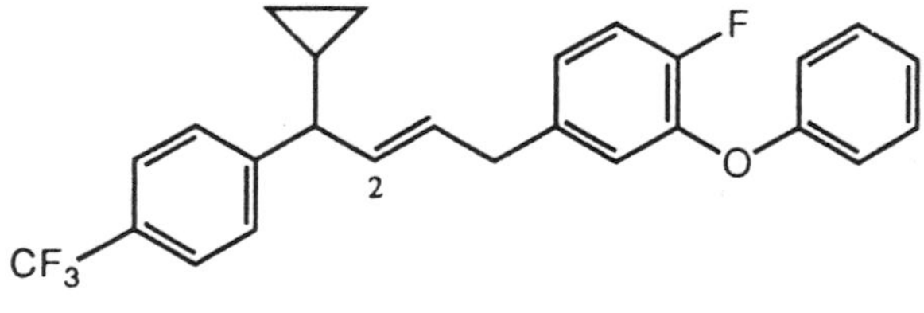

(*143*)

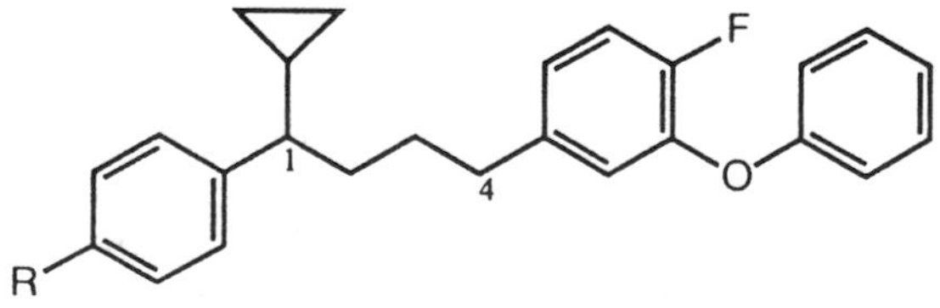

(*144*)

Elliott and co-workers have studied the effect on activity of changing the gem-dimethyl group in analogs of the general type *136* to a cyclopropyl group, as in *145*, [137, 143]. There is a general increase in activity against the housefly and mustard beetle in going from *136* to *145* when A is an E -CH=CHCH$_2$- group. For the corresponding ethers (when A is -CH$_2$OCH$_2$-) the change is less favorable and one observes either a small increase or a decrease in activity against these two insect species (cf. *133* versus *147*, Table 9). With the alkenes, good activity is obtained when R^1 is chloro or ethoxy. The introduction of a 4-fluoro substituent (R^2 = F) gives an increase in activity against the mustard beetle but little change against the housefly. The best analog reported in this study was *146*, which has activity similar to bioresmethrin (Table 1) against these two insect species [143]. We also found that the corresponding ethers *147–150* have good activity and that the introduction of a 4-fluoro substituent generally gives an increase in activity against *Heliothis virescens* and the housefly (Table 9). The alkanes (*145*; A = -(CH$_2$)$_3$-) in this series are less active than the corresponding 2-alkenes [137, 144]. The related ester *151* is inactive [137]. Introduction of gem-difluoro atoms into the cyclopropane ring of the ether (*147*) gives a 1.5–9-fold increase in activity. These 2,2-difluorocyclopropane ethers, such as *152* (R = H), are considerably more active than the corresponding 2,2-dichlorocyclopropane esters (e.g., *92*, Table 7) against several insect species [145]. The introduction of a 4-fluoro substituent in these analogs [e.g., *152* (R = F)] gave little change in activity against the insect species tested and the related pentafluorobenzyl ethers were inactive.

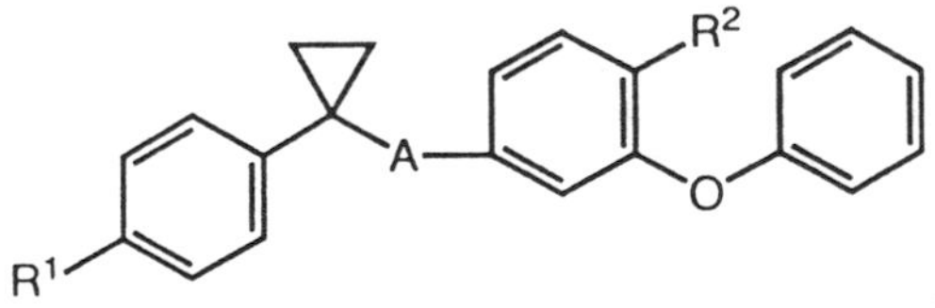

(*145*)

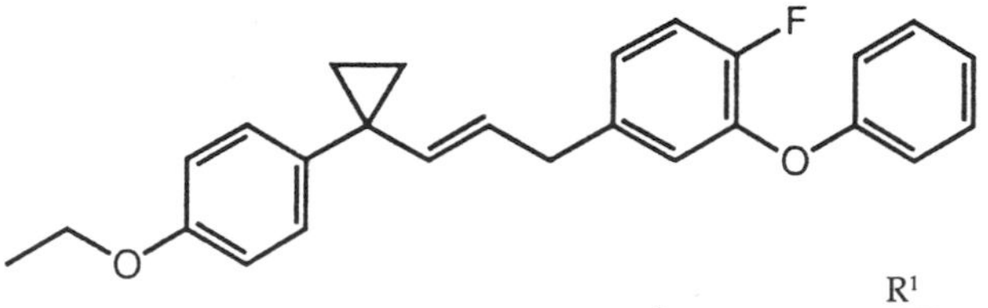

(*146*)

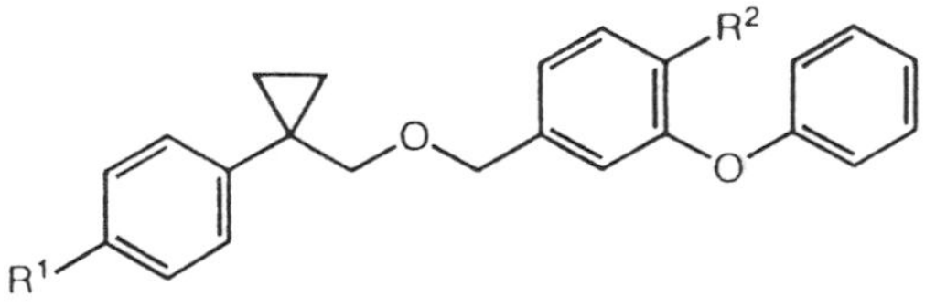

R^1	R^2	
CH$_3$CH$_2$O·	H	(*147*)
CH$_3$CH$_2$O·	F	(*148*)
Cl·	H	(*149*)
Cl·	F	(*150*)

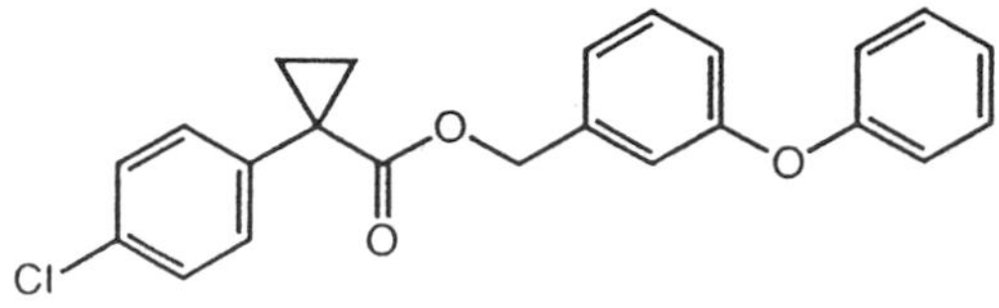

(*151*)

(152)

Both gem-dimethyl groups in etofenprox (133) are not required for activity, and the analog 153 is threefold to fourfold more active than etofenprox against a number of rice insect pests [135]. The corresponding 4-fluoro analog 154 is similar in activity to 153 against most insect species although it is twice as active against *Nilaparvata lugens* [135, 146]. Both 153 and 154 are more active than MTI-800 (135) against *N. lugens* but are less active against *Chilo partellus*. The (R) isomer of 153 is more active than the corresponding (S) isomer even though this (R) isomer has the trifluoromethyl group oriented in the opposite direction to the isopropyl group in esfenvalerate (23) [146]. The monotrifluoromethyl analog 153 is much more active than the corresponding monomethyl analog of etofenprox. The alkane-linked analogs of 153 are reported to be less active than the ethers except against lepidopterous species and mites [135].

R=H   (153)

R=F   (154)

Some silicon analogs of etofenprox and MTI-800 show good activity, although 155 and 157 are considerably less active than the corresponding carbon analogs 133 and 134, respectively (Table 9). Both 156 and 158 show about one-third of the activity of the corresponding carbon analogs against the tobacco cutworm, *Spodoptera litura* [147]. The analog 155 is reported to be generally less active than etofenprox [148, 149] and 158 is about four times less active than MTI-800 (135) against *Trichoplusia ni* [148]. In these analogs, the introduction of a 4-fluoro group moderately enhances the activity of the ether 155 but not that of the alkane-linked analog 157 [147]. The analog of 157 with (2-methyl[1,1'-biphenyl]-3-yl)methyl as the "alcohol" moiety is inactive [148]. The ether 155 also has very low fish toxicity; the corresponding (6-phenoxy-2-pyridyl)methyl ether is less active as an insecticide and is much more toxic to fish [149]. The alkane-linked analog 158 has very low fish toxicity, even much lower than MTI-800 [148]. Some simple trimethylstanniomethyl ethers such as *159* are reported to show surprisingly good activity against the Asiatic rice borer, *Chilo suppressalis*, and the housefly. They are much more active than the corresponding carbon analogs [150].

R=H	(*155*)
R=F	(*156*)

R=H	(*157*)
R=F, silafluofen	(*158*)

(*159*)

In conclusion, it is evident from this overview that structure-activity relationships in pyrethroids are complex. Attempts at quantitative structure-activity correlations in this field have been only modestly successful. It is difficult to rationalize suitable conformations of all the diverse active structures fitting a single binding site. Attempts have been made to deduce the shape of a common active conformation for all pyrethroids at the target site [1], but it is likely that there are several different binding sites for the different types of pyrethroids.

V. EFFECT OF TEMPERATURE ON INSECTICIDAL ACTIVITY

The insecticidal toxicity of pyrethrins and DDT decreases considerably with increasing temperature, especially posttreatment temperature within the normal range of 10 to 30°C, but this large negative temperature coefficient does not always occur with pyrethroids. The effect of temperature on pyrethroid toxicity varies considerably among different analogs and is influenced by the method of application, the temperature range selected, and the target insect species. Considerable differences have even been observed among different resistant and susceptible strains of the same insect species [151–153]. The influence of temperature on insecticidal activity is particularly important when pyrethroids are used in crops, such as cotton, in hot climates. The fall off in toxicity at high temperature may be responsible, in some cases, for poor field performance. It is not possible to predict the temperature response of a particular insect species and, thus, field efficacy at different temperatures may vary considerably from results obtained in laboratory bioassays performed at 25°C. Toxicity of a pyrethroid may increase or decrease with increasing temperature or it may show a negative correlation over one range of temperature and a positive correlation over another range. The knockdown activity of pyrethroids also varies with temperature, often showing a positive correlation [151, 154]. However, the recovery from knockdown may also be more rapid at higher temperatures [151].

With topical application of photostable pyrethroids to third instar larvae of *Heliothis virescens* there is a marked difference of the temperature coefficients for toxicity above 30°C between 3-phenoxybenzyl and α-cyano-3-phenoxybenzyl esters. We found that permethrin (*18*), *48*, *59*, and *60* showed a considerable negative temperature-activity correlation throughout the range 15–to 37°C, whereas the corresponding α-cyano analogs, cypermethrin (*19*), fenvalerate (*22*), *64*, and fluvalinate (*65*), respectively, showed a negative temperature-activity correlation between 15 and 30°C, but above 30°C the decrease in activity leveled off so there was little change in activity from 31 to 37°C. This leveling off effect was particularly dramatic for fluvalinate (*65*) [46, 155, 156]. Field tests confirmed that fluvalinate maintains its effectiveness in crop applications under conditions of high temperature (28–38°C). In another study, other researchers found similar leveling off effects between 27 and 38°C with flucythrinate (*54*), fenvalerate (*22*), and deltamethrin (*20*) against *H. virescens*, whereas there was a considerable continued decrease in activity for permethrin (*18*) at higher temperatures ([82]; see also [157]). Thus, in these studies against *H. virescens*, 3-phenoxybenzyl esters showed a large negative temperature coefficient of toxicity throughout the range 7–38°C, whereas α-cyano-3-phenoxybenzyl esters showed a negative correlation from 7°C to ~ 27°C and then, in general, little further decrease in activity at higher temperatures (up to 38°C). Thus, α-cyano-3-phenoxybenzyl esters of this type may be more consistently active than compounds such as permethrin in very hot field conditions.

Under similar laboratory conditions other scientists have obtained somewhat different results. It has been reported that, in the temperature range 16–38°C, the toxicity of fenvalerate and deltamethrin against

Heliothis virescens increased as the temperature increased, whereas both permethrin and cypermethrin showed a negative temperature coefficient throughout this range. The effect was more dramatic for permethrin than for cypermethrin [49, 158]. With other lepidopterous species, considerable variation in the temperature-activity response has been reported. Against the cabbage looper, *Trichoplusia ni* [49, 154], and the Egyptian cotton leafworm, *Spodoptera littoralis* [153], almost all of the pyrethroids tested gave an overall negative temperature-activity coefficient. Against the fall armyworm, *Spodoptera frugiperda*, the temperature coefficient was either negative, positive, or variable within the range 16–38°C depending on the pyrethroid under study [49]. Many nonlepidopterous insect species have shown overall negative temperature-activity coefficients, but there can be considerable variation in the pattern of response with different pyrethroids. A small reversal of the negative coefficient was often observed at high temperatures [51, 158, 159]. Against the housefly, most pyrethroid esters showed greater kill at 18°C than at 32°C [151, 152]. However, the nonester pyrethroid MTI-800 (*135*) showed only a small loss in activity against the housefly from 15 to 25°C and there was little further change in toxicity in the range 25–35°C [134].

Early studies with DDT had demonstrated that although the temperature coefficients for toxicity were large and usually negative for posttreatment temperatures in the range 10–30°C, reversal of the temperature effects can occur at low and high temperatures outside of this range [160, 161]. Extremes of temperature can stress insects and reduce their tolerance to an insecticide. Thus, in some cases, the reversal of the temperature coefficient from negative to positive, observed with some pyrethroids in the range 30–38°C, could be due in part to increased stress on the insect at the higher temperatures [154, 159, 162].

VI.   COMMERCIAL APPLICATIONS

A.   Agrochemical Markets

In the 1930s, pyrethrum was used as a dust and a spray for the control of a number of agricultural insect pests, especially in vegetable and fruit crops. With the shortage of pyrethrum during World War II and the subsequent introduction of DDT and other more potent insecticides with greater persistence and lower cost, the use of pyrethrum in agricultural and horticultural crops essentially disappeared. The first four commercial photostable pyrethroids fenvalerate (*22*), permethrin (*18*), cypermethrin (*19*), and deltamethrin (*20*) were introduced in the late 1970s. They are broad-spectrum, nonsystemic insecticides with high activity against a wide range of important agricultural pests by both contact and ingestion. Pyrethroids rapidly penetrate the insect cuticle and are much more active topically than orally against insects [163]. Good spray coverage is essential for effective control of less exposed foliar pests and, thus, pyrethroids are often applied by boom-spraying or mist-blowing to maximize the amount of contact with the target. Pyrethroids are effective in the field at application rates of less than 200 g/ha. The most active compounds, such as deltamethrin, α-cypermethrin, and λ-cyhalothrin are used at rates as low as 5–25 g/ha. They are essentially immobile on leaf surfaces with little movement of the compound into or within the plant and are usually nonphytotoxic to crops. They persist on crops for approximately 7–30 days and, because the residues are metabolized to polar materials, they do not accumulate in the food chain.

The commercialization of 18–20 came at a very fortuitous time. Insect resistance to organophosphates and carbamates had become a major problem in some areas, and these pyrethroids made it possible to control pests that were becoming increasingly difficult to control with existing insecticides. The early success of pyrethroids was largely due to their efficiency in controlling lepidopterous pests with fewer treatments per season than required by organophosphate insecticides, and their competitive prices. The most important insect pests are those belonging to the family Noctuidae, which cause damage to a wide range of crops, but especially to cotton. Heavy pesticide use is required in cotton due to the diversity of insect pest species and the severity of infestations. The high activity of pyrethroids against bollworms and leafworms in cotton, along with their cost-effectiveness and favorable toxicological and environmental properties, led to rapid market acceptance by growers, and they soon became the most widely used class of compounds for insect control in cotton. Species such as the tobacco budworm, *Heliothis virescens*, the bollworm, *Helicoverpa zea*, and the Old World bollworm, *Helicoverpa armigera*, were readily controlled. Some species such as the armyworms, *Spodoptera littoralis* and *Spodoptera exigua*, and the pink bollworm, *Pectinophora gossypiella*, required somewhat higher rates and more precisely timed applications that did *H. virescens* [164–166]. The insecticidal activity in laboratory bioassay varies considerably even between closely related lepidopterous species. These differences can be even greater in the field, due to behavioral differences, the influence of temperature on activity, and the presence of resistant field strains.

Almost all other crop areas are markets for pyrethroids, especially the fruit, vegetable, and horticultural sectors. Many lepidopterous pest species in these markets are controlled at low rates, although the diamondback moth, *Plutella xylostella*, is as difficult to control with pyrethroids as it is by any other class of insecticide. Many important insect pests of the order Coleoptera are well controlled by pyrethroids, although the boll weevil, *Anthonomus grandis grandis*, is an exception. In general, this species requires high rates, frequent applications, and careful timing of the sprays. Efficacy in the order Homoptera varies considerably according to the species. Most aphid species can be readily controlled, provided they are contacted by the spray. Lack of control in aphids and whitefly species is often due to the insect's behavior and the difficulty in contacting the insect rather than to lack of intrinsic insecticidal activity. Pyrethroids, generally, are not very effective against scale insects, especially in citrus orchards. This may be due in part to poor penetration into the insect.

A number of important beneficial sublethal effects of pyrethroids are observed in some field applications [164]. At sublethal concentrations effects may be observed on the growth, development, and fecundity of the insect. The affected insects can become hyperactive and tend not to remain on treated foliage ("irritancy" or "repellency") [167, 168], exposed larvae and adults may be inhibited from feeding, and adult females may be inhibited from laying eggs on treated plants [157]. Fenvalerate, for example, shows high feeding deterrent activity against *Spodoptera litura* at well below the lethal dose [163]. Such repellency and antifeeding effects can enhance the efficacy of the pyrethroid under field conditions because the affected insects may die for reasons other than acute toxicity, such as desiccation. These effects can also greatly suppress the transmission of viruses by aphids [167].

Since 1976, more than 10 new photostable pyrethroid active ingredients have been registered for agricultural use along with several additional products, including α-cypermethrin, λ-cyhalothrin, esfenvalerate, β-cyfluthrin, and τ-fluvalinate, which contain partially or fully resolved active isomers of the older products [169]. This has led to decreasing prices, putting the older products into essentially commodity status. Pyrethroid sales grew rapidly in the early 1980s, increasing from 7% of the world insecticide market in 1980 to a level of 21% (worth approximately $1600 million) in 1991. However, the share of market has remained in the range 20–25% over the past 4 years, 1987–1991 [170]. Today, the leading products are deltamethrin, fenvalerate, and cypermethrin. Cotton remains the major crop application, consuming 40% of the worldwide pyrethroid usage. This percentage has decreased from a high of 46% in 1987, due to resistance problems and increased applications in other sectors such as fruit, vegetables, cereals, and rice.

The early photostable pyrethroids had a number of limitations, including the lack of useful plant systemic or translaminar properties, generally poor activity against phytophagous mites, high toxicity to nontarget animals such as bees and fish, and physical property limitations, such as low vapor activity and strong absorption to soil particles, that made them ineffective against insect pests in the soil. Due to the lipophilic nature of pyrethroids it is unlikely that effective systemic analogs will be discovered. However, the other limitations have been overcome by the introduction of specific pyrethroids as discussed in Section IV. For example, tefluthrin (105) is more mobile and more stable in soil than most other pyrethroids and shows increased efficacy against soil insect pests such as corn rootworms. Several nonester pyrethroids, such as etofenprox (133) have low fish toxicity and high activity against planthoppers and leafhoppers, which allows their use in flooded rice. Treatment of rice for control of the brown planthopper, *Nilaparvata lugens*, frequently leads to increases in the pest population due to both poor activity of the pyrethroid and stimulation of the reproduction of the pest. This resurgence of *N. lugens* is not a problem with certain new products such as etofenprox.

Many early pyrethroids showed poor selectivity between target pests and beneficial species in certain applications. The effect of these broad-spectrum insecticides on beneficial predators and parasites has to be considered carefully to avoid unwanted effects. The selectivity between target species and beneficial arthropods varies widely. For example, application of the early pyrethroids permethrin, cypermethrin, deltamethrin, and fenvalerate, for insect control in orchard crops such as apples and pears is incompatible with integrated mite control programs. Use of these compounds can lead to a dramatic resurgence of populations of spider mites (family Tetranychidae) after a short period. The cause of this pyrethroid-induced resurgence is complicated. These four compounds are much more toxic to the predatory phytoseiid mites than to the target spider mites. At the recommended field rates, their use causes considerable predator mortality with little effect against the pest mites [171–173]. Integrated mite control programs depend on conservation of the predatory mites to aid in the regulation of the spider mite populations. Although the differential toxicity between the spider mites and their phytoseiid predators is an important factor, other indirect effects on spider mite behavior, reproduction, development, and diapause are also involved. For example, in response to pyrethroid residues spider mites (and also phytoseiid mites) are irritated and repelled, and exhibit hyperactivity, avoidance, run-off, spin-down, and aerial dispersal behaviors. This repellent activity induces the spider mites to disperse to plant areas with reduced or no pyrethroid residues or to leave the treated plants. In mite outbreaks, the result is that the temporary removal of predators and competitors allows the dispersed spider mites to rapidly proliferate. However, if the pyrethroid has sufficiently greater toxicity for the target mite species than the predators and high enough irritant and repellent activity, control may occur as the residual mites leave the treated crop. Other indirect effects on the mites also contribute to the outcome of pyrethroid treatment on the mite population [174, 175].

Thus, the miticidal action of a pyrethroid involves contributions from both direct toxicity and repellent (irritation) activity. Laboratory assays, like the slide-dip technique, that only measure direct mortality do not reflect field results. In order to obtain useful information, the assay must measure both the toxicity and the repellent activity of the compound against the spider mite [176]. Such an assay was used to obtain the data in the tables in this chapter [46]. There are large variations between pyrethroids in their direct toxicity and repellency to various mite species. Some pyrethroids, such as fenpropathrin (11), acrinathrin (37), flucythrinate (54), τ-fluvalinate (66), and bifenthrin (96) are reported to give effective control of spider mites in field crops. This is presumably due to a combination of their acceptable differential toxicity against the target mite over their predators and, perhaps more importantly, their ability to produce very high mite run-off and other indirect effects on the mite population dynamics. τ-Fluvalinate, which causes very rapid dispersal of spider mites and loss of fecundity, is one of the few pyrethroids that has not caused spider mite resurgence [176–178]. Selection in the laboratory and the field for suitable pyrethroid-resistant strains of predatory phytoseiid mites also offers some potential for reestablishment of natural control of spider mites in integrated pest management programs in orchards [174].

B. Noncrop Markets

1. *Household–Industrial*

One of the earliest uses of pyrethrum was for rapid knockdown and subsequent kill of flying insects. Even today, a major use of pyrethrum is as a nonresidual contact insecticide to control flies, mosquitoes, and cockroaches in household and public health sectors. In the hyperexcited state associated with knockdown, biting flies and mosquitoes are repelled and disoriented, and crawling insects, such as cockroaches, are flushed from their harborages, increasing their exposure to the insecticide. Differences in the speed of knockdown produced by different pyrethroids may be due, in part, to differences in their rates of penetration and detoxification within the insect. The structural requirements for knockdown differ from those for kill. Pyrethroids with high knockdown activity are normally less lipophilic than those with high lethal activity. Thus, the optimum polarity for knockdown is greater than that for kill and the more polar compounds may penetrate more rapidly to the site of action for knockdown. It is likely that excessive sensory hyperactivity in the peripheral nervous system is largely responsible for early knockdown symptoms from which affected insects may or may not recover.

In the household market, the consumer expects to see flying and crawling insects "killed" instantly and prefers one spray to be effective against all insect species. Consequently, mixtures are used containing rapid knockdown agents such as tetramethrin (7) and bioallethrin combined with compounds, such as bioresmethrin and (1R)-phenothrin (10), which show excellent kill (but poor knockdown at the recommended use rate) and a broader spectrum of activity [179]. Synergized pyrethrum is, generally, more effective under practical conditions than blends of tetramethrin with resmethrin or bioresmethrin. Pyrethrum is repellent to insects even at low levels, and its excellent flushing action is probably responsible for its superior performance [179]. Photostable pyrethroids, such as permethrin (18) and deltamethrin (20), are used for surface residual applications to control cockroaches, flies, and mosquitoes in dwellings [180].

In recent years, several new pyrethroids have been evaluated for this market. Prallethrin (121), the propargyl analog of allethrin, was introduced in 1990 by Sumitomo Chemical Co. Prallethrin is more active than bioallethrin in both knockdown and killing activity [126]. The pyrethroid imiprothrin (120) is another recent introduction, and empenthrin (122) is noteworthy for its relatively high vapor pressure, allowing it to be used as a fumigant in enclosed areas [39, 181].

2. Synergism

Both the knockdown and killing effects of pyrethrum and some pyrethroids can be potentiated manyfold, especially against the housefly, by the addition of synergists that have little or no toxicity alone [4]. In the late 1930s and 1940s, a number of synergists, such as sesamin, were discovered and the importance of the methylenedioxyphenyl group in these molecules was recognized. The first highly effective synergist for the pyrethrins to be found was piperonyl butoxide (160) [182], followed soon thereafter by the dicarboximide MGK 264 (161) [183]. Combination of these synergists with expensive pyrethrum allows a reduction in the pyrethrum concentration, which permits the mixture to remain competitive in cost with other classes of insecticides in the household market. Piperonyl butoxide is the most widely used synergist and is formulated with the pyrethrins (and pyrethroids) with a 5–20-fold higher concentration of the synergist.

piperonyl butoxide   (160)

MGK 264   (161)

The mode of action of pyrethroid synergists is not completely understood, but they appear to enhance toxicity by inhibiting the oxidative and/or hydrolytic metabolism of the insecticide within the insect. By interfering with *in vivo* detoxification of the pyrethroid, the synergist increases the effectiveness and also the likelihood that insects knocked down will subsequently die rather than recover. Pyrethrin I (1), synergized with 160, produces a response almost identical to that of bioresmethrin in the housefly, supporting the conclusion that detoxification limits the potency of 1. Which synergist is optimal for a pyrethroid depends on the importance of various metabolic routes in the insect species being treated. The insecticidal activity of a pyrethroid is not necessarily increased by a particular synergist against all insect species. For example, in the housefly, an important detoxification route for 1 involves the microsomal mixed-function oxidase system, and 160 is an extremely effective synergist. However, in the mustard beetle, *Phaedon cochleariae*, this oxidase system appears to be less important and very little synergism is observed [15]. The synergistic effect is also influenced by the formulation, pyrethroid:synergist ratio, timing and method of application, and the particular synergist and pyrethroid being studied. The synergistic factors vary widely with the alcohol moiety of the ester, being very high with the natural pyrethrolone esters (such as 1) and very low with 5-benzyl-3-furylmethyl esters like resmethrin (8) [184]. Deltamethrin (20) and other photostable pyrethroids are generally synergized to a much smaller degree than the pyrethrins or compounds such as tetramethrin (7), which are much more susceptible to metabolic breakdown.

3. Subterranean Termite Control

Termites are a major threat to timber in buildings in tropical and some temperate parts of the world. With the removal of the highly effective cyclodiene organochlorine insecticides from the market, because of their undesirable toxic effects, three pyrethroids (permethrin, cypermethrin, and fenvalerate) have been registered for termite control along with the organophosphorus insecticide chlorpyrifos. Although all four compounds are used commercially, the major product in this market is chlorpyrifos [185]. These compounds are applied to the soil beneath or surrounding building foundations. They protect the structures from subterranean termites by both toxicity and repellency [186]. Repellency appears to be the major effect of pyrethroids, whereas contact toxicity during tunneling is important for chlorpyrifos. Thus, the pyrethroids deter the termites from penetrating treated soil. Both types of action are effective in protecting structures when the compounds are properly applied to the soil. Recently, the silicon pyrethroid silafluofen (*158*) has been evaluated as a termaticide. It is very effective in protecting wood and is more stable than the pyrethroid esters under alkaline conditions.

4. Other Applications

Pyrethroids are used in all of the main areas of public health. Applications include the control of major disease vectors, such as malarial mosquitoes, tsetse fly, and black fly, and use in the urban sector to control pests such as cockroaches, fleas, bedbugs, and lice [187]. Thermal vaporization of pyrethrins has been used since the early days of pyrethrum commercialization. A suitable slowly burning solid mixture produces an aerosol of the pyrethrin or pyrethroid in the smoke. *S*-Bioallethrin (*6*) and its isomeric mixtures are used extensively in mosquito coils and electrically heated mats to protect people in the tropics from the biting of mosquitoes. Tetramethrin and resmethrin are less effective than allethrin and bioallethrin in this application, despite their higher topical insecticidal activity and their better activity when used in aerosol formulations [25]. Large-scale thermal fogging or spraying of pyrethroids is often used in urban public health applications to control nuisance and disease-carrying insect pests.

In the veterinary field, pyrethroids are used to control external parasites of livestock. The principal targets are cattle ticks and ectoparasites on sheep. Although most pyrethroids are not very active against ticks, a number of the analogs are as active as the organophosphorus insecticides [1]. In particular, flumethrin (*30*) is a highly effective selective tickicide. It controls all stages of both one-host and multihost ticks in dip, spray, and pour-on applications with an excellent safety margin for cattle and sheep [69].

VII. TOXICITY TO NONTARGETS

A. Mammals

Although the selectivity between target and nontarget organisms is high for most pyrethroids, their low toxicity is often overstated in the literature. They are not entirely "harmless" to mammals and several of the commercial pyrethroids, such as bifenthrin (96), fenpropathrin (14), flucythrinate (54), and tefluthrin (105), have quite high acute oral toxicities to rats (LD_{50}, 20–80 mg/kg) [1, 188]. Many pyrethroids have high intrinsic toxicity to mammals as demonstrated by their very high intravenous toxicities in rats [189] and dogs, although such measurements bear little relation to practical applications. Three exceptions are bioresmethrin, (1R)-phenothrin, and permethrin, which show remarkably low intravenous toxicities [189]. The "natural is safe" idea that prevails in popular opinion and in much of the literature is nonsense; many natural products are extremely toxic to mammals. Even pyrethrins I (1) and II (2) have very high intravenous toxicities in female rats (the approximate lethal doses are 5 and 1 mg/kg, respectively) [190]. Nevertheless, pyrethroids are, overall, a remarkably safe class of insecticides and they are used at low application rates.

The intrinsic toxicity to mammals and other nontargets is not usually observed due to the physical and chemical properties of the pyrethroids and the metabolic differences between mammals and insects. Pyrethroids and pyrethrins are able to penetrate rapidly to the sites of action in insects, whereas in mammals they are absorbed slowly, and the absorbed material is rapidly metabolized. Mammalian metabolism efficiently converts the compounds to predominately polar metabolites, which are rapidly eliminated in the feces and urine before sensitive sites are reached. Thus, many pyrethroids have low to moderate oral toxicities to mammals and some, such as acrinathrin (37), bioresmethrin, cycloprothrin (93), empenthrin (122), etofenprox (133), phenothrin (10), and tetramethrin (7) are essentially nontoxic orally [188]. These analogs are much less toxic to mammals than the natural pyrethrins. The measured oral LD_{50} value for a pyrethroid against rats or mice depends dramatically on the type of carrier and formulation used, the species and strain of the test animal, its sex, age, and degree of fasting at the time of dosing, and the isomer ratio of the sample. More lipophilic carriers increase the observed toxicity, and reported values for acute oral toxicity can vary fivefold to tenfold for a particular pyrethroid depending on the dosage form.

In the chrysanthemate type of pyrethroid, the *cis*-cyclopropane-carboxylates are more toxic to vertebrates than the corresponding trans isomers. The introduction of an α-cyano group in 3-phenoxybenzyl esters reduces the rate of metabolism and increases the mammalian toxicity [189]. Dermal toxicities of pyrethroids to mammals are almost always of a low order. The pyrethroids are one of the best studied classes of insecticides and numerous subchronic and chronic studies have shown their toxicology in mammals to be of minimal concern. Birds are even less sensitive to pyrethroids than mammals.

In instances of occupational acute pyrethroid poisoning in humans, the prognosis is generally good with adequate therapeutic treatment. In a study of 573 cases of acute occupational or accidental (ingestion) poisoning in China, it was found that the vast majority of patients recovered in 1–6 days [191]. Most of these cases involved the inappropriate handling of these insecticides. Under normal field conditions there is a large margin of safety for humans.

Most pyrethroids produce an abnormal cutaneous sensation (paresthesia) on contact with exposed human skin, particularly on the face and arms. One feels an uncomfortable tingling or cold burning sensation that occurs 30 min to 3 hr after dermal exposure and persists from a few hours to a day or so [192–194]. No inflammation or abnormal neurological findings are observed and there is no permanent damage. This effect is presumably due to the repetitive firing of sensory nerve endings in the skin. Although this appears to be a general property of pyrethroids, some compounds, such as the (6-phenoxy-2-pyridyl)methyl esters *106* and *110–116*, have a more pronounced effect than others [116, 121].

B. Fish

The acute toxicity of most pyrethroids to fish is extremely high under laboratory conditions in pure water [195]. They are also very toxic to aquatic insects and crustaceans, although mollusks are generally tolerant to pyrethroids [196, 197]. In order to accurately assess the environmental hazard associated with pyrethroid run-off after agricultural applications, one needs to quantitate the bioavailability. The high lipophilicity and low water solubility of these compounds ensures that they readily partition from the water into the gills of fish and then rapidly enter the bloodstream even at low aqueous concentrations. However, under field conditions, this is much less of a problem than laboratory experiments would predict. Because pyrethroids are strongly adsorbed by suspended particulate matter and organic matter in general in an aqueous natural environment, their concentration in the water is greatly diminished. This means the compounds are only present in the aqueous phase for a short time and, therefore, present a low hazard under practical conditions with respect to spray drift or run-off. It also explains why there is little toxicity to fish and aquatic invertebrate populations under normal field conditions. Any toxic effects observed are usually localized and transitory.

As discussed in Section VIA, the application of early photostable pyrethroids to flooded rice crops was limited by their very high toxicity to fish and their relatively low activity against the brown planthopper, *Nilaparvata lugens*. Many of these compounds promoted resurgence of some strains of this important pest. This resurgence appears to be caused by a combination of low toxicity against *N. lugens* and the stimulation of reproduction of the insect [198]. Several of the more recent pyrethroids, such as etofenprox (*133*), show good activity against such rice insect pests and have very low toxicity to fish [134, 135].

C. Bees

The honey bee, *Apis mellifera* (order Hymenoptera), the alfalfa leafcutting bee, *Megachile rotundata*, and the alkali bee, *Nomia melanderi*, are important pollinators of wild and cultivated plants. It is necessary to minimize the incidental poisoning of these beneficial insects with pyrethroids. Under laboratory conditions, the acute oral and contact toxicities of pyrethroids to the honey bee are usually high or extremely high, with topical LD_{50} values in the range of 0.01 to 0.3 μg/bee [196]. However, pyrethroid residues on plant surfaces are much less toxic due to adsorption of the compound on the plant material and "repellent" effects. Extensive field trials at the low recommended application rates have shown that pyrethroids are of low hazard to the honey bee under practical outdoor conditions [196]. Even deltamethrin has little detrimental effect on honey bee survival in field experiments on flowering crops; thus, pyrethroids have been used extensively in such crops as oilseed rape and alfalfa.

Although pyrethroids are adsorbed by plant material, limiting their availability to foraging bees, some compounds, such as permethrin and bifenthrin, cause high mortalities when honey bees are *confined* with freshly treated cotton leaves [199]. Field applications of pyrethroids are associated with subsequent suppression of bee foraging activity in the treated crop. This is usually ascribed to repellent effects of the compounds on the treated plants. Honey bees cease foraging for periods of a few hours to several days following the application of pyrethroids. Conditioning experiments with bees have demonstrated that exposure to pyrethroids reduces their learning response to odor [200]. Thus, the inhibition of foraging activity following crop treatment could be due to a sublethal odor learning dysfunction.

τ-Fluvalinate (66) is a unique pyrethroid in that it is essentially nontoxic to honey bees (topical LD_{50}, 18.4 µg/bee; oral LC_{50}, 1000 ppm in nectar) [46, 84, 199, 201, 202]. In addition, foliar residues of 66 are nontoxic and nonrepellent to bees [199, 203, 204] and do not inhibit pollination of plants by bees. Even at a high rate, τ-fluvalinate had the least impact of all pyrethroids in a study on the odor learning response of honey bees [200]. It is also an effective miticide [46]. These observations led to the development of τ-fluvalinate, formulated in PVC resin (Apistan™ strip), for the control of the parasitic mite, *Varroa jacobsoni*, in honey bee colonies. This ectoparasitic mite, which originated in Asia, is now a serious economic threat to beekeeping worldwide. The female *V. jacobsoni* feed on the hemolymph of the bees, especially the drone and worker larvae in operculated cells [205, 206]. Slow-release formulations of τ-fluvalinate are also effective in controlling infestation by the parasitic Asian honey bee broad mite, *Tropilaelaps clareae*, which is a major impediment to honey bee production in Southeast Asia [207, 208].

D. Other Beneficial Arthropods

The effects of pyrethroids on beneficial terrestrial nontarget organisms have been extensively reviewed [196]. Pyrethroids are broad-spectrum insecticides. As most insect orders are very susceptible, one would expect some effects on beneficial predator and parasitic arthropod species. In pest management programs it is desirable that an insecticide has a minimal effect on the natural populations of beneficial entomophagous species. As discussed in Section VIA, pyrethroids can exacerbate insect pest problems by promoting a resurgence of the pest population or by increasing the economic significance of the pest. In some cases, this is due, at least in part, to greater effects of the pyrethroid on the predatory and parasitic arthropods than on the target insect species, or to stimulatory effects on the reproduction of the pest. The effects of any pyrethroid on field populations of different insect species is quite complicated. In some situations, effective control of a pest requires a careful study of the life history of both the entomophagous and the pest species present [196]. In cereals, for example, extensive studies on predatory and parasitic arthropods indicate that there are probably no long-term adverse effects from the use of pyrethroids. However, in cotton, severe effects are often observed on predatory arthropods after multiple applications.

VIII. METABOLISM AND DEGRADATION

The metabolism and environmental fate of pyrethrins and pyrethroids have been investigated in considerable detail [209–211]. The two main routes of degradation, i.e., photochemical and biological, are often superimposed, and the sites and rates of metabolism depend on the organism and the pyrethroid under investigation. The pyrethrins are unstable in air and light, and metabolic detoxification in the insect is a major limiting factor for their insecticidal activity. An understanding of the sites of photodegradation and biodegradation of the pyrethrins aided in the discovery of photostable pyrethroids [210]. Photochemical studies have been carried out with many pyrethroids under a variety of conditions, such as in organic and aqueous solvents and on plant and inert surfaces. The introduction of the α-cyano group into 3-phenoxybenzyl esters tends to reduce the photostability of the analog [46, 111]. In the nonester pyrethroids, such as etofenprox (133), the nature of the central linkage has a large influence on the photostability [212].

Most metabolic studies in mammals have been carried out in rats, mice, dogs, and in in vitro preparations, such as isolated liver fractions, from these animals. The rates of metabolism and excretion of a pyrethroid and its metabolites have a significant effect on the toxicity of the compound. Even though pyrethroids are very lipophilic, they are not stored to any significant extent in mammalian tissues. They are rapidly converted to less lipophilic metabolites in mammals by ester cleavage, oxidation, hydroxylation, and conjugation. For example, pyrethroid esters administered orally to mammals are generally rapidly metabolized by oxidation of the acid and alcohol moieties and cleavage of the central ester linkage, and conjugation of the resulting products (carboxylic acids, alcohols, and phenols) with glucuronic acid, sulfuric acid, or amino acids such as glycine and taurine. These polar metabolites are then readily excreted in the urine and feces. Almost complete elimination of the applied dose occurs relatively rapidly, which limits the toxicity of the compounds to mammals. Some minor liphophlic conjugates of the acid moiety, such as cholesterol esters, can be isolated in certain instances [213].

With the pyrethrins and close analogs such as allethrin, initial metabolism by mammals primarily involves oxidation of the intact molecule by the microsomal mixed-function oxidase system. Ester cleavage is not a significant reaction. For permethrin and other cyclopropane-containing pyrethroids, ester cleavage is a major reaction. The trans isomers are usually metabolized faster than the corresponding cis isomers. The introduction of an α-cyano group into 3-phenoxybenzyl esters reduces the susceptibility of the molecule to both hydrolytic and oxidative metabolism. However, even α-cyano pyrethroids such as deltamethrin, cis-cypermethrin, and fenvalerate are metabolized and efficiently eliminated from mammals at a reasonably rapid rate. The metabolism of fluvalinate by rats is unusual in that much more of the oral dose is excreted unchanged in the feces than with most other pyrethroids [214, 215]. Fenvalerate, on the other hand, is metabolized in a fashion similar to the cyclopropane pyrethroids. Excretion is fairly rapid and metabolism is extensive with ester cleavage and 4-hydroxylation in the alcohol moiety being the major initial reactions [211].

Metabolism of pyrethroids in insects is complex. Oxidation at one or more sites in the molecule along with ester hydrolysis, followed by secondary oxidations, can generate a large number of metabolites. The relative significance of different metabolic processes can vary substantially among different pyrethroids, as well as their stereoisomers, and among different insect species [216]. The rates of metabolism in insects appear, in general, to be somewhat slower than in mammals.

Metabolism of commercial pyrethroids by plants has also been studied. A combination of photodecomposition and metabolism by the plant is observed on leaf surfaces in sunlight. As in mammals, ester cleavage and hydroxylation of both alkyl groups and aromatic rings are important initial metabolic processes, but the subsequent oxidation of the primary alcohol metabolites to carboxylic acids is less important in plants. Subsequent conjugation of the metabolites with sugars or amino acids occurs readily in plants [211].

Most pyrethroids bind strongly to soil and are rapidly and completely degraded in most soil types under both aerobic and anerobic conditions. Soil degradation of commercial pyrethroids has been extensively investigated. The half-life in soil varies widely with different analogs and soil types. But most pyrethroids bind irreversibly to soil and are completely degraded, forming carbon dioxide as the major end product, particularly under aerobic conditions [211].

IX. MODE OF ACTION

Considerable research has been directed at attempts to determine the mode of action of pyrethroids in insects, but the actual cause of death by these nerve poisons remains obscure [1, 217–219]. Care must be used in interpreting the results obtained from animals other than insects, as their relevance to insect target sites is questionable in many cases. In addition, experiments on isolated nerve preparations are not good models for toxic effects on whole animals. The symptoms of pyrethroid poisoning leave little doubt that the nervous system of insects is a primary site of toxic action, and pyrethroids act at almost every part of the insect nervous system including the central and peripheral nerves. The progressive series of symptoms in insects is similar to that observed, at much higher doses, in mammals; it includes hyperactivity, lack of coordination, tremor, and convulsions, followed by paralysis and eventually death.

Changes in nerve membrane ionic permeability are mediated by discrete molecular structures, called ionic channels, that are formed by proteins embedded in the lipid matrix of the membrane. Control of sodium ion permeability through the nerve membrane is vital to nerve function, and the resting membrane is electrically polarized. Pyrethroids act stereoselectively on a small fraction of the fast voltage-dependent sodium ion channels in excitable nerve membranes in several parts of the insect nervous system. The major initial effect is to delay the closing of the sodium ion channel activation gate. Prolongation of the transient inward sodium ion movement through the nerve membrane produces membrane depolarization, which eventually causes a block of nerve impulses and results in the toxic symptoms. Other cation ion channels may also be affected, and some additional sites may be important with different pyrethroids [218, 219].

In insects, pyrethroids have been classified into two types based on their symptoms and electrophysiological effects, i.e., presence or absence of repetitive firing in the peripheral and central nervous systems [217–219]. Type-I pyrethroids, including allethrin, tetramethrin, bioresmethrin, and most non-α-cyano esters, produce marked hyperactivity followed by incoordination, prostration, convulsions, tremors, and paralysis in whole insects and also induce repetitive discharges in nerve cords. Type-II pyrethroids, which include most α-cyano-3-phenoxybenzyl esters, cause some restlessness, followed by ataxia, prostration, convulsions, tremors, and paralysis in whole insects. In nerve cords, they block the production of action potentials without inducing repetitive after discharges. They maintain the sodium ion channels in a modified open state, causing depolarization of nerve axons and terminals. However, the differences between the two types are only quantitative and somewhat artificial, especially as there is a continuum from type I to type II with some compounds (such as permethrin and fenvalerate) showing intermediate effects [154]. Both types affect the sodium ion channels of the nerve membrane in a similar fashion. The major difference is the time that the modified sodium ion channels remain open, being very short for type-I and much longer for type-II pyrethroids. Only limited success has been achieved in attempts to correlate the actions of pyrethroids on isolated nerve fibers with the knockdown and lethal effects on whole insects. There is poor correlation between the electroneurophysiological data obtained from nerve preparations and the insect toxicities for different analogs. The discrepancies can be very large.

In mammals there are also two different toxic syndromes at very high doses. The T-syndrome, induced by type-I pyrethroids, has the prominent symptomology of whole-body tremors. The CS-syndrome, induced by type-II pyrethroids, is characterized by choreoathetosis (sinuous writhing) and profuse salivation. As with insects, the distinction between these syndromes is somewhat artificial, and some pyrethroids produce tremors and salivation (TS-syndrome) [189, 194, 220]. In addition, there is a large range of other symptoms, all of which are indicative of strong excitatory action on the nervous system. As in insects, the pyrethroids appear to act in mammals primarily on the voltage-dependent sodium ion channels in excitable nerve membranes. Although it is often difficult to achieve systemic poisoning in mammals by oral dosing of pyrethroids, due to slow absorption and rapid metabolism, neurotoxic effects are readily observed when the nervous system is directly accessible, as in intravenous administration [189].

In general, insects *do not die rapidly* after treatment with pyrethroids. For example, larvae of *Heliothis virescens* and *Spodoptera littoralis* when treated topically with cypermethrin were observed to die slowly at all doses tested. Even at high doses, the death or recovery could not be predicted with certainty until at least 3–4 days after treatment [221, 222]. In these studies with *S. littoralis*, using LD_{95} doses and continuous long-term recordings of nerve activity in the intact treated insects, death did not occur until up to 5 days after the initial dosing. Thus, actual death may take much longer than the time frame of most bioassay experiments, so that mortality counts should not be made until at least 72 hr after treatment in lepidopterous species [221, 222] (see Table 1). Of course, in a field situation most of the treated insects that have been knocked down would probably die from other causes such as dehydration or attacks by predators.

This time factor in pyrethroid poisoning of insects is sometimes overlooked in mode of action studies. The initial symptoms and electrophysiological effects observed on the nervous system during the first few hours after treatment are transient lesions that may have little bearing on the death or recovery of the insect. Death occurs as a result of irreversible effects on the insect after prolonged action on the nervous system. Insects are not as dependent as mammals on continuous nervous control of respiration and circulation; thus, the point of death is more difficult to define in insects. The overstimulation of the insect's nervous system by a pyrethroid does not explain the lethality, so that other events evidently determine whether or not the poisoning is fatal. Although the voltage-sensitive membrane bound sodium ion channel is probably the major initial site of action, it is not the only site involved in poisoning [1, 217, 218]. For example, insect neurosecretory neurons are very sensitive to pyrethroids, and repetitive activity can also be induced in these cells resulting in the abnormal release of neurohormones. Disruption of the balance of neurohormones within the insect could be a major factor in the toxicity. The insect eventually dies due to a complex series of secondary effects, sometimes including a dramatic loss of body fluid [221, 223], due, perhaps, to release of a diuretic hormone from the neurosecretary cells into the hemolymph.

X.   INSECT RESISTANCE

One of the main factors that will determine the future commercial use of pyrethroids is the degree and rate of development of resistance by economically important insect pests and how successfully resistance management strategies are implemented. The use of pyrethrum and photolabile analogs, such as allethrin and bioresmethrin, against household and public health pests resulted in relatively low levels of resistance, but soon after the widespread commercial use of photostable pyrethroids in agriculture and public health, high levels of resistance were observed. The initial high resistance seen with houseflies and mosquitoes appears to be due to cross-resistance induced by previous intense selection pressure with DDT. The mechanism of this resistance in the adult housefly was characterized as decreased target site sensitivity in the nervous system and has become known as knockdown resistance (*kdr*). It is associated with a recessive gene that confers knockdown resistance to both DDT and pyrethroids [224]. One early example of the rapid development of resistance to photostable pyrethroids was their use in cattle ear-tags to control the horn fly, *Haematobia irritans*. The slow release of pyrethroids in the plastic eartags rapidly selected for horn fly resistance, so that by 1984 a practical level of control could no longer be obtained [225, 226]. This example shows the danger of using persistent slow-release formulations. In general, insecticide persistence and/or frequent application encourage the development of insect resistance.

It is essential to control the development of economically significant levels of resistance. The heavy use of photostable pyrethroids in crops such as cotton has resulted in a number of well-documented cases of resistance, especially by lepidopterous species. Often resistance is caused by a number of mechanisms, including delayed or reduced penetration through the insect cuticle, reduced target site (nerve) sensitivity (*kdr*), enhanced capacity of the insect to metabolically detoxify the pyrethroid, and altered behavioral patterns. The resistance is usually due to a combination of such factors [227]. Interestingly, insecticide resistance often imparts a considerable cost to the resistant population compared with the susceptible strain. Deficiencies in fitness, vigor, behavior, or reproductive potential can occur in resistant strains. For example, there are biological constraints associated with pyrethroid resistance in *Heliothis virescens*. There is a rapid decline in resistance in laboratory and field strains in the absence of pyrethroid pressure, showing that the resistant strains are less "fit" than the susceptible insects [228]. However, resistant strains of insect species are not always at a competitive disadvantage [229].

Considerable effort has been applied to the development of strategies to manage resistance. Pyrethroid resistance in *Heliothis* and *Helicoverpa* spp. was anticipated from the first widespread commercial applications in cotton, and in recent years, significant resistance has built up due to strong selection pressures. For example, control failures have been reported in Australia, Colombia, Guatemala, India, Indonesia, Pakistan, Thailand, Turkey, and the United States, especially against the tobacco budworm, *Heliothis virescens*, and the Old World bollworm, *Helicoverpa armigera*. To prolong the utility of these cost-effective and environmentally compatible insecticides, it is essential to limit their use, so that susceptible genotypes in the pest population are preserved. Increasing the dosage and/or shortening the interval between applications usually encourages the development of resistance, as does retreatment following a control failure. Thus, the aim of resistance management is to conserve susceptibility in the population. This usually involves, in part, minimizing selection pressure by limiting the overall use of pyrethroids combined with careful timing of applications.

The ability to develop a successful resistance management strategy depends on identifying and understanding the various factors influencing resistance-induced field control failures. The development of resistant strains of insects is a dynamic and complicated phenomenon involving biochemical, physiological, genetic, and ecological factors [227, 230]. These factors vary with the insect species, population, and geographic location. It is essential to have a good knowledge of the population dynamics of the pest and to develop rapid and reliable tests for monitoring field populations, so that changes in the frequency and distribution of resistant insects can be observed.

A good case study is the response to the sudden development of very high resistance by the Old World bollworm, *Helicoverpa armigera*, in Australian cotton crops in the fall of 1983. Following these serious control failures, three resistance mechanisms were identified [230, 231]. They were a strongly decreased nerve sensitivity (*super-kdr* mechanism), an enhanced metabolism factor which could be overcome with piperonyl butoxide, and reduced cuticular penetration. Of these, nerve insensitivity appeared to be the major cause of the high-order resistance (>100-fold). Increased metabolism had some influence, and penetration resistance was relatively unimportant. The *super-kdr* mechanism normally confers such a high order of resistance that control of an insect species is difficult. A resistance-management program was instituted in 1983 that restricted pyrethroid use on all crops in the problem areas to only one of the four or five yearly generations of *H. armigera*. Pyrethroids were applied in cotton only during the period of peak squaring, flowering, and boll formation, and no applications of pyrethroids were allowed after this time. This compromise strategy of partial withdrawal of pyrethroid use resulted in considerable changes in both the level and type of resistance. By 1990, 7 years later, the *super-kdr* nerve insensitivity mechanism was no longer a major factor in field populations. The relative importance of enhanced metabolism and penetration resistance had increased, along with the appearance of a new low-order *kdr*-type mechanism of little importance. Obviously, the resistance to pyrethroids in this species is sensitive to use strategy. Careful and restricted application resulted in a reduction of the selection pressure on the *super-kdr* mechanism, and its frequency declined to an undetectable level, leaving a high frequency of enhanced metabolism and penetration mechanisms, but an overall much lower order of pyrethroid resistance in these areas [231, 232].

In cotton fields in the United States, the first control failure attributed to pyrethroid-resistant genotypes of *Heliothis virescens* occurred in west Texas in 1985. In the following years, control failures resulting from decreased susceptibility to pyrethroids were observed throughout the southwest and midsouth. Nerve insensitivity (*kdr*) appears to be the predominant mechanism for the widespread resistance to pyrethroids by *H. virescens* in the cotton-growing areas of the United States [233–235]. In response to this problem, resistance-monitoring assays and resistance-management strategies were developed and implemented. The extent of resistance was monitored by the use of a discriminating dose assay using a treated glass vial technique [233]. It was found that the occurrence of resistance reached high levels early in the season, probably due to the emergence of overwintering resistant insects, followed by a steep decline late in spring and early in summer. This steep decline in the proportion of resistant individuals is most likely due to decreased fecundity and mating success of the resistant insects compared with susceptible insects. The occurrence of this decline in the frequency of resistance is the key to a successful management program. Resistance appears to be successfully managed by avoiding the use of pyrethroids early in the season and restricting their application to the critical mid-season period. Later in the season, the frequency of resistance usually increases due to the selection pressure caused by pyrethroid use. Decisions to use pyrethroids, especially late in the season, are based on monitoring results. Avoiding early season use takes advantage of the lower reproductive success of the resistant insects. Thus, identifying the costs to the insect associated with resistance and finding how to take advantage of them are very important factors in successful resistance management.

   These examples demonstrate the importance of understanding resistance mechanisms and dynamics, as well as the need for rapid and reliable techniques for early detection and monitoring of the frequency and distribution of resistance insects in field populations. Unfortunately, the agronomic practice of extending cotton production until late season, in an attempt to maximize yield, requires heavy insecticide use and thereby encourages the development of insect resistance. Therefore, another component of resistance management involves using crop practices that reduce the need for late-season insecticide applications, such as early planting and reduced inputs of water and fertilizer.

XI. CONCLUSION

The discovery of photostable pyrethroids is the most significant commercial example of the use of a natural product as a model for the development of superior insecticides. Although it took 50 years from the elucidation of the general structural features of the pyrethrins to the discovery of photostable pyrethroids, after major breakthroughs in the early 1970s the field developed very rapidly. Pyrethroids soon became major agricultural products due to their superior efficacy against lepidopterous pests of cotton. Synthetic analogs are now known that have a wide range of activities and use patterns. The commercial pyrethroids are much more insecticidally active than the natural pyrethrins and are among the most active insecticides known. They control a wider range of insect and acarid pests at lower rates than most other insecticides. Although their effects on beneficial insects and other wildlife must be considered in various applications, the pyrethroids are, overall, an environmentally sound class of insecticides with a wide margin of safety to mammals. Even though pyrethroids used in agriculture are sufficiently photostable to be effective in the field, they are readily degraded by metabolizing systems in mammals, as well as by soil microorganisms, and they do not accumulate in the food chain.

The future utility of this class of insecticides in agriculture will be heavily influenced by the extent to which important insect pest populations develop economically significant levels of resistance and how well this resistance can be managed. It is essential to maintain susceptible genotypes in the field insect populations. Detailed evaluation of the dynamics of a particular pest situation combined with limited use of the appropriate pyrethroid, in carefully timed applications, should allow these compounds to remain a significant factor in the agricultural insecticide market for decades to come.

ACKNOWLEDGMENTS

I would like to thank Gerardus B. Staal and David C. Cerf for the bioassay results presented in the tables and William L. Collibee for drawing the structures.

REFERENCES

1. K. Naumann, *Chemistry of Plant Protection, Vol. 4, Synthetic Pyrethroid Insecticides: Structures and Properties* (W. S. Bowers, W. Ebing, D. Martin, R. Wegler, eds.), Springer-Verlag, Berlin (1990).
2. K. Naumann, *Chemistry of Plant Protection, Vol. 5, Synthetic Pyrethroid Insecticides: Chemistry and Patents* (W. S. Bowers, W. Ebing, D. Martin, R. Wegler, eds.), Springer-Verlag, Berlin (1990).
3. J. H. Davis, *The Pyrethroid Insecticides* (J. P. Leahey, ed.), Taylor and Francis, London, p. 1 (1985).
4. M. Matsui and I. Yamamoto, *Naturally Occurring Insecticides* (M. Jacobson and D. G. Crosby, eds.), Marcel Dekker, New York, p. 3 (1971).
5. L. Crombie, *Pestic Sci,* 11:102 (1980).

6. M. Elliott, *Pesticides and Alternatives* (J. E. Casida, ed.). Elsevier, Amsterdam, p. 345 (1990).
7. R. J. Soreng and E. A. Cope, *Baileya, 23*:145 (1991).
8. C. C. McDonnell, R. C. Roark, and G. L. Kennan, *Insect Powder*, United States Department of Agriculture, Department Bulletin No. 824 (issued 1920), U.S. Government Printing Office, Washington, D.C. (1922).
9. C. C. McDonnell, R. C. Roark, and F. B. LaForge, *Insect Powder*, United States Department of Agriculture, Department Bulletin No. 824 (revised 1926), U.S. Government Printing Office, Washington, D.C., (1926).
10. G. A. McLaughlin, *Pyrethrum, The Natural Insecticide*, (J. E. Casida, ed.) Academic Press, New York, p. 3 (1973).
11. C. B. Gnadinger, *Pyrethrum Flowers*, 2nd ed., McLaughlin Gormley King Co., Minneapolis, MN (1936); C. B. Gnadinger, *Pyrethrum Flowers, Suppl. to 2nd ed.* (1936–1945), McLaughlin Gormley King Co., Minneapolis, MN (1945).
12. S. W. Head, *Pyrethrum, The Natural Insecticide* (J. E. Casida, ed.), Academic Press, New York, p. 25 (1973).
13. M. Elliott, N. J. Jones, and D. A. Pulman, *J. Chem. Soc. Perkin I*, 2470 (1974).
14. De Visiani, *Ann. Chim. Applicata Medicina, Ser. 3, 19*:84 (1854).
15. M. Elliott and N. F. Janes, *Pyrethrum, The Natural Insecticide* (J. E. Casida, ed.), Academic Press, New York, p. 55 (1973).
16. H. Staudinger and L. Ruzicka, *Helv. Chem. Acta, 7*:448 (1924).
17. M. S. Schechter, N. Green, and F. B. LaForge, *J. Amer. Chem. Soc, 71*:3165 (1949).
18. Y. Katsuda, *J. Pestic. Sci, 7*:317 (1982).
19. J. W. Wickham, *Pestic. Sci, 7*:273 (1976).
20. T. Kato, K. Ueda, and K. Fujimoto, *Agric. Biol. Chem, 28*:914 (1964).
21. M. Elliott, A. W. Farnham, N. F. Janes, P. H. Needham, and B. C. Pearson, *Nature (Lond.), 213*:493 (1967).
22. M. Elliott, A. W. Farnham, N. F. Janes, and P. H. Needham, *Pestic. Sci, 5*:491 (1974).
23. K. Ueda, L. C. Gaughan, and J. E. Casida, *J. Agric. Food Chem, 22*:212 (1974).
24. Y. Katsuda, T. Chikamoto, H. Ogami, H. Hirobe, and T. Kunishige, *Agric. Biol. Chem, 33*:1361 (1969).
25. Y. Nishizawa, *Bull. Wld. Hlth. Org, 44*:325 (1971).
26. M. Elliott, N. F. Janes, K. A. Jeffs, P. H. Needham, and R. M. Sawicki, *Nature, (Lond.), 207*:938 (1965).
27. M. Elliott, *Bull. Wld. Hlth. Org, 44*:315 (1971)
28. M. Elliott, *Synthetic Pyrethroids* (M. Elliott, ed.), ACS Symposium Series, No. 42, American Chemical Society, Washington, D.C., p. 1 (1977).
29. M. Elliott and N. F. Janes, *Chem. Soc. Rev, 7*:473 (1978).
30. M. Elliott and N. F. Janes, Proceeding of 10th International Congress of Plant Protection 1983, Brighton, England, Vol. 1, pp. 216–223, (c. 1981), [1983].

31. K. Fujimoto, N. Itaya, Y. Okuno, T. Kadota, and T. Yamaguchi, *Agric. Biol. Chem.,* 37:2681 (1973).
32. A. Fujinami, *Japan Pesticide Information,* No. 37:30 (1980).
33. T. Matsuo, N. Itaya, Y. Okuno, T. Mizutani, N. Ohno, and S. Kitamura, U.S. Patent, 3,835,176 (1974).
34. T. Matsuo, N. Itaya, T. Mizutani, N. Ohno, K. Fujimoto, Y. Okuno, and H. Yoshioka, *Agric. Biol. Chem.,* 40:247 (1976).
35. M. Matsui and T. Kitahara, *Agric. Biol. Chem.,* 31:1143 (1967).
36. J. Farkaš, P. Kouřím, and F. Šorm, *Coll. Czech. Chem. Commun.,* 24:2230 (1959).
37. L. Velluz, J. Martel, and G. Nominé, *C. R. Acad. Sci., Paris. Ser. C,* 268:2199 (1969).
38. J. Lhoste and F. Rauch, *Pestic. Sci.,* 7:247 (1976).
39. T. Matsuo, T. Nishioka, M. Hirano, Y. Suzuki, K. Tsushima, N. Itaya, and H. Yoshioka, *Pestic. Sci.,* 11:202 (1980).
40. M. Elliott, A. W. Farnham, N. F. Janes, P. H. Needham, D. A. Pulman, and J. H. Stevenson, *Nature (Lond.),* 246:169 (1973).
41. M. Elliott, A. W. Farnham, N. F. Janes, P. H. Needham, and D. A. Pulman, *Nature (Lond.),* 248:710 (1974).
42. H. Yoshioka, *Rev. Plant Protect. Res.,* 11:39 (1978).
43. N. Ohno, K. Fujimoto, Y. Okuno, T. Mizutani, M. Hirano, N. Itaya, T. Honda, and H. Yoshioka, *Agric. Biol. Chem.,* 38:881 (1974).
44. N. Ohno, K. Fujimoto, Y. Okuno, T. Mizutani, M. Hirano, N. Itaya, T. Honda, and H. Yoshioka, *Pestic. Sci.,* 7:241 (1976).
45. P. Ackermann, F. Bourgeois, and J. Drabek, *Pestic. Sci.,* 11:169 (1980).
46. C. A. Henrick, B. A. Garcia, G. B. Staal, D. C. Cerf, R. J. Anderson, K. Gill, H. R. Chinn, J. N. Labovitz, M. M. Leippe, S. L. Woo, R. L. Carney, D. C. Gordon, and G. K. Kohn, *Pestic. Sci.,* 11:224 (1980).
47. C. A. Henrick, R. J. Anderson, and G. B. Staal, *Pesticide Chemistry: Human Welfare and the Environment* (J. Miyamoto and P. C. Kearney, eds.-in-chief), Vol. 1, IUPAC Symposium Series, Pergamon Press, Oxford, p. 107 (1983).
48. M. Elliott, N. F. Janes, and C. Potter, *Ann. Rev. Entomol.,* 23:443 (1978).
49. T. C. Sparks, M. H. Shour, and E. G. Wellemeyer, *J. Econ. Entomol.,* 75:643 (1982).
50. C. Y. Brempong-Yeboah, T. Saito, T. Miyata, and Y. Tsubaki, *J. Pestic. Sci.,* 7:47 (1982)
51. C. F. Hinks, *Canad. Entomol.,* 117:1007 (1985).
52. C. A. Henrick, R. J. Anderson, R. L. Carney, B. A. Garcia, and G. B. Staal, *Recent Advances in the Chemistry of Insect Control* (N. F. Janes, ed.), Royal Society of Chemistry, London, p. 133 (1985).
53. A. J. Hopfinger and R. D. Battershell, *Advances in Pesticide Science* (H. Geissbühler ed.), Part 2, Pergamon Press, Oxford, p. 196 (1979).
54. G. G. Briggs, M. Elliott, A. W. Farnham, N. F. Janes, P. H. Needham, D. A. Pulman, and S. R. Young, *Pestic. Sci.,* 7:236 (1976).

55. M. G. Ford, *Pestic. Sci.,* 10:39 (1979).
56. T. Sugiyama, A. Kobayashi, and K. Yamashita, *Agric. Biol. Chem.,* 39:1483 (1975).
57. T. Kitahara, K. Fujimoto, and M. Matsui, *Agric. Biol., Chem.,* 38:1511 (1974).
58. H. Greuter, P. Bissig, P. Martin, V. Flück, and L. Gsell, *Pestic. Sci.,* 11:148 (1980).
59. R. G. Bolton, *Pestic. Sci.,* 7:251 (1976).
60. M. Elliott, A. W. Farnham, N. F. Janes, P. H. Needham, and D. A. Pulman, *Pestic. Sci.,* 7:499 (1976).
61. M. Elliott, A. W. Farnham, N. F. Janes, P. H. Needham, and D. A. Pulman, *Pestic. Sci.,* 7:492 (1976).
62. M. Elliott, A. W. Farnham, N. F. Janes, P. H. Needham, and D. A. Pulman, *Pestic. Sci.,* 6:537 (1975).
63. P. E. Burt, M. Elliott, A. W. Farnham, N. F. Janes, P. H. Needham, and D. A. Pulman, *Pestic. Sci.,* 5:791 (1974).
64. M. Elliott, N. F. Janes, and B. P. S. Khambay, *Pestic. Sci.,* 17:708 (1986).
65. E. Bosone, F. Corda, F. Gozzo, A. Menconi, P. Piccardi, and V. Caprioli, *Pestic. Sci.,* 17:621 (1986).
66. P. D. Bentley, R. Cheetham, R. K. Huff, R. Pascoe, and J. D. Sayle, *Pestic. Sci.,* 11:156 (1980).
67. V. K. Stubbs, C. Wilshire, and L. G. Webber, *Aust. Vet. J.,* 59:152 (1982).
68. A. R. Jutsum, M. D. Collins, R. M. Perrin, D. D. Evans, R. A. H. Davies, and C. N. E. Ruscoe, Proceedings 1984 British Crop Protection Conference—Pest and Diseases, Vol. 2, Brighton, 421–428 (1984).
69. W. Stendel and R. Fuchs, *Vet. Med. Rev,* 2:115 (1982).
70. H. J. Schnitzerling, J. Nolan, and S. Hughes, *Exp. Appl. Acarology,* 6:47 (1989).
71. J. R. Tessier, *Recent Advances in the Chemistry of Insect Control* (N. F. Janes, ed.), Royal Society of Chemistry, London, p. 26 (1985); J. J. Heller, F. Alibert, G. Soubrier, and L. Roa, *Med. Fac. Landbouww. Univ. Gent,* 57(3a):931 (1992).
72. L. O. Ruzo, L. C. Gaughan, and J. E. Casida, *Pestic. Biochem. Physiol.,* 15:137 (1981).
73. L. M. Cole, L. O. Ruzo, E. J. Wood, and J. E. Casida, *J. Agric. Food Chem.,* 30:631 (1982).
74. L. O. Ruzo and J. E. Casida, *J. Agric. Food Chem.,* 29:702 (1981).
75. S. N. Irving and T. E. M. Fraser, *J. Agric. Food Chem.,* 32:111 (1984).
76. Y. Minamite, H. Hirobe, H. Ohgami, and Y. Katsuda, *J. Pestic. Sci.,* 3:437 (1978).
77. K. Ozawa, S. Ishii, K. Hirata, and M. Hirose, *J. Pestic. Sci.,* 11:175 (1986).
78. D. G. Brown and R. W. Addor, *Advances in Pesticide Science* (H. Geissbühler ed.), Part 2, Pergamon Press, Oxford, p. 190 (1979).
79. K. Tsushima, N. Matsuo, N. Itaya, T. Yano, and M. Hatakoshi, *Pesticide Chemistry: Human Welfare and the Environment* (J. Miyamoto and P. C. Kearney, eds.-in-chief), Vol. 1, IUPAC Symposium Series, Pergamon Press, Oxford, p. 91 (1983).

80. M. Miyakado, N. Ohno, Y. Okuno, M. Hirano, K. Fujimoto, and H. Yoshioka, *Agric. Biol. Chem.,* 39:267 (1975).

81. M. Elliott, A. W. Farnham, N. F. Janes, D. M. Johnson, and D. A. Pulman, *Pestic. Sci.,* 11:513 (1980).

82. W. K. Whitney, Proceedings 1979 British Crop Protection Conference— Pest and Diseases, Vol. 2, Brighton, 387–394 (1979).

83. G. Nisha and M. Kalyanasundaram, *Pestic. Sci.,* 35:21 (1992).

84. J. Heath, *Med. Fac. Landbouww. Rijksuniv. Gent,* 50(2b):665 (1985).

85. R. J. Anderson, K. G. Adams, and C. A. Henrick, *J. Agric. Food Chem.,* 33:508 (1985).

86. H. M. Ayad and T. N. Wheeler, *J. Agric. Food Chem.,* 32:85 (1984).

87. T. N. Wheeler, *J. Agric. Food Chem.,* 32:1125 (1984).

88. A. S. Hirwe, R. L. Metcalf, and I. P. Kapoor, *J. Agric. Food Chem.,* 20:818 (1972).

89. G. Holan, D. F. O'Keefe, C. Virgona, and R. Walser, *Nature, (Lond.),* 272:734 (1978).

90. G. Holan, W. M. Johnson, C. T. Virgona, and R. A. Walser, *J. Agric. Food Chem.,* 34:520 (1986).

91. G. Holan, W. M. P. Johnson, D. F. O'Keefe, G. L. Quint, K. Ribs, T. H. Spurling, R. Walser, C. T. Virgona, C. Frelin, M. Lazdunski, G. A. R. Johnston, and S. C. Chow, *Recent Advances in the Chemistry of Insect Control* (N. F. Janes, Ed.), Royal Society of Chemistry, London, p. 114 (1985).

92. G. Holan, W. M. P. Johnson, D. F. O'Keefe, K. Rihs, D. R. J. Smith, C. T. Virgona, R. Walser, and J. M. Haslam, *Pesticide Chemistry: Human Welfare and the Environment* (J. Miyamoto and P. C. Kearney, eds.-in-chief), Vol. 1, IUPAC Symposium Series, Pergamon Press, Oxford, p. 119 (1983).

93. M. Elliott, A. W. Farnham, M. G. Ford, N. F. Janes, and P. H. Needham, *Pestic. Sci.,* 3:25 (1972).

94. M. Elliott, R. L. Elliott, N. F. Janes, B. P. S. Khambay, and D. A. Pulman, *Pestic. Sci.,* 37:691 (1986).

95. M. Elliott, R. L. Elliott, N. F. Janes, and B. P. S. Khambay, *Pestic. Sci.,* 17:701 (1986).

96. E. L. Plummer and D. S. Pincus, *J. Agric. Food Chem.,* 29:1118 (1981).

97. Y. Minamite, Y. Tsuji, H. Hirobe, H. Ohgami, and Y. Katsuda, *J. Pestic. Sci.,* 7:349 (1982).

98. M. Elliott, A. W. Farnham, N. F. Janes, P. H. Needham, and D. A. Pulman, *Mechanism of Pesticide Action* (G. K. Kohn, ed.). ACS Symposium Series No. 2, American Chemical Society, Washington, D. C., p. 80 (1974).

99. H-T. Tu, W. T. Brady, and S. J. Norton, *J. Agric. Food Chem.,* 33:751 (1985).

100. E. L. Plummer, *Pesticide Synthesis Through Rational Approaches* (P. S. Magee, G. K. Kohn, and J. J. Menn, eds.), ACS Symposium Series No. 255, American Chemical Society, Washington, D. C., p. 297 (1984).

101. J. F. Engel, E. L. Plummer, R. R. Stewart, W. A. VanSaun, R. E. Montgomery, P. A. Cruickshank, W. N. Harnish, A. A. Nethery, and G. A. Crosby, *Pesticide Chemistry: Human Welfare and the Environment* (J. Miyamoto and P. C. Kearney, eds.-in-chief), Vol. 1, IUPAC Symposium Series, Pergamon Press, Oxford, p. 101 (1983).

102. E. L. Plummer and R. R. Stewart, *J. Agric. Food Chem.*, 32:1116 (1984).

103. J. F. Engel, C. A. Staetz, S. T. Young, and G. A. Crosby, *Recent Advances in the Chemistry of Insect Control* (N. F. Janes, ed.), Royal Society of Chemistry, London, p. 162 (1985).

104. Y. Nakada, S. Muramatsu, M. Asai, S. Ohno, and Y. Yura, *Agric. Biol. Chem.*, 42:1357 (1978).

105. Y. Nakada, S. Ohno, M. Yoshimoto, and Y. Yura, *Agric. Biol. Chem.*, 42:1365 (1978).

106. Y. Nakada, S. Muramatsu, M. Asai, H. Tsuji, and Y. Yura, *Agric. Biol. Chem.*, 42:1767 (1978).

107. M. Elliott, A. W. Farnham, N. F. Janes, and B. P. S. Khambay, *Pestic. Sci.*, 13:407 (1982).

108. M. Elliott, N. F. Janes, B. P. S. Khambay, and D. A. Pulman, *Pestic. Sci.*, 14:182 (1983).

109. M. H. Bresse, *Pestic. Sci.*, 8:264 (1977).

110. M. Elliott, A. W. Farnham, N. F. Janes, and D. M. Soderlund, *Pestic. Sci.*, 9:112 (1978).

111. F. Barlow, A. B. Hadaway, L. S. Flower, J. E. H. Grose, and C. R. Turner, *Pestic. Sci.*, 8:291 (1977).

112. S. D. Lindell, J. D. Elliott, and W. S. Johnson, *Tetrahedron Lett.*, 25:3947 (1984).

113. M. Elliott, A. W. Farnham, N. F. Janes, and B. P. S. Khambay, *Pestic. Sci.*, 22:231 (1988).

114. I. Hammann and R. Fuchs, *Pflanzenschutz-Nachrichten Bayer*, 34:121 (1981).

115. W. Behrenz, A. Elbert, and R. Fuchs, *Pflanzenschutz-Nachrichten Bayer*, 36:127 (1983).

116. C. A. Henrick, R. L. Carney, G. B. Staal, and D. Cerf, unpublished results, Zoecon Corporation, Palo Alto, CA (1982).

117. W. Behrenz and K. Naumann, *Pflanzenschutz-Nachrichten Bayer*, 35:309 (1982).

118. W. Behrenz, J. Hartwig, B. Homeyer, and K. Naumann, *Abstracts of Papers*, vol. 1, 7th International Congress of Pesticide Chemistry, Hamburg, Germany, IUPAC, 01A-06 (1990).

119. E. McDonald, N. Punja, and A. R. Jutsum, Proceedings of the 1986 British Crop Protection Conference—Pests and Diseases, Vol. 1, Brighton, 199–206 (1986); K. A. Simmons, A. C. Lew, I. R. Silverman, and S. F. Ali, *J. Agric. Food Chem.*, 40:1432 (1992).

120. A. R. Jutsum, R. F. S. Gordon, and C. N. E. Ruscoe, Proceedings 1986 British Crop Protection Conference—Pests and Diseases, Vol. 1, Brighton, 97–106 (1986).

121. S. K. Malhotra, J. C. VanHeertum, L. L. Larson, and M. J. Ricks, *J. Agric. Food Chem.,* 29:1287 (1981).
122. K. Ozawa, S. Ishii, K. Hirata, and M. Hirose, *J. Pestic. Sci.,* 11:169 (1986).
123. T. Ohsumi, M. Hirano, N. Itaya, and Y. Fujita, *Pestic. Sci.,* 12:53 (1981).
124. T. Ohsumi, C. Takayama, T. Motoki, T. Yano, M. Hirano, and N. Itaya, *J. Pestic. Sci.,* 12:659 (1987).
125. M. Hirano, N. Itaya, I. Ohno, Y. Fujita, and H. Yoshioka, *Pestic. Sci.,* 10:291 (1979).
126. T. Matsunaga, K. Yoshida, G. Shinjo, S. Tsuda, Y. Okuno, and H. Yoshioka, *Pesticide Chemistry: Human Welfare and the Environment* (J. Miyamoto and P. C. Kearney, eds.-in-chief), Vol. 2, IUPAC Symposium Series, Pergamon Press, Oxford, p. 231 (1983).
127. N. Matsuo, T. Yano, H. Yoshioka, S. Kuwahara, T. Sugai, and K. Mori, *Pesticide Chemistry: Human Welfare and the Environment* (J. Miyamoto and P. C. Kearney, eds.-in-chief), Vol. 1, IUPAC Symposium Series, Pergamon Press, Oxford, p. 279 (1983).
128. N. Matsuo, T. Yano, and N. Ohno, *Agric. Biol. Chem.,* 49:3029 (1985).
129. P. E. Berteau and J. E. Casida, *J. Agric. Food Chem.,* 17:931 (1969).
130. M. J. Bull, J. H. Davies, R. J. G. Searle, and A. C. Henry, *Pestic. Sci.,* 11:249 (1980).
131. K. Nanjyo, N. Katsuyama, A. Kariya, T. Yamamura, S-B. Hyeon, A. Suzuki, and S. Tamura, *Agric. Biol. Chem.,* 44:217 (1980).
132. G. Holan, W. M. P. Johnson, K. Rihs, and C. T. Virgona, *Pestic. Sci.,* 15:361 (1984).
133. T. G. Cullen, D. P. Suarez, A. C. Lew, C. E. Williams, M. Reed, F. L. Marek, and G. L. Meindl, *Abstracts of Papers, Part 1, Fourth Chemical Congress of North America, New York,* American Chemical Society, Washington, D. C., Agro 72 (1991).
134. T. Udagawa, S. Numata, K. Oda, S. Shiraishi, K. Kodaka, and K. Nakatani, *Recent Advances in the Chemistry of Insect Control* (N. F. Janes, ed.), Royal Society of Chemistry, London, p. 192 (1985); T. Udagawa, *Japan Pestic. Inform.* No. 53:9 (1988); T. Udagawa, *Japan Pestic. Inform.* No. 61:7 (1992).
135. M. J. Bushell, *Recent Advances in the Chemistry of Insect Control II* (L. Crombie, ed.), Royal Society of Chemistry, Cambridge, p. 125 (1990); see also R. F. S. Gordon, M. J. Bushell, R. Pascoe, and T. Enoyoshi, Proceedings 1992 British Crop Protection Conference—Pest and Diseases, Vol. 1, Brighton, 81–88 (1992).
136. M. Elliott, A. W. Farnham, N. F. Janes, and B. P. S. Khambay, *Pestic. Sci.,* 23:215 (1988).
137. A. E. Baydar, M. Elliott, A. W. Farnham, N. F. Janes, and B. P. S. Khambay, *Pestic. Sci.,* 23:231 (1988).
138. A. Svendsen, L-E. K. Pedersen, and P. D. Klemmensen, *Pestic. Sci.,* 17:93 (1986).
139. G. A. Meier, S. E. Burkart, T. G. Cullen, J. F. Engel, C. M. Langevine,

S. Sehgel, and S. M. Sieburth, *Abstracts of Papers, Part 1, 199th National Meeting of the American Chemical Society, Boston,* American Chemical Society, Washington, D. C., Agro 13 (1990); G. A. Meier, T. G. Cullen, S. Sehgel, J. F. Engel, S. E. Burkart, S. M. Sieburth, and C. M. Langevine, *Synthesis and Chemistry of Agrochemicals III* (D. R. Baker, J. G. Fenyes, and J. J. Steffens, eds.), ACS Symposium Series, No. 504, American Chemical Society, Washington, D. C., p. 258 (1992).

140.   G. A. Meier, T. G. Cullen, J. F. Engel, S. M. Sieburth, J. A. Dybas, A. W. Fritz, C. M. Langevine, G. L. Meindl, L. W. Stratton, J. H. Strickland, and D. P. Suarez, *Abstracts of Papers, 203rd National Meeting of the American Chemical Society, San Francisco,* American Chemical Society, Washington, DC, Agro 29 (1992).

141.   T. G. Cullen, S. M. Sieburth, S. Y. Lin, J. F. Engel, G. A. Meier, C. M. Langevine, J. H. Strickland, F. L. Marek, G. L. Meindl, and S. E. Burkart, *Abstracts of Papers, Part 1, Fourth Chemical Congress of North America,* New York, American Chemical Society, Washington, D.C., Agro 73 (1991); T. G. Cullen, S. M. Sieburth, J. F. Engel, G. A. Meier, A. C. Lew, S. E. Burkart, F. L. Marek, and J. H. Strickland, *Synthesis and Chemistry of Agrochemicals III* (D. R. Baker, J. G. Fenyes, and J. J. Steffens, eds.), ACS Symposium Series No. 504, American Chemical Society, Washington, D.C., p. 271 (1992).

142.   G. A. Meier, T. G. Cullen, J. F. Engel, S. M. Sieburth, C. M. Langevine, G. L. Meindl, and D. P. Suarez, *Abstracts of Papers, 203rd National Meeting of the American Chemical Society, San Francisco,* American Chemical Society, Washingon, D.C., Agro 30 (1992).

143.   A. E. Baydar, M. Elliott, A. W. Farnham, N. F. Janes, and B. P. S. Khambay, *Pestic. Sci.,* 23:247 (1988).

144.   A. W. Farnham, A. Ifill, N. F. Janes, T. Javed, and B. P. S. Khambay, *Pestic. Sci.,* 28:25 (1990).

145.   G. Holan, W. M. P. Johnson, K. E. Jarvis, C. T. Virgona, and R. A. Walser, *Pestic. Sci.,* 17:715 (1986).

146.   K. Tsushima, T. Yano, T. Takagaki, N. Matsuo, M. Hirano, and N. Ohno, *Agric. Biol. Chem.,* 52:1323 (1988).

147.   Y. Yamada, T. Yano, and N. Itaya, *J. Pestic. Sci.,* 12:683 (1987).

148.   S. M. Sieburth, S. Y. Lin, J. F. Engel, J. A. Greenblatt, S. E. Burkart, and D. W. Gammon, *Recent Advances in the Chemistry of Insect Control II* (L. Crombie, ed.), Royal Society of Chemistry, Cambridge, p. 142 (1990).

149.   A. Teruoka, T. Ohtsuka, Y. Hayase, Y. Fujita, S. Katayama, K. Ohba, A. Mizutani, T. Takahashi, T. Ishiguro, and Y. Hayashi, *J. Pestic. Sci.,* 14:1 (1989).

150.   K. Tsushima, T. Yano, K. Umeda, N. Matsuo, M. Hirano, and N. Ohno, *Pestic. Sci.,* 25:17 (1989).

151.   J. G. Scott and G. P. Georghiou, *Pestic. Biochem. Physiol.* 21:53 (1984).

152.   Y. J. Ahn, T. Shono, and J-I. Fukami, *Pestic. Biochem. Physiol.* 28:301 (1987).

153. M. R. Riskallah, *Experientia, 40*:188 (1984).
154. S. J. Toth Jr., and T. C. Sparks, *J. Econ. Entomol., 83*:342 (1990).
155. C. A. Henrick, *Pyrethroid Insecticides: Chemistry and Action* (J. Mathieu, ed.). Tables Rondes Roussel Uclaf, Paris, No. 37, p. 23 (1980).
156. C. A. Henrick, G. B. Staal, D. Cerf, and A. Kilkenny, unpublished results, Zoecon Corporation, Palo Alto, CA (1979).
157. K. Wettstein, *Med. Fac. Landbouww. Rijksuniv. Gent. 48*(2):331 (1983).
158. T. C. Sparks, A. M. Pavloff, R. L. Rose, and D. F. Clower, *J. Econ. Entomol., 76*:243 (1983).
159. E. Grafius, *J. Econ. Entomol., 79*:588 (1986).
160. M. Das, *Ann. Appl. Biol., 49*:39 (1961).
161. M. Das and A. H. McIntosh, *Ann. Appl. Biol. 49*:267 (1961).
162. S. J. Toth, Jr., and T. C. Sparks, *J. Econ. Entomol., 81*:115 (1988).
163. M. Hirano, *Pestic. Sci., 27*:353 (1989).
164. C. N. E. Ruscoe, *Pestic. Sci., 8*:236 (1977).
165. J. J. Hervé, *The Pyrethroid Insecticides* (J. P. Leahey, ed.), Taylor and Francis, London, p. 343 (1985).
166. J. R. Cox, *Pestic. Outlook, 1*:26 (1990).
167. D. P. Highwood, Proceedings 1979 British Crop Protection Conference— Pest and Diseases, Vol. 2, Brighton, 361–369 (1979).
168. G. L. Gist and C. D. Pless, *Florida Entomol., 68*:450, 456, 462 (1985).
169. Agrochemical Service, *Agrochemical Products Section Parts 1 and 2*, County NatWest WoodMac, London, p. 38 (1992).
170. Anon., *Pyrethroids Feature, Agrow*, No. 157, PJB Publications, Surrey, p. 22 (1992).
171. S. C. Hoyt, P. H. Westigard, and E. C. Burts, *J. Econ. Entomol., 71*:431 (1978).
172. S. W. Wong and R. B. Chapman, *Aust. J. Agric. Res, 30*:497 (1979).
173. M. T. Aliniazee and J. E. Cranham, *Environ. Entomol., 9*:436 (1980).
174. D. R. Penman and R. B. Chapman, *Expl. Appl. Acarology, 4*:265 (1988); J. D. Fitzgerald and M. G. Solomon, Proceedings 1992 British Crop Protection Conference—Pest and diseases, Vol. 3, Brighton, 1199–1204 (1992).
175. U. Gerson and E. Cohen, *Exp. Appl. Acarology, 6*:29 (1989); S. Y. Li, R. Harmsen, and H. M. A. Thistlewood, *Exp. Appl. Acarology, 15*:259 (1992).
176. D. R. Penman, R. B. Chapman, and M. H. Bowie, *J. Econ. Entomol., 79*:1183 (1986).
177. J. M. Holland and R. B. Chapman, *Proc. N.Z. Weed Pest Control Conf., 43*:95 (1990).
178. J. M. Holland and R. B. Chapman, *Proc. N.Z. Weed Pest Control Conf., 44*:248 (1991).
179. R. Winney, *Int. Pest Control, 18*:11 (1976).
180. P. R. Chadwick, *Pestic. Sci., 10*:32 (1979).
181. S. Tsuda, K. Yoshida, and Y. Okuno, Soap Cosmetics, *Chem. Spec, 59*:36 (1983).
182. H. Wachs, *Science, 105*:530 (1947).

183. R. H. Nelson, R. A. Fulton, J. H. Fales, and A. H. Yeomans, *Soap Sanitary Chem.,* 25:120 (1949).
184. P. E. Burt, M. Elliott, A. W. Farnham, N. F. Janes, P. H. Needham, and J. H. Stevenson, *Crop Protection Agents—Their Biological Evaluation* (N. R. McFarlane, ed), Academic Press, London, p. 393 (1977).
185. J. W. M. Logen and D. S. Buckley, *Pestic. Outlook,* 2:33 (1991).
186. N.-Y. Su and R. H. Scheffrahn, *J. Econ. Entomol.,* 83:1918 (1990).
187. S. W. Carter, *Pestic. Sci.,* 27:361 (1989).
188. C. R. Worthing and R. J. Hance (eds.), *The Pesticide Manual,* 9th ed., British Crop Protection Council, Surrey (1991).
189. R. D. Verschoyle and W. N. Aldridge, *Arch. Toxicol.,* 45:325 (1980).
190. R. D. Verschoyle and J. M. Barnes, *Pestic. Biochem. Physiol.,* 2:308 (1972).
191. F. He, S. Wang, L. Liu, S. Chen, Z. Zhang, and J. Sun, *Arch. Toxicol.,* 63:54 (1989).
192. S. A. Flannigan, S. B. Tucker, M. M. Key, C. E. Ross, E. J. Fairchild II, B. A. Grimes, and R. B. Harrist, *Br. J. Indust. Med.,* 42:363 (1985).
193. S. A. Flannigan and S. B. Tucker, *Contact Dermatitis,* 13:140 (1985).
194. W. N. Aldridge, *Crit. Rev. Toxicol.,* 21:89 (1990).
195. K. Haya, *Environ. Toxicol. Chem.,* 8:381 (1989).
196. C. Inglesfield, *Pestic. Sci.,* 27:387 (1989).
197. J. R. Clark, L. R. Goodman, P. W. Borthwick, J. M. Patrick Jr., G. M. Cripe, P. M. Moody, J. C. Moore, and E. M. Lores, *Environ. Toxicol. Chem.,* 8:393 (1989).
198. S. Chelliah and E. A. Heinrichs, *Environ. Entomol.,* 9:773 (1980).
199. G. D. Waller, B. J. Estesen, N. A. Buck, K. S. Taylor, and L. A. Crowder, *J. Econ. Entomol.,* 81:1022 (1988).
200. K. S. Taylor, G. D. Waller, and L. A. Crowder, Apidologie, 18:243 (1987).
201. Anon, *Am. Bee J.,* 124:358 (1984).
202. S. R. Duff and B. Furgala, *Am. Bee J.,* 132:476 (1992).
203. M. Barnavon, *Déf. Vég.,* 39(236):8 (1985).
204. B. J. Estesen, N. A. Buck, G. D. Waller, K. S. Taylor, and A. Mamood, *J. Econ. Entomol.,* 85:700 (1992).
205. S. Cobey and T. Lawrence, *Am. Bee J.,* 128:112 (1988).
206. J. S. Pettis, W. T. Wilson, H. Shimanuki, and P. D. Teel, *Apidologie,* 22:1 (1991).
207. Y. Lubinevski, Y. Stern, Y. Slabezki, Y. Lensky, H. Ben-Yossef, and U. Gerson, *Am. Bee J.,* 128:48 (1988).
208. D. M. Burgett and C. Kitprasert, *Am. Bee J.,* 130:51 (1990).
209. D. A. Otieno and G. Pattenden, *Pestic. Sci.,* 11:270 (1980).
210. J. E. Casida and L. O. Ruzo, *Pestic. Sci.,* 11:257 (1980).
211. J. P. Leahey, *The Pyrethroid Insecticides* (J. P. Leahey, ed.), Taylor and Francis, London, p. 263 (1985).
212. T. J. Class, J. E. Casida, and L. O. Ruzo, *J. Agric. Food Chem.,* 37:216 (1989).
213. N. Isobe, H. Kaneko, K. Shiba, K. Saito, S. Ito, N. Kakuta, A. Saito, A. Yoshitake, and J. Miyamoto, *J. Pestic. Sci.,* 15:159 (1990).

214. G. B. Quistad, L. E. Staiger, G. C. Jamieson, and D. A. Schooley, *J. Agric. Food Chem.,* 31:589 (1983).
215. L. E. Staiger and G. B. Quistad, *J. Agric. Food Chem.,* 32:1130 (1984).
216. D. M. Soderlund, J. R. Sanborn, and P. W. Lee, *Progress in Pesticide Biochemistry and Toxicology* (D. H. Hutson and T. R. Roberts, eds.), Vol. 3, John Wiley and Sons, New York, p. 401 (1983).
217. T. A. Miller and V. L. Salgado, *The Pyrethroid Insecticides* (J. P. Leahey, ed.), Taylor and Francis, London, p. 43 (1985).
218. D. B. Sattelle and D. Yamamoto, *Advances in Insect Physiology* (P. D. Evans and V. B. Wigglesworth, eds.), Vol. 20, Academic Press, London, p. 147 (1988).
219. D. M. Soderlund and J. R. Bloomquist, *Ann. Rev. Entomol.,* 34:77 (1989).
220. H. P. M. Vijverberg and J. van den Bercken, *Crit. Rev. Toxicol.,* 21:105 (1990).
221. M. P. Broderick and L. D. Leake, *Progress and Prospects in Insect Control* (N. R. McFarlane, ed.), BCPC Monograph No. 43, British Crop Protection Council, Surrey, p. 273 (1989).
222. M. J. Firko and J. L. Hayes, *Southwest. Entomol.,* 15:(Suppl.) 83 (1991).
223. J. E. Casida and S. H. P. Maddrell, *Pestic. Biochem. Physiol.,* 1:71 (1971).
224. R. M. Sawicki, *Insecticides* (D. H. Hutson and T. R. Roberts, eds.), John Wiley and Sons, New York, p. 143 (1985).
225. S. E. Kunz and C. D. Schmidt, *J. Agric. Entomol.,* 2:358 (1985).
226. D. C. Sheppard and N. C. Hinkle, *J. Agric. Entomol.,* 2:317 (1985); D. C. Sheppard and J. A. Joyce, J. Econ. Entomol., 85:1587 (1992).
227. L. B. Brattsten, C. W. Holyoke, Jr., J. R. Leeper, and K. F. Raffa, *Science,* 231:1255 (1986).
228. C. Campanhola, B. F. McCutchen, E. H. Baehrecke, and F. W. Plapp, Jr., *J. Econ. Entomol.,* 84:1404 (1991).
229. R. Weinzierl, Illinois Agricultural Pesticides Conference, summaries of presentations, Urbana, IL, 56–74 (1988).
230. R. M. Sawicki, *Pestic. Sci.,* 26:401 (1989).
231. R. V. Gunning, C. S. Easton, M. E. Balfe, and I. G. Ferris, *Pestic. Sci.,* 33:473 (1991).
232. J. C. Daly and J. H. Fisk, Bull. Entomol. Res., 82:5 (1992); see also P. G. Cox and N. W. Forrester, *J. Econ. Entomol.,* 85:1539 (1992).
233. F. W. Plapp, Jr., J. A. Jackman, C. Campanhola, R. E. Frisbie, J. B. Graves, R. G. Luttrell, W. F. Kitten, and M. Wall, *J. Econ. Entomol.,* 83:335 (1990).
234. F. W. Plapp, Jr., *Southwest. Entomol.,* 15(Suppl.):69 (1991).
235. A. R. McCaffery, R. T. Gladwell, H. El-Nayir, C. H. Walker, J. N. Perry, and M. J. Miles, *Southwest. Entomol.,* 15(Suppl.):143 (1991).

3

Juvenoids

Clive A. Henrick

Sandoz Agro, Inc., Palo Alto, California

I. INTRODUCTION

The commercial development of juvenoids (analogs of the natural juvenile hormones) in the mid-1970s was the result of 50 years of basic research in insect physiology aimed at understanding the puzzle of insect metamorphosis. This work began in the early part of this century with the studies of a Polish biologist Stefan Kopeć on the metamorphosis of the gypsy moth, *Lymantria dispar*. In a carefully thought-out series of microsurgical experiments he demonstrated the presence of a neuroendocrine factor (the "brain" hormone) that was produced by the brain and transmitted through the hemolymph to trigger metamorphosis [1, 2]. Later work in the 1940s demonstrated that this brain hormone acts on the prothoracic gland to cause the release of the molting hormone α-ecdysone (*1*) which, as its active metabolite β-ecdysone (*2*), initiates the various molts [3]. Pioneering research by V. B. Wigglesworth, an insect physiologist working at Cambridge University in England, on the hemipteran bug, *Rhodnius prolixus*, established the existence in the 1930s of an "inhibitory" hormone. In 1940 he suggested the name juvenile hormone (JH) for this material, which acts to suppress adult differentiation and maintain the immature nature of the developing insect [4]. Wigglesworth traced the source of the juvenile hormone to the corpora allata, two tiny endocrine glands at the

147

base of the insect brain. These three hormones regulate the processes of molting and metamorphosis which allow the animal to grow.

α-ecdysone (R=H)   (1)

β-ecdysone (R=OH)   (2)

 The structural elucidation of the juvenile hormone was hindered by the extremely small quantities present in insect larvae. A breakthrough occurred in the early 1950s when Carroll Williams found a rich source of stored JH in the abdomen of the adult male cecropia moth, *Hyalophora cecropia* [5]. He made an ether extract of macerated cecropia abdomens and obtained the legendary "golden oil," which was an impure but highly active concentrate of JH. To his dismay, this oil became colorless after initial purification steps. It was not until 1966 that the first juvenile hormone, JH I [(3); Figure 1], was finally identified from these extracts by Herbert Röller and co-workers at the University of Wisconsin. This work was published in 1967 [6].
 As early as 1956, the understanding of the complexities of insect metamorphosis had progressed sufficiently to enable Williams to speculate that once the juvenile hormone of insects was identified and synthesized it could prove to be an effective ("third-generation") insecticide [5, 7]. The elucidation of the structure of JH I (3) led to immediate attempts by many research groups to exploit this discovery as the basis for a new generation of selective environmentally sound insecticides. As JH is involved in the regulation of physiological processes in insects, such as molting and metamorphosis, which do not have a counterpart in vertebrates, it was anticipated that juvenoids would disrupt such processes leading to the death of the insect without detrimental effects against most other classes of animals.
 The term juvenoid is used in this chapter to include all chemicals showing the same qualitative physiological activity against insect species as the natural JH (a more restrictive definition requires that a juvenoid be capable of fully restoring the normal JH-mediated functions in an insect in which the corpora allata have been surgically removed).

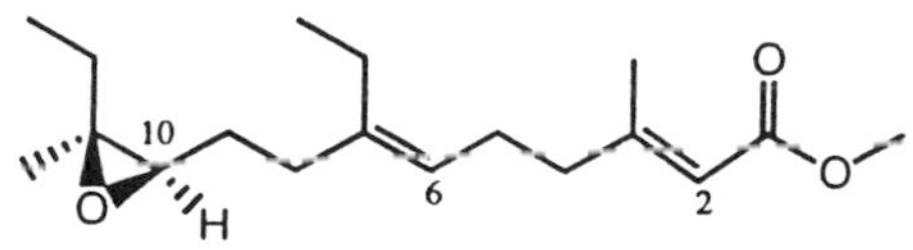

3, JH I

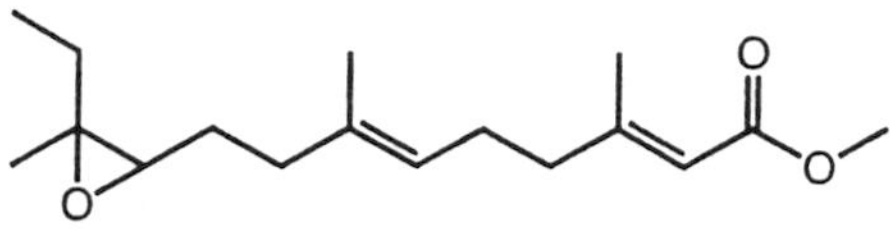

4,   JH II

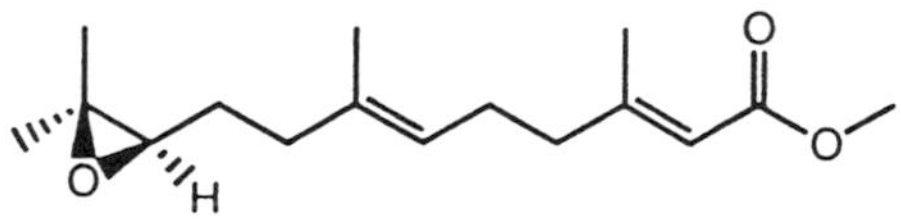

5,   JH III

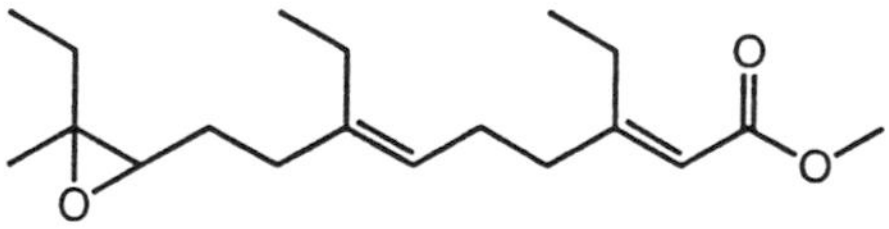

6,   JH 0

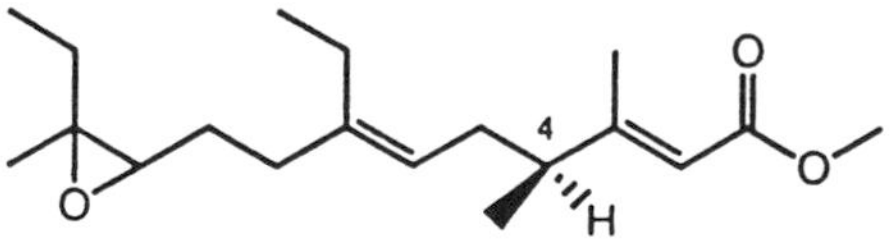

7,   4-methyl-JH I

Figure 1   Natural juvenile hormones.

II. ENDOCRINE CONTROL OF POSTEMBRYONIC DEVELOPMENT

The distinguishing feature of insect growth and development is the process of molting, during which the old exoskeleton is discarded and a new one is formed. Postembryonic growth and development from a young larva into the reproducing adult proceeds through a series of immature stages that are separated by periodic molts. This molting is essential for animals in which growth is restricted by an exoskeleton. There are two principal types of metamorphosis in insects: simple metamorphosis (Hemimetabola) and complete metamorphosis (Holometabola). In hemimetabolous insects, there are several nymphal molts followed by a nymphal–adult molt. In holometabolous insects, there are several (usually four to six) larval molts followed by one larval–pupal molt and one pupal–adult molt. The insect retains its juvenile characteristics after larval (or nymphal) molts but undergoes metamorphosis at the nymphal–adult molt (Hemimetabola) or at the larval–pupal and pupal–adult molts (Holometabola).

The four components of the neuroendocrine system that interact to regulate development in insects are the neurosecretory cells in the brain, the associated head glands innervated by the brain the corpora cardiaca and the corpora allata, and the prothoracic glands in the prothorax. At each molt, when the larva reaches a critical size, the brain is triggered to release a prothoracicotropic hormone (the "brain" hormone) that is produced by groups of neurosecretory cells located in the brain and axonally transported to the corpora cardiaca (which act as neurohemal organs) from which it is released into the hemolymph [8]. This neuropeptide hormone acts on the prothoracic gland causing the release of the molting hormone α-ecdysone (1) into the hemolymph. The majority of 1 is rapidly converted by 20-hydroxylation into the much more active β-ecdysone by many insect tissues, especially the Malpighian tubules, the fat body, and the midgut [9]. The molting hormone, or more specifically its active metabolite, β-ecdysone (2), is a regulator of gene expression and initiates all molting processes and induces metamorphosis. The juvenile hormone is synthesized and secreted by the corpora allata and acts in concert with the molting hormone to determine the outcome of each molt as either larval, pupal, or adult. The steroidal molting hormone is bound in the target cells by a specific receptor that mediates the action at the transcription level in the cell nucleus, but the cellular programming is dependent on the titre of the JH in the hemolymph. The JH permits growth but suppresses adult differentiation and favors maintenance of larval structures. JH also appears to act at the genetic level and be involved in the transcription of specific mRNAs. A high titre of JH must be present in the hemolymph for the immature larva to grow and pass through a series of larval molts induced by successive waves of molting hormone secretion. However, the titre of JH must be low during the last larval stage leading to the larval–pupal molt and essentially zero in the pupae (of holometabolous species) in order for the cells to be programmed for metamorphosis to the sexually mature adult [10, 11]. Although the prothoracic glands secrete α-ecdysone in the normal life of the insect, there may be additional sources of pupal ecdysteroids (other than the prothoracic glands) in some insect species, especially under stressful conditions [12].

The JH of insects influences a wide range of physiological processes in both the developing and the mature insect and the corpora allata in most insects resume activity in the adult stage [13–16]. In addition to regulating metamorphosis in essentially all insect species, JH is required in the adult of many species for several reproductive functions such as ovarian development, yolk synthesis and maturation of eggs in females, pheromone function, and accessory reproductive gland development in males. Regulation of diapause is also under JH control, and JH is involved in the regulation of caste determination and phase polymorphism in the groups of insects in which these phenomena occur. It also appears to have a role in the regulation of molting by affecting the synthesis of α-ecdysone in the prothoracic gland.

The endocrine system of insects is regulated by a complex arrangement of hormonal interactions and feedback mechanisms that modulate endocrine activity both temporarily and quantitatively. There is considerable variability between different insect species in the biosynthesis, titre regulation, and transport of JH [17, 18]. The titre of the JH in the hemolymph is regulated by an overall balance between the rate of its synthesis, its uptake and release by tissues, and its clearance by degradation and excretion. The biosynthesis of JH by the corpora allata is subject to dominant regulation from the brain. The brain regulatory mechanisms of JH biosynthesis are complex and may be neural, neurosecretory, or a combination of both pathways. Neurohormones controlling JH biosynthesis in the corpora allata may be released into neurohemal organs such as the corpora cardiaca and act via the hemolymph on the corpora allata, or they may be delivered directly to the corpora allata by neurosecretory axons. Both stimulatory and inhibitory mechanisms that control the activity of the corpora allata have been detected. In some species, allatotropins have been identified that stimulate an increase in biosynthesis of JH [19]. Allatostatins [20, 21] and allatinhibins [22] have also been detected that rapidly and reversibly inhibit JH biosynthesis. The brain may produce both types of neuropeptides although both stimulation and inhibition of corpora allata activity are under nerve control in some species. One of these mechanisms may predominate, depending on the insect species.

The regulation of the endocrine activity of the corpora allata represents only one aspect of the overall control of JH titres within the insect. Other factors, such as metabolism, excretion, biosynthesis from JH acid in other tissues, and tissue uptake and release are also involved. The primary route of *in vivo* metabolism of JH appears to be ester hydrolysis with secondary metabolism by epoxide hydration, oxidation, and conjugation. High hemolymph JH esterase activity is usually associated with low or declining JH levels, and the high metabolic activity in the last larval instar probably has the role of removing the last traces of JH from the insect. Although JH is biosynthesized in the corpora allata, extra-allatal sites are also involved. For example, JH acids are secreted by the corpora allata of certain male lepidopterous species and later enzymatically esterified in the accessory sex glands. JH acids are also secreted by the corpora allata of prepupal stages of *Manduca sexta* and methylated in the developing imaginal discs [23, 24]. Thus, the accessory sex glands in the adult male *Hyalophora cecropia* have the ability to synthesize and sequester JH and this was the source of the JH discovered by Williams [5] (discussed in Section I).

The JH does not appear to be stored in the corpora allata and is released as it is synthesized [17]. Upon release into the hemolymph it associates with various binding proteins in a noncovalent fashion forming dissociable hormone–protein complexes. These complexes enhance the dispersion of JH, lower the excretion rate, reduce metabolism, and provide a reservoir of JH that is available to target tissues. Thus, these binding proteins are also involved in the transport and regulation of the JH titre in the hemolymph.

The details of the precise mode of action of JH (and juvenoids) in regulating metamorphosis, reproduction, and other processes are not known, but the general actions of JH and juvenoids have been extensively reviewed [13–15, 25]. The modes of action of JH and juvenoids may be different for different insect species.

III. NATURAL JUVENILE HORMONES

Following the discovery by Williams in 1956 that the abdomens of adult male *Hyalophora cecropia* contained high JH biological activity, several research groups attempted to isolate and identify a natural juvenile hormone from this source. Finally, in 1967 Röller and his co-workers published the successful isolation and structural identification of the first natural juvenile hormone, JH I (3) (Figure 1), from *H. cecropia* [6, 26]. This was followed soon after with the publication, by a competing research group, of a closely related second juvenile hormone, JH II (4), from the same species [27]. The third natural hormone to be identified was JH III (5), which was isolated from *in vitro* organ cultures of corpora cardiaca–corpora allata complexes of the adult female tobacco hornworm, *Manduca sexta* [28]. The fourth hormone, JH O (6), was isolated from the developing embryos of *M. sexta* [29], although racemic 6 had been synthesized and bioassayed several years prior to its identification as a natural JH [30]. Soon after, a fifth hormone, 4-methyl-JH I ("iso-JH O"; 7), was isolated from the eggs of the same species [31]. Several research groups have developed methods for the identification and quantification of these homologous natural JHs in biological samples, [24, 32–34]. These methods include radioimmunoassay and sophisticated physicochemical methods using various derivatization procedures to enhance JH detectability. Because of the minute levels of JH present in most insect tissues and the difficulty of purifying these very small amounts of lipophilic JHs, the literature contains many misidentifications of the various JHs. The JHs I, II, and III (3–5) have been identified from a variety of insect species from several orders, and all five of the hormones, 3–7 have been isolated from various stages of *M. sexta*. However, from the very limited number of insect species investigated to date it appears that, in general, nonlepidopterous species contain or biosynthesize only JH III, whereas lepidopterous species produce JH I, II, and III [24, 33]. Although there have been a number of reports in the literature of the presence of JH I and/or JH II in nonlepidopterous species, most of these studies did not unequivocally identify the material and there is little evidence for the existence of higher homologs of JH III in any order other than Lepidoptera [33].

To this group of five homologous JHs (3–7) may tentatively be added methyl farnesoate (8) which has been detected (along with JH III; 5) in embryos of the cinereous cockroach, *Nauphoeta cinerea* [35]. However, the physiological importance of 8 in insects is not clear. In addition, 8 has been isolated from several crustaceans. For example, it was identified in *Libinia emarginata* along with traces of JH III (almost racemic) [36]. The latter was probably an artifact arising from air oxidation of the co-occurring methyl farnesoate. (R)-JH III has been isolated from the plant *Cyperus iria* (Cyperaceae) [37].

Furthermore, methyl 6,7;10,11-bisepoxyfarnesoate (JHB$_3$) (9) is biosynthesized *in vitro* by the ring gland of third instar larvae of *Droso-*

(8)

JHB$_3$ (9)

phila melanogaster, whereas JH III (5) is only a minor product of the larval ring gland *in vitro* [38]. Although this finding indicates that the bis-epoxide (9) has a direct role in modulating morphogenesis in higher Diptera, 9 is much less active than JH III in bioassays on this species.

The natural juvenile hormones are all chiral compounds. JH I (3), II (4), O (6), and 4-methyl-JH I (7) contain asymmetric carbon atoms at C-10 and C-11. The hormone 7 also has an asymmetric center at C-4, and JH III (5) has a chiral center at C-10. The absolute configuration of JH I has been established as 10R, 11S [39], and the natural JH III has been shown to have the 10R configuration [28]. These are the only two natural JHs whose absolute configurations have been rigorously established. The epoxy groups in 4, 6, and 7 are assumed to have the (10R, 11S) configuration, but this has not been proven. The absolute configuration at C-4 in 7 has been established as S [40]. Mori and co-workers have synthesized the pure (R)-enantiomer of JH III, the (10R, 11S)-enantiomers of JH I, II, and O, and the (4S, 10R, 11S)-isomer of 4-methyl-JH I [41].

The morphogenetic activities of synthetic racemic samples of the JHs 3–7 are presented in Table 1 [42]. In general, JH I shows the highest biological activity against most insect species, although JH O is more active than JH I against *Tenebrio molitor*. In this chapter all bioassay data in the tables are only for the morphogenetic effects of JHs and juvenoids. Effects on embryogenesis and other developmental stages are not tabulated. All of the biological results in the tables were obtained in our laboratories at Zoecon Corporation (subsequently Sandoz Argo, Inc.) using synchronized sensitive stages of the six insect species. The activities are expressed as ID$_{50}$ or IC$_{50}$ values (dose or concentration required to produce 50% inhibition of metamorphosis) as measured on a graded scale. The procedures for bioassay on the yellowfever mosquito, *Aedes aegypti* (last larval instars), the greater wax moth, *Galleria mellonella* (fresh pupae), the yellow mealworm, *Tenebrio molitor* (fresh pupae), the housefly, *Musca domestica* (full-grown larvae), the pea aphid, *Acyrthosiphon pisum* (second and third instar nymphs), and the tobacco budworm, *Heliothis virescens* (larvae) have been described in detail [43, 44].

Table 1 Racemic Samples of the Natural Juvenile Hormones. ID_{50} and IC_{50} Values on Sensitive Synchronized Instars

No.	Compound	Aedes aegypti (ppm)	Galleria mellonella (μg/pupa)	Tenebrio molitor (μg/pupa)	Musca domestica (μg/prepupa)	Acyrthosiphon pisum (% active ingredient in spray)	Heliothis virescens (ppm in medium)
3[a]	JH I	0.15	0.060	0.70	>100	—	24
4[a]	JH II	0.26	0.13	4.3	>100	0.035	>100
5[b]	JH III	0.35	12	4.5	>100	>0.1	>100
6[a]	JH O	1.0	1.0	0.15	>100	—	>100
7[a]	4-MeJH I	—	80	—	—	—	—

[a] (2E, 6E)-cis-10,11-isomer.
[b] (2E, 6E)-isomer.

IV. EARLY ANALOGS AND DISCOVERY OF COMMERCIAL JUVENOIDS

Several years before the establishment of the structure of JH I (*3*),
Schmialek detected weak JH activity in the sesquiterpenoids farnesol
and farnesal, which he had isolated from the feces of the yellow mealworm, *Tenebrio molitor* [45]. This important discovery marked the beginning of juvenoid chemistry and analogs with better activity, such
as farnesyl methyl ether (*10*) and N,N-diethylfarnesylamine (*11*), were
soon prepared [46, 47]. Then, with remarkable foresight, Bowers et al.
in 1965 predicted the general structural features of the natural JH and
prepared racemic *5*, which showed a large increase in activity over *10*
against most insect species [48]. This compound was subsequently identified as the natural hormone JH III (Section III). Also during the mid-
1960s, Law and co-workers prepared the "Law–Williams mixture" by
passing dry hydrogen chloride into solutions of farnesoic acid in
alcohols such as methanol and ethanol [49]. These mixtures contained
a number of components, but most of the JH activity against *Pyrrhocoris
apterus* was found to be due to the corresponding 7,11-dichloro esters
12 and *13* [50]. These esters show low activity against many other insect species, including *T. molitor* (Table 2).

(*10*)

(*11*)

R= · CH$_3$ (*12*)

R= · CH$_2$CH$_3$ (*13*)

      In the mid-1960s Sláma and Williams also observed JH activity in
paper of North American origin (used to line the cages) against the
European linden bug, *Pyrrhocoris apterus*, which Sláma had brought
with him from Prague [51]. Bowers and co-workers soon isolated (+)-
juvabione (*14*) from the wood of the balsam fir, *Abies balsamea*, and
showed it to be one of the active materials of this "paper factor" [52].
Subsequently, it was found that several closely related juvenoids contribute to the JH activity of the "paper factor" ([53]; see also [54]).
These compounds are also very species specific and are highly active

Table 2 Early Analogs and Commercial Juvenoids. ID_{50} and IC_{50} Values

No.	Compound[a]	Aedes aegypti (ppm)	Galleria mellonella (µg/pupa)	Tenebrio molitor (µg/pupa)	Musca domestica (µg/prepupa)	Acyrthosiphon pisum (% active ingredient in spray)	Heliothis virescens (ppm in medium)
13		6.6	>100	>100	72	0.033	>100
15		0.0032	0.082	0.0050	2.0	0.020	>100
16	Methoprene	0.00017	5.7	0.0040	0.0035	0.0054	0.77
17	Kinoprene	0.23	0.64	1.3	>100	0.00095	3.0
18	Hydroprene	0.0078	0.040	0.25	18	0.0039	0.30
19	Triprene	0.00040	0.080	0.10	0.35	0.00083	0.41
20	Epofenonane	0.057	0.037	0.0024	54	—	2.2
23		0.019	>100	70	0.16	—	>100
24	CGA 045128	0.0000036	0.0000001	—	0.045	—	—
25	Fenoxycarb	0.000010	0.000020	—	0.090	—	0.018
26	(Z) S-21149	0.0000013	0.00086	—	0.0044	—	—
	(E) S-21149	0.00000028	0.00022	—	0.044	—	—
27	(E) S-21150	0.000028	0.008	—	0.016	—	—
28	Pyriproxyfen	0.0000039	0.00068	—	0.00033	—	—

[a] All compounds are racemic.

only against some hemipterous species of the family Pyrrhocoridae, such as *Pyrrhocoris apterus* and *Dysdercus cingulatus*.

(+)-juvabione   (*14*)

(*15*)

Other work that greatly impacted early analog synthesis was the discovery of the JH activity of aryl terpenoid ethers by Bowers [55]. During an attempt to increase the activity of juvenoids by the addition of insecticide synergists he observed that some 3,4-methylenedioxy-phenolic compounds, such as piperonyl butoxide and sesamex, showed weak JH activity against *Tenebrio molitor*. He then prepared 6,7-epoxygeranyl 3,4-methylenedioxyphenyl ether (*15*) and related ethers and found that they had remarkably high morphogenetic activity. These aryl geranyl ethers were the first group of juvenoids to be discovered that show a significant improvement in JH activity over JH I (*3*) against a broad range of insect species (Table 2). Although this class of juvenoid has been extensively investigated [56], no commercial products of this type have been developed.

Then when the structures of JH I (*3*) and JH II (*4*) were published in the late 1960s, many research groups prepared analogs based on these terpenoid-like structures. This work also led directly to the founding of Zoecon Corporation in August 1968 to carry out research on insect hormones (and pheromones) with the specific aim of discovering new, environmentally safe, selective insecticides [57]. We at Zoecon found that the least stable groups in JH I *in vivo* were the epoxide group and especially the methyl ester function. Hydrolysis of the ester and/or hydration of the epoxide group results in loss of biological activity. Therefore, we carried out modifications in the structure of *3* with the aim of eliminating the need for the epoxide function and the labile methyl ester group. In addition, it was observed that saturation of the 6-ene double bond did not remove the biological activity. For these 6,7-dihydro compounds the presence of a terminal epoxide or a double bond at the 10,11-position reduces the JH activity against some insect species compared with that of the corresponding 10,11-dihydro compounds. Furthermore, introduction of an additional 4E double bond to form a (2E,4E)-dienoic ester enhances the activity. For this class of juvenoid the presence of either a 10,11-epoxide or 10,11-olefin function generally decreases the activity relative to that of the corresponding 10,11-dihydro or the 11-methoxy analogs. These studies led to the discovery of the alkyl and alkynyl (2E,4E)-3,7,11-trimethyl-2,4-dodecadienoates [43, 56]. These very potent juvenoids show activity against a broad spectrum of species (Table 2).

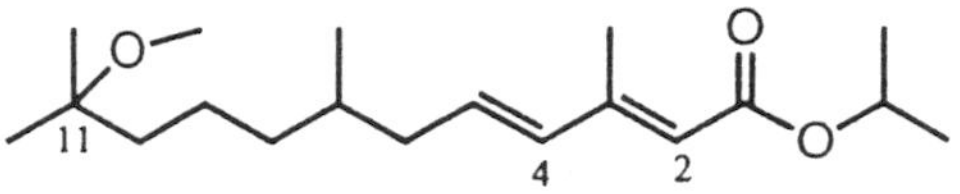

methoprene   (16)

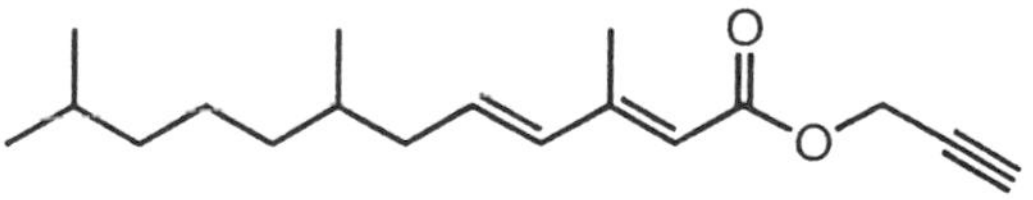

kinoprene   (17)

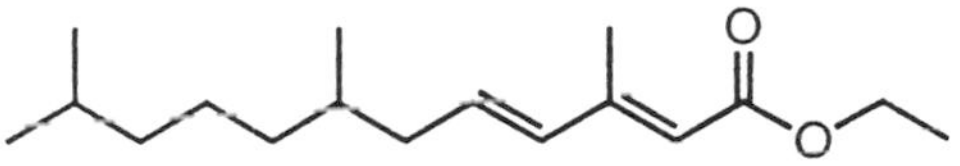

hydroprene   (18)

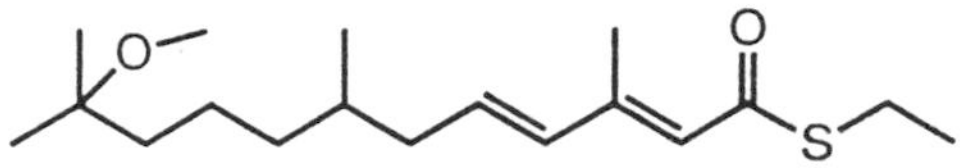

triprene   (19)

Three commercial products for insect control, methoprene (16), kinoprene (17), and hydroprene (18) were developed from this series. Hydroprene (18) was prepared in October 1970, followed by methoprene (16) in March 1971. Methoprene was the first juvenoid to be commercially developed and received its first full registration in March 1975 from the U.S. Environmental Protection Agency for the control of floodwater mosquitoes. Methoprene was the first "biorational" insecticide, a term coined by Zoecon scientists in 1974 to describe this approach to developing environmentally acceptable insecticides based on an understanding of the physiology of the target insect species [58, 59]. The biological activity and the various commercial applications of methoprene have been extensively reviewed [14, 16, 56, 57, 60]. It is a remarkable compound with a unique combination of properties. It is nontoxic to mammals (acute oral LD_{50} to rats is >34,600 mg/kg) and has negligible effects on most nontarget organisms. It is somewhat unstable in air and sunlight without special formulations and is readily metabolized in vertebrates. Its volatility is a considerable advantage in household applications. Methoprene has been used worldwide as the standard juvenoid since its discovery. It is highly effective in controlling insect species from many orders, including Diptera, Lepidoptera, Homoptera, Coleoptera, Siphonaptera, and Hymenoptera. The 2-propynyl ester kinoprene (*17*), which is very active against many homopterous species, was commercialized in December 1975 for the control of aphids and whiteflies in greenhouses on ornamental plants and vegetable seed crops. Hydroprene (*18*) is highly active against many insect species, especially of the orders Lepidoptera, Coleoptera, Homoptera, and Blattodea. Since 1984, it has been marketed for controlling cockroach populations. The *S*-ethyl thioate triprene (*19*) shows very high activity against a wide range of insect pests (Table 2), especially of orders Lepidoptera and Homoptera, but it was not commercialized.

epofenonane
(Ro 10-3108)   (*20*)

                                    (*21*)

Although geranyl phenyl ethers such as *15* have interesting JH activity, they are unstable in sunlight [61]. The search for juvenoids with higher stability under field conditions led to the study of various citronellyl phenyl ethers and related analogs. This work resulted in the discovery of epofenonane (*20*). This juvenoid is less active than the corresponding 2-ene compound (*21*) against some species (e.g., *Aedes aegypti*) but is equally active as *21* against *Tenebrio molitor* and *Tribolium castaneum* and shows high JH activity against many lepidopterous, homopterous, and coleopterous species [61–64]. It possesses much higher foliar stability in sunlight than most other juvenoids known up to this time including *16–18*. The increased field stability of *20* is presumably due to lack of the 2-ene double bond and, at least in part, to the increased steric shielding of the epoxide function by the ethyl branches. Epofenonane has a relatively broad spectrum of JH activity but is only moderately active against dipterous species (Table 2). It was not developed commercially.

Pro-drone™ (R = · CH₃)   (*22*)

R = · CH₂CH₃   (*23*)

Work on aryl terpenoids led Schwarz and co-workers to the discovery of the juvenoids *22* and *23* [56, 65, 66]. These two compounds are much more stable toward microbial degradation than methoprene and are selectively active against only a few orders (e.g., Diptera). Although they show good activity against many fly and mosquito species, they show low activity against *Tenebrio molitor* (Table 2) and *Oncopeltus fasciatus*. The juvenoid *22* [predominantly the (S)-isomer; Pro-drone™] was developed for the control of colonies of the red imported fire ant, *Solenopsis invicta*. Applied as a bait, it causes the eventual death of the colonies by the disruption of reproduction, caste differentiation, and social organization [67].

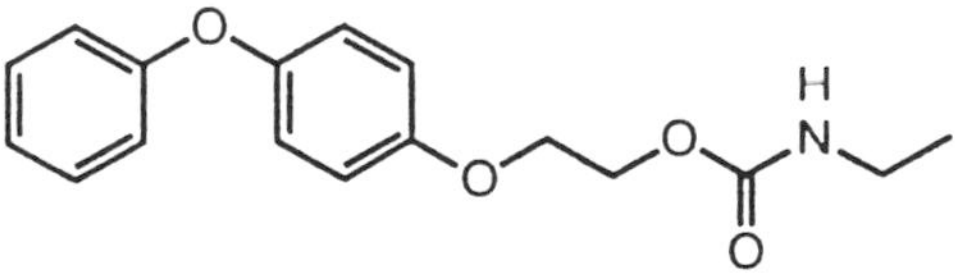

CGA 045128   (*24*)

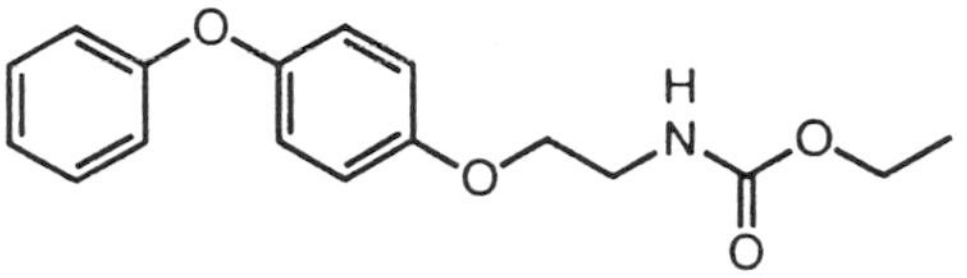

fenoxycarb, (Ro 13 5223)   (*25*)

The 2,4-dodecadienoates *16–18* are biodegradable compounds of low persistence in air and light. Their lack of foliar stability makes them unsuitable for crop applications. A major breakthrough in the use of juvenoids in field applications was the discovery of the phenoxy compounds *24* (CGA 045128) [68] and fenoxycarb (*25*) [69] by researchers at Ciba Geigy and Hoffman–La Roche (Dr. R. Maag), respectively. Both of these juvenoids (and many related analogs discussed in Section V) show very high activity against a wide range of insect species [56, 68, 69] (Table 2). Fenoxycarb has been commercialized for a variety of uses. Its high activity and foliar stability allow it to be used against homopterous and multivoltine lepidopterous species in orchard and vine crops. It has also been registered for the control of imported fire ant populations and in household indoor applications for the control of fleas and cockroaches.

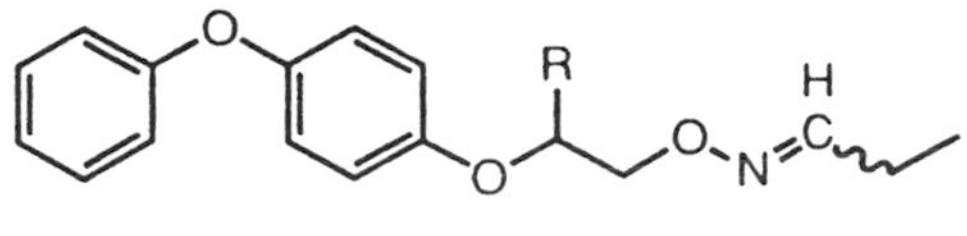

S-21149 (R=H)   (*26*)

S-21150 (R= · CH₃)   (*27*)

pyriproxyfen (S-31183)   (28)

More recently, researchers at Sumitomo Chemical Company discovered several oxime ethers [e.g., 26 (S-21149) and 27 (S-21150)] which show remarkably high JH activity, especially against dipterous species [70] (Table 2). However, these juvenoids have not been commercially developed. In addition, a series of heterocyclic derivatives were found and one of these juvenoids, pyriproxyfen (28), is being developed for household applications [71]. This juvenoid shows excellent activity against a wide range of insect species (Table 2).

juvocimene II   (29)

Several other natural products isolated from plants have been shown to have some JH activity against a limited range of insect species. For example, juvocimene II (29) was isolated from the oil of sweet basil, *Ocimum basilicum*, and shown to have very good activity against the large milkweed bug, *Oncopeltus fasciatus* [72]. However, the apparent high species selectivity of these natural juvenoids isolated from plants [52–54, 72–75] makes it doubtful that their presence in the plant offers any useful protection against insect pests as has sometimes been implied [75].

The morphogenetic activities of most of the juvenoids discussed in this section are presented in Table 2. The more recent compounds 24–28 show outstanding activity against many insect species.

V. STRUCTURE-ACTIVITY RELATIONSHIPS

Structure-activity relationships of juvenoids have been extensively reviewed [42, 56, 68, 76–79]. A remarkably wide range of chemical structures can produce a JH response in at least some insect species, including many compounds with little or no structural resemblance to the natural JHs. Many thousands of analogs have been prepared over the past 25 years, and this work has resulted in the discovery of a surprisingly large number of active structural types. Structure-activity can be quite complex, partially due to the diversity of physiological responses. The JH activity of some juvenoids is specific to a particular order or even family of insect species, whereas other juvenoids show activity against a broad range of species. Some insect species (e.g., *Tenebrio molitor*) respond to a large range of structures, whereas others are only sensitive to very specific compounds. For a particular juvenoid, large differences in sensitivity are often observed between the two common laboratory species, *Oncopeltus fasciatus* (Heteroptera) and *T. molitor* (Coleoptera). Structure-activity relationships obtained with one group of insects are frequently not applicable for another and generalizations, even within one order, are difficult to make. However, there are often relatively small differences in activity between species of the same genus. For a specific type of juvenoid, changes in functional groups that produce an increase in activity in one species can cause a decrease in another. In addition, different insect species can differ greatly in their overall level of sensitivity to juvenoids [15]. For example, insects of the family Pyrrhocoridae (Heteroptera), such as *Dysdercus cingulatus*, are very sensitive to juvenoids. This family responds to a diverse range of juvenoids, including many that are inactive against other groups of insects [77]. Some families in the orders Hymenoptera, Diptera, and Coleoptera respond poorly to all known juvenoids.

The natural JHs and many of the known juvenoids are very flexible molecules in solution and can adopt a large number of conformations. Thus, the active conformation that these compounds presumably adopt, which allows effective binding with the receptor protein involved in the JH response, is difficult to ascertain. In essentially all cases studied, the JH binding site recognizes stereochemical features with very high specificity. One of the possible geometric and/or optical isomers is usually responsible for all of the morphogenetic activity [79]. In general, the optimum size of a juvenoid corresponds to an aliphatic carbon chain of 14–17 units (~21 Å), but the optimum length varies for different insect species. The overall size and shape of the molecule, its lipophilicity, the branching along the carbon skeleton at critical positions, the character and location of olefinic and aromatic centers, and the dimensions and type of terminal functional groups and their relative hydrophobicities all greatly influence the biological activity. The rigidity of the molecule influences the possible conformations that the juvenoid can adopt and can have a major effect on the activity and species specificity [56, 76, 80]. Several research groups have attempted to derive binding models for different insect species and to fit different types of juvenoids into a common model. In particular, Iwamura and co-workers have extensively studied the steric dimension and hydrophobicity of the whole molecule, and the contour maps of electrostatic potentials for functional groups, in a number of different types of juvenoids in attempts to find an inclusive receptor model [81–84]. It was found in these studies that the optimum overall molecule dimension and the position-specific functional group interaction sites can vary considerably for different types of juvenoids, even against the same insect species [82]. Some success was achieved in examining the electrostatic potential contour maps of various juvenoids. A negative potential peak was observed at a common distance from the molecular edge within the receptor model [83, 84]. Thus, molecular recognition may involve an electrostatic interaction at a specific site on the JH receptor surface.

The observed JH activity of a topically applied juvenoid may depend to some extent in the rates of penetration through the cuticle, transport to the receptor site, absorption by various tissues, metabolism, and excretion. Overall, it appears that the persistence of a juvenoid within the insect plays only a minor role in the activity observed in the laboratory, but it was concluded in one study that the high activity of *28* against *Manduca sexta* is due to its high penetration rate and its relative stability in the hemolymph as well as its high intrinsic JH activity [85].

Various developmental stages of the same insect species can show quite different sensitivities to a given juvenoid. The action of juvenoids on embryogenesis varies widely, with almost every group of insect species showing the highest sensitivity to a different compound [15]. Most structure-activity studies in the literature report the effects of juvenoids on inhibition of insect metamorphosis. However, structural features that are optimum for the effect on metamorphosis are often quite different from those that are optimum for the action on embryogenesis. In this section, all bioassay data relate to the effects on inhibition of metamorphosis for synchronized sensitive stages of the insect species. In practical applications, the insect population is usually not synchronized, but additional effects on other developmental stages may contribute to control of the target species.

A. Farnesoates and 2,4-Dodecadienoates

All eight of the racemic geometric isomers of JH I (*3*) have been prepared and bioassayed and the natural (*2E, 6E*)-*cis*-10,11 isomer is the most active. The [*E*]-configuration at C-2 is very important for high activity in all farnesoate-type analogs with the corresponding 2Z isomers showing much lower activity. The *E* stereochemistry at C-6 is also important for high activity, but the stereochemistry of the oxirane ring is of secondary importance. The *trans*-10,11 isomer also shows good activity [30, 56]. In these compounds the oxirane ring is important for activity; the corresponding 10,11-olefins show somewhat lower JH activity. Increasing the length of a terminal C-11 alkyl substituent to *n*-butyl tends to remove the activity differences between the 10,11-epoxide and the corresponding 10,11-olefin. In this type of JH I analog, the effect on activity of epoxidizing the 10,11-double bond can be positive or negative depending on the insect species [56]. It is noteworthy that juvenoids have been discovered that are much more active than JH I against essentially all insect species.

(40)

(41)

(42)

Although the presence of an *E* 2-ene double bond is necessary for high morphogenetic activity in farnesoates and related structures, many variations in the chain are allowable. Saturation of the 6-ene double bond can give an increase in activity depending on the species (Table 3), and in these 6,7-dihydro derivatives various substituents at C-11, such as methyl, chloro, hydroxy, acetoxy, and alkoxy can give active analogs [56]. Even the 6,7;10,11-tetrahydro derivative *36* shows considerable JH activity. The 2*E*,4*E*-dienoates (e.g., *18*) show, in general, much higher activity than the corresponding 2,10-dienoates (e.g., *35*) and the 2,6,10-trienoates (e.g., *34*). In the 2*E*,4*E*-dienoates, the 11-methoxy analogs (e.g., *39*) and the 10,11-dihydro analogs (e.g., *18*) are much more active than the corresponding 10,11-olefins (e.g., *37*) and 10,11-epoxy derivatives (e.g., *38*) (Table 3). In general, the JH activity of farnesoates is very dependent on the isoprenoid nature of the carbon chain, and analogs with 3,7-dialkyl branching show the highest activity. However, the C-3 methyl group is more important in the 2,4-dienoates *16–19* than it is in the 10,11-epoxyfarnesoates [42, 56]. Replacing the methyl branches at C-7 and C-11 in JH III (*5*) with ethyl groups to give JH I (*3*) gives a large increase in activity against *Galleria*

Table 3 Ethyl Farnesoate and Related Analogs. ID$_{50}$ and IC$_{50}$ Values

No.[a]	Substituents	Aedes aegypti (ppm)	Galleria mellonella (μg/pupa)	Tenebrio molitor (μg/pupa)	Musca domestica (μg/prepupa)	Acyrthosiphon pisum (% active ingredient in spray)	Heliothis virescens (ppm in medium)
30	10,11-Epoxy-2,6-diene	0.14	26	5.5	3.4	—	>100
31	10,11-Epoxy-2-ene	0.29	6.7	14	3.7	—	42
32	11-Methoxy-2,6-diene	>0.1	84	11	>100	—	>100
33	11-Methoxy-2-ene	0.040	33	46	2.1	—	100
34	2,6,10-Triene	0.023	~300	4.4	6.4	—	>300
35	2,10-Diene	0.20	6.8	12	2.8	0.029	>100
36	2-Ene	0.062	3.3	2.6	1.4	0.035	100
18[b]	2,4-Diene	0.0078	0.040	0.25	18	0.0039	0.30
37	2,4,10-Triene	0.10	0.62	1.1	11	—	1.7
38	10,11-Epoxy-2,4-diene	0.11	1.2	1.7	1.1	—	4.7
39	11-Methoxy-2,4-diene	0.0020	0.17	8.9	5.0	0.000056	0.20
40		0.094	0.050	0.10	2.5	0.0037	1.4
41		0.053	5.0	4.4	31	—	13
42		0.0030	0.0029	0.0085	0.027	—	0.0085

[a] All compounds are racemic.
[b] Hydroprene.

mellonella (Table 1). In contrast, replacing the methyl branches at C-7 and C-11 in hydroprene (*18*) with ethyl groups to give *40* did not produce an increase in activity against most insect species (Table 3). Homologation of the methyl group at C-3 in *18* to an ethyl group to give *41* produces a decrease in activity. However, the cyclopentene analog (*42*) is very active against all of the insect species in Table 3 [80]. The juvenoid *42* possesses a fixed 3-*s-trans*-diene conformation, and the very high morphogenetic activity of *42* and related analogs [76, 80, 86] suggests that the 2,4-dienoates (such as *18*) adopt this conformation when bound at the active site. The natural JHs may assume a similar active conformation in which the carbon atoms one through six are almost coplanar [56].

$R^1 = H;\ R^2 = H$ (*43*)
$R^1 = H;\ R^2 = \cdot CH_2CH_3$ (*44*)
$R^1 = \cdot OCH_3;\ R^2 = H$ (*45*)
$R^1 = \cdot OCH_3;\ R^2 = \cdot CH_2CH_3$ (*46*)

(*47*)

$R^1 = H;\ R^2 = \cdot CH_2CH_3$ (*48*)
$R^1 = H;\ R^2 = \cdot CH_2CH_2CH_3$ (*49*)
$R^1 = \cdot OCH_3;\ R^2 = \cdot CH_2CH_3$ (*50*)
$R^1 = \cdot OCH_3;\ R^2 = \cdot CH_2CH_2CH_3$ (*51*)
$R^1 = \cdot OH;\ R^2 = \cdot CH_2CH_3$ (*52*)

The nature of the unsaturated carbonyl functional group in the farnesoate type of structure is very important for high activity [56]. For esters, the activity decreases rapidly when the alcohol moiety of the ester contains more than four carbon atoms. The ethyl esters are the most active against many insect species, but some dipterous and coleopterous species are particularly sensitive to isopropyl esters such as methoprene (*16*) (Table 1). *N*-Ethyl and *N,N*-diethyl amides (e.g., *43–46*) show very good activity against several orders, as do some *S*-ethyl thiocarboxylates (e.g., *19* and *47*) and some α, β; γ, δ-unsaturated ketones (e.g., *48–52*, Table 4). The hydroxy ketone (*52*) shows especially good activity against the two lepidopterous species in Table 4 (*Galleria mellonella* and *Heliothis virescens*). In the farneoates and 2,4-dodecadienoates, many different substituents at C-10 and C-11 have been investigated. The effect of such substitution is variable and depends on the type of substituent and on the insect species. For the 2,4-dodecadienotes, an 11-methoxy or 11-ethoxy group often gives an increase in activity, especially for the isopropyl esters, cf. *16* [56].

Table 4 N-Alkylamides, S-Alkyl Esters, Unsaturated Ketones and Other Analogs. ID$_{50}$ and IC$_{50}$ Values

No.	Structure[a]	Aedes aegypti (ppm)	Galleria mellonella (μg/pupa)	Tenebrio molitor (μg/pupa)	Musca domestica (μg/prepupa)	Acyrthosiphon pisum (% active ingredient in spray)	Heliothis virescens (ppm in medium)
43		0.060	0.063	0.0046	0.23	0.032	3.0
44		0.17	0.33	0.0068	0.21	0.027	3.6
45		0.026	0.11	0.00033	0.0077	0.03	0.62
46		0.27	0.067	0.0010	0.030	>0.1	0.48
47		0.0022	0.045	0.025	2.4	0.00043	0.55
19[b]		0.00040	0.080	0.10	0.35	0.00083	0.41
48		0.024	0.070	0.0032	0.54	—	0.050
49		0.038	0.90	0.0053	3.6	0.029	2.6
50		0.0015	0.084	0.00046	0.25	0.055	0.036
51		0.028	2.6	0.00055	0.72	—	0.48
52		0.064	0.0061	0.00024	0.18	—	0.032
55[c]		0.00066	28	—	0.0046	—	43
56		0.26	0.52	>10	6.6	—	—
60		>0.1	0.016	0.70	57	—	2.7
61		>0.1	0.026	0.55	4.2	—	0.013

[a] All compounds are racemic.
[b] Triprene.
[c] Cyklopren.

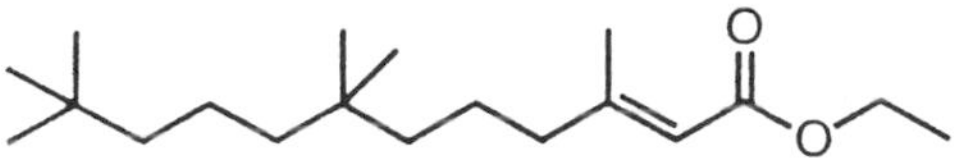

(53)

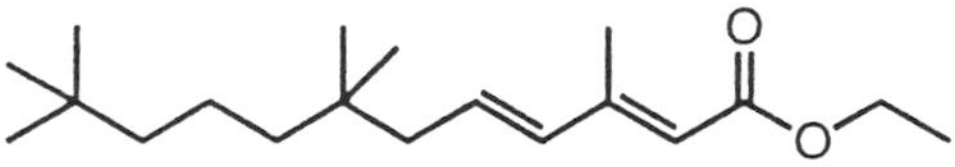

(54)

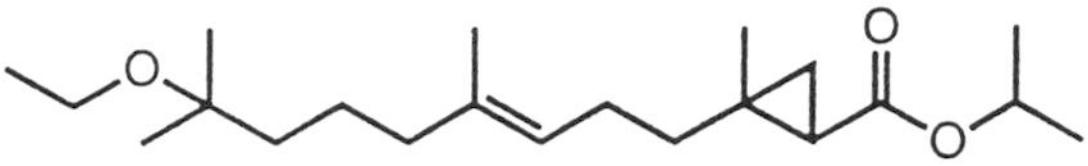

cyklopren   (55)

In this farnesoate type of juvenoid the introduction of an additional substitutent at C-7 (such as chloro, methyl, or a 6,7-epoxy group) gives analogs (e.g., 13, 53, 54) that show low activity against the species listed in Table 1 [56]. However, many such 7,7-disubstituted analogs show high activity against hemipterous species of the families Pyrrhocoridae, Lygaeidae, and Reduviidae. Replacement of the E 2-ene double bond in the farnesoates by a cyclopropane ring gives analogs that show a considerable increase in activity against some species. For example, cyklopren (55) shows excellent activity against *Aedes aegypti* and *Musca domestica* [56] (Table 4).

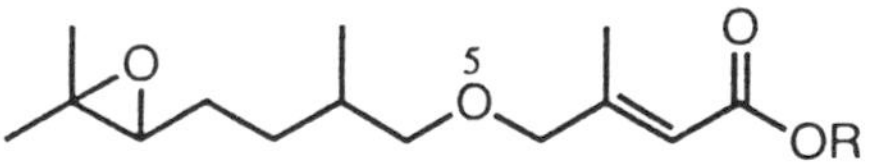

R = · CH₂CH₃   (56)
R = · CH(CH₃)₂   (57)

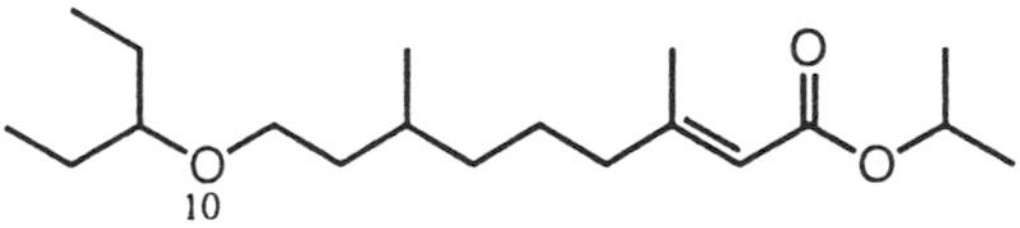

R = $\cdot$ CH$_2$CH$_3$ (56)
R = $\cdot$ CH(CH$_3$)$_2$ (57)

(58)

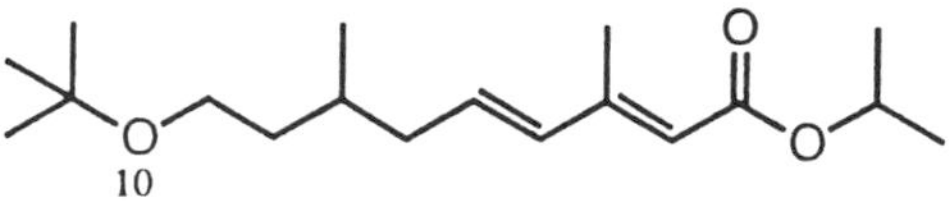

(59)

Jarolím and co-workers have extensively investigated the effects on JH activity of replacing carbon atoms in the farnesoate chain with oxygen atoms [87]. The JH activity of these analogs is quite species specific and varies considerably with the position of the oxygen atom in the chain. The 5-oxa analogs 56 and 57 and some 10-oxa analogs (e.g., 58 and 59) give high activity against a number of insect species. For example, the 5-oxa juvenoids 56 and 57 are very active against certain fly species, but 56 shows low activity against the species in Table 4 including *Musca domestica*. The 10-oxa ester (58) is also active against a number of fly species and this juvenoid is quite active against *M. domestica*. The 10-oxa analog *59* is more active than methoprene (*16*) against several cyclorrhaphous fly species.

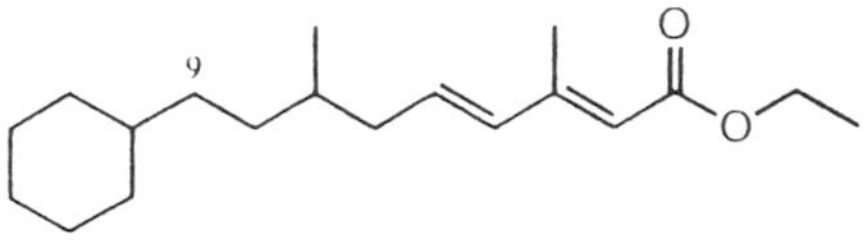

(*60*)

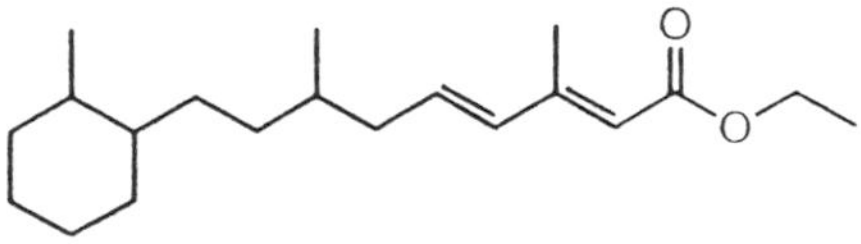

(*61*)

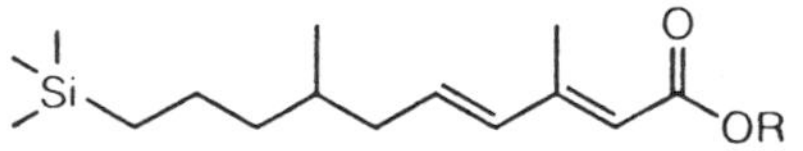

R = · CH_2CH_3 (*62*)
R = · $CH(CH_3)_2$ (*63*)

Another group of juvenoids that show high species specificity are analogs with a cycloalkyl group attached to the end of the carbon chain. The effects of a terminal cyclohexyl group on JH activity has been studied against a wide range of insect species [88, 89]. The 9-cyclohexyl-2,4-dienoates (e.g., *60* and *61*) are, overall, the most active juvenoids of this type and are especially active against several lepidopterous species (mainly noctuids). However, they are relatively inactive against some other orders such as Diptera (Table 4). For comparison, methoprene (*16*) shows very good activity against dipterous species (Table 2) and is also, overall, the most active juvenoid of those tested against most noctuid species (except *Heliothis virescens*) [88–90]. The silicon analogs *62* and *63* have been prepared and bioassayed. The isopropyl ester *63* is considerably less active than methoprene (*16*) against several insect species. However, the ethyl ester *62* shows increased activity over hydroprene (*18*) against two insect species and decreased activity against two other species [91].

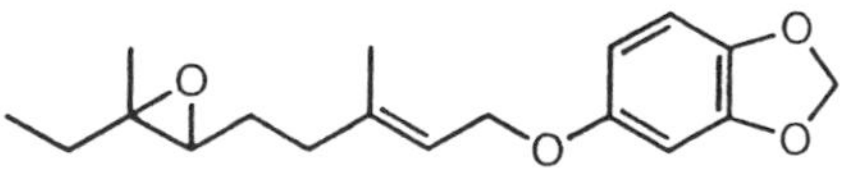

(*64*)

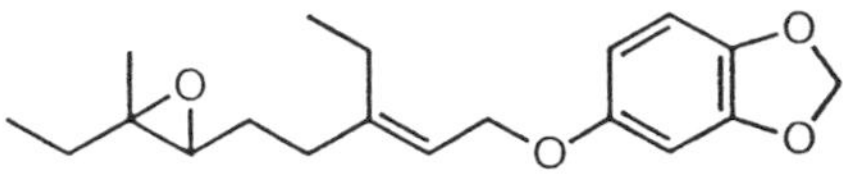

(*65*)

B. Terpenoid Phenyl Ethers

Aryl terpenoid ethers and related structures were extensively investi
gated in the late 1960s and early 1970s following the discovery of the
juvenoid 15 [55]. The structure-activity relationships for these analogs
have been reviewed in detail [56]. The higher homologs 64 and 65 are
more active than 15 against many insect species, but there are cases in
which 15 is the more active analog. The (E)-2-ene isomer of 15 is much
more active than the corresponding (Z)-2-ene isomer. The juvenoids 15,
64, and 65 have been evaluated against a large number of insect species, and good activity has been found against some species of the orders Coleoptera, Diptera, and Lepidoptera [56].

The effects on JH activity of substitution in the phenyl ring of
these ethers have been extensively studied. Substitution of appropriate groups in the 4-position of the phenyl ring generally gives analogs
with the highest activity, and these substituent effects are often very
species specific [56]. For example, a p-nitro or p-methoxycarbonyl group
gives analogs with high activity against several hemipteran families but
very poor activity against many other insect species. The 4-ethylphenyl
ether 66 (R-20458) shows high activity against a number of insect species, including *Aedes aegypti* and *Tenebrio molitor* (Table 5) [56, 92]. The
JH effect of 66 against fly species varies considerably from high to very
low activity, depending on the species.

$R = \cdot CH_2CH_3$ (R-20458)	(66)
$R = \cdot Cl$	(68)

Compounds of this type with a 6,7-epoxy group are usually much
more active than the corresponding analogs (e.g., 67, Table 5) with a
6,7-olefin function. Structure-activity observations suggest that the role
of the 6,7-epoxy group in 66 is equivalent to that of the 10,11-epoxy
group in the natural JHs (e.g., 5). The 4-chlorophenyl ether 68 is quite
active against a number of insect species, being much more active than
15 or 66 against *Eurygaster integriceps* [93]. However, 68 is less active
than 15 or 66 against the species in Table 5.

(69)

Table 5 Terpenoid Phenyl Ethers. ID_{50} and IC_{50} Values

No.	Structure[a]	Aedes aegypti (ppm)	Galleria mellonella (μg/pupa)	Tenebrio molitor (μg/pupa)	Musca domestica (μg/prepupa)	Acyrthosiphon pisum (% active ingredient in spray)	Heliothis virescens (ppm in medium)
15		0.0032	0.082	0.0050	2.0	0.020	>100
66[b]		0.00070	0.072	0.0025	3.0	0.043	~100
67		2.3	>200	0.52	>100	0.013	—
68		0.094	>100	0.19	5.6	—	—
69		0.33	>100	0.00062	>100	>0.01	24
71		>10	>100	>100	>100	>0.1	>100
72		>1	>100	>100	>100	>0.1	—
75		0.054	2.2	0.019	3.9	—	>100
20[c]		0.057	0.037	0.0024	54	—	2.2

[a] All compounds are racemic.
[b] R-20458.
[c] Epofenonane.

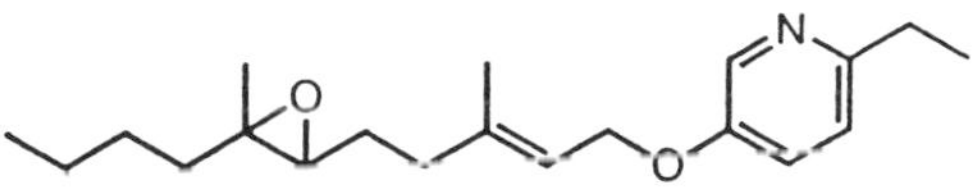

(70)

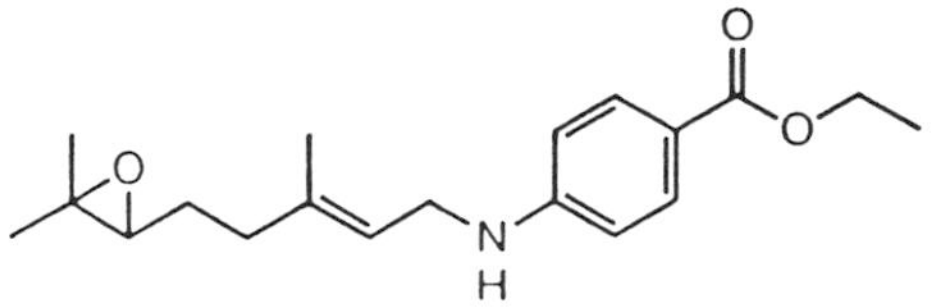

(71)

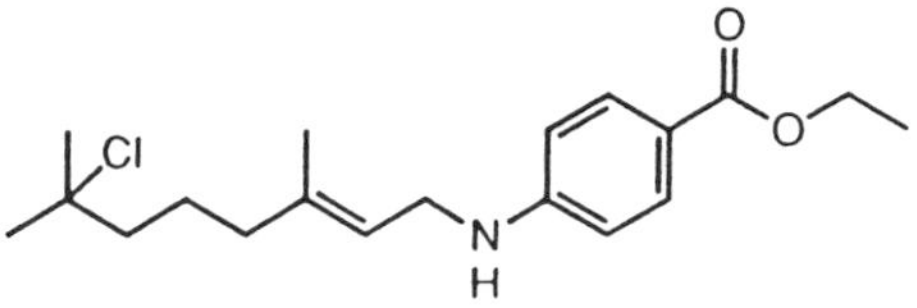

(72)

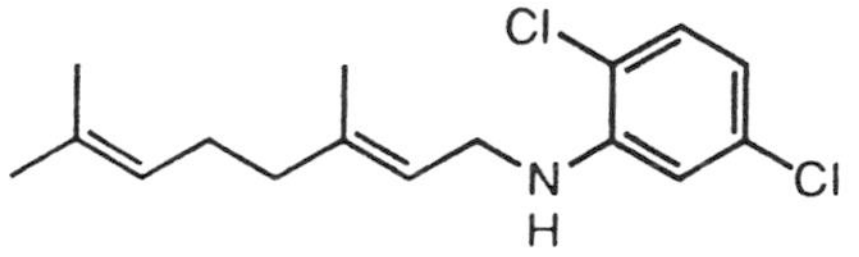

(73)

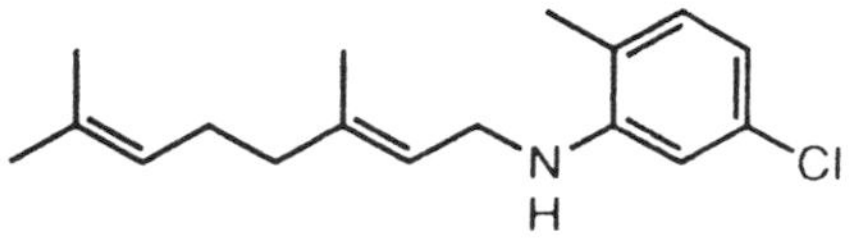

(74)

Replacement of the 6,7-epoxy function in these ethers by a 7-alkoxy group increases species specificity and gives analogs with a narrow spectrum of activity. The 7-ethoxy compound 69 is very active against *Tenebrio molitor*, being more active than 15 or 66 (Table 5) [56, 94]. However, 69 is relatively inactive against many other species. Some 6-ethyl-3-pyridyl terpenoid ethers (e.g., 70) also show high JH activity. The juvenoid 70 is very active against a wide range of coleopterous and lepidopterous pests of stored products [95]. Anilino analogs, such as 71 and 72, show very high species specificity. They are inactive against the insect species in Table 5, but show high activity against some hemipterous species. For example, 71 and 73 are very active against *Rhodnius prolixus* [96]. The juvenoids 73 and 74 also show high activity against the related species *Panstrongylus megistus*, an important vector of Chagas' disease in Brazil [97]. The presence of a 3-chloro group in the phenyl ring considerably increases the JH activity against these two hemipterous species.

The 6,7-epoxycitronellyl ethers, such as 75, show somewhat lower activity than the corresponding geranyl ethers (e.g., 66) against most insect species (Table 5). The juvenoid 75 shows generally poor activity against lepidopterous species and very variable activity against dif-

(75)

ferent fly species [56]. As discussed in Section IV, the homolog *20* (epofenonane) shows interesting JH activity against many lepidopterous, coleopterous, and homopterous species and has much higher foliar stability than the corresponding geranyl ether [61–64]. However, it has relatively low activity against dipterous species (Table 5).

(76)

A large number of related aryl terpenoid analogs have also been prepared, but only a few of these compounds show interesting JH activity [56]. Most of these analogs show activity against a narrow range of insect species. As discussed in Section IV, the juvenoids 22 and 23 show good activity against many dipterous species [65, 66] (Table 2) and are active in disrupting colonies of the red imported fire ant, *Solenopsis invicta* [67]. The 3-methoxyphenyl derivative *76* is one of the most active juvenoids against *Rhodnius prolixus*, being about 300 times more active than JH I [98]. It also has high activity against the related species *Panstrongylus megistus* (Reduviidae) [97]. In this type of juvenoid, the presence on the aryl ring of an alkoxy or chloro group *meta* to the terpenoid chain is important for activity against these two heteropterous species.

C.   4-Phenoxyphenoxy Analogs and Related Structures

Each of the three isoprene or homoisoprene units of JH I (*3*) can be replaced by an appropriately substituted benzene or heterocyclic ring to afford very active juvenoids [56, 68]. A considerable number of 4-phenoxyphenyl and 4-benzylphenyl compounds have been prepared (e.g., *77–81*, Table 6) and many of them show excellent morphogenetic activity. They often are quite species specific in their JH effects. For example, the juvenoid *77* is quite active against *Aedes aegypti* and yet is inactive against *Musca domestica* (Table 6). This juvenoid also shows good activity against a number of lepidopterous species [68]. The compound *78* (CGA 13353) shows very good activity against many homopterous and lepidopterous species but is inactive against *Tenebrio molitor* and *M. domestica* (Table 6) [99, 100]. The related juvenoid *79* (CGA 34301) is also active against homopterous species such as both armored and soft-scale species, and pear psyllids [101, 102].

Table 6 4-Phenoxyphenoxy Analogs and Related Structures. ID_{50} and IC_{50} Values

No.	Structure[a]	Aedes aegypti (ppm)	Galleria mellonella (µg/pupa)	Tenebrio molitor (µg/pupa)	Musca domestica (µg/prepupa)	Heliothis virescens (ppm in medium)
77		0.0012	3.5	0.43	>100	4.9
78[b]		0.017	~10	>100	>100	1.9
79[c]		0.026	4.5	0.38	25	140
80		>0.1	>100	0.28	6.5	>100
81		0.021	0.71	0.00053	~100	19
24[d]		0.0000036	0.0000001	—	0.045	—
25[e]		0.000010	0.000020	—	0.090	0.018
86		0.00028	0.00065	—	0.0059	—
87		0.00029	0.00086	—	0.0036	—
88		0.00016	0.00062	—	0.052	—

[a] All compounds are racemic.
[b] CGA 13353.
[c] CGA 34301.
[d] CGA 045128.
[e] Fenoxycarb.

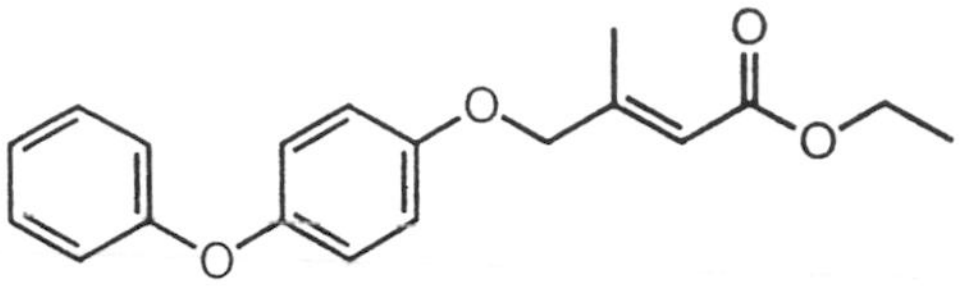

(77)

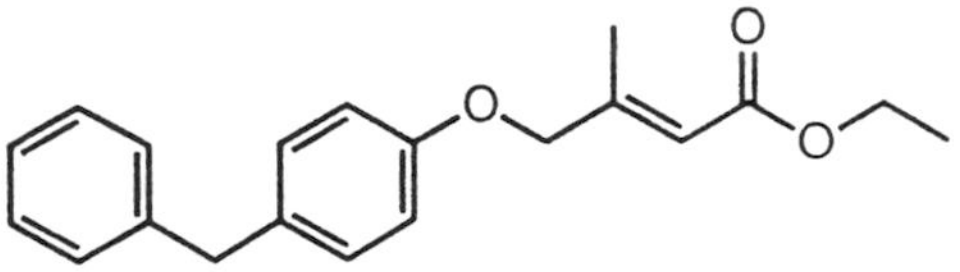

CGA 13353   (78)

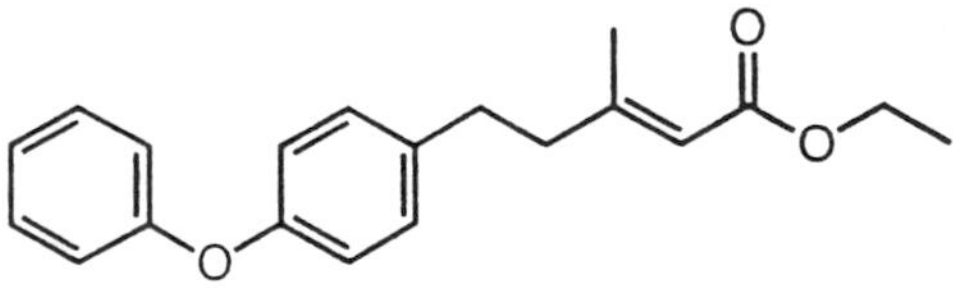

CGA 34301   (79)

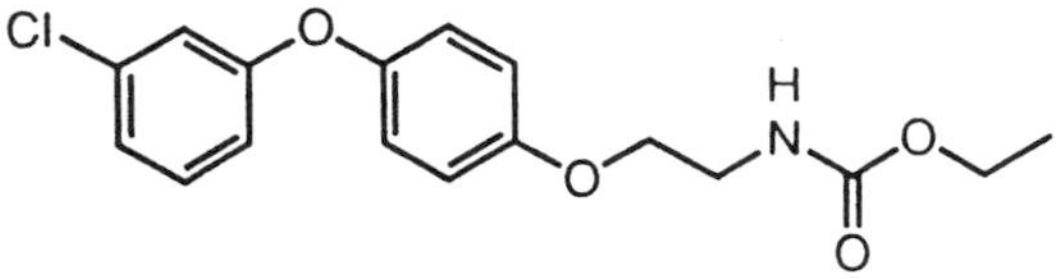

Ro 16 1294   (82)

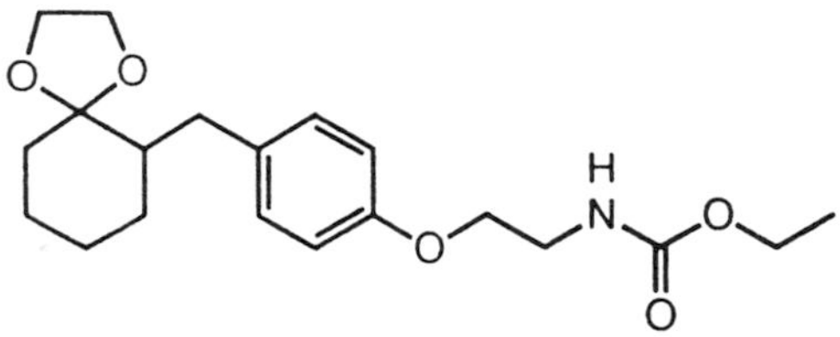

(83)

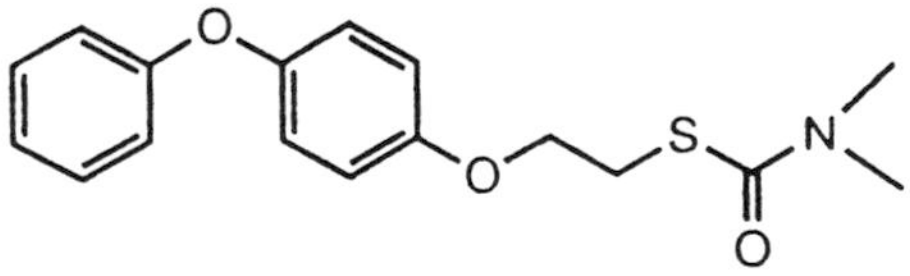

(84)

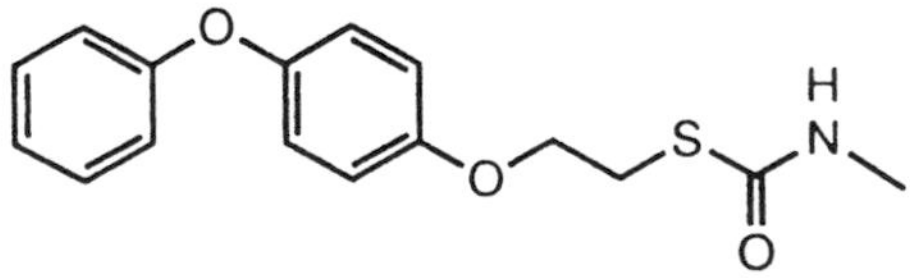

(85)

A major breakthrough in juvenoid chemistry was the discovery of the photostable carbamates CGA 045128 (24) and fenoxycarb (25) (Ro 13 5223) (see Section IV) by chemists at Ciba-Geigy AG [68] and R. Maag Ltd. [69], respectively. Both of these juvenoids show excellent activity against a wide range of insect species. They are not as active as methoprene (16) against the housefly, *Musca domestica*, but are more active than 16 against many other species (Table 2). These analogs are stable enough in sunlight to be useful against many lepidopterous and homopterous insect pests in orchard and vine crops, and fenoxycarb (25) has been commercialized for such field applications. Many variations in this structural type have been investigated and some of the analogs show excellent JH activity [57, 68]. The analog 82 (Ro 16-1294) is very active against mosquito species under field conditions [103], and the juvenoid 83 shows very good activity on a number of homopterous species [104]. The thiocarbamate 84 has high activity against mosquito species [105] and the analog 85 has good activity against *Galleria mellonella* [106].

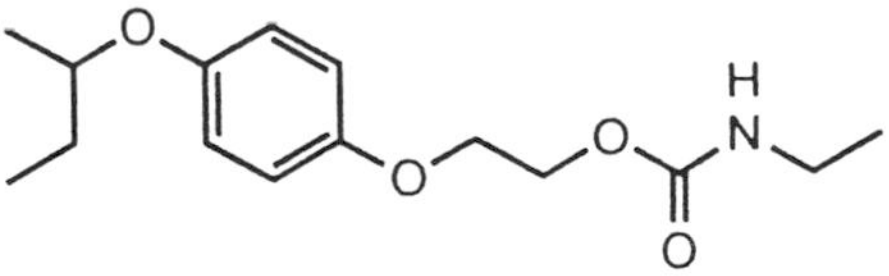

(86)

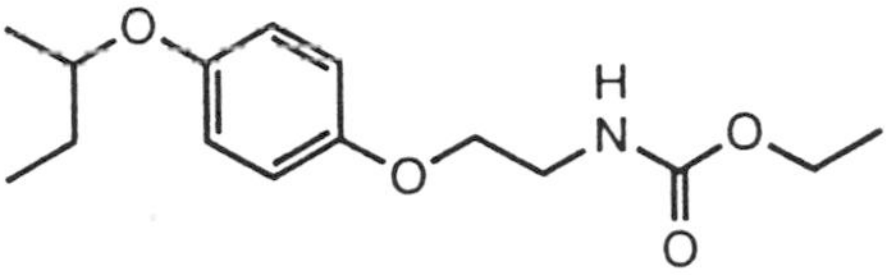

(87)

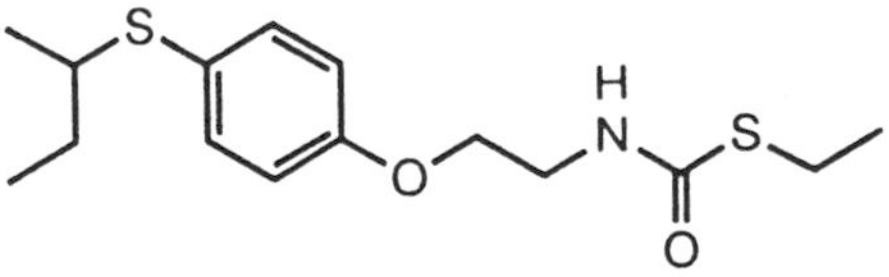

(88)

When the terminal phenyl group in 24 or 25 is replaced with an aliphatic group, some of the analogs have high JH activity. In particular, the 1-methylpropyl analogs 86 and 87 are much more active than the corresponding phenyl analogs 24 and 25, respectively, against *Musca domestica* (Table 6) [57]. Although 87 is much less active than fenoxycarb (25) against *Aedes aegypti*, it is nevertheless much more persistent in water contaminated with sewage in sunlight and it outperforms 25 after 14 days' exposure under these conditions [57]. The related analog 88 is very active against cockroaches, being 10 times more active than both fenoxycarb and (S)-hydroprene against the German cockroach, *Blattella germanica*, in a diet incorporation assay based on both morphogenetic and reproductive inhibition [57].

The juvenoid 89 (CGA 29170; dofenapyn) with a terminal acetylene group has high activity against several insect species and also shows considerable ovicidal activity on spider mite species (Tetrany-

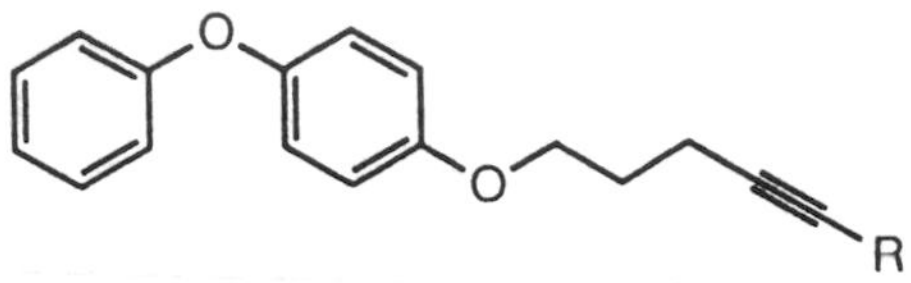

R = H, dofenapyn, CGA 29170   (*89*)

R = Cl   (*90*)

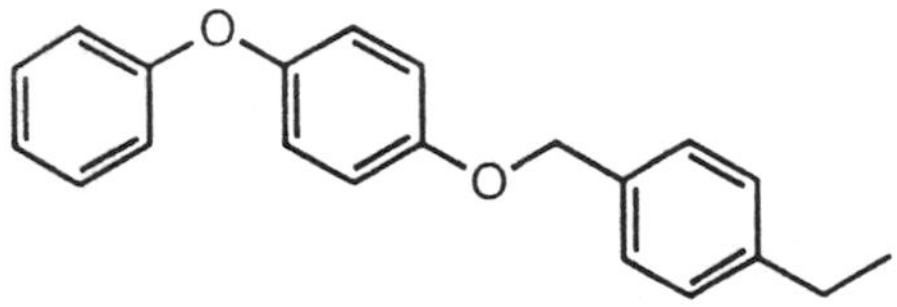

CGA 35452   (*91*)

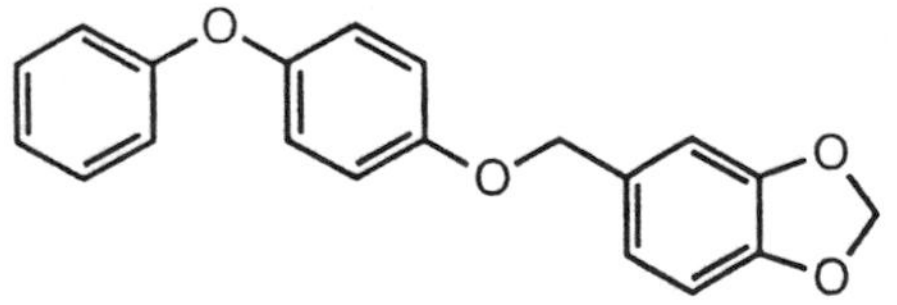

CGA 28772   (*92*)

chidae) [68]. The closely related structure *90* ("JH-286") has good activity against many insect species and shows useful field efficacy against several homopterous species [107]. It is particularly interesting that analogs such as *91* and *92*, in which all three sections of JH I have been replaced by benzene rings, show considerable JH activity [68].

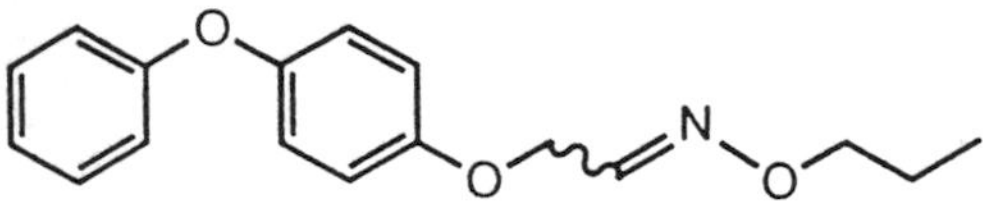

(*93*)

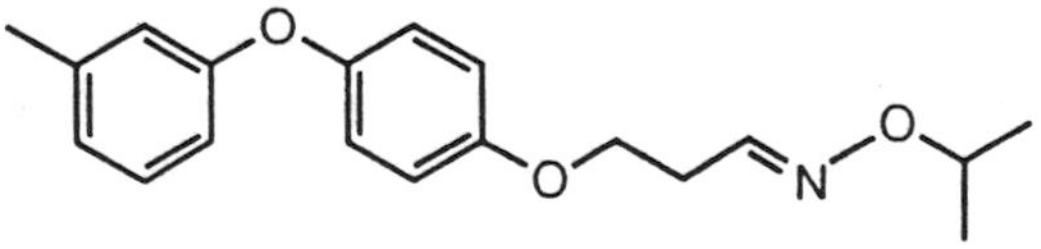

(*94*)

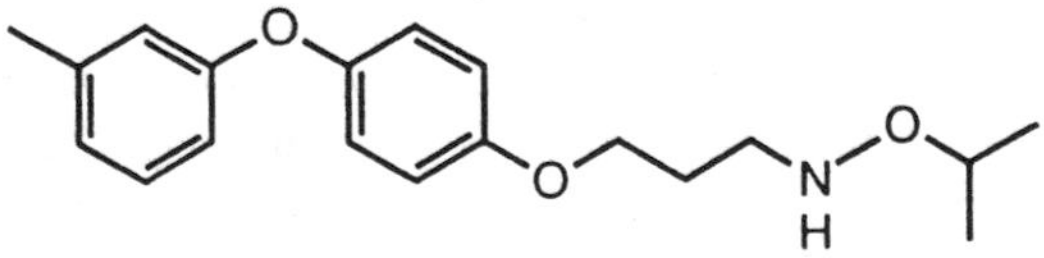

(*95*)

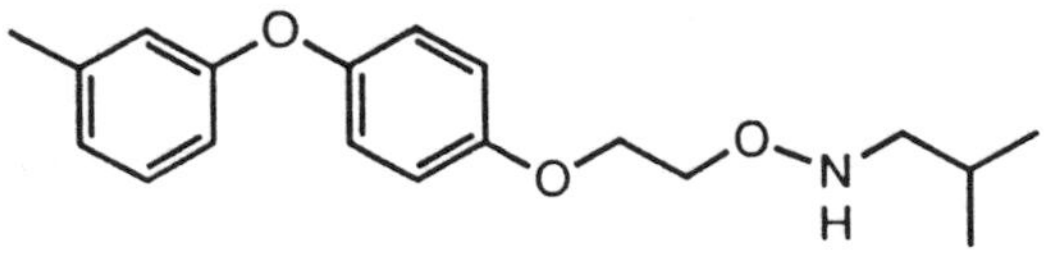

(*96*)

(97)

(98)

The propionaldehyde oxime ethers *26* (S-21149) and *27* (S-21150) show remarkably high JH activity against the larvae of dipterous species such as the common mosquito, *Culex pipiens pallens*, the yellow-fever mosquito, *Aedes aegypti*, and the housefly, *Musca domestica* [70, 108, 109]. There is little difference in activity reported between *E* and *Z* isomers of the oximes [70], but the data in Table 2 show considerable differences in activity for the isomers of *26*. The *R* isomer of *27* is more active than the corresponding *S* isomer [70]. It has been reported that *27* is more active than *26* against *M. domestica* and that *27* is much more active than methoprene at inhibiting adult emergence [108] (cf. Table 2). The juvenoids *26* and *27* are considerably more active than methoprene (*16*) against most mosquito species [109]. The juvenoids *26* and *27*, and closely related analogs, are also very active against *Galleria mellonella*, being considerably more active than methoprene or hydroprene [110] (Table 2). The reversed oxime ethers like *93* also show good activity against *C. pipiens pallens*. However, *93* is not as active as *26* or *27* and shows a level of activity similar to methoprene against this species [111]. Certain other variations in this structural type give analogs with increased activity. For example, the oxime ether *94* is more active than methoprene against *C. pipiens pallens* [112] and *Spodoptera litura* [113]. The corresponding alkoxyamines *95* and *96* likewise show higher activity than methoprene against *C. pipiens pallens* [81]. Replacing the terminal phenoxy group in *94* with a suitably branched alkyl substituent gives analogs (e.g., *97*) with excellent activity against *C. pipiens pallens* [84], although *97* is less active than methoprene against *Musca domestica* [113]. Interestingly, some structurally related ethers (e.g., *98*) have high activity against *C. pipiens pallens* [114].

The pyridine derivative pyriproxyfen (*28*) and the closely related analog *99* are examples from a group of heterocyclic juvenoids discovered by Sumitomo Chemical Company [71]. Pyriproxyfen is very active against a wide range of insect species, especially in the orders Diptera [115–117], Lepidoptera [118], and Homoptera [119]. It is also as

effective as fenoxycarb *25* in suppressing populations of the red imported fire ant, *Solenopsis invicta* [120]. Pyriproxyfen also shows strong ovicidal effects against 1-day-old eggs of some lepidopterous species [121].
A number of related heterocyclic analogs have been reported in the
patent literature and many of them have very high JH activity. For
example, the pyrimidine analogs *100* and *101* [122], and the 1,3-thiaz-
ole analogs 102 and 103 [123] show excellent JH activity (Table 7). The
acetylenic analog *104* also has good activity [124], although the corresponding (*E*)-olefin *105* is somewhat more active against *Galleria
mellonella* [125].

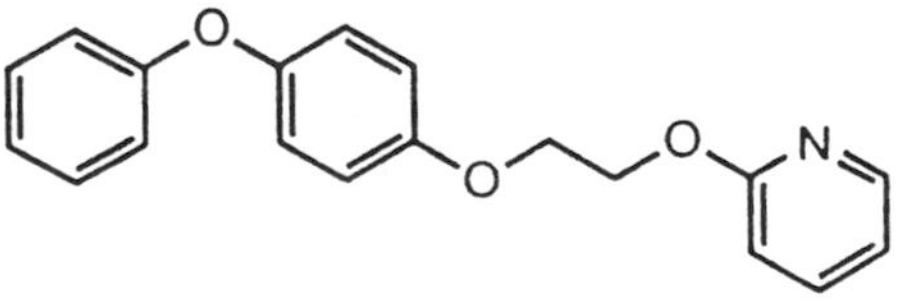

(*99*)

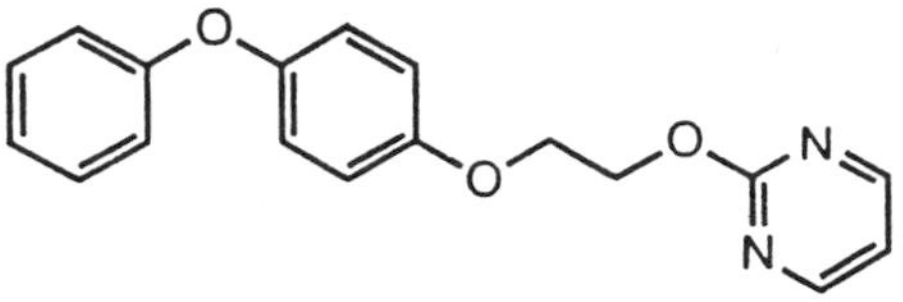

(*100*)

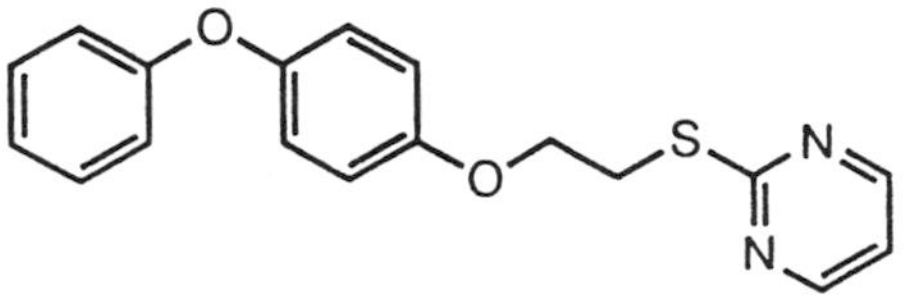

(*101*)

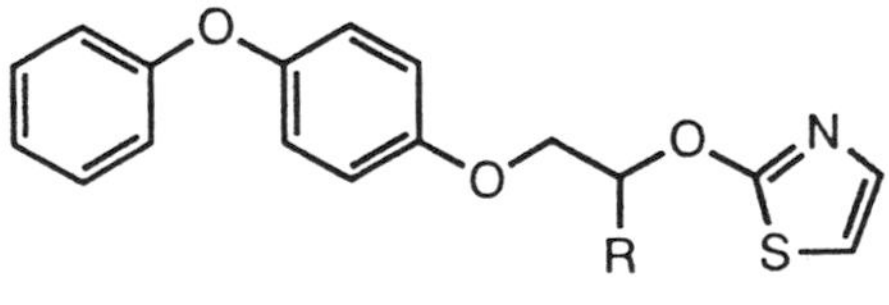

R = · CH₃   (*102*)

R = H   (*103*)

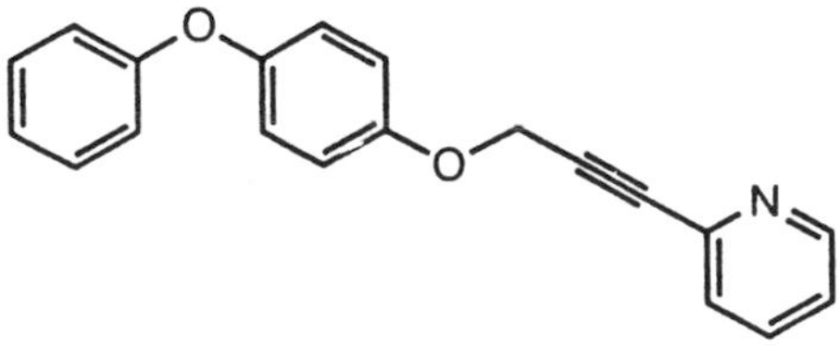

(*104*)

Table 7 4-Phenoxyphenoxy Heterocyclic Analogs. ID_{50} and IC_{50} Values

No.	Structure	Aedes aegypti (ppm)	Galleria mellonella (μg/pupa)	Musca domestica (μg/prepupa)
28[a]		0.0000039	0.00068	0.00033
99		0.0000024	0.000056	0.0038
100		0.000008	0.0062	0.00081
101		0.000021	0.001	0.0023
102		0.0000026	0.0000058	0.00019
103		0.00000037	0.0000032	0.0031
104		0.000039	0.052	0.064
105		0.000030	0.00095	0.058

[a] Pyriproxyfen (S31183).

(*105*)

D. General Analogs

There are some other structural types, not covered in Sections VA–VC, that produce a morphogenetic response in at least a few insect species [56]. For example, the aliphatic bisthiocarbamate *106* (R-31026) shows the same high morphogenetic activity as *66* (Table 5) against *Tenebrio molitor* and much higher ovicidal activity than *66* against the saltmarsh caterpillar, *Estigmene acrea* [126]. The compound *106* is also active on newly formed pupae of *Tribolium castaneum* and *T. confusum*, but the larvae of these two species are not affected; hence, the action of *106* is different from that of a typical juvenoid [127]. The analog *106* is very specific in its JH activity action and shows little or no morphogenetic activity against most other insect species [56, 126].

R-31026   (*106*)

(S)-   (*107*)

(S)-   (*108*)

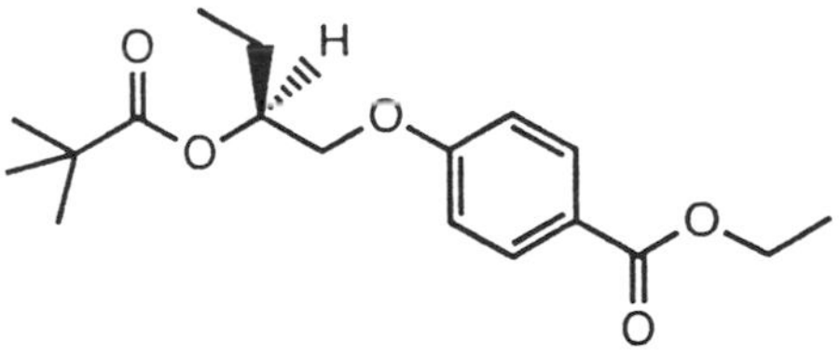

(S)-ETB   (109)

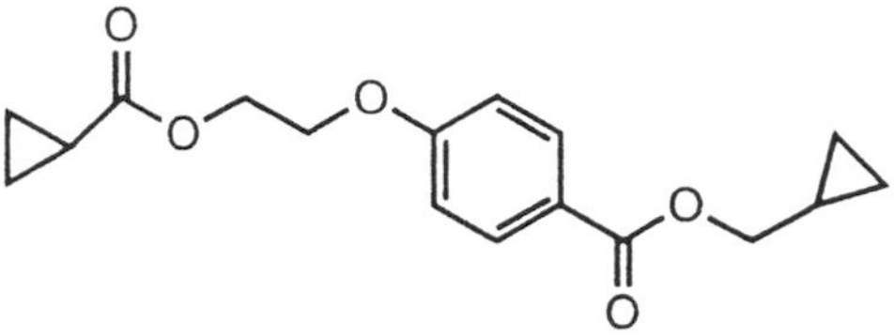

(110)

The peptidic juvenoids discovered by Sláma and co-workers in the early 1970s are an example of completely selective action against only one family of insects. Peptide derivatives of ethyl 4-aminobenzoate, such as 107 and 108, are extremely active but only against bugs of the family Pyrrhocoridae [13, 56, 128]. Only analogs derived from the corresponding L-amino acid are active.

The juvenoid 109 ("ETB") shows JH activity against a number of lepidopterous species but is inactive against *Musca domestica* and *Tenebrio molitor* [56]. This compound also exhibits anti-JH activity over a narrow dose range against *Manduca sexta* [129, 130]. Both the JH activity (at high doses) and the anti-JH activity (at lower doses) are due entirely to the (S)-enantiomer [131, 132]. The related compound 110 shows interesting JH activity, being as active as hydroprene (18) against *Heliothis virescens* and *Tenebrio molitor*. This compound (110) is very active as an ovicide against eggs of *Samia cynthia* and even has ovicidal activity against the two-spotted spider mite [56].

Pyran derivatives like 111 have been found to be very active at inducing morphogenetic aberrations in the cotton stainer, *Dysdercus intermedius*. The effects include disturbances in morphogenesis and gonad development, and the induction of supernumerary larval–adult intermediates. The compound 111 is active specifically against insects of the family Pyrrhocoridae and its effects are comparable to those of other juvenoids active on this family [133].

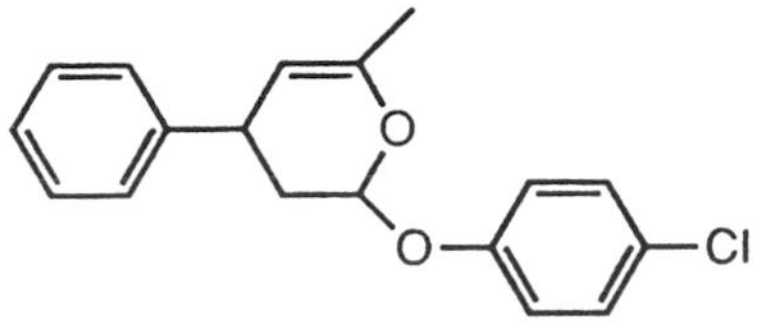

(111)

NC-170   (*112*)

Another interesting highly selective juvenoid is the pyridazin-3 (2H)-one *112* (NC-170). This compound strongly inhibits metamorphosis in leafhoppers like *Nephotettix cincticeps* when topically applied to mid-penultimate larvae [134]. Affected insects develop into supernumerary larval–adult mosaic individuals which subsequently die. This juvenoid has similar effects against the small brown planthopper, *Laodelphax striatellus*, when applied to the early final stadium [135]. The compound *112* is selective in its JH-like effects to leafhoppers and planthoppers and has no JH activity on nonhopper species such as those in Table 6. Interestingly, methoprene (*16*), fenoxycarb (*25*), and pyriproxyfen (*28*) are almost completely inactive against leafhoppers, and even kinoprene (*17*) requires a high concentration to affect metamorphosis.

E.   Optical Isomerism

There are usually considerable differences observed in the biological activity for the different optical isomers of the natural JHs and juvenoids [79, 136]. In almost all cases studied of compounds containing asymmetric carbon atoms, one optical isomer is responsible for essentially all of the morphogenetic activity, and in no case has synergism or antagonism been observed for the other stereoisomers. Thus, it is clear that a chiral receptor is involved in the action of JH and juvenoids containing asymmetric centers.

The natural (10*R*)-enantiomer of JH III (*5*) shows a much higher level of morphogenetic activity than does the 10*S* enantiomer [137]. When samples of high enantiomeric purity of the (10*R*, 11*S*)- and (10*S*, 11*R*)-stereoisomers of JH I were tested by topical application on allatectomized fourth instar larva of the silkworm, *Bombyx mori*, it was found that the natural (10*R*, 11*S*)-JH I (*3*) was 12,000 times more active than the (10*S*, 11*R*)-enantiomer [138]. Similarly, the (10*R*)-JH III (*5*) was much more active than the (10*S*)-enantiomer in this bioassay [138].

As discussed in Section II, larval hemolymph JH-binding proteins are involved in the transport of JH, its protection from JH degrading enzymes, and the regulation of JH titre. These JH-binding proteins also display a high degree of enantioselectivity, with the natural stereoisomers being more strongly bound than their enantiomers. However, the differences in binding affinity between the natural isomers and their enantiomers are much less than the differences observed in JH activity on the insect species [137–139]. The relative binding of juvenoids to these JH-binding proteins does not correlate with their observed morphogenetic activity, as highly active compounds like methoprene (*16*), hydroprene (*18*), and the aryl ether *66* do not generally bind with high affinity to these proteins [140]. With JHs, this binding is optimum with the natural enantiomer of the principal hormone present in a particular species.

The chiral centers present in most active juvenoids are not present in the natural JHs *3–7*. For example, the 2,4-dodecadienoates *16–19* contain an asymmetric carbon atom C-7 and essentially all of the JH activity is due to the (*S*)-enantiomers (Figure 2) [136]. The 6,7-epoxycitronellyl phenyl ethers like 75 have asymmetric carbon atoms at C-3 and C-6. A study of the activity of the 3*S* and 3*R* isomers of *75* (but unresolved at C-6) showed that the 3*R* configuration appears to be the stereochemistry required for the JH response. Similarly, in the aryl terpenoids like *23* the (*S*)-enantiomer is more active than the (*R*)-enantiomer [136]. These results suggest that if the juvenoids of types *16*, *23*, and *75* bind at the same chiral receptor, then the 2,4-dodecadienoates like *16* bind in a reverse orientation compared to the aryl terpenoids *23* and *75* (Figure 2) [136]. Therefore, the ester moiety of *16* presumably could occupy the same space as the ether or epoxide function of *23* and *75*, respectively. However, such conclusions must take into account the fact that juvenoids like methoprene *16* are much more active than *23* and *75* against most insect species, and thus, *23* and *75* may bind poorly at the active site.

Several of the 4-phenoxyphenoxy and related analogs described in Section VC contain asymmetric carbon atoms [79]. For example, the oxime ether *27* contains a chiral carbon atom at C-2 and essentially all of the JH activity appears to be due to the (*R*)-enantiomer [70]. With pyriproxyfen (*28*) the (*S*)-isomer appears to be more active than the (*R*)-enantiomer [71]. Also in a study of the optical isomers of the 4-(1-methylpropoxy)phenoxy analog *86*, it appears that all of the JH activity is due to the (*S*)-enantiomer [79].

In reviewing the morphogenetic activity of the various juvenoids discussed in this section it is clear that broad structure-activity generalizations are difficult to make. This is partially due to the very large variations in sensitivity of different taxonomic groups of insects to dif-

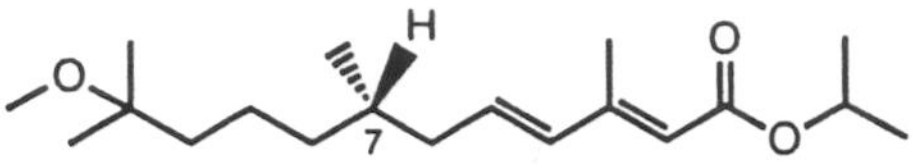

(S)-16

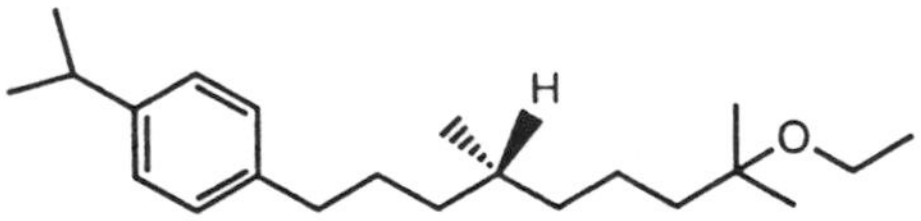

(S)-23

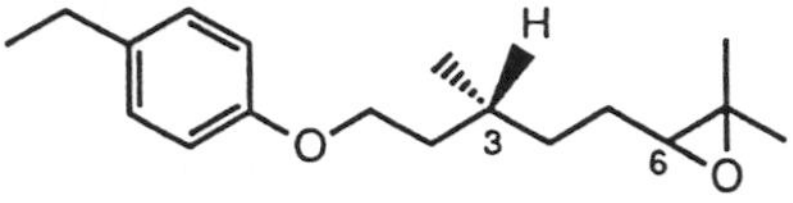

(R)-75

Figure 2 Biologically active stereoisomers.

ferent juvenoids and to the remarkably diverse range of structural types
of compounds that can produce a JH response in at least some insect
species. The JH binding site recognizes stereochemical features with
very high specificity and the ability of insect species to respond selectively to a variety of compounds containing different chiral centers
suggests that the JH receptor for a particular species may have several
adjacent chiral binding sites. Highly active analogs like methoprene,
that have asymmetric centers at positions where the natural JHs are
achiral, may bind more effectively at the receptor due to supplemen-
tary chiral interactions.

VI. COMMERCIAL APPLICATIONS

The juvenile hormones influence a wide range of important processes in both the developing and the mature insect [13–16]. Applications of JHs or juvenoids at sensitive stages in an insect's life can disrupt development, reproduction, behavior, and general physiology. Sensitive periods include the early stages of embryogenesis, the end of the last larval or nymphal instar, the early pupal period (in Holometabola), and, in a relatively few species, the adults. In some homopterous species, the penultimate nymphal instar is the most sensitive stage to juvenoids. The most important periods of JH sensitivity are metamorphosis and the beginning of embryogenesis. The best-studied lethal effect is that on metamorphosis. As the natural JH titre in an insect during the last larval or nymphal stadium must decline to a low level for metamorphosis to take place, the presence of a juvenoid at this time can lethally derange the development. Several different types of morphogenetic effects can be distinguished [25, 69]. Application of juvenoids to the last larval instar can produce effects ranging from a prolongation of the instar or the formation of almost perfect supernumerary larvae, through nymphal–adult intermediates (Hemimetabola) or larval–pupal and pupal–adult intermediates (Holometabola), to adults with minor deformities. The production of supernumerary larvae or nymphs occurs in relatively few species and only with carefully timed applications of large doses or when exposure to the juvenoid is continuous. The various mosaic intermediates usually fail to undergo normal ecdysis and soon die. In some homopterous species, like scale insects, mealybugs, and psyllids, treatment of the younger instars can result in death even at the early molts.

The second major lethal effect of juvenoids is interference with normal embryonic development [14–16, 77]. Application of juvenoids to freshly laid eggs of many insect species results in an ovicidal effect due to disruption of embryogenesis. The period of sensitivity is restricted to the very early stages of embryonic development. For example, hydroprene (18) and the analog 60 are very effective as ovicides against newly laid eggs of lepidopterous species [141, 142]. In addition, derangement of the reproductive system is observed in a limited number of insect species when late nymphs or young adult females are treated with juvenoids. The sensitive periods for effects on reproduction vary with the insect species [14, 16, 77]. Treatment during these periods results in the deposition of eggs that fail to hatch or sterilization of both sexes due to defective reproductive organs. The reduced fertility in sensitive female adults may result from direct physiological effects on the

JH-regulated reproductive system or from direct ovicidal activity on the eggs in the gravid female. More attention should be given to the evaluation of juvenoids that have high activity against insect eggs and/or the fecundity of adults. The ovicidal activity of photostable juvenoids like fenoxycarb (25) offers some promise for the control of lepidopterous pests like *Heliothis virescens* on cotton [143].

Applications of juvenoids can terminate the reproductive diapause in some adult insect species. Juvenoids may also have useful effects on polymorphism in insects. The compounds can affect the formation of castes in social insects, seasonal polymorphism in aphids, and phase polymorphism in locusts. The extent and character of all of these various responses to juvenoid application depends on the insect species, the time of application, the dose, the mode of application, the type of juvenoid, and the stage of development. Abnormal developmental effects resulting from the application of juvenoids are mostly irreversible and often lethal to the insect. The effects of a juvenoid are not the result of excessive amounts of JH but rather occur when cells are exposed to JH at abnormal times in their development. The application of juvenoids to insects when the cells are not in a sensitive stage has little or no effect.

Juvenoids are rather slow in their mode of action and do not stop the development of early larval stages of most insect species. In many plant-feeding insects, the growing immature larvae are the most destructive stage to the crop. Furthermore, in order to disrupt development, the juvenoid must be present at a sensitive stage, which may be short. Some sensitive resting stages, such as pupae and adults in reproductive diapause, are often inaccessible. Juvenoids are usually not useful for the control of insect species that are highly migratory, polyphagous, or that only have a few generation cycles per year. Many lepidopterous species are of this type. The better target insect species are those that go through several short life cycles per season and in which the damage caused by feeding larvae in the first generation can be tolerated (e.g., the summerfruit tortrix moth, *Adoxophyes reticulana*). Other suitable targets include those in which the younger instars die at the early molts after treatment (e.g., scale insects and psyllids); insects in which only the adult stage is a pest and the larvae do not cause significant damage (e.g., mosquitoes and most fly species); and insects that are in confined environments (e.g., stored product facilities, buildings, and greenhouses) where one can prevent small populations of the pest from increasing to economically significant levels, and where limited reinfestation can occur [60]. Reinfestation by untreated insects is a general problem with the use of juvenoids because JH action is based on delayed or indirect effects on developmental processes. Therefore, the best target species are those that are relatively sedentary or live in restricted environments. Juvenoids are most effective against species that are susceptible to both morphogenetic and reproductive effects.

A. Noncrop Markets

1. *Mosquitoes and Flies*

Mosquitoes are an excellent target for juvenoids because the larvae cause no damage, the larval developmental cycle is short, and, in field flooding applications, a synchronous population is produced from the diapausing eggs. We at Zoecon Corporation selected control of mosquito populations as our initial commercial target and methoprene received the first full commercial registration for a juvenoid in March 1975 for the control of floodwater mosquitoes [57]. Methoprene is particularly effective against mosquitoes such as *Aedes aegypti* and *Aedes nigromaculis*. It is applied as a microencapsulated, slow-release flowable formulation (Altosid™ SR 10 IGR). Treatment of fourth larval instars of these mosquito species results in failure of the adults to emerge from the pupal case [14]. Low-field persistence of technical methoprene (half-life of ~24 hr) requires the use of special formulation techniques. The microencapsulated Altosid™ SR 10 formulation extends field persistence to 4–7 days, allowing excellent control of the floodwater mosquito species found in irrigated pastures, coastal tidal marshpools, and rice fields [144]. Aedine and Anopheline mosquitoes are the most susceptible to methoprene, whereas Culicine mosquitoes are less sensitive. Additional longer-lasting formulations, such as Altosid™ XR extended residual briquets, have been commercialized. These charcoal-based materials release effective levels of (*S*)-methoprene over a period of up to 150 days to control *Culex* spp. which breed in standing water [60].

In these applications, methoprene has been found to be very safe to the environment with little or no effect on nontarget species such as mammals, fish, crustacea, birds, protozoa, annelids, molluscs, amphibia, damsel flies, mayflies, and water beetles [145]. Methoprene is also active against most strains of floodwater mosquitoes that are tolerant to or have developed resistance to conventional insecticides [144]. As discussed in Section V, several other juvenoids like *24–28* (Table 2), *55* (Table 4), *66* (Table 5), and *99–105* (Table 7) have excellent activity against mosquito species, but none of these compound have been commercialized for this application.

Another good target for juvenoids is the control of a number of fly pests of livestock by incorporation of the compounds into the food or mineral supplements fed to the animals. Several of these fly species develop entirely in the manure of the animal and, thus, this feed-through treatment brings the larvae into direct contact with the juvenoid. Methoprene is highly active against dipterous species and a successful product was registered in May 1975, based on a feed-through application in cattle, for the control of the horn fly, *Haematobia irritans*. A mineral block, containing 0.02% methoprene, fed to cattle completely inhibits the emergence of adult horn flies from the feces [146]. More recently, the Inhibitor™ sustained-release bolus has been developed for beef and dairy cattle. It provides sustained release of methoprene sufficient to prevent the emergence of adult horn flies from the manure of treated cattle for up to 6.5 months [147, 148].

The extremely low mammalian toxicity of methoprene and the rapid and complete metabolism in the livestock to simple products (mostly acetate) are a great advantage in this type of feed-through treatment [149]. Methoprene released from sustained-release boluses administered to cattle is compatible with nontarget beneficial insects in the manure. No effect was observed on reproduction of two dung beetle species when they were reared on the manure from a steer that had received a methoprene bolus, and no effect was found on two manure-inhabiting insect predators of the horn fly [148].

A more difficult application of juvenoids is the control of housefly larvae in the manure of poultry. Administration of methoprene to poultry as a food additive does not give consistent fly control in the manure at economic rates. This is partly because the target is the less susceptible housefly, *Musca domestica*, and partly because the larvae migrate from the manure to pupation sites that are free of methoprene [60]. Pyriproxyfen (*28*) is considerably more effective than methoprene in controlling the housefly in the manure by feed-through treatment in poultry [117]. More stable juvenoids like pyriproxyfen may also be effective by direct application to the manure [116, 117]. Several other juvenoids including *22, 23, 26, 55, 64,* and *66* likewise have good activity against several dipterous species which develop in manure (e.g., [65]), but commercial development of these compounds for this application has not been pursued.

Methoprene was also developed commercially for the control of sciarid flies, like *Lycoriella auripila*, in mushroom culture. Incorporation of methoprene into the mushroom-growing medium prevents emergence of the adult flies [150]. Several juvenoids like methoprene and *66* are very active against *Simulium* spp., but practical control of these blackfly species under field conditions is difficult. These flies breed in rapidly moving water and a prolonged sustained release of the juvenoid would be required. Methoprene is also very active against a number of midge species and excellent control can be obtained under field conditions [151].

2. Household Pests—Fleas and Cockroaches

Another very good application for juvenoids is the control of fleas (Siphonaptera) in household situations. Methoprene (16) is extremely effective in preventing larval development and adult emergence of fleas. The LC_{50} of (S)-methoprene for preventing adult emergence of the oriental rat flea, *Xenopsylla cheopis*, is 0.00011 ppm, when incorporated into the larval-rearing medium [152, 153]. A higher concentration of methoprene is required to prevent cocoon formation. The cat flea, *Ctenocephalides felis*, which is the most frequently encountered species on dogs and cats in the United States, is also very sensitive to methoprene [154, 155]. Excellent residual effects with methoprene are observed in larval substrates such as carpets and animal bedding. The treatment of the fur of cats and dogs with methoprene at 2–10 mg/kg of body weight effectively inhibits the egg hatch of the cat flea [156]. This activity has led to a number of very successful flea products containing methoprene or (S)-methoprene, beginning with Precor™ IGR in 1980. When a quick knockdown of adult fleas is desired, an adulticide such as pyrethrins or permethrin is added to the juvenoid. However, the adulticide is not necessary for good control of the flea population. The ability of methoprene to move from one surface to another in household situations aids considerably in the effectiveness of the treatment. Hydroprene (18) is much less active than methoprene against fleas [152].

Fenoxycarb (25) is also very active in the larval-rearing medium in preventing larval development and emergence of adults of the oriental rat flea [153] and the cat flea [154]. However, it is not as active as methoprene against fleas. The juvenoid 25 is also an effective ovicide against eggs of the cat flea [157]. Commercial products containing fenoxycarb have been recently developed for domestic indoor flea control. Pyriproxyfen (28) is as effective as [S]-methoprene in flea control and is photostable enough to persist in the outdoor environment for at least 3 weeks. It shows promise for the control of the cat flea outdoors in home yards [158].

Hydroprene (18) is active at very low rates against the last instar nymphs of both sexes of cockroach species, producing sterile adults [159]. The exposed cockroaches molt into abnormal adults with twisted wings and other deformities. The sterility is irreversible, even after very brief exposure to hydroprene. The sensitive period with the

German cockroach, *Blattella germanica*, is the early part of the last nymphal instar, approximately 5–17 days before the molt to the adult. Therefore, treatment at this stage with hydroprene eliminates the reproductive potential, and the population declines as the abnormal sterile adultoids die. The sterility appears to be due to defective metamorphosis of the internal reproductive organs of both sexes. Although this is a slow control method, long-term cockroach control with conventional insecticides is rarely successful. To achieve a more rapid initial population reduction, a conventional insecticide like a pyrethroid is applied simultaneously with the juvenoid. Combination products containing hydroprene and an insecticide have been marketed since 1984 (Gencor™ IGR) for the long-term control of cockroach populations [160]. Retreatment on a schedule of 4–6-month intervals gives the best results. The excellent efficacy of hydroprene in this application is due to its very high intrinsic activity, nonrepellency, and volatility. The volatility of hydroprene allows it to translocate between many household surfaces and penetrate otherwise inaccessible cockroach harborages [159, 161]. Interestingly, methoprene is much less active than hydroprene against the German cockroach [159].

Fenoxycarb (*25*) is very active in sterilizing cockroaches during the last nymphal stadium and has also been commercialized for the control of cockroaches in indoor applications [162]. In contrast to hydroprene, fenoxycarb at high doses produces some mortality at the molt of younger nymphs of *Blattella germanica* [163] and has ovicidal activity when applied to young oothecae [164]. However, fenoxycarb produces these effects at much higher doses than that required for the sterilization of *B. germanica*. Although fenoxycarb is superior to hydroprene when scored for morphogenetic inhibition of cockroaches, the reverse is true for reproductive inhibition. In all of our large-scale chamber and apartment tests against cockroaches conducted at Zoecon Corporation and Sandoz Agro, Inc., hydroprene is much more effective in sterilization and population reduction than less-volatile compounds such as fenoxycarb. The lower volatility of fenoxycarb requires thorough area treatment. In indoor situations, the volatility of hydroprene (and methoprene in flea control) leads to rapid redistribution of the active ingredient, resulting in superior efficacy. Translocation effects with fenoxycarb are insignificant, whereas hydroprene has considerable sterilization activity against *B. germanica* through volatilization from all treated surfaces when applied at recommended use rates [159, 161].

The analog *88* is the most active of all juvenoids tested against *B. germanica* in our laboratory bioassay, based on both morphogenetic and reproductive inhibition, being 10 times more active than fenoxycarb and (*S*)-hydroprene [57]. However, in full-scale chamber tests it is inferior to hydroprene in population control. As with fenoxycarb, this is presumably due to the lower volatility of *88* compared to hydroprene and on its less complete effect on inhibition of reproduction. Pyriproxyfen (*28*) also has good activity against *B. germanica* [165] but is not as effective as hydroprene in the control of cockroach populations in apartments [166].

3. Stored Products

Insect infestations cause considerable losses in many parts of the world in postharvest storage [167]. Juvenoids offer considerable promise for the long-term control of stored product insect pests in confined environments, especially when continuous reinfestation by large numbers of untreated insects is not a problem. Treatment with juvenoids prevents the development of damaging populations by eliminating reproduction of the small number of insects originally infesting the stored product. Methoprene and hydroprene are highly effective against many of the lepidopterous and coleopterous pests of stored commodities when added to these products [167, 168]. These juvenoids remain effective for several years in stored grain and grain products.

The first successful commercial application of a juvenoid to a stored product was the use of methoprene for the protection of stored tobacco from damage by the larvae of both the cigarette beetle, *Lasioderma serricorne*, and the tobacco moth, *Ephestia elutella*. A single treatment of 10 ppm of racemic methoprene (Kabat™ tobacco protector) applied to tobacco immediately prior to compaction into storage containers gives protection for up to 4 years [169]. This treatment is completely effective in preventing adult emergence and eliminates the need for fogging and fumigation with toxic agents such as phosphine.

Coleopterous species are not uniform in their response to juvenoids. A problem in the application of these compounds for the protection of some stored products has been the relative ineffectiveness of many juvenoids against beetles of the genus *Sitophilus* (grain weevils). For example, the rice weevil, *Sitophilus oryzae*, and the granary weevil, *S. granarius*, are not controlled at economical rates by the application of most juvenoids. The relatively low activity against these species in whole grain is due partly to the poor penetration of some juvenoids into the grain kernels in which the developing larvae feed [170] and, more importantly, to the lower intrinsic activity of most juvenoids against *Sitophilus* spp. [14]. Fenoxycarb (*25*) has considerable potential as a protectant of stored grain. This juvenoid is more effec-

tive than methoprene and most other juvenoids against *Sitophilus* spp.
Fenoxycarb gives outstanding long-term protection against coleopterous
pests in stored wheat, although the control of *S. granarius* requires higher rates than for other beetle pests of stored grain [171]. It also has a
long half-life in grain. This compound likewise gives excellent control
of both lepidopterous and coleopterous species in stored rough rice
[172]. The juvenoid *24* shows promise in inhibiting the development of
damaging populations of *Tribolium confusum* in stored products [173].

Even though *Sitophilus* spp. are relatively insensitive to
methoprene, the lesser grain borer, *Rhyzopertha dominica*, is very sensitive to this juvenoid [174]. Methoprene is much more active than
fenoxycarb against *R. dominica* [175]. At application rates of 1–10 ppm,
methoprene gives complete control of *R. dominica* in stored wheat, barley, and corn for more than 12 months [174]. This very high activity
of methoprene, in preventing the production of progeny in the lesser
grain borer, complements the activity of conventional organophosphorus insecticides against other members of the pest complex in cereal
grains. Thus, combinations of methoprene and fenitrothion are effective against the entire pest complex in wheat. Methoprene is sold (as
Diacon™ IGR) for the control of *R. dominica* and the sawtoothed grain
beetle, *Oryzaephilus surinamensis*, including organophosphorus resistant
strains [176], especially in stored wheat. Methoprene is also used commercially as an effective protectant of many stored commodities, including grains, nuts, spices, dried fruit, and oil seeds. The remarkable absence of mammalian toxicity and residue problems with methoprene
provides a wide safety margin in these applications.

4. Social Insects

The action of juvenoids on colonies of social insect pests has been extensively investigated. The absence of repellency of juvenoids and their
slow mode of action enables them to be used in bait formulations. The
baits are taken into the colonies by the foraging workers and the
juvenoid is redistributed, allowing it to disrupt a number of processes
that are essential for survival of the colony. The mode of action is complex and may include effects on morphogenesis, brood care, fertility
and mortality of queens, and especially caste differentiation. These effects can lead to the slow decline and eventual death of the colony [14,
16].

The application of baits enables juvenoids to be used for the control of several ant species. An example is the use of methoprene
(Pharorid™ ant growth regulator) to eradicate colonies of the Pharoah

ant, *Monomorium pharaonis*, in indoor situations like hospitals and commercial buildings. Methoprene acts by interfering with the development of immature ants in the colony and by sterilizing the egg-laying queens. Complete control takes up to 20 weeks [177]. The very low mammalian toxicity and the nonrepellancy of the methoprene baits are important characteristics for this application. Many other ant species, such as the Argentine ant, *Iridomyrmex humilis*, frequently change food preferences; this makes development of a consistently attractive bait formulation difficult.

A number of juvenoids, including *22, 25, 28, 69,* and *90* have been shown to be effective in disrupting colonies of the red and black imported fire ants, *Solenopsis invicta* and *S. richteri*, leading to the eventual death of the colonies. The effective juvenoids cause death of the developing larvae, reduction or cessation of egg production by the queen through degeneration of the ovaries, and a shift in caste differentiation from worker to sexual forms. Death of the colony results from the lack of worker replacement coupled with natural mortality of the existing worker ants [67, 178]. The juvenoids Pro-drone™ (*22*) and fenoxycarb (*25*) (Logic™ fire ant bait) have been commercialized for this application. Fenoxycarb is one of the most effective juvenoids for the areawide management of *S. invicta* [178, 179]. Baits containing pyriproxyfen (*28*) have been found to be equal in effectiveness to those containing fenoxycarb in suppressing field populations of *S. invicta* [180].

Juvenoids have potential for long-term control of subterranean termite populations. The major effect of application of a juvenoid is usually the destabilization of the termite colony by the formation of excess soldiers and presoldiers from pseudergates normally destined to become workers. Considerable research still needs to be carried out on delivery systems (impregnation of structural wood, baiting, etc.) and assessing the long-term efficacy under practical conditions. Fenoxycarb has considerable activity against the subterranean termite species, *Coptotermes formosanus* and *Reticulitermes virginicus*, when applied using small blocks of decayed wood impregnated with the juvenoid [181]. Large numbers of superfluous presoldiers and/or nonfunctional soldier–worker intercastes are produced, and morphological abnormalities are observed in several stages of the termites. Methoprene shows considerable potential for use in the bait-block technique for the control of *C. formosanus* [182]. This juvenoid causes defaunation and starvation, induction of presoldier and intercaste development, and ecdysis inhibition. Pyriproxyfen does not show promising activity against *C. formosanus* [182].

5. Sericulture

A minor commercial application of juvenoids is in sericulture to improve the yield of silk. Carefully timed administration of methoprene (Manta™ IGR) at the beginning of the fifth instar larvae of the silkworm, *Bombyx mori*, produces giant larvae. This results in a substantial increase in the size of the cocoon and the length of the cocoon filament. Silk production can be increased as much as 5–15%, depending on the dose and the timing of the application, with only minimal increase in food consumption [183, 184].

B. Agrochemical Markets

1. Glasshouse Pests

Insect species with short developmental cycles and high reproductive potential, like many homopterous species, are excellent targets for juvenoids, especially in confined environments such as greenhouses. Economic damage often requires a population buildup through several generations, and many of the life stages are well exposed and susceptible to the effects of juvenoids. Kinoprene (*17*) shows considerable activity against many homopterous species [42, 185–187]. This juvenoid was commercialized in 1975 for the control of aphids and whiteflies in greenhouses on ornamental plants and vegetable seed crops. Application of kinoprene results in a gradual reduction of the pest population due to its ovicidal, morphogenetic, and sterilizing effects on these insect species. The penultimate instar of homopterous species is the most sensitive to the morphogenetic effects of kinoprene, but this juvenoid is unusual in that at higher rates (2.5 times) it produces direct mortality at all developmental stages [185–187]. Kinoprene is relatively specific for homopterous species (Table 2) and has little effect on most beneficial predators and parasites. Kinoprene lacks sufficient foliar stability for most field applications. Methoprene and hydroprene also show excellent activity against many homopterous insect species, but they have not been registered for this application. The juvenoid *78* has high activity on many homopterous species [100], but this compound was not developed commercially. On the green peach aphid, *Myzus persicae*, hydroprene is more potent than *78* [99]. Not all aphid species are sensitive to the action of juvenoids [188]. The effects of juvenoids on reproduction of aphids can result from direct action on the development of the embryo inside the adult female or from incomplete metamorphosis of the genital pore [14].

2. Field and Orchard Crops

Most major agricultural insect pests are not good targets for control by juvenoids. The important lepidopterous species do their damage in the larval stages, are present in the field for only one to three generations per season, and build their populations rapidly. To be effective, an insecticide must be active against the eggs and/or early larval stages and must have a rapid action to prevent the development of the later, more damaging larval stages. Although there are possibilities for the control of some of these lepidopterous species, such as *Heliothis virescens*, by the ovicidal effects of photostable juvenoids, relatively little field research has been carried out in this area [143]. One may need to treat large areas to limit reinfestation.

Juvenoids such as epofenonane (*20*), CGA 045128 (*24*), fenoxycarb (*25*), and pyriproxyfen (*28*) possess sufficient foliar stability and activity to be useful against multivoltine lepidopterous species in orchard and vine crops, especially when damage produced by the first generation is not a major problem. They also have considerable potential for the control of many homopterous species, such as scale insects, mealybugs, psyllids, and whiteflies. Homopterous species have multiple generations per season and often several stages of development are very sensitive to juvenoids. Scale insects are an ideal target for juvenoids due to their lack of mobility and the delayed damage to the host plant.

Epofenonane (*20*) was extensively evaluated against lepidopterous and homopterous species, and although good field results were obtained [61–64], this compound was not developed further. However, since 1985, fenoxycarb (*25*) has been commercialized (Insegar™) for a number of field applications against lepidopterous species attacking fruit trees and vines. Fenoxycarb has very high morphogenetic activity against the summerfruit tortrix moth, *Adoxophyes reticulana*, being considerably more active than epofenonane [69]. Field tests in apple orchards, against *Adoxophyes orana* and other tortricid larvae, demonstrate that applications to last larval instars of the first generation in spring give year-long control of the leafroller complex [189]. In addition to the morphogenetic effects against these multivoltine lepidopterous species, fenoxycarb shows excellent ovicidal activity on freshly laid eggs of the codling moth, *Cydia pomonella*, the plum fruit moth, *Grapholita funebrana*, the grape berry moth, *Lobesia botrana*, and the vine moth, *Eupoecilia ambiguella* [190–192]. The ovicidal activity against eggs of *Adoxophyes orana* is poor [190]. Thus, both the morphogenetic and ovicidal effects of fenoxycarb are useful in the control of lepidopterous pests in orchards. Carefully timed applications are necessary for good efficacy [191]. For example, a single application of fenoxycarb in plum orchards at the start of the second flight reduced damage from larvae of *G. funebrana* below the tolerance level via the ovicidal action of the juvenoid [193]. Fenoxycarb is used in integrated pest-management programs in orchards because of its selectivity and very low mammalian toxicity. Fenoxycarb shows high ovicidal activity against freshly laid eggs of the spruce budworm, *Choristoneura fumiferana* (Tortricidae), and fifth instar larvae display lethal morphogenetic effects. However, the most sensitive stage is the adult female; treated females lay eggs that fail to hatch. Untreated adult females that mate with treated males also lay infertile eggs. Thus, this juvenoid has considerable potential to control field populations of forest pests such as *C. fumiferana* [194]. In forestry, there is no necessity for the immediate elimination of the feeding stages of the insect pest, and overall population reduction may be sufficient in many cases.

Fenoxycarb has also been commercialized for the control of a number of homopterous pests, especially certain scale species attacking olive, citrus, and other fruit trees. It is used for the control of soft scales such as the black scale, *Saissetia oleae*, and the Florida wax scale, *Ceroplastes floridensis*, which are major pests of citrus and olive [195]. Although fenoxycarb is very effective in controlling soft scales, it is less active against armored scales like the California red scale, *Aonidiella aurantii*. It has little or no effect on some beneficial parasites of soft scales [196], but it does harm predatory beetles to some extent [197]. Fenoxycarb is suitable for integrated pest management in pears for control of the pear psylla, *Cacopsylla pyricola*, without causing damaging effects on larval or adult anthocorids [198].

The juvenoid CGA 045128 (24) has similar morphogenetic and ovicidal activities to fenoxycarb against orchard lepidopterous pests [192]. Pyriproxyfen (28) has good ovicidal activity against newly laid eggs of the codling moth, *Cydia pomonella* [121] and the Egyptian cotton leafworm, *Spodoptera littoralis* [199]. As with other juvenoids, the response is much poorer with 1–2-day-old eggs, and even older eggs are insensitive. This juvenoid is very effective against many homopterous species. It has high ovicidal activity against the sweetpotato whitefly, *Bemisia tabaci*, in both preinfestation and postinfestation treatments [119, 199]. Male and female first instar nymphs of the California red scale, *Aonidiella aurantii*, are highly susceptible to pyriproxyfen [119, 197]. This juvenoid is more active than methoprene or fenoxycarb in its ability to kill or sterilize first and second instar nymphs of *A. aurantii*. Pyriproxyfen is also very active against the Florida wax scale, *Ceroplastes floridensis*. It controls the early nymphal stages and the young females [119, 197]. Therefore, in contrast to fenoxycarb, pyriproxyfen controls both armored and soft-scale pests of citrus. It has no adverse effect on *Aphytis holoxanthus*, a beneficial ectoparasite of some armored-scale insects [119]. However, pyriproxyfen has harmful effects on some beneficial neuropterous species and predatory beetles [197]. Pyriproxyfen has very strong sterilizing activity against the green peach aphid, *Myzus persicae*, and is effective in controlling this pest in field tests [119]. Pyriproxyfen has no effect on the larvae of *Thrips palmi* and only controls this pest at the pupal stage. However, the juvenoid has little or no effect on the hemipteran predatory *Orius* sp. In spray tests in which both pest and predator are present, the population of *T. palmi* decreases rapidly due to the selective action of pyriproxyfen. Therefore, it is possible to control this pest species by only affecting metamorphosis at the pupal stage in the presence of this predator [200]. Although pyriproxyfen shows promise for controlling a number of agricultural insect pests, it has not yet been commercialized for such applications.

VII. TOXICITY TO NONTARGETS

Juvenoids generally have low toxicity against nontarget organisms. All of the commercial juvenoids are remarkably nontoxic to vertebrates. Both methoprene and hydroprene have acute oral LD_{50} values for rats of >34,600 mg/kg. Fenoxycarb has an acute oral LD_{50} for rats of >10,000 mg/kg and the value for pyriproxyfen is >5000 mg/kg [201]. The acute percutaneous LD_{50} values for rats for all of these compounds are >2000 mg/kg [201]. These juvenoids pose a very low hazard to wildlife, including birds, fish, and bees. They do not bioaccumulate in fish and do not persist in the environment. The environmental fate of methoprene has been thoroughly investigated. It undergoes rapid degradation and metabolism to innocuous products in plants, animals, aquatic microorganisms, and soil microbes [149, 202, 203]. This juvenoid, in particular, shows negligible harmful effects on the environment and is one of the safest materials known for use in insect control. Even the photostable juvenoids dissipate fairly rapidly in soil and on plants.

In most applications there is a need to consider the effects of these juvenoids on nontarget species, especially important beneficial entomophagous arthropods. A considerable advantage of juvenoids lies in their stage-specific mode of action that renders them safe for all insects that are not in a sensitive developmental stage. In addition, juvenoids often show very high selectivity among different taxonomic groups of insects. The combination of stage and species specificity frequently allows selective control of insect pests without harmful effects on beneficial predatory and parasitic arthropods. However, in some cases, the beneficial insect species are adversely affected. Some juvenoids have sufficient selectivity to allow their use in integrated pest-management programs. Although topical applications of juvenoids to last instar larvae of the honey bee, *Apis mellifera*, can produce deformities and mortality, field applications cause no measurable effects on brood development or worker bee population density in *A. mellifera* colonies [10, 204–206].

Due to the application of methoprene as a mosquito larvicide, detailed studies have been carried out on the effects of this juvenoid on nontarget aquatic organisms. Methoprene has negligible effects on most nontarget species [145, 207, 208]. Any observed effects are usually transitory. However, under certain conditions, methoprene may be detrimental to the development and survival of some crab species [208]. Applications to coastal marshland and freshwater environments can have harmful effects on small crustaceans, but recovery of the population is rapid [209]. Interestingly, hydroprene causes premature metamorphosis of the cyprid larvae of the acorn barnacle, *Balanus galeatus*, at very low concentrations in sea water [210]. In this barnacle, the act of settling triggers the hormonally controlled metamorphosis of the larvae. Treatment with hydroprene induces metamorphosis to the adults before the larvae can attach themselves to a substrate, and these adults are unable to feed and subsequently die. Methoprene is *not* active against *B. galeatus*. Fenoxycarb and pyriproxyfen also have a good margin of safety to fish, wildlife, and dominant macroinvertebrates prevailing in aquatic mosquito breeding sites [211–213]. However, fenoxycarb is more harmful than methoprene to some nontarget organisms in these aquatic habitats [214, 215]. As discussed in Section VIA, methoprene is compatible with nontarget beneficial insect species in feed-through applications in cattle to control the horn fly breeding in the manure [148].

An area that has been extensively studied is the effect of juvenoids on beneficial predatory and parasitic arthropods in integrated pest-management programs. Although epofenonane (*20*) was not developed commercially, it showed considerable promise for selective use in orchards. This juvenoid has a relatively broad spectrum of activity but is only moderately active against dipterous species. This selectivity results in the juvenoid being safe for most beneficial insect species belonging to the orders Diptera and Hymenoptera [61, 205]. Negligible adverse effects were observed with epofenonane in tests against several predators and parasites of important lepidopterous and homopterous pests in several crops [63, 205, 216]. Likewise, the juvenoids CGA 13353 (*78*) and CGA 34301 (*79*) are selective against some lepidopterous and/or homopterous pests. They have little or no harmful effects on beneficial insect species and show promise for use in integrated control programs in greenhouse and field applications [100, 102]. Both kinoprene (*17*) and hydroprene (*18*) can have adverse effects on hymenopterous parasitoids of aphid and scale insect species [217, 218]. Although hydroprene is more active than CGA 13353 (78) against *Myzus persicae*, *78* is less harmful than hydroprene against the egg and last larval instar of the aphid's predator, *Aphidoletes aphidimyza* [99].

Fenoxycarb (*25*) has adverse effects on the development of some parasitoids and predators belonging to the orders Diptera, Coleoptera, and Neuroptera. For example, fenoxycarb produces high mortality of parasitoid tachinid fly larvae [219]. It also shows highly deleterious morphogenetic effects on the common predators associated with the Egyptian cotton leafworm, *Spodoptera littoralis*, such as the larvae of the ladybird beetle, *Coccinella undecimpunctata*, and the aphid lion, *Chrysopa vulgaris* [220]. In integrated pest-management programs against soft-scale insect species in citrus, both fenoxycarb and pyriproxyfen do not affect parasitic wasps but are moderately harmful to predatory beetles and neuropterous species [119, 196, 197]. The analog *83* shows good control of several homopterous species but has harmful effects against some important parasitoids and predators [104]. Even in these cases where harmful effects are observed with juvenoids on beneficial arthropods, the effects are usually much less severe than those seen with conventional broad-spectrum insecticides.

VIII. INSECT RESISTANCE

Although Williams suggested that insects would be unable to develop resistance to their own juvenile hormone [5], there is little reason not to expect the development of resistance to juvenoids under intense selection pressure. Juvenoids are metabolized in insects by the same oxidative, hydrolytic, and other inactivation processes that degrade other foreign compounds such as conventional insecticides. Furthermore, insects regulate the titre of their natural JHs in the hemolymph partly by metabolism. Thus, it is hardly surprising that insect species of several orders have shown cross-resistance to juvenoids and that resistance to these compounds can be induced under laboratory conditions with constant selection pressure [221].

Several insecticide-resistant strains of the housefly, *Musca domestica*, show various degrees of cross-resistance to methoprene (*16*), hydroprene (*18*), *66*, and several other juvenoids in laboratory tests [222, 223]. The enhanced capacity of the larvae of the insecticide-resistant strains to metabolize the juvenoids via microsomal mixed-function oxidases, appears to be the most important mechanism for this cross-resistance [222, 223]. An insecticide-resistant strain of the mosquito, *Anopheles gambiae*, also shows cross-resistance to methoprene [224].

Resistance to methoprene in the northern house mosquito, *Culex pipiens pipiens*, may be induced by selection pressure under laboratory conditions [225]. This development of resistance can be accompanied by a reduction in reproductive potential of *C. pipiens pipiens* [225]. Enhanced oxidative metabolism of methoprene is an important mechanism in the resistant strains, but other factors such as reduced uptake may contribute to the resistance [226]. Selection pressure with hydroprene induces resistance in laboratory colonies of *Tribolium confusum* [225]. The *Met* mutant of *Drosophila melanogaster* is highly resistant to JH III and methoprene. With JH III, the resistance appears to be due to reduced binding affinity to a cytosolic binding protein in the JH target tissue. This is the first reported example of a JH-resistance mechanism involving decreased target-site sensitivity [227]. This resistant strain is at a large competitive disadvantage to susceptible strains. In the absence of methoprene selection pressure, the *Met* mutant flies were noncompetitive with a wild-type strain [228].

Despite these various laboratory results, resistance to juvenoids has not yet developed in any application. Although this could be partly due to the limited use of juvenoids, most of these compounds have low environmental persistance and only specific stages in the insect's development are sensitive to the action of juvenoids. Because not all of the insects are sensitive at the same time in nonsynchronous field populations, these compounds allow preservation of susceptible genotypes in the pest population. Due to the resulting low selection pressure, resistance is unlikely to develop rapidly to methoprene and most other juvenoids in many of their applications.

IX. CONCLUSION

Juvenoids were originally investigated with the aim of developing environmentally sound and selective insecticides based on the disruption of the endocrine system of insects. The hope that juvenoids would become a major class of commercial insecticides has not materialized. However, a number of juvenoids with high biological activity have been discovered. Several of them, namely, methoprene, hydroprene, kinoprene, fenoxycarb, and pyriproxyfen have found commercial applications for the control of mosquitoes, flies, ants, fleas, cockroaches, and certain stored product and greenhouse pest. In addition, fenoxycarb is used in integrated pest control programs in orchard and vine crops. Although limited in their use patterns, due to their insect species selectivity and stage-specific slow mode of action, these juvenoids are very effective in controlling certain insect pests while being remarkably safe for humans and almost all other nontarget organisms.

ACKNOWLEDGMENTS

I would like to thank Gerardus B. Staal and David C. Cerf for the bioassay results presented in the tables and William L. Collibee for drawing the structures.

REFERENCES

1. S. Kopeć, *Bull. Int. Acad. Sci. Cracovie*, B:57 (1917).
2. S. Kopeć, *Biol. Bull.*, 42:323 (1922).
3. C.M. Williams, *Biol. Bull.*, 93:89 (1947); C.M. Williams, *Biol. Bull.*, 94:60 (1948); C.M. Williams, *Biol. Bull.*, 103:120 (1952).
4. V.B. Wigglesworth, *J. Exp. Biol.*, 17:201 (1940).
5. C.M. Williams, *Nature (Lond.)*, 178:212 (1956).
6. H. Röller, K.H. Dahm, C.C. Sweely, and B.M. Trost, *Angew. Chem., Int. Ed. Engl.*, 6:179 (1967).
7. C.M. Williams, *Sci. Am.*, 217:13 (1967).
8. W.E. Bollenbacher and N.A. Granger, *Comprehensive Insect Physiology Biochemistry and Pharmacology* (G.A. Kerkut and L.I. Gilbert, eds.), Vol. 7, Pergamon Press, Oxford, p. 109 (1985).
9. J. Koolman, *Insect Biochem.*, 12:225 (1982).
10. A. Retnakaran, J. Granett, and T. Ennis, *Comprehensive Insect Physiology Biochemistry and Pharmacology* (G.A. Kerkut and L.I. Gilbert, eds.), Vol. 12, Pergamon Press, Oxford, p. 529 (1985).
11. A. KrishnaKumaran, *Morphogenetic Hormones of Arthropods* (A.P. Gupta, ed.), Vol. 1, Rutgers University Press, New Brunswick, NJ, p. 182 (1990).
12. S. Sakurai, J.T. Warren, and L.I. Gilbert, *Arch. Insect Biochem. Physiol.*, 18:13 (1991).
13. K. Sláma, M. Romaňuk, and F. Šorm, *Insect Hormones and Bioanalogues*, Springer-Verlag, New York (1974).
14. G.B. Staal, *Ann. Rev. Entomol.*, 20:417 (1975).
15. F. Sehnal, *The Juvenile Hormones* (L.I. Gilbert, ed.), Plenum, New York, p. 301 (1976).
16. J.P. Edwards and J.J. Menn, *Chemie der Pflanzenschutz- und Schädlingsbekämpfungsmittel* (R. Wegler, ed.), Vol. 6, Springer-Verlag, Berlin, p. 185 (1981).

17. R. Feyereisen, *Comprehensive Insect Physiology Biochemistry and Pharmacology* (G.A. Kerkut and L.I. Gilbert, eds.), Vol. 7, Pergamon Press, Oxford, p. 391 (1985).
18. W.G. Goodman, *Morphogenetic Hormones of Arthropods* (A.P. Gupta, ed.), Vol. 1, Rutgers University Press, New Brunswick, NJ, p. 83 (1990).
19. H. Kataoka, A. Toschi, J.P. Li, R.L. Carney, D.A. Schooley, and S.J. Kramer, *Science*, 243:1481 (1989).
20. A.P. Woodhead, B. Stay, S.L. Seidel, M.A. Khan, and S.S. Tobe, *Proc. Natl. Acad. Sci.* 86:5997 (1989).
21. G.E. Pratt, D.E. Farnsworth, and R. Feyereisen, *Mol. Cell. Endocrin.*, 70:185 (1990).
22. G. Bhaskaran, K.H. Dahm, P. Barrera, J.L. Pacheco, K.E. Peck, and M. Muszynska-Pytel, *Gen. Comp. Endocrin.*, 78:123 (1990).
23. F.C. Baker, L.W. Tsai, C.C. Reuter, and D.A. Schooley, *Insect Biochem.*, 17:989 (1987).
24. F.C. Baker, *Morphogenetic Hormones of Arthropods* (A.P. Gupta, ed.), Rutgers University Press, New Brunswick, NJ, p. 389 (1990).
25. W. Vogel., P. Masner, O. Graf, and S. Dorn, *Experientia*, 35:1254 (1979).
26. H. Röller and K.H. Dahm, *Recent Prog. Horm. Res.*, 24:651 (1968).
27. A.S. Meyer, H.A. Schneiderman, E. Hanzmann, and J.H. Ko, *Proc. Natl. Acad. Sci.*, 60:853 (1968).
28. K.J. Judy, D.A. Schooley, L.L. Dunham, M.S. Hall, B.J. Bergot, and J.B. Siddall, *Proc. Natl. Acad. Sci.*, 70:1509 (1973).
29. B.J. Bergot, G.C. Jamieson, M.A. Ratcliff, and D.A. Schooley, *Science*, 210:336 (1980).
30. R.J. Anderson, V.L. Corbin, G. Cotterrell, G.R. Cox, C.A. Henrick, F. Schaub, and J.B. Siddall, *J. Am. Chem. Soc.*, 97:1197 (1975).
31. B.J. Bergot, F.C. Baker, D.C. Cerf, G. Jamieson, and D.A. Schooley, *Juvenile Hormone Biochemistry* (G.E. Pratt and G.T. Brooks, eds.), Elsevier, Amsterdam, p. 33 (1981).
32. D.A. Schooley, *Analytical Biochemistry of Insects* (R.B. Turner, ed.), Elsevier, Amsterdam, p. 241 (1977).
33. D.A. Schooley, F.C. Baker, L.W. Tsai, C.A. Miller, and G.C. Jamieson, *Biosynthesis, Metabolism and Mode of Action of Invertebrate Hormones* (J. Hoffmann and M. Porchet, eds.), Springer-Verlag, Berlin, p. 373 (1984).
34. D.A. Schooley and F.C. Baker, *Comprehensive Insect Physiology Biochemistry and Pharmacology* (G.A. Kerkut and L.I. Gilbert, eds.), Vol. 7, Pergamon Press, Oxford, p. 363 (1985).
35. E. Brüning, A. Saxer, and B. Lanzrein, *Int. J. Invertebr. Reprod. Dev.*, 8:269 (1985); E. Brüning and B. Lanzrein, *Int. J. Invertebr. Reprod. Dev.*, 12:29 (1987).
36. H. Laufer, D. Borst, F.C. Baker, C. Carrasco, M. Sinkus, C.C. Reuter, L.W. Tsai, and D.A. Schooley, *Science*, 235:202 (1987).
37. Y.C. Toong, D.A. Schooley, and F.C. Baker, *Nature (Lond.)*, 333:170 (1988).
38. D.S. Richard, S.W. Applebaum, T.J. Sliter, F.C. Baker, D.A. Schooley, C.C. Reuter, V.C. Henrich, and L.I. Gilbert, *Proc. Natl. Acad. Sci.*, 86:1421

(1989); D.S. Richard and L.I. Gilbert, *Experientia, 47*:1063 (1991); H. Duve, A. Thorpe, K.J. Yagi, C.G. Yu, and S.S. Tobe, *J. Insect Physiol., 38*:575 (1992).

39. D. J. Faulkner and M.R. Petersen, *J. Am. Chem. Soc., 93*:3766 (1971); A.S. Meyer, E. Hanzmann, and R.C. Murphy, *Proc. Natl. Acad. Sci., 68*:2312 (1971); K. Nakanishi, D.A. Schooley, M. Koreeda, and J. Dillon, *Chem. Commun.*, 1235 (1971); D.J. Faulkner and M.R. Petersen, *J. Am. Chem. Soc., 95*:553 (1973).

40. T. Koyama, K. Ogura, F.C. Baker, G.C. Jamieson, and D.A. Schooley, *J. Am. Chem. Soc., 109*:2853 (1987).

41. K. Mori and M. Fujiwhara, *Israel J. Chem., 31*:223 (1991); and references cited therein.

42. C.A. Henrick, W.E. Willy, and G.B. Staal, *J. Agric. Food Chem., 24*:207 (1976).

43. C.A. Henrick, G.B. Staal, and J.B. Siddall, *J. Agric. Food Chem. 21*:354 (1973).

44. C.A. Henrick, W.E. Willy, B.A. Garcia, and G.B. Staal, *J. Agric. Food Chem., 23*:396 (1975).

45. P. Schmialek, *Z. Naturforschg. B: Anorg. Chem., Org. Chem., Biochem. Biophys., Biol., 16*:461 (1961).

46. P. Schmialek, *Z. Naturforschg. B: Anorg Chem., Org. Chem., Biochem., Biophys., Biol., 18*:516 (1963).

47. A. Krishnakumaran and H.A. Schneiderman, *J. Insect Physiol., 11*:1517 (1965); P.A. Cruickshank, *Mitt. Schweiz. Entomol. Ges., 44*:97 (1971).

48. W.S. Bowers, M.J. Thompson, and E.C. Uebel, *Life Sci., 4*:2323 (1965).

49. J.H. Law, C. Yuan, and C.M. Williams, *Proc. Natl. Acad. Sci., 55*:576 (1966).

50. M. Romaňuk, K. Sláma, and F. Šorm, *Proc. Natl. Acad. Sci., 57*:349 (1967).

51. K. Sláma and C.M. Williams, *Proc. Natl. Acad. Sci., 54*:411 (1965).

52. W.S. Bowers, H.M. Fales, M.J. Thompson, and E.C. Uebel, *Science, 154*:1020 (1966).

53. J.F. Manville, *Can. J. Chem., 53*:1579 (1975); J.F. Manville, *Can. J. Chem., 54*:2365 (1976); J.F. Manville and C.D. Kriz, *Can. J. Chem., 55*:2547 (1977); J.F. Manville, L. Greguss, K. Sláma, and E. von Rudloff, *Collect. Czech. Chem. Commun., 42*:3658 (1977).

54. P. Bhan, B.S. Pande, R. Soman, N.P. Damodaran, and S. Dev, *Tetrahedron, 40*:2961 (1984); K. Kawai, C. Takahashi, A. Numata, S. Chernysh, A. Nesin, and H. Numata, *Appl. Entomol. Zool., 28*:118 (1993).

55. W.S. Bowers, *Science, 164*:323 (1969).

56. C.A. Henrick, *Insecticide Mode of Action* (J.R. Coats, ed.), Academic Press, New York, p. 315 (1982).

57. C.A. Henrick, *Insect Chemical Ecology* (I. Hrdý, ed.), Academia Praha and SPB Academic Publishing BV, The Hague, Netherlands, p. 429 (1991).

58. C. Djerassi, C. Shih-Coleman, and J. Diekman, *Science, 186*:596 (1974).

59. J.J. Menn, *J. Agric. Food Chem., 28*:2 (1980); J.J. Menn and C.A. Henrick, *Phil. Trans. R. Soc. Lond., B295*:57 (1981).

60. J.J. Menn, C.A. Henrick, and G.B. Staal, *Regulation of Insect Development and Behavior* (F. Sehnal, A. Zabża, J.J. Menn, and B. Cymborowski, eds.), Vol. 2, Wrocław Tech. Univ. Press, Wrocław, Poland p. 735 (1981).

61. R.C. Zurflüh, *The Juvenile Hormones* (L.I. Gilbert, ed.), Plenum, New York, p. 61 (1976).

62. W.W. Hangartner, M. Suchý, H.-K. Wipf, and R.C. Zurflüh, *J. Agric. Food Chem.*, *24*:169 (1976).

63. H. Schooneveld, J.P. Van Der Molen, and J. Wiebenga, *Entomol. Exp. Appl.*, *19*:227 (1976).

64. W. Vogel, P. Masner, and M.L. Frischknecht, *Mitt. Schweiz. Entomol. Ges.*, *49*:245 (1976).

65. M. Schwarz, R.W. Miller, J.E. Wright, W.F. Chamberlain, and D.E. Hopkins. *J. Econ. Entomol.*, *67*:598 (1974); J.E. Wright, G.E. Spates, and M. Schwarz, *J. Econ. Entomol.*, *69*:79 (1976).

66. R.E. Lowe, M. Schwarz, A.L. Cameron, and D.A. Dame, *Mosq. News*, *35*:561 (1975).

67. W.A. Banks, C.S. Lofgren, and J.K. Plumley, *J. Econ. Entomol.*, *71*:75 (1978).

68. F. Karrer and S. Farooq, *Regulation of Insect Development and Behavior* (F. Sehnal, A. Zabża, J.J. Menn, and B. Cymborowski, eds.), Vol. 1, Wrocław Tech. Univ. Press, Wrocław, Poland, p. 289 (1981).

69. P. Masner, S. Dorn, W. Vogel, M. Kälin, O. Graf, and E. Günthart, *Regulation of Insect Development and Behavior* (F. Sehnal, A. Zabża, J.J. Menn, and B. Cymborowski, eds.), Vol. 2, Wrocław Tech. Univ. Press, Wrocław, Poland, p. 809 (1981); S. Dorn, M.L. Frischknecht, V. Martinez, R. Zurflüh, and U. Fischer, *Z. Pflanzenkr. Pflanzenschutz*, *88*:269 (1981); S. Grenier and A.-M. Grenier, *Ann. Appl. Biol.*, *122*:369 (1993).

70. T. Ohsumi, M. Hatakoshi, H. Kisida, N. Matsuo, I. Nakayama, and N. Itaya, *Agric. Biol. Chem.*, *49*:3197 (1985).

71. S. Nishida, N. Matsuo, M. Hatakoshi, and H. Kisida, U.S. Patent 4,751,225 (1988).

72. R. Nishida, W.S. Bowers, and P.H. Evans, *J. Chem. Ecol.*, *10*:1435 (1984).

73. M. Jacobson, R.E. Redfern, and G.D. Mills, Jr., *Lloydia*, *38*:473 (1975); however, see M.P. Cooke, Jr., *J. Org. Chem.*, *44*:2461 (1979); F. Orsini, F. Pelizzoni, G. Sello, and G.B. Serini, *Gazz. Chim. Italiana*, *112*:277 (1982).

74. P. Bhan, R. Soman, and S. Dev, *Agric. Biol. Chem.*, *44*:1483 (1980).

75. R. Nishida, W.S. Bowers, and P.H. Evans, *Arch. Insect Biochem. Physiol.*, *1*:17 (1983).

76. C.A. Henrick, G.B. Staal, and J.B. Siddall, *The Juvenile Hormones* (L.I. Gilbert, ed.), Plenum, New York, p. 48 (1976).

77. K. Sláma, *Comprehensive Insect Physiology Biochemistry and Pharmacology* (G.A. Kerkut and L.I. Gilbert, eds.), Vol. 11, Pergamon Press, Oxford, p. 357 (1985).

78. Z. Wimmer and M. Romaňuk, *Collect. Czech. Chem. Commun.*, *54*:2302 (1989).

79. C.A. Henrick and G.B. Staal, *Stereoselectivity of Pesticides* (E.J. Ariëns, J.J.S. van Rensen, and W. Welling, eds.), Elsevier, Amsterdam, p. 303 (1988).
80. C.A. Henrick, J.N. Labovitz, V.L. Graves, and G.B. Staal, *Bioorg. Chem.*, *7*:235 (1978).
81. A. Niwa, H. Iwamura, Y. Nakagawa, and T. Fujita, *J. Agric. Food Chem.*, *38*:514 (1990).
82. T. Hayashi, H. Iwamura, and T. Fujita, *J. Agric. Food Chem.*, *38*:1965 (1990).
83. T. Hayashi, H. Iwamura, and T. Fujita, *J. Agric. Food Chem.*, *38*:1972 (1990).
84. T. Hayashi, H. Iwamura, and T. Fujita, *J. Agric. Food Chem.*, *39*:2029 (1991).
85. M. Hatakoshi, I. Nakayama, and L.M. Riddiford, *Appl. Ent. Zool.*, *22*:641 (1987).
86. L. Novák, J. Rohály, G. Gálik, J. Fekete, L. Varjas, and C. Szántay, *Liebigs Ann. Chem.*, 509 (1986).
87. V. Jarolím, *Regulation of Insect Development and Behavior* (F. Sehnal, A. Zabža, J.J. Menn, and B. Cymborowski, eds.), Vol. 1, Wrocław Tech. Univ. Press, Wrocław, Poland, p. 261 (1981).
88. F. Sehnal, M. Romaňuk, and L. Streinz, *Acta Entomol. Bohemoslov.*, *73*:1 (1976).
89. F. Sehnal, V. Skuhravý, R. Hochmut, and V. Landa, *Acta Entomol. Bohemoslov.*, *73*:373 (1976).
90. F. Sehnal, M.M. Metwally, and I. Gelbič, *Z. Angew. Entomol.*, *81*:85 (1976).
91. I. Ujváry, A. Kis-Tamás, L. Varjas, and L. Novák, *Acta Chimica Hungarica*, *113*:165 (1983).
92. F.M. Pallos, J.J. Menn, P.E. Letchworth, and J.B. Miaullis, *Nature* (Lond.), *232*:486 (1971).
93. C.A. Kontev, J. Žďárek, K. Sláma, F. Sehnal, and M. Romaňuk, *Acta Entomol. Bohemoslov.*, *70*:377 (1973).
94. R. Sarmiento, T.P. McGovern, M. Beroza, G.D. Mills Jr., and R.E. Redfern, *Science*, *179*:1342 (1973).
95. K.J. Kramer, H.E. McGregor, and K. Mori, *J. Agric. Food Chem.*, *27*:1215 (1979); Y. Ichikawa, M. Komatsu, T. Takigawa, K. Mori, and K.J. Kramer, *Agric. Biol. Chem.*, *44*:2709 (1980).
96. J.W. Patterson and M. Schwarz, *J. Insect Physiol.*, *23*:121 (1977).
97. R. Pinchin, A.M. De Oliveira Filho, M.J. Figueiredo, C.A. Muller, B. Gilbert, A.P. Szumlewicz, and W.W. Benson, *J. Econ. Entomol.*, *71*:950 (1978); A.M. Oliveira Filho, R. Pinchin, M.J. Figueiredo, and C.A. Muller, *Insect Sci. Applic.*, *5*:127 (1984).
98. J.F. Grove, R.C. Jennings, A.W. Johnson, and A.F. White, *Chem. Ind.* (London), 346 (1974).
99. F. El-Gayar, *Entomophaga*, *21*:297 (1976).
100. R. Scheurer, V. Flück, and M.A. Ruzette, *Mitt. Schweiz Entomol. Ges.*, *48*:315 (1975).

101. R. Scheurer and M.A. Ruzette, Z. Angew. Entomol., 77:218 (1974).
102. R. Scheurer, M.A. Ruzette, and V. Flück, Z. Angew. Entomol., 78:313 (1975).
103. C.H. Schaefer, W.H. Wilder, E.F. Dupras, Jr., and R.J. Stewart, J. Agric. Food Chem., 33:1045 (1985).
104. K. Novák, J. Zelený, and F. Sehnal, Insect Chemical Ecology (I. Hrdý, ed.), Academia Praha and SPB Academic Publishing BV, The Hague, Netherlands, p. 457 (1991); J. Kuldová, Z. Wimmer, and I. Hrdý, Insect Chemical Ecology (I. Hrdý, ed.), Academia Praha and SPB Academic Publishing BV, The Hague, Netherlands, p. 461 (1991).
105. H. Kisida, M. Hatakoshi, N. Itaya, and I. Nakayama, Agric. Biol. Chem., 48:2889 (1984).
106. M. Hatakoshi, H. Kisida, I. Fujimoto, N. Itaya, and I. Nakayama, Appl. Ent. Zool. 19:523 (1984).
107. P. Massardo, F. Bettarini, P. Piccardi, and A. Longoni, Pestic. Sci., 14:461 (1983).
108. M. Hatakoshi, T. Osumi [sic], H. Kisida, N. Itaya, and I. Nakayama, Jpn. J. Sanit. Zool., 36:327 (1985).
109. M. Hatakoshi, T. Ohsumi, H. Kisida, N. Itaya, and I. Nakayama, Jpn. J. Sanit. Zool., 37:99 (1986).
110. M. Hatakoshi, T. Ohsumi, I. Fujimoto, H. Kisida, N. Itaya, and I. Nakayama, Appl. Ent. Zool., 22: 638 (1987).
111. A. Nakayama, H. Iwamura, A. Niwa, Y. Nakagawa, and T. Fujita, J. Agric. Food Chem., 33:1034 (1985).
112. A. Niwa, H. Iwamura, Y. Nakagawa, and T. Fujita, J. Agric. Food Chem., 36:378 (1988).
113. T. Hayashi, H. Iwamura, T. Fujita, N. Takakusa, and T. Yamada, J. Agric. Food Chem., 39:2039 (1991).
114. A. Niwa, H. Iwamura, Y. Nakagawa, and T. Fujita, J. Agric. Food Chem., 37:462 (1989).
115. K. Iwanaga and T. Kanda, Appl. Ent. Zool., 23:186 (1988); D. Amalraj, V. Vasuki, C. Sadanandane, M. Kalyanasundaram, B.K. Tyagi, and P.K. Das, Indian J. Med Res., 87:19 (1988).
116. M. Hatakoshi, H. Kawada, S. Nishida, H. Kisida, and I. Nakayama, Jpn. J. Sanit. Zool., 38:271 (1987).
117. R.W. Miller, J. Agric. Entomol., 6:77 (1989).
118. M. Hatakoshi, N. Agui, and I. Nakayama, Appl. Ent. Zool., 21:351 (1986); M. Hatakoshi, I. Nakayama, and L.M. Riddiford, J. Insect Physiol., 34:373 (1988).
119. B.A. Peleg, J. Econ. Entomol., 81:88 (1988); M. Hatakoshi, Y. Shono, H. Yamamoto, and M. Hirano, Appl. Ent. Zool., 26:412 (1991); I. Ishaaya, S. Yablonski, Z. Mendelson, and A.R. Horowitz, Phytoparasitica, 20:79 (1992); I. Ishaaya and A.R. Horowitz, J. Econ. Entomol., 85:2113 (1992).
120. W.A. Banks and C.S. Lofgren, J. Entomol. Sci., 26:331 (1991).
121. V.Y. Yokoyama and G.T. Miller, J. Econ. Entomol., 84:942 (1991).

122. S. Nishida, N. Matsuo, M. Hatakoshi, and H. Kisida, U.S. Patent 4,879,292 (1989).
123. S. Nishida, N. Matsuo, M. Hatakoshi, and H. Kisida, U.S. Patent 4,970,222 (1990).
124. F. Bettarini, P. Massardo, P. Piccardi, F. Reggiori, and A. Longoni, U.S. Patent 4,607,035 (1986).
125. C.A. Henrick, B.A. Garcia, G.B. Staal, and D.C. Cerf, unpublished results, Sandoz Agro, Inc., CA (1986).
126. F.M. Pallos, P.E. Letchworth, and J.J. Menn, *J. Agric. Food Chem.*, 24:218 (1976).
127. I. Ishaaya, *Pestic. Sci.*, 13:204 (1982).
128. T.H. Babu and K. Sláma, Science, 175:78 (1972); J. Hlaváček, K. Poduška, F. Šorm, and K. Sláma, *Collect Czech. Chem. Commun.*, 41:317 (1976); J. Hlaváček, K. Poduška, F. Šorm , and K. Sláma, *Collect Czech. Chem. Commun.*, 41:1257 (1976); J. Hlaváček, J. Koudelka, and J. Järv, *Bioorg. Chem.*, 21:7 (1993).
129. G.B. Staal, *Pontif. Acad. Sci. Scr. Varia*, 41:353 (1977).
130. G.B. Staal, C.A. Henrick, B.J. Bergot, D.C. Cerf, J.P. Edwards, and S.J. Kramer, *Regulation of Insect Development and Behavior* (F. Sehnal, A. Zabża, J.J. Menn, and B. Cymborowski, eds.), Vol. 1, Wrocław Tech. Univ. Press, Wrocław, Poland, p. 323 (1981).
131. D.H.S. Horn, R.H. Nearn, J.B. Siddall, G.B. Staal, and D.C. Cerf, *Aust. J. Chem.*, 36:1409 (1983).
132. F.C. Baker, C.A. Miller, L.W. Tsai, G.C. Jamieson, D.C. Cerf, and D.A. Schooley, *Insect Biochem.*, 16:741 (1986).
133. H.G. Kallenborn, G.C. Mosbacher, and C. Künast, *J. Appl. Ent.*, 104:385 (1987); C. Künast, W. Himmele, and H. Theobald, *Z. Pflanzenk. Pflanzenschutz*, 95:285 (1988).
134. T. Miyake, H. Haruyama, T. Ogura, T. Mitsui, and A. Sakurai, *J. Pestic. Sci.*, 16:441 (1991).
135. T. Miyake, H. Haruyama, T. Mitsui, and A. Sakurai, *J. Pestic. Sci.*, 17:75 (1992); see also T. Miyake and T. Ogura, *J. Pestic. Sci.*, 17:S231 (1992).
136. C.A. Henrick, R.J. Anderson, G.B. Staal, and G.F. Ludvik, *J. Agric. Food Chem.*, 26:542 (1978).
137. H. Kindle, M. Winistörfer, B. Lanzrein, and K. Mori, *Experientia*, 45:356 (1989).
138. S. Sakurai, T. Ohtaki, H. Mori, M. Fujiwhara, and K. Mori, *Experientia*, 46:220 (1990).
139. M.G.Peter, *Chirality and Biological Activity* (B. Holmstedt, H. Frank, B. Testa, eds.), Proceedings of International Symposium, Tubingen, Germany, April 1988, A.R. Liss, New York, pp. 111–117 (1990).
140. R.C. Peterson, M.F. Reich, P.E. Dunn, J.H. Law, and J.A. Katzenellenbogen, *Biochem.*, 16:2305 (1977); S. Turunen and G.M. Chippendale, *Insect Biochem.*, 11:429 (1981); H. Van Mellaert, S. Theunis, and A. De Loof, *Insect Biochem.*, 15:655 (1985).

141. S. Matolín and I. Gelbič, *Acta Ent. Bohemoslov.*, 72:360 (1975).
142. I. Gelbič and S. Matolín, *Acta Ent. Bohemoslov.*, 81:321 (1984).
143. P. Masner, M. Angst, and S. Dorn, *Pestic. Sci.*, 18:89 (1987).
144. C.H. Schaefer and W.H. Wilder, *J. Econ. Entomol.*, 66:913 (1973).
145. T. Miura and R. M. Takahashi, *J. Econ. Entomol.*, 66:917 (1973).
146. R.L. Harris, E.D. Frazer [sic], and R.L. Younger, *J. Econ. Entomol.*, 66:1099
     (1973); R.L. Harris, W.F. Chamberlain, and E.D. Frazar, *J. Econ. Entomol.*,
     67:384 (1974).
147. J.A. Miller, M.L. Beadles, J.S. Palmer, and M.O. Pickens, *J. Econ.
     Entomol.*, 70:589 (1977); J.A. Miller, F.W. Knapp, R.W. Miller, and C.W.
     Pitts, *Southwest Entomol.*, 4:195 (1979).
148. G.T. Fincher, *Environ. Entomol.*, 20:77 (1991).
149. G.B. Quistad, L.E. Staiger, B.J. Bergot, and D.A. Schooley, *J. Agric. Food
     Chem.*, 23:743 (1975); G.B. Quistad, L.E. Staiger, and D.A. Schooley, *J.
     Agric. Food Chem.*, 23:750 (1975).
150. P.F. White, *Entomol. Exp. Appl.*, 26:332 (1979).
151. M.S. Mulla, R.E. Norland, T. Ikeshoji, and W.L. Kramer, *J. Econ.
     Entomol.*, 67:165 (1974).
152. W.F. Chamberlain and J.D. Becker, *Southwest Entomol.*, 2:179 (1977); W.F.
     Chamberlain, *Pest Control*, 47:22 (1979).
153. W.F. Chamberlain, J. Maciejewska, and J.J. Matter, *J. Econ. Entomol.*,
     81:1420 (1988).
154. L.M. El-Gazzar, P.G. Koehler, R.S. Patterson, and J. Milio, *J. Med.
     Entomol.*, 23:651 (1986).
155. B.A. Moser, P.G. Koehler, and R.S. Patterson, *J. Econ. Entomol.*, 85:112
     (1992).
156. A. Olsen, *Int. Pest Control*, 27:10 (1985); see also K.G. Palma, S.M. Meola,
     and R.W. Meola, *J. Med. Entomol.*, 30:421 (1993).
157. A.A. Marchiondo, J.L. Riner, D.E. Sonenshine, K.F. Rowe, and J.H.
     Slusser, *J. Med. Entomol.*, 27:913 (1990).
158. K.G. Palma and R.W. Meola, *J. Med. Entomol.*, 27:1045 (1990).
159. G.B. Staal, C.A. Henrick, D.L. Grant, D.W. Moss, M.C. Johnston, R.R.
     Rudolph, and W.A. Donahue, *Bioregulators for Pest Control* (P.A. Hedin,
     ed.), ACS Symposium Series, No. 276, American Chemical Society,
     Washington, DC, p. 201 (1985).
160. G.W. Bennett, J.W. Yonker, and E.S. Runstrom, *J. Econ. Entomol.*, 79:1032
     (1986).
161. T.H. Atkinson, P.G. Koehler, and R.S. Patterson, *J. Med. Entomol.*, 29:364
     (1992).
162. R.J. Brenner, P.G. Koehler, and R.S. Patterson, *J. Econ. Entomol.*, 81: 1404
     (1988); J. E. King and G. W. Bennett, *J. Econ. Entomol.*, 82:833 (1989).
163. J.E. King and G.W. Bennett, *J. Econ. Entomol.*, 81:225 (1988).
164. J.E. King and G.W. Bennett, *J. Med. Entomol.*, 27:642 (1990).
165. H. Kawada, I. Kojima, and G. Shinjo, *Jpn. J. Sanit. Zool.*, 40:195 (1989).
166. P.G. Koehler and R. S. Patterson, *J. Econ. Entomol.*, 84:917 (1991).

167. J.T. Snelson, *Grain Protectants*, Australian Centre for International Agricultural Research, Canberra, p. 239 (1987).
168. R.G. Strong and J. Diekman, *J. Econ. Entomol.*, 66:1167 (1973).
169. M.A. Manzelli, *J. Econ. Entomol.*, 75:721 (1982).
170. K.J. Kramer and H.E. McGregor, *J. Econ. Entomol.*, 71:132 (1978).
171. J.P. Edwards, J.E. Short, and L. Abraham, *J. Stored Prod. Res.*, 27:31 (1991).
172. R.R. Cogburn, *J. Econ. Entomol.*, 81:722 (1988).
173. H. Smet, M. Rans, and A. De Loof, *J. Stored Prod. Res.*, 25:165 (1989).
174. L.S. Mian and M.S. Mulla, *J. Econ. Entomol.*, 75:599 (1982).
175. P.R. Samson, R.J. Parker, and E.A. Hall, *J. Stored Prod. Res.*, 26:215 (1990).
176. P.J. Collins and D. Wilson, *Pestic. Sci.*, 20:93 (1987).
177. J.P. Edwards and B. Clarke, *Int. Pest Control*, 20:5 (1978).
178. W.A. Banks, L.R. Miles, and D.P. Harlan, *Florida Entomol.*, 66:172 (1983).
179. S.A. Phillips, Jr. and H.G. Thorvilson, *J. Econ. Entomol.*, 82:1646 (1989).
180. W.A. Banks and C.S. Lofgren, *J. Entomol. Sci.*, 26:331 (1991).
181. S.C. Jones, *J. Econ. Entomol.*, 77:1086 (1984).
182. M.I. Haverty, N.-Y. Su, M. Tamashiro, and R. Yamamoto, *J. Econ. Entomol.*, 82:1370 (1989).
183. C.-F. Chang and S. Tamura, *Appl. Entomol. Zool.*, 6:143 (1971); C.-F. Chang, S. Murakoshi, and S. Tamura, *Agric. Biol. Chem.*, 36:692 (1972); S. Murakoshi, C.-F. Chang, and S. Tamura, *Agric. Biol. Chem.* 36:695 (1972).
184. M. Nihmura, S. Aomori, K. Mori, and M. Matsui, *Agric. Biol. Chem.*, 36:889 (1972); Y. Ozawa, K. Mori, and M. Matsui, *Agric. Biol. Chem.*, 37:2373 (1973).
185. G.B. Staal, S. Nassar, and J.W. Martin, *J. Econ. Entomol.*, 66:851 (1973).
186. S.G. Nassar, G.B. Staal, and N.I. Armanious, *J. Econ. Entomol.*, 66:847 (1973).
187. R.J. Bauernfeind and R.K. Chapman, *J. Econ. Entomol.*, 77:211 (1984).
188. G. Singh and H.S. Sidhu, *Insect Sci. Applic.*, 11:69 (1990).
189. R.H. De Reede, P. Alkema, and L.H.M. Blommers, *Entomol. Exp. Appl.*, 39:265 (1985).
190. P.-J. Charmillot, K. Vernez, B. Bloesch, M. Berret, and D. Pasquier, *Mitt. Schweiz. Entomol. Ges.*, 58:393 (1985).
191. P.-J. Charmillot, M. Baillod, B. Bloesch, E. Guignard, Ph. Antonin, M.L. Frischknecht, H. Hoehn, and A. Schmid, *Rev. Suisse Vitic. Arboric. Hortic.*, 19:183 (1987).
192. P.-J. Charmillot, *Entomol. Exp. Appl.*, 51:59 (1989).
193. P.-J. Charmillot and B. Bloesch, *Rev. Suisse Vitic. Arboric. Hortic.*, 19:87 (1987).
194. B.J. Hicks and R. Gordon, *Can. Ent.*, 124:117 (1992).
195. A.A. Eisa, M.A. El-Fatah, A. El-Nabawi, and A.A. El-Dash, *Phytoparasitica*, 19:49 (1991); A.A. Eisa, M.A. El-Fatah, A. El-Nabawi, and A.A. El-Dash, *Anz. Schädlingskde., Pflanzenschutz, Umweltschutz*, 64:16 (1991).

196. B.A. Peleg, *Entomophaga, 28*:117, 367 (1983).
197. I. Bar-Zakay, *Phytoparasitica, 20*:80 (1992).
198. M.G. Solomon and J.D. Fitzgerald, *J. Hortic. Sci., 65*:535 (1990).
199. K.R.S. Ascher and M. Eliyahu, *Phytoparasitica, 16*:15 (1988).
200. K. Nagai, *Appl. Ent. Zool., 25*:199 (1990).
201. C.R. Worthing and R.J. Hance (eds.), *The Pesticide Manual*, 9th ed. British Crop Protection Council, Surrey (1991).
202. G.B. Quistad, L.E. Staiger, and D.A. Schooley, *J. Agric. Food Chem., 22*:582 (1974); D.A. Schooley, B.J. Bergot, L.L. Dunham, and J.B. Siddall, *J. Agric. Food Chem., 23*:293 (1975); G.B. Quistad, L.E. Staiger, and D.A. Schooley, *J. Agric. Food Chem., 23*:299 (1975).
203. D.A. Schooley and G.B. Quistad, *Pesticide and Xenobiotic Metabolism in Aquatic Organisms* (M.A.Q. Khan, J.L. Lech, and J.J. Menn, eds.), ACS Symposium Series, No. 99, American Chemical Society, Washington, DC, p. 161 (1979).
204. L. Gerig, *Schweiz. Landwirtsch Forsch., 14*:355 (1975).
205. M.L. Frischknecht and P.J. Müller, *Mitt. Schweiz. Entomol. Ges., 49*:239 (1976).
206. G.M. Copijn, J. Beetsma, and P. Wirtz, *Proc. K. Ned. Akad. Wet. Ser. C. Biol. Med. Sci., 82*:29 (1979).
207. P.G. Hester, C.B. Rathburn, Jr., and A.H. Boike, Jr., *Proc. Florida Anti-Mosq. Assoc., 51*:16 (1980).
208. L.S. Mian and M.S. Mulla, *Residue Reviews, 84*:27 (1982).
209. T.P. Breaud, J.E. Farlow, C.D. Steelman, and P.E. Schilling, *Mosq. News, 37*:704 (1977); L. Bircher and E. Ruber, *J. Am. Mosq. Control Assoc., 4*:520 (1988); C.L. McKenney, Jr. and E. Matthews, *Environ. Pollution, 64*:169 (1990).
210. E.D. Gomez, D.J. Faulkner, W.A. Newman, and C. Ireland, *Science, 179*:813 (1973); M. Ramenofsky, D.J. Faulkner, and C. Ireland, *Biochem. Biophys. Res. Commun., 60*:172 (1974).
211. M.S. Mulla, H.A. Darwazeh, B. Kennedy, and D.M. Dawson, *J. Am. Mosq. Control Assoc., 2*:314 (1986).
212. T. Miura, C.H. Schaefer, and R.J. Stewart, "Impact of Carbamate Insect Growth Regulators on the Selected Aquatic Organisms: A Preliminary Study," Proceedings and Papers of the Annual Conference of the California Mosquito and Vector Control Association, 54th, pp. 36–38 (1986).
213. C.H. Schaefer and T. Miura, *J. Econ. Entomol., 83*:1768 (1990).
214. T. Miura and R.M. Takahashi, *J. Am. Mosq. Control Assoc., 3*:476 (1987).
215. B.M. Lee and G.I. Scott, *Bull. Environ. Contam. Toxicol. 43*:827 (1989).
216. M. Frischknecht, W. Jucker, M. Baggiolini, and A. Schmid, *Z. Pflanzenkr. Pflanzenschutz, 85*: 334 (1978).
217. J. McNeil, *Science, 189*:640 (1975).
218. A.I. Abd. El-Kareim, B. Darvas, and F. Kozár, *J. Appl. Entomol., 106*:270 (1988).
219. S. Grenier and G. Plantevin, *J. Appl. Ent., 110*:462 (1990).

220. M.T. El-Ibrashy, *Insect Sci. Applic.*, *8*:743 (1987).
221. T.C. Sparks and B.D. Hammock, *Pest Resistance to Pesticides* (G.P. Georghiou and T. Saito, eds.), Plenum Press, New York, p. 615 (1983).
222. D.C. Cerf and G.P. Georghiou, *Nature* (*Lond.*), *239*:401 (1972); D.C. Cerf and G. P. Georghiou, *Pestic. Sci.*, *5*:759 (1974).
223. F.W. Plapp, Jr. and S.B. Vinson, *Pestic. Biochem. Physiol.*, *3*:131 (1973); S.B. Vinson and F.W. Plapp, Jr., *J. Agric. Food Chem.*, *22*:356 (1974).
224. A.B.H. Kadri, *J. Med. Ent.*, *12*:10 (1975).
225. T.M. Brown, D.H. DeVries, and A.W.A. Brown, *J. Econ. Entomol.*, *71*:223 (1978).
226. T.M. Brown and G.H.S. Hooper, *Pestic. Biochem. Phyisol.*, *12*:79 (1979); T.M. Brown and A.W.A. Brown, *Ent. Exp. Appl.*, *27*:11 (1980).
227. L. Shemshedini and T.G. Wilson, *Proc. Natl. Acad. Sci.*, *87*:2072 (1990).
228. C. Minkoff III and T.G. Wilson, *Genetics*, *131*:91 (1992).

4

Avermectins and Milbemycins

Gabe I. Kornis

Upjohn Company, Kalamazoo, Michigan

I. INTRODUCTION

The milbemycins and avermectins are 16-membered macrocycles with very potent anthelmintic and insecticidal activity. In 1975, Mishima et al., [1] at the Sankyo Company in Japan isolated the milbemycins as a complex mixture from a fermentation broth, and although they recognized the acaricidal properties of the crude extract, for reasons never explained they missed or perhaps ignored their important anthelmintic activity. The avermectins were isolated from the fermentation broth produced by a culture originating in Japan, and their anthelmintic properties were discovered by Merck workers [2a, 2b, 2c] in 1979 using an *in vivo* assay. Recognizing the structural similarities between the avermectins and the milbemycins, the Merck workers synthesized milbemycins from avermectins and patented their anthelmintic properties [3a, 3b]. Site-specific hydrogenation of avermectin B_1 yielded ivermectin, the most potent semisynthetic anthelmintic known to date. The great commercial success of ivermectin, and later Milbemycin D (Figure 1) led to a widespread search for other naturally occurring anthelmintics, and several companies were successful in finding and, in some cases, marketing compounds with great structural similarities to the milbemycins. These will be described in the isolation section.

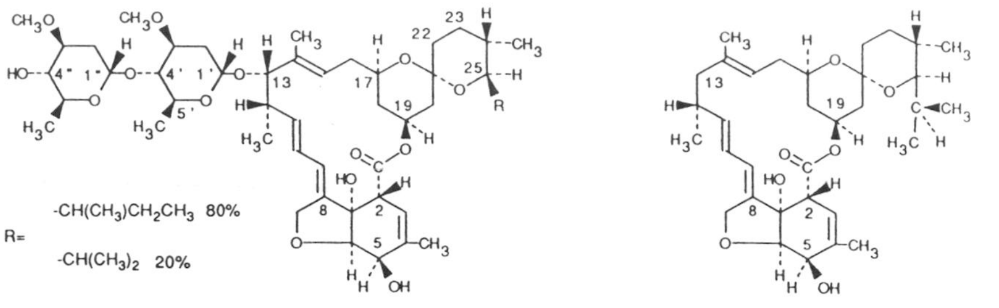

(a)   (b)

Figure 1   Chemical structures of (a) ivermectin and (b) milbemycin D.

The patent and scientific literature relating to the milbemycins and avermectins has expanded tremendously, and a computerized search carried out in November 1992 yielded well over 3000 references. There are many excellent review articles covering this field; foremost is the book edited by W. C. Campbell [4] on avermectins and published in 1989. Other excellent reviews are by Fisher and Mrozik [5, 6] and by Davies and Green [7–9].

II. ISOLATION

The *Streptomyces hygroscopicus* culture produced a complex mixture from which, after extensive purification, 13 entities were isolated and named milbemycins; they were divided into the α and β series. Mutation of the organism provided other milbemycins, designated by capital letters. A common feature of all milbemycins is a 16-membered macrocycle, a spiroketal unit formed by two 6-membered rings, and the hydroxyl functionality at C_7 (Figure 2). In the α series, the southern portion consists of a hexahydrobenzofuran ring, and the various members of the series are distinguished by the following substituents. At R_4 the functionality is methyl or in two cases 2-pyrrolylcarbonyloxy; at R_5, hydroxyl or methoxyl; at R_{22}, hydrogen or axial β-hydroxyl; at R_{23}, hydrogen or α-hydroxycarbonyl-alkyl groups, and at R_{25}, methyl or ethyl moieties. The milbemycins designated with capital letters such as D, F, G, J, and K have a methyl or 2-pyrrolylcarbonyloxy group at R_4; a hydrogen, hydroxyl, methoxyl, or ketone grouping at R_5, and at R_{25} a methyl, ethyl, or isopropyl group. The milbemycins β_1, β_2, and H differ from the previously described series by having an opened tetrahydropyranyl ring and have the following substituents: hydrogen, β methoxyl, or ketone at R_5; hydroxymethylene or methyl at R_8; and

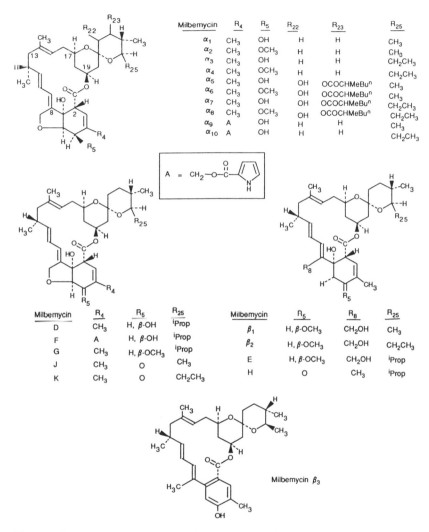

Milbemycin	R_4	R_5	R_{22}	R_{23}	R_{25}
α_1	CH_3	OH	H	H	CH_3
α_2	CH_3	OCH_3	H	H	CH_0
α_3	CH_3	OH	H	H	CH_2CH_3
α_4	CH_3	OCH_3	H	H	CH_2CH_3
α_5	CH_3	OH	OH	$OCOCHMeBu^n$	CH_3
α_6	CH_3	OCH_3	OH	$OCOCHMeBu^n$	CH_3
α_7	CH_3	OH	OH	$OCOCHMeBu^n$	CH_2CH_3
α_8	CH_3	OCH_3	OH	$OCOCHMeBu^n$	CH_2CH_3
α_9	A	OH	H	H	CH_3
α_{10}	A	OH	H	H	CH_2CH_3

$$A = CH_2-O-\overset{O}{\underset{}{C}}-\text{(pyrrole)}$$

Milbemycin	R_4	R_5	R_{25}
D	CH_3	H, β-OH	^{i}Prop
F	A	H, β-OH	^{i}Prop
G	CH_3	H, β-OCH_3	^{i}Prop
J	CH_3	O	CH_3
K	CH_3	O	CH_2CH_3

Milbemycin	R_5	R_8	R_{25}
β_1	H, β-OCH_3	CH_2OH	CH_3
β_2	H, β-OCH_3	CH_2OH	CH_2CH_3
E	H, β-OCH_3	CH_2OH	^{i}Prop
H	O	CH_3	^{i}Prop

Milbemycin β_3

Figure 2 Chemical structures for various milbemycins.

methyl, ethyl, or isopropyl at R_{25}. Finally, milbemycin β_3 is the only member of the family with an aromatic ring in the southern hemisphere; it could be an artifact formed during purification.

The avermectins are a complex group of macrolides produced by *Streptomyces avermitilis* from which eight entities were isolated (Figure 3). In common with the α-milbemycins, they have a 16-membered macrocycle, a spiroketal group, a hexahydrobenzofuran ring in the

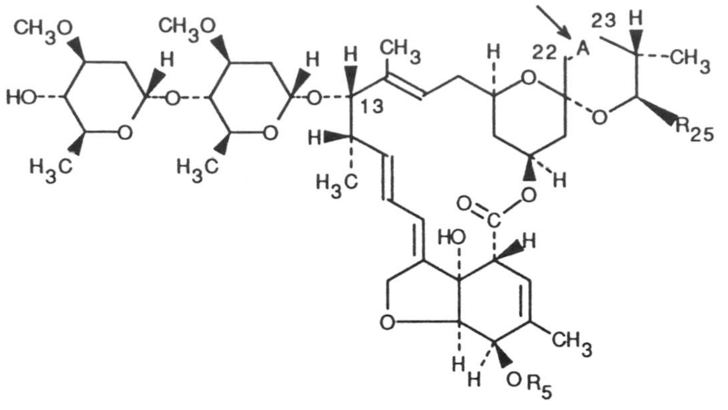

Figure 3 Chemical structures for the avermectins.

AVERMECTIN	R_5	A	R_{25}
A_{1a}	CH_3	HC=CH	^sBu
A_{1b}	CH_3	HC=CH	^iPr
A_{2a}	CH_3	CH_2-CH(OH)	^sBu
A_{2b}	CH_3	CH_2-CH(OH)	^iPr
B_{1a}	H	HC=CH	^sBu
B_{1b}	H	HC=CH	^iPr
B_{2a}	H	CH_2-CH(OH)	^sBu
B_{2b}	H	CH_2-CH(OH)	^iPr

southern hemisphere, and a tertiary hydroxyl at C_7. The major difference is that the avermectins have the α-L-oleandrosyl-α-L-oleandrosyloxy disaccharide at the C_{13}-α position, and a double bond at C_{22} in some cases. The nomenclature of the avermectins is straightforward. In the A series, at C_5 there is a methoxyl group, whereas in the B series, this is replaced with an hydroxyl. In the 1 series there is a double bond between C_{22} and C_{23}, whereas in the 2 series, the double bond is hydrated with the hydroxyl group at C_{23}. Finally, at C_{25}, the small letter "a" denotes a secondary butyl group, while the small letter "b" indicates an isopropyl group attached to C_{25}. After isolation, the ratio of compounds a and b is about 80:20 and, because their biological activities are almost indistinguishable, they are not normally separated. Ivermectin (Figure 1), one of the commercial products from the avermectin family, is an 80:20 mixture of 22,23-dihydroavermectin B_{1a} and B_{1b} and is obtained by the site-specific hydrogenation of the naturally occurring mixture of avermectin B_{1a} and B_{1b}.

Many other milbemycin type compounds have been isolated following the great commercial success of ivermectin. The excellent review by Davies and Green [8] provides full details; a summarized version is presented here. Workers at American Cyanamid utilizing a culture of *Streptomyces cyaneogriseus* isolated over 20 macrocycles closely related to the milbemycins. The major difference is the presence of an alkene grouping at C_{25}. The structures of four members of this class of compounds are shown in Figure 4. LL-F 28249α, also called nemadectin, is being developed as an anthelmintic agent. A second product in this series is named moxidectin; it differs from the naturally occurring nemadectin by replacement of the C_{23} hydroxyl with the methyl ether of the oxime (Figure 4). Ramsay et al. [10] at Glaxo have isolated from *Streptomyces thermoarchaensis* milbemycin-type compounds identical to the LL-F 28249 series and designated as the S541 factor; the structures are also shown in Figure 4. Hood et al. [11] at Beecham also isolated several milbemycin analogs from various *Streptomyces* strains distinguished by acyl substituents on the C_4 methyl group, and a hy-

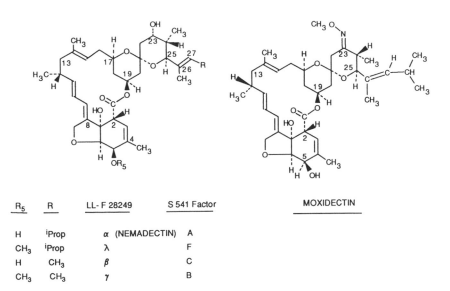

R_5	R	LL-F 28249	S 541 Factor	MOXIDECTIN
H	iProp	α (NEMADECTIN)	A	
CH₃	iProp	λ	F	
H	CH₃	β	C	
CH₃	CH₃	γ	B	

Figure 4 Chemical structures for some milbemycin analogs.

droxyl at C_{22}. Haxell et al. [12] at Pfizer utilized *Streptomyces hygroscopicus* ATCC 53718 to obtain milbemycin analogs oxygenated at C_{13}, C_{22}, and C_{23}, called the N787-182 complex (For further details, see also the structure determination section.) Also at Pfizer, Dutton et al. [13] produced interesting novel avermectins by mutational biosynthesis. They discovered that the organism *Streptomyces avermitilis* mutant strain ATCC 53568 lacked the ability to form isobutyric and S-2-methylbutyric acids, and, therefore, could not produce avermectins. By supplying alternative carboxylic acids, a very large number of analogs were obtained with alkyl, alkene, cycloalkyl, and cycloheteroaryl substituents at C_{25}. Chen et al. [14] were also able to induce *Streptomyces avermitilis* to produce avermectin homologs that carry the 2-pentyl and 2-hexyl groups at C_{25} by externally supplying sodium 2-methyl pentanoate and sodium 2-methyl hexanoate, respectively. Omura et al. [15] mutated the parent strain (K139) of *Streptomyces avermitilis* and produced strains that yielded exclusively the B_{1a} and B_{2a} components of avermectin. Řezanka et al. [16] grew *S. avermitilis* in the presence of phenoxyacetic and phenoxypropionic acids, substances known to inhibit fatty acid biosynthesis, and this resulted in the enhanced production of avermectins with a secondary butyl group at C_{25}. This was due to the diversion of the common precursor, 2-methylbutyryl-Co-A, from fatty acid biosynthesis to that of avermectins. Nakagawa and co-workers [17] in continuing efforts on the milbemycins at Sankyo used *Syncephalastrum racemosum* to selectively hydroxylate the methyl group at C_{24} of milbemycin α_4, α_3, D, and 5-keto-milbemycin-α_4-5-oxime. (See Figure 2 for structures.) Similar procedures have been utilized to substantially change the type and proportions of products obtained by fermentation. Undoubtedly, many more milbemycin- and avermectin-type compounds will be discovered in the future.

III. STRUCTURE DETERMINATION

In 1975, Mishima [1] published an article on the structural features of the β-milbemycins. The primary hydroxy group apparent from spectral studies in milbemycin β_1 was treated with *p*-bromophenylisocyanate to yield a crystalline urethane derivative whose structure was determined by x-ray analysis. Treatment of milbemycin α_3 by Mishima [18] with the same reagent also afforded a crystalline derivative suitable for x-ray analysis. To distinguish between various members of the milbemycin family, spectroscopic methods were used extensively. The diene chromophore and the pyrrole moiety in the α series were suggested by the ultraviolet (UV) spectrum and were distinct from the β series. Infrared spectra indicated the presence of a hydroxyl group and the lactone. With the structure known from the x-ray analysis, the fragmentation patterns obtained from mass spectrometry (MS) could be explained as taking place via a retro Diels–Alder reaction, ester elimination, allylic fission, and also cleavage α to the oxygen atom. This was confirmed by combining ^{13}C nuclear magnetic resonance (NMR) experiments with biosynthetic studies. Mass spectral studies were also useful in distinguishing the methyl from the ethyl homologs, and the hydroxyl versus the methoxyl group at C_5. Both ^1H and ^{13}C NMR were useful in distinguishing between close analogs in the milbemycin family. The absolute configuration by Mishima et al. [19] was established by measuring the specific rotation of milbemycin D, the C_5 acetate, and their epimers, and by investigating the circular dichroism of the *p*-diethylaminobenzoates of the epimers. These results led to the assignment of the R configuration at C_5 and were in agreement with the relative and absolute configuration of milbemycin D prepared from one of the avermectins.

The structural elucidation of the avermectin class of compounds was published by Albers-Schonberg et al. [20]. Based on MS and NMR measurements, the similarity to the milbemycins was noted soon after isolation, facilitating further structural work. The presence of the secondary butyl and isopropyl group at C_{25} was established with the aid of NMR, as was the presence of a methoxyl group at C_5 in the A series. The location of the C_{22} double bond in the A series and of the hydroxyl at C_{23} was established by NMR after removal of the disaccharide. The monosaccharide, the aglycone, and methyl oleandroside were obtained by acidic methanolysis, or sulfuric acid hydrolysis of the avermectins. The C_7 hydroxyl group was also pinpointed by NMR and the attachment of the disaccharide at C_{13} was suggested by NMR experiments performed on the aglycone. This was later ascertained chemically by treating avermectin A_{2a} with ozone, followed by reduction with sodium borohydride, which gave fragments 1 and 2 (Figure 5). Treatment with methanol/HCl yielded α- and β-methyl oleandroside and the two epimers of 2-methylpentanediol (3). Springer and co-workers [21] combined the power of single-crystal x-ray diffraction techniques and NMR experiments to establish the absolute stereochemistry and conformation of avermectin B_{2a} aglycone (4) as shown in Figure 5. Neszmélyi et al. [22] studied the solution conformation of the dis-

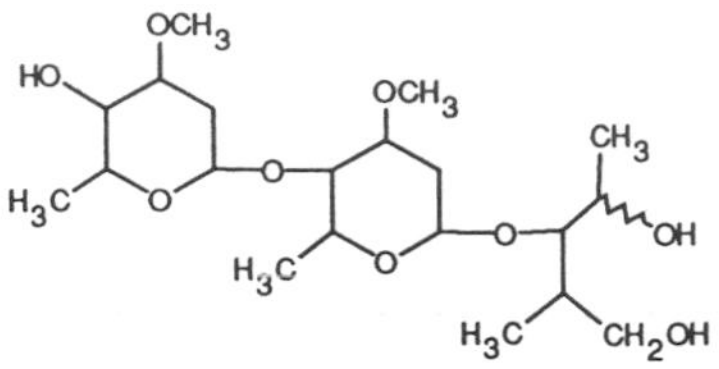

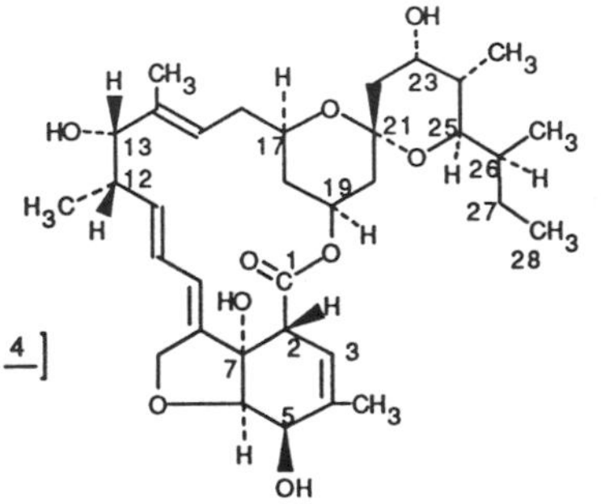

Figure 5 Degradation products of Avermectin A_{2a}.

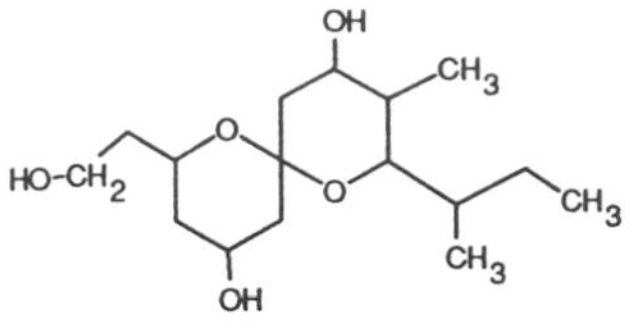

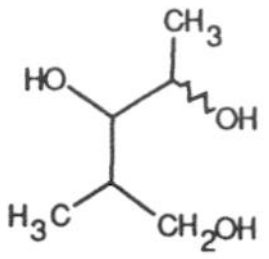

accharide of avermectin B_{1a} by utilizing NMR data and conformational energy calculations. They concluded that the conformation of avermectin B_{1a} in solution and in the crystalline state is similar and the oleandrose ring closest to the flat macrocycle is perpendicular to its plane, whereas the more distant oleandrose ring is almost parallel to the macrocycle. In an article published in 1990, Diez-Martin and coworkers [23] using more modern NMR techniques, have shown that in avermectin B_{1a} the chemical shifts assigned to the C_{22} and C_{23} atoms (olefinic) should be interchanged.

The structural elucidation of the newer milbemycins was based mainly on the work performed at Sankyo and Merck. Baker et al. [24], by using UV, MS, and NMR, identified VM44857 (Figure 6) as being similar to milbemycin D, except for the substituent at C_{25}. Extensive use of the nuclear Overhauser effect (NOE) established the 1,3-diaxial relationship between protons at positions 17 and 19, and also between the two diaxial protons at positions 18 and 20; this was utilized to prove the conformation of the C_{17}–C_{21} tetrahydropyran ring. The configuration of the important side chain at C_{25}, as shown in Figure 6, was also established with the aid of NOE. The structure of LL-F 28249-α (Fig-

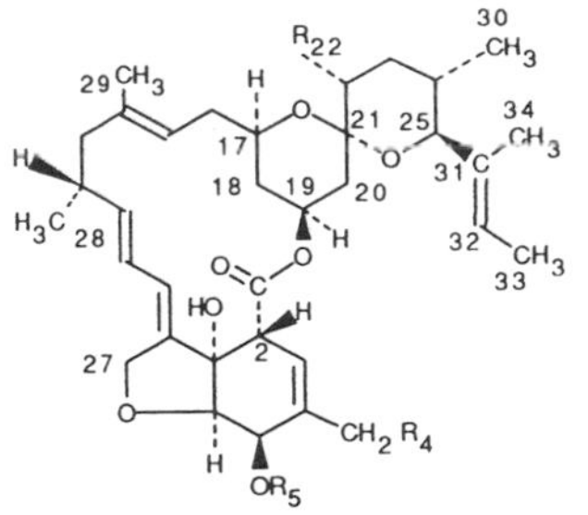

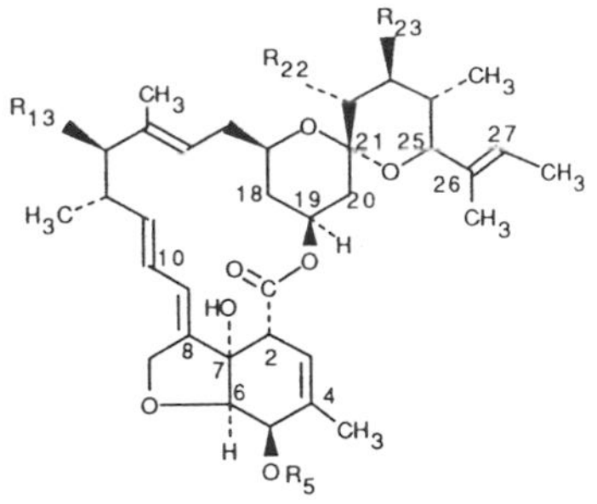

COMPOUND	R₄	R₅	R₂₂
VM 44857	H	H	H
VM 44864	H	CH₃	OH
VM 44865	A	CH₃	OH
VM 44866	H	H	OH

COMPOUND	R₅	R₁₃	R₂₂	R₂₃
UK-78,629	CH₃	OCOCH(CH₃)₂	OH	H
UK-78,624	H	OCOCH(CH₃)₂	OH	H
UK-80,694	H	OCOCH(CH₃)₂	OH	OCOCH(CH₃)₂
UK-80,695	H	OCOCH(CH₃)₂	H	H

$$A = \quad O-\overset{\overset{\displaystyle O}{\|}}{C}-\underset{\underset{\displaystyle CH_3}{|}}{C}=C-CH_3$$

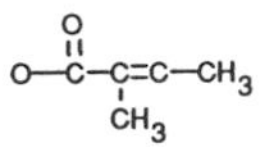

Figure 6 Some chemical structures for the VM and UK series.

ure 4) was determined by Carter and co-workers [25] using high-resolution MS and NMR spectroscopy, in conjunction with x-ray crystallographic studies of LL-F 28249-γ. Rajan and Stockton [26], using sophisticated two-dimensional NMR spectroscopy, fully supported the structural assignment of Carter and co-workers [25] and concluded that the LL-F 28249 family of macrocycles in the crystalline forms is dominated by intramolecular forces and that the crystal and solution conformations are similar. In a cooperative effort from three Pfizer research centers, Haxel and co-workers [27] published an article on the structural elucidation of C_{13}-β-acyloxymilbemycins, also called the N787-182 complex and designated by the prefix UK (Figure 6). By comparing the UV, NMR, and MS data of UK-78629 to those of LL-F 28249-γ and VM44864, it became immediately apparent that all three compounds possess the same skeleton between C_1 and C_{28}. The NMR spectrum also indicated the presence of an isobutyryl group attached at C_{13}; the β stereochemistry was established by comparison to the avermectin family of macrocycles. The hydroxyl at C_{22} was also shown to be present in VM44864, and the stereochemistry was established by two-dimensional (2D) Cosy NMR experiments. The C_{26}–C_{27} double bond was assumed to have the same geometry as other closely related milbemycins. The structures of the other 11 compounds in the series was established by spectroscopic means. UK-78624, UK-80694, and UK80695 showed potent in vitro nematicidal and insecticidal activity.

IV. STRUCTURE-ACTIVITY RELATIONSHIPS

Two comprehensive reviews by Fisher and Mrozik [5] and Davies and Green [7] are available and only the salient points will be highlighted; more recent findings will be discussed in greater detail. As mentioned previously, the family of avermectins comprises eight compounds, designated by the letters A, B, a, and b and the numbers 1 and 2. As the anthelmintic activity of the a series (sec. butyl at C_{25}) is almost identical to the b series (isopropyl at C_{25}), they will not be discussed separately. Compounds in the A series (methoxyl at C_5) are generally less potent than the ones in the B series (hydroxyl at C_5). There is no substantial difference in activity between the 1 series (double bond at C_{22}) or the 2 series (hydroxyl at C_{23}), and reduction of the C_{22} double bond of avermectin B_1 yields the commercial anthelmintic ivermectin (Figure 1), which fortuitously combines the beneficial properties of the B_1 and B_2 series, such as equal efficacy with oral and parenteral administration, and an increased spectrum. The insensitivity to biological efficacy of the C_{22} double bond is in direct contrast to the C_3–C_4 double bond, which, when moved into the C_2–C_3 position, leads to an almost total loss of anthelmintic activity. The role of the disaccharide in promoting efficacy has been thoroughly investigated. Acylation of 4″-hydroxyl of avermectin B_1 produced compounds with anthelmintic activity comparable to, but not better than, the naturally occurring compounds. This is in contrast to the 4″-epiaminomethyl derivative, which is much more active than avermectin B_1, a commercial acaricide. Removal of one of the sugars lowered the activity twofold to fourfold, while removal of both sugars to give the aglycone lowered the activity to only 1/30th of avermectin B_1. Interestingly, removal of the oxygen atom linking the disaccharide to the macrocycle in ivermectin restores most of the anthelmintic and acaricidal activity and gives rise to the milbemycin series of compounds. Derivatization of the hydroxyl groups at C_5, C_7, and C_{23} is detrimental to activity. Since the introduction of ivermectin to the marketplace in 1981, Merck chemists have continued probing the biological spectrum of this fascinating class of compounds. Mrozik et al. [28] have synthesized several avermectin aglycons with substituents at C_{13} and tested their activity against the two-spotted spider mite on bean plants, neonate southern armyworms and on the parasite nematode *Trichostrongylus colubriformis* in the jird *Meriones unguiculatus*. Binding constants were also measured by displacement of [22,23-^{3}H$_2$]-ivermectin from an avermectin receptor prepared from the free living nematode *Caenorhabditis elegans*. Although none of the derivatives were superior to avermectin B$_1$, ivermectin, or milbemycin D, derivatives with halogen long-chain alkyl ethers, or methoxime group substituted at C_{13} showed potent anthelmintic and insecticidal activity. Stereochemistry-wise, β-substituents had a slight advantage over α-substituents. Perhaps encouraged by this slight advantage in potency, Jones et al. [29] developed an efficient procedure for synthesizing both the α and β C_{13} hydroxyl derivatives of the aglycone of ivermectin. This was utilized by Blizzard and co-workers [30] in the synthesis of the 13 epimers of avermectin B$_1$ and B$_2$, which were tested in sheep and shown to retain the full anthelmintic potency of the natural products. Furthermore, based on mouse LD$_{50}$ values, these epi compounds possess an improved therapeutic ratio in mammals as compared to the naturally occurring isomers. It is of interest that 13-epi ivermectin was first prepared and patented by Frei and Mereyala [31] at the Ciba-Geigy company. Blizzard et al. [32] also carried out the conversion of avermectin B$_1$ to 19-epi-avermectin B$_1$ (*7*) in six steps (Figure 7). This involved treatment of the protected avermectin B$_1$ (*5*) with base which caused the opening of the lactone and migration of the double bond to the 2,3 position to yield *6*. Relactonization under Mitsunobu conditions gave the epilactone, which after treatment with base to isomerize the double bond to the 3,4 position, followed by deprotection, yielded the desired *7*, which was about 40-fold less active than avermectin B$_1$. In an accompanying publication, Hanessian and Chemca [33] described the synthesis of 19-epi-avermectin A$_{1a}$ from avermectin B$_{1a}$ by a somewhat similar procedure.

Shih et al. [34] transposed the olefinic side chain of nemadectin (LL-F 28249α) to the C_{25} position in avermectin B$_1$, starting with the C_{22} avermectin aldehyde derivative *8* (Figure 8). Although the product *9* retained activity in the brine shrimp assay, it was less potent than avermectin B$_1$ or nemadectin. Meinke et al. [35] continued exploring chemically the 4″ position in avermectin B$_1$ and ivermectin, a position known to lend itself to substantial modification without loss of biological activity. Thus, the 5-O-protected avermectin was oxidized to the 4″ ketone, which was then treated with trimethylsilyldiazomethane to yield the epoxide *10* and the oxepinyl epoxide *11*. Reaction with vari-

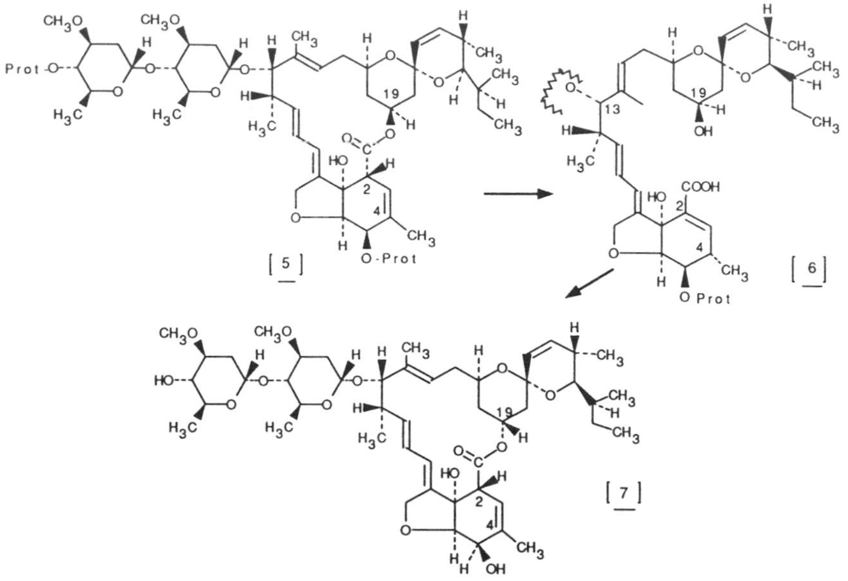

Figure 7 Synthesis of 19-epi-avermectin B$_1$.

ous nucleophiles provided structures of type *12, 13,* and *14.* From the many compounds prepared, *12* (R=CN or 2-imidazoylthio) showed anthelmintic activity comparable to that of ivermectin, when tested against six gastrointestinal helminths in experimentally infected sheep at a dose of 0.1 mg/kg. Blizzard et al. [36] studied the effect of changing the spatial relationship between the disaccharide and the aglycone by introducing a spacer between them. Compounds *15* and *16* approached the activity of ivermectin as measured in the brine shrimp immobilization assay. Analogs with spacers such as ester groupings or longer ether chains had substantially diminished anthelmintic activity. Frazer-Reid and co-workers [37] at Duke University were able to migrate the 3,4 double bond of avermectin B$_{1a}$ into the exocylic 4,4a position (*17*). When tested in experimentally infected sheep against gastrointestinal helminths, activity comparable to ivermectin was observed against *Haemonchus contortus* but not against *Ostertagia circumcincta.* One of the most successful synthetic changes effected in the field of avermectin chemistry was replacement of the 4''-α-hydroxyl in avermectin B$_1$ with the β-*N*-methylamino group (*18*) leading to a 1600-fold enhancement of activity in certain insect species, when compared to abamectin. This compound was named emamectin [6].

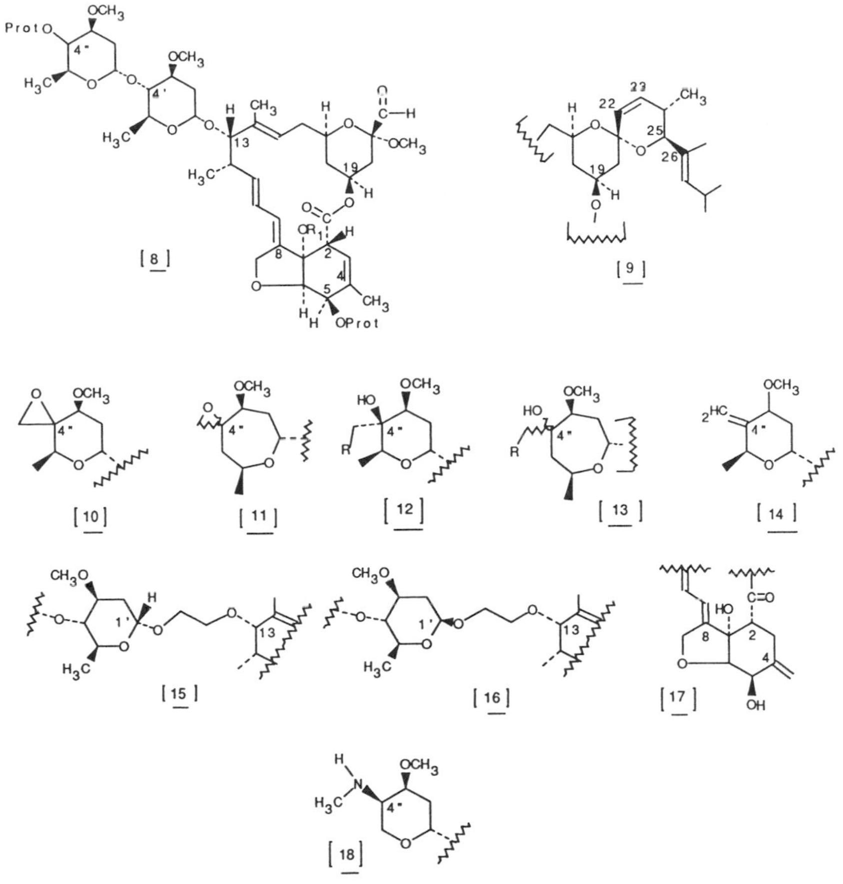

Figure 8 Semisynthetic analogs of the avermectins and milbemycins.

V. SYNTHESIS

One may arbitrarily divide the synthesis of naturally occurring compounds into two classes. In the first class, the aim is to construct simplified analogs containing the pharmacophore and to maximize biological activity at a minimum cost; this activity is mostly carried out in the pharmaceutical industry. In the second class, the primary aim is to construct by total synthesis a molecule identical in all respects to the naturally occurring substance; biological activity of the intermediates is of secondary importance. This work is mainly carried out in academia. In this section, only a few examples related to the first class will be discussed.

A. Partial Synthesis

In a cooperative effort between Oxford University and imperial chemical industries (ICI), Hughes et al. [38] synthesized the simplified milbemycin β_1 analog *19* (Figure 9). Kornis et al. [39] also prepared, by total synthesis, simplified milbemycin (*20a*) and ivermectin (*20b*) analogs. Schow et al. [40] synthesized the milbemycin H analog *21*, and Scherkenbeck et al. [41] synthesized the simple open-chain milbemycin E analog *22*. None of these analogs approached the potency of the naturally occurring compounds.

   One may conclude that while modest structural modifications of the naturally occurring compounds have produced analogs possessing good biological activity, fully synthetic analogs have been rather disappointing.

B. Total Synthesis

Naturally occurring compounds, which combine intricate molecular architecture, profuse functionality and remarkably potent biological activity, have always been irresistible targets for the synthetic chemist, and, of course, the milbemycins and avermectins have been no exception. A cursory literature search on the total synthesis of these com-

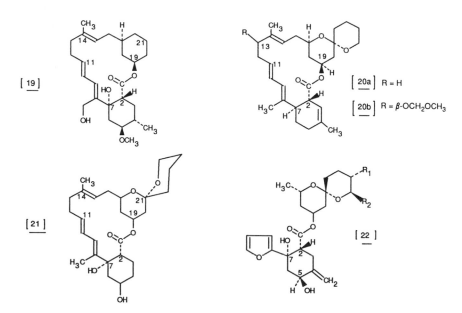

Figure 9 Fully synthetic analogs of the avermectins and milbemycins.

pounds yielded over 400 publications, and it is beyond the scope of this chapter to cover the topic. The reader partial to synthetic efforts is referred to the excellent reviews by Fisher and Mrozik [5, 6], Davies and Green [7–9], Blizzard et al. [42], and Ley and Armstrong [43].

VI. TOXICOLOGY

Two reviews, by Lankas and Gordon [44] and Fisher and Mrozik [6], have appeared recently. It is obvious that a compound that has been administered to about 2 million people at a dose of 0.2 mg/kg cannot be very toxic. Some of the oral LD_{50} values in milligrams per kilogram are as follows: rat: 50; mouse: 25; dog: 80; rhesus monkey >24. Subchronic oral toxicity studies with rats (3 months), dogs (3 months), and rhesus monkeys (2 weeks) have demonstrated that monkeys tolerated ivermectin better than rodents, with a no-effect level as high as 1.2 mg/kg/day. Negative results with ivermectin were obtained in the following genotoxicity studies: Ames, DNA synthesis, and mouse lymphoma. These predictive tests were confirmed in developmental and reproductive toxicity studies carried out with mice, rats, and rabbits. The toxic dose to the pregnant female animal was closely related to the appearance of abnormalities in the offspring, indicating lack of toxicity to the embryo. The toxicity of abamectin is quite similar to that of ivermectin. When abamectin was administered to rats at a dose of up to 2.0 mg/kg for 105 weeks, there was no increase in tumor incidence. Abamectin in the feed at up to 8 mg/kg/day, administered to mice during a 94-week study, did not produce tumors. Dermal penetration ranged between 0.17% and 0.55% of the applied dose.

The toxicity of milbemycin D was investigated by Matsunuma et al. [45]. The LD_{50} value for 5–7-week-old male mice treated by intraperitoneal injection was 667.9 mg/kg. For rats of both sexes, the LD_{50} value was over 5000 mg/kg when treated subcutaneously. In subacute tests, rats of both sexes were treated orally at 250, 50, 10, and 2 mg/kg/day for 13 weeks. No toxic effects were apparent.

VII. USAGE IN AGRICULTURE

A. Animal Health

The use of ivermectin and abamectin in agriculture has been thoroughly reviewed, mainly by Merck researchers, and for publications up to 1988 the best source is the book edited by W. C. Campbell [4]. A second book, written primarily for the veterinarian and the parasitologist by Oakley [46], gives special emphasis on the gastrointestinal, respiratory, and vascular parasites of cats and dogs. In this review, a brief summary of older work is presented, followed by a more detailed description of reports published after 1988.

1. *Cattle*

Benz et al. [47] have reviewed the utility of ivermectin in cattle and other species. The route of administration can be subcutaneous, oral, or topical. The dose for the injectable formulation is 0.2 mg/kg body weight and is effective on a very large number of adult and fourth-stage nematodes and lungworms. Administration at 3, 8, and 13 weeks after the cattle were moved to the pasture gives excellent control. Ivermectin is also very effective on ectoparasites—mites: *Sarcoptes scabiei* var. *bovis* and *Psoroptes ovis*; ticks: *Boophilus microplus*, *B. decoloratus*; lice: *Haematopinus eurysternus* and others; grubs: *Hypoderma bovis* and *H. lineatum*. The pharmacokinetics of ivermectin administered intravenously were studied by Wilkinson et al. [48] and described as a three-compartment open model with elimination from the central compartment.

Ivermectin can also be administered to cattle orally, as a solution or as a paste, and the spectrum of activity on gastrointestinal nematodes and lungworms is similar to ivermectin administered subcutaneously, although not as effective on ectoparasites. The topical formulation contains 0.5% w/v of ivermectin and is highly effective on gastrointestinal nematodes, lungworms, and as an ectoparasiticide. Recently, Eagleson and Allerton [49] in Australia conducted 15 controlled feed lot trials with topically applied ivermectin on 339 cattle in 3 different states, at a dose rate of 0.5 mg/kg. The treatment was efficacious, and no signs of toxicity were evident. Soll et al. [50] applied ivermectin topically at 0.5 mg/kg to 12 cattle infected with *Sarcoptes scabiei*. After 14 days, no mites were found in scrapings from the treated animals, and clinical signs of mange were absent. Bisset et al. [51] also applied ivermectin topically on weaner cattle, and at slaughter 14 days after treatment, the efficacy on gastrointestinal nematodes was between 95% and 100%, however, the sizes of the *Cooperia spp.* and *Trichuris ovis* burdens did not differ from those in the control group.

Controlled-release delivery of ivermectin has also been developed. Pope et al. [52] reported that ivermectin was administered to cattle at a controlled zero order rate for 35 days via an osmotic pump. Taylor and Kenny [53] introduced orally into calves a bolus designed to release ivermectin for 120 days at 8 mg/day, this being equivalent to 0.04 mg/kg/day for the expected weight of 200 kg at the end of the delivery. Ticks (*Ixodes ricinus*) were put on the calves and observed routinely. There was no difference between treatment groups for female mortality, but ivermectin reduced the numbers of egg-laying and engorged females, and the number of larvae on the treated group was 3% of that of the control group.

Many trials with ivermectin have been carried out on all continents, most of them providing excellent results. One adverse effect was noted by Keck et al. [54] in an area of France where *Hypoderma lineatum* predominates over *Hypoderma bovis*. After ivermectin treatment for *Hypoderma,* oesophagitis, edema and ataxia occurred in 22% of the treated cattle. This was due to allergic inflammatory reaction to lysed *H. lineatum* larvae. Soll et al. [55] found ivermectin to be highly efficacious against eyeworm infections. Fifty cattle with natural infections of *Thelazia rhodesii* were injected subcutaneously at a dose rate of 0.2 mg/kg. After 8 days, the reduction in numbers of *T. rhodesii* was greater than 99% as compared to the controls.

Although there is no doubt on the efficacy of ivermectin on a wide spectrum of endoparasites and ectoparasites, its commercial success depends on a favorable cost-to-benefit ratio. The following studies were designed to address this point. Wohlgemuth et al. [56], using 953 cattle over a 2-year period, found that treated calves averaged at least 20 lbs more than calves in the control group. A significant economic benefit was achieved. Bauck et al. [57] in 1989 compared ivermectin with a topical organophosphate on growth rate and feed efficiency in a feedlot in Western Canada. Ivermectin-treated calves gained 0.08 kg/day more, required 0.23 kg less feed/kg gain, and required 11 fewer days occupation compared to the calves treated with an organophosphate. This translated into a net benefit of $7.04 per head of cattle. Soll et al. [58] found that 400 ivermectin-treated weaner cattle originating from a *Parafilaria bovicola* endemic area had an 80% reduction in *P. bovicola*-caused lesion area and affected tissue, when compared to the untreated controls. The benefit to cost ratio was 4:1 for the ivermectin treatment. Bauck et al. [59] found a significant weight gain in a trial involving 400 cattle which had been treated with ivermectin when compared to cattle that were treated with fenthion.

Abamectin, with a chemical structure identical to ivermectin except for having an extra double bond at C_{22} has also been registered for use on cattle in Australia. Its formulation and biological spectrum is very similar to that of ivermectin. As will be discussed later, abamectin is also being used as an acaricide.

2. *Sheep*

There are oral and injectable ivermectin formulations available for treating sheep. The recommended dose for both is 0.2 mg/kg body weight, affording excellent control against most gastrointestinal nematodes and lungworms found in sheep [47]. Ivermectin is also effective on *Haemonchus contortus, Ostertagia circumcincta,* and *Trichostrongylus colubriformis,* which are resistant to structurally unrelated anthelmintics. Ivermectin provides good ectoparasitic activity on *Oestrus ovis, Psoroptes ovis,* and *Sarcoptes scabiei.* In a recent publication, Sharma et al. [60] suggested a dose rate of 0.4 mg/kg for the control of *Dictyocaulus filaria.* Hughes and Levot [61] found ivermectin, abamectin, and 4"-deoxy-4"-epimethylamino avermectin B_1 to be highly effective on the sheep blowfly *Lucilia cuprina;* the LC_{50} values were between 0.0021 and 0.0028 mg/ Soll et al. [62] conducted a large trial in South Africa and found ivermectin to be highly efficacious on *Psoroptes ovis.* Easterly et al. [63] found ivermectin not to be effective on lungworm infections in free-ranging bighorn sheep. Reinemeyer et al. [64] in a survey of 126 sheep producers in Tennessee found the great majority used ivermectin for parasite control.

3. Pigs

Ivermectin is administered to pigs subcutaneously at the dose of 0.3 mg/kg in the same formulation as used for cattle and sheep. It provides excellent control on most gastrointestinal roundworms, kidney worms and lungworms, lice, and mites. Ivermectin is somewhat slow to act on lice and mites and appropriate precautions must be taken against reinfestation. Palmer [65] reported that ivermectin-treated pigs had reduced white-spot lesions on the liver at slaughter. Hiepe et al. [66], in a massive effort, eradicated clinically manifest mange by injecting 109,000 pigs within 24 hr. This was followed by immediate treatment of newly arrived animals and other prophylactic measures. To totally eradicate the *Sarcoptes suis* population, two treatments with ivermectin are recommended with a 14-day interval. Several publications dealt with the cost/benefit ratio of using ivermectin. Thus, Arends et al. [67] found that untreated sows had litters weighing 4.1 kg less than treated sows; further, treated sows consumed 1.9 kg less feed per weaned piglet, and piglets from treated sows were 5.8 kg heavier at slaughter than the controls. Stewart et al. [68] calculated from a trial of 90 pigs that the production cost for ivermectin-treated pigs was $1.53 per pig less than for the nontreated group.

4. Horses

The use of ivermectin in horses has been thoroughly reviewed by Campbell et al. [69], and most of the information presented in this section was summarized from that source. Ivermectin is administered by veterinarians to horses as an oral paste or as a clear liquid. The recommended dose is 0.2 mg/kg. In early trials, ivermectin was injected subcutaneously; however, adverse effects and even mortality occurred, leading to withdrawal of this formulation. It is believed these effects were not due to a species-specific toxicity of ivermectin but rather due to bacteria at the site of injection. As in most other species, ivermectin is highly efficacious on most nematodes and mites present in or on horses. Ivermectin was judged to be highly active against immature (L3 and L4) and adult *Parascaris equorum*, as judged by the elimination of eggs. In one study, naturally infected horses were killed 14 days after treatment; no larvae were found in the lungs. Several studies were carried out with experimentally infected horses to ascertain ivermectin activity on immature *P. equorum*. L3 larval stage was investigated by treating 11 days after infection, and necropsy took place on day 14. There was 100% clearance, whereas in the controls, many larvae were found in the lungs. The L4 larval stage was investigated by treating 28 days after the infection and necropsy took place 14 days after treatment. The number of ascarids in the intestines was greatly reduced. Ivermectin was 100% effective against adult *Strongylus vulgaris* and the migratory fourth-stage larvae. Activity on the larvae could best be demonstrated by necropsy 5 weeks after treatment. There was evidence that ivermectin had some prophylactic properties. Ivermectin was highly active on small strongyles as shown by egg counts in the feces. This was confirmed by necropsy on experimentally infected animals. Benzimidazole-resistant strongyles were also eliminated. Several studies have shown the high efficacy of ivermectin against *Dictyocaulus arnfieldi*, *Strongyloides westeri*, *Gastrophilus* spp., and the microfilaria of *Onchocerca* spp. Efficacy was also reported on summer sores and sarcoptic mange. In more recent work, Bell and Holste [70] compared routes of administration on 100 naturally infected horses. Fourteen days after treatment the fecal ova counts were down to zero, whether ivermectin was administered as a liquid by oral drench, nastrogastric tube, or as a paste. Nilsson et al. [71] advocated two treatments of ivermectin per season for total control and appreciable weight gain.

5. *Dogs and Cats*

a. IVERMECTIN   Ivermectin can be administered orally to dogs as tablets or as chewable treats. The dose is 0.006 mg/kg administered on a monthly basis and is used to inhibit the establishment of heartworms. The literature has been reviewed by Campbell [72]. Although microfilariae of *Dirofilaria immitis* are exceedingly sensitive to ivermectin, the adult worm appears to be immune, despite the fact that labeled studies have shown that the drug is distributed throughout the worm. Although not registered for use, ivermectin has shown outstanding activity against *Toxascaris leonina, Ancylostoma caninum, Trichuris vulpis,* and *Sarcoptes scabiei.* The safety of ivermectin when administered to dogs has been a contentious subject, although it is generally believed that long-haired collie breeds are especially sensitive. The subject has been reviewed at some length by Pulliam and Preston [73]. Rohrer and Evans [74] attempted to measure the amount of ivermectin transported across the blood-brain barrier of sensitive collies. No differences were found in ivermectin binding to plasma from sensitive and nonsensitive collies. Paul et al. [75] concluded that administering a dose three times over the recommended dose of chewable ivermectin for 12 months to sensitive collies did not produce any toxic effects. Clark et al. [76] compared the bioavailability of ivermectin tablets and chewable ivermectin treats by using tritium-labeled material. No differences were found.

b.   MILBEMYCIN D   Mainly due to patent reasons, the development of milbemycin D was slow indeed. Based on work performed in Japan, milbemycin D was registered against heartworm in the dog. Tagawa et al. [77] reported that milbemycin D was 100% effective at 0.1 mg/kg against *Ancylostoma caninum.* Horie and Noda [78] treated orally 278 dogs with milbemycin D and found that *Toxocara canis* was eliminated at 0.1 mg/kg, *Ancylostoma caninum* at 0.025 mg/kg, and *Trichuris vulpis* at 1.0 mg/kg. Sakamoto et al. [79] investigated the long-term effects of milbemycin D, dosed at 5 mg/kg on dogs. Twenty-six factors in the blood and nine factors in the urine were measured, followed by postmortem and histopathological examinations. There were no abnormal findings. Fukase et al. [80] reported that at a dose between 0.05 and 0.1 mg/kg, milbemycin D was efficacious in eliminating *Toxocara cati* and *Ancylostoma tubaeforme* in cats. Several workers have published on the preventive effect of milbemycin D against *Dirofilaria immitis.* Shiramizu et al. [81,82] treated 25 young dogs every 5 weeks, and adult dogs monthly at a dose of 1 mg/kg. No side effects were evident. Horie et al. [83] treated 34 dogs at the reduced dose of 0.1 mg/kg and observed a total prophylactic effect. Tagawa et al. [84] mimicked natural infection by inoculating beagles with larvae on 5–10 occasions over a 4–7-month period. None of the treated dogs (1 mg/kg) showed the presence of adult *D. immitis.* Sasaki [85] treated dogs at 1 and 5 mg/kg with or without circulating microfilariae. The adverse effects were not dose related and were severe in infected dogs or in dogs with circulating microfilariae present at the time of treatment. The dogs that were not infected at the time of treatment showed only mild adverse effects, or none at all. Takiyama et al. [86] summarized the efficacy of milbemycin D, and advocated a dose of 1 mg/kg administered once a month during the period of transmission. Adverse effects in dogs due to treatment with milbemycin D have been reported. Sasaki [87] noted a shock-like reaction in 8% of the *D. immitis* uninfected and infected dogs after milbemycin D treatment at 0.1–5 mg/kg. Clinical signs were depression, weak pulse, staggering, prostration, and a remarkable decrease in blood pressure. Electrocardiogram and hematological readings were also altered. The shock reaction occurred 1.5–4 hr after treatment, and the dogs recovered 1–4 hr after the shock took place, without supportive treatment. Dogs that displayed shock after the first treatment were reinjected; no adverse effects were noted. Kitagawa et al. [88] reported that 10% of microfilaria positive dogs developed dirofilarial hemoglobinuria after treatment with milbemycin D. Symptoms occurred 3–24 hr after administration, and recovery took place 20–117 hr later. Between 10 and 39 adult heartworms were present in the dogs, and it was theorized that suppression of cardiac functions initiates the movement of heartworms from the pulmonary arteries toward the venae cavae. Due to the known sensitivity of collies to ivermectin, their response to milbemycin D was studied. Sasaki et al. [89] treated orally 5 collies at 1.0, 2.5, and 5 mg/kg with milbemycin D for 10 consecutive days. Dogs on the high dose displayed neurological signs such as arrhythmia, salivation, mydriasis, and lethargy. None of the dogs died, and all recovered 24 hr after administration of the drug. Kitagawa et al. [90] investigated the concentration of milbemycin D in the plasma of rough-coated collies, shetland sheep dogs, and Japanese mongrels. At the 5.0-mg/kg dose, neurological signs were apparent in the collies and shelties, and the milbemycin D concentration in the plasma was 400 ng/ml. Although the mongrels were not affected, increasing the dose to 12.5 mg/kg led to similar neurological symptoms and high plasma levels of milbemycin D. Intravenous administration at 0.5 mg/kg caused no differences in pharmacokinetics between the collie–sheltie groups and the mongrel group. It was concluded that the difference between the breeds is due to differences in bioavailability.

6. *Miscellaneous Animals*

Ivermectin is an effective endoparasiticide and ectoparasiticide on a very large number of animals. Although only registered for some domestic animals, as discussed above, it has been extensively used "off label" on many exotic and not so exotic species. Soll [91] has reviewed such use and a summary is presented here. Most endoparasites inhabiting mice such as *Heligmosomoides polygyrus* (*Nematospiroides dubius*), *Syphacia obvelata*, *Strongyloides* spp., and *Aspiculuris tetraptera* are effectively eliminated by ivermectin at a dose between 0.3 and 2 mg/kg. Less promising results were obtained on *Toxocara canis* and *Trichinella* spp. Mites and triatomid bugs have also been eradicated from mice with ivermectin. Although rats were used extensively for toxicological studies, the effect of ivermectin on parasites harbored by the rat is not well known. Ivermectin has eliminated *H. contortus* and *T. colubriformis* from the jird; this has been used by Conder et al. [92] to test for new anthelmintics with a target parasite in a nontarget model, with good reproducibility and predictive value for the sheep. Richardson [93] has used ivermectin for the control of *Trixacarus caviae* in guinea pigs. Other laboratory animals, for example, monkeys, have also been treated. Dschobava and Tscherner [94] reported good control of *Prosarcoptes faini* in a colony of Hamadryas baboons. Joseph et al. [95] obtained good control of pulmonary acariasis with a single injection of ivermectin at 0.2 mg/kg in rhesus macaques. Kutzer et al. [96] found ivermectin to be highly effective on *Dictyocaulus viviparus*, *Varestrongylus sagitatus*, and other strongylid nematodes in red deer. Nordkvist et al. [97] carried out a study on about 50 reindeer in Sweden and found ivermectin to provide excellent control on *Oedemagena tarandi*, *Cephenomyia trompe*, *Dictyocaulus viviparus*, and others. During the winter the treated animals lost less weight than the control group. Several workers have reported on the effectiveness of ivermectin for treating camels. At 0.2 mg/ kg, ivermectin controlled many gastrointestinal nematodes and sarcoptic mange mites Robin et al. [98]. Ivermectin has been registered for this use in a limited number of countries Soll [91]. Control on lice was also achieved in llamas by Fowler [99]. Ivermectin was successfully used in zoological gardens on such varied animals as antelopes, tapirs, camels, ponies, wild boars, zebras, elephants, and bears [100, 101]. Other exotic species treated with ivermectin were snakes [102], leopard frogs [103], and budgerigars [104].

7. Milbemycin-like Compounds

Two other compounds, first discussed in the structure determination section and related to the milbemycins, are being developed for animal health use by American Cyanamid. Carter et al. [25] claimed good activity for nemadectin on endoparasites and ectoparasites of many domestic animals. Moxidectin, a synthetic derivative of nemadectin (see Figure 4) has been evaluated as an insecticide by Webb [105] and as an anthelmintic on horses [106] and cattle [107–110]. Based on published results, the efficacy of these compounds is promising; however, they are almost certain to be co-resistant with ivermectin and milbemycin D.

B. Plant Health

The major component of the avermectin mixture obtained by fermentation is avermectin B_1, also called abamectin. It was developed as an insecticide and acaricide by Merck and is being sold under various trade names. It is an extremely potent agent with application rates varying between 0.005 and 0.025 lbs of active ingredient per acre. The literature up to about 1988 has been thoroughly reviewed by Dybas [111] and a synopsis will be presented here. More emphasis will be given to recent publications.

The effectiveness of abamectin on mites attacking citrus was recognized early. In several trials, good control on rust mites was obtained at rates of 0.0125–0.05 lbs/acre; however, the control was of short duration. The combination of abamectin with paraffinic oil increased substantially the intrinsic potency of abamectin and controlled the mite population for 16 weeks. French and Villarreal [112] found abamectin in combination with oil to be superior to other marketed miticides for the control of citrus rust mite. Good control was also achieved on the broad mite using the same dose and oil as an adjuvant. Control of red mites was less satisfying, even when oil was added to the formulation; this was attributed to the restricted movement of abamectin into the leaf, and its instability under field conditions [113]. Morse et al. [114] reported abamectin to be nontoxic to predacious mites and advocated its use in an integrated pest-management program.

Abamectin has been utilized in greenhouses and in the field on ornamentals. Good activity was seen against mobile forms of the two-spotted spider mite but not on eggs. When applied at 4.5 ppm on roses or chrysanthemums, excellent activity and long-term residual control was obtained. Abamectin is unstable to sunlight, and its long residual can be explained by its ability to permeate the leaf and be stored in the mesophyll, from where it is consumed by the mites. At the rate of 0.01–0.02 lbs/acre, abamectin provided good protection on foliage and flowers from leaf miner adults and early instar larvae; however, it was not toxic to the eggs. In a laboratory study, Harris et al. [115] compared the effectiveness of nematode control versus spraying with abamectin on chrysanthemums burdened with *Liriomyza trifoli*. There was no difference, and he advocated control with nematodes instead of abamectin, for environmental reasons. Several trials were made to establish the effect of abamectin on thrips, which are common to ornamentals [111]. In a laboratory study, abamectin was applied at 40 ppm to western flower thrips, resulting in about 50% mortality; it was more toxic to *Ecinothrips americanus*, which were placed on previously treated impatiens leaves. Although 90% mortality was seen on adult and second instar immatures, the residual activity was quite poor. Some reduction in the hatch was also apparent but insufficient to be viable commercially. Control of *Frankliniella occidentalis* with abamectin on roses, carnations, and dahlias at various rates was also poor. Immaraju et al. [116] in a 1992 publication claims to have discovered the first documented case of resistance to abamectin on a field population of *F. occidentalis*. Resistance ratio values were 18-fold in one test area, and 798-fold in a second area. Similar levels of resistance to several other insecticides were also apparent.

The use of abamectin on corn, cotton, and soybeans has been researched. Laboratory studies on cotton plants have demonstrated that abamectin is highly effective on spider mites, including mites that are resistant to regularly used miticides. Several trials in cotton fields have confirmed the findings from the laboratory. An application of 0.008 lbs/acre gave 90% control 13 days after abamectin was applied, and residual control up to 40 days after application. At the higher rate of 0.015 lbs/acre, abamectin controlled 98% of the mite population, and 100% control was apparent after 21 days; the aphid population was not controlled. The recommended range of abamectin on cotton for control of mites is 0.01–0.02 lbs/acre. This range will provide better than 80% reduction of the spider mites within 7–10 days after treatment and over 90% mortality after 14 days. The slow action is due to lack of ovicidal activity. Although the mites are controlled soon after application, the eggs hatch normally, and several days of feeding are required before knockdown occurs. The long residual activity is achieved mainly when abamectin is applied in combination with an emulsifiable oil. This enhances translaminar action, and the leaf thus protects abamectin from the rapid degradation that normally occurs under field conditions. The toxicity of abamectin against lepidoptera is species dependent. The LC_{90} larval values reported for tobacco hornworm, corn earworm and southern armyworm are 0.02, 1.5, and 6.0 ppm, respectively. Under laboratory conditions, toxicity at low dose was achieved against the pink bollworm moth. Corbitt et al. [117] investigated the action of abamectin on various larval stages. When topically applied, the relative toxicity of abamectin decreased from the third to the fourth and fifth larval instars of *Spodoptera littoralis*, but a 200-fold increase in toxicity was observed from the fifth to the sixth instar. Injection increased the toxicity 20-fold, suggesting varying rates of penetration. When placed on treated cotton, abamectin was most toxic to *Heliothis virescens* and least toxic to *S. littoralis*. Further, abamectin was eightfold more active against third instar *Heliothis armigera* than against *S. littoralis*; the reason was attributed to different feeding rates. In a direct comparison between the effect of abamectin and milbemycin D on gypsy moth *Lymantria dispar*, Deecher et al. [118] found that at 5 ppm after 2 hr on an artificial diet, 50% of the third instars were paralyzed by milbemycin D but were not affected by abamectin; however, paralysis did occur after 24 hr. Generally, abamectin was more toxic to third instars than milbemycin D. Abamectin is also used on vegetable crops. At a range of 0.01–0.02 lbs/acre, good control of *Liriomyza trifolii* was reported on pepper, tomato, and celery; the addition of a surfactant enhanced control by penetration. Using the same range, satisfactory control was achieved on the two-spotted spider mite, tomato russet mite, and tomato pinworm. Integrated pest-management programs emphasize the use of an insecticide/acaricide that does not harm beneficial arthropods. Abamectin-treated foliage had no effect on *Diglyphus intermedius*, *D. begini*, *Chrysonotomyia punctiventris*, *Chrysocharis parksi*, *C. ainsliei*, and *Halticoptera circulus*. In a more recent publication, Zhang and Sanderson [119] found that at concentrations between 0.08 and 16 ppm abamectin had little effect on two predatory mites, *Phytoseiulus persimilis* and *Acari tetranychidae*; however, at the higher rates, reproduction was decreased.

Abamectin is registered for use on the red imported fire ant and is formulated as a bait containing 0.011% w/w abamectin in soybean oil on corn grits. The bait is fed by worker ants to the queen, causing permanent sterility. Hara and Hata [120] in a recent article found abamectin to be very effective against the big-headed ant, *Pheidole megadcephalia*. Cochran [121] found abamectin to be effective against both susceptible and pyrethroid-resistant German cockroaches.

C. Human Health

It is common knowledge that many drugs developed as anthelmintics for animal use become the drugs of choice for the treatment of humans, especially in tropical countries. Ivermectin is no exception, and it can safely be said that it revolutionized the treatment of onchocerciasis, for which it is registered. It is distributed free of charge in many countries, under the name mectizan. Onchocerciasis is caused by *Onchocerca volvulus* and is transmitted via an insect vector, *Simulium* species, in which the microfilariae develop into the infective C-3 larvae. These larvae grow into adult worms after about 12 months and persist for many years under the skin, producing millions of microfilariae. More than 20 million people suffer from the disease, which causes debilitating dermatitis and sclerosing keratitis which eventually leads to blindness. Disease of the lymph nodes is frequently observed, and there is a fourfold increase in mortality in adults who have lost their eyesight. People who are severely infected with *O. volvulus* have symptoms of a cell-mediated immune response, which are more severe than those observed in individuals who are less infected. It is thought that onchocerciasis is due to an elevated cell-mediated response to the antigen released by the worm [122]. The number of publications on the beneficial effects of ivermectin for people suffering from onchocerciasis has grown at a rapid pace, and a literature survey using the key words "ivermectin and onchocerciasis" yielded over 300 publications. Fortunately, Campbell [123] has written an excellent review on the use of ivermectin in humans and has covered the literature to about 1991. A summary of earlier work is presented here, and publications that appeared after that date will be covered in more detail.

Although ivermectin may have utility on human intestinal nematodes and other filarial infections, at present it is only approved for the treatment of onchocerciasis, and this is the only disease discussed in this section. Egerton et al. [124] was the first to describe the efficacy of ivermectin on *Onchocerca cervicalis* in horses, and this observation prompted trials in humans. Initial dosing was very low, at 0.005 mg/kg given to *O. volvulus*-infected men. The dose was raised in increments up to 0.2 mg/kg, and it was established that after a single oral dose of ivermectin, there was a reduction of microfilariae in the skin; the low numbers lasted for about a year. No serious adverse effects were apparent, and damage to the eyes was absent. These very encouraging results led to many more trials in various parts of the world, starting with a few patients in a hospital setting, and progressing to very large-scale community-based programs. Many studies were undertaken to elucidate the effect of ivermectin on *O. volvulus*; none of them, however, yielded a definitive answer. Although it is clear that there is a direct effect of ivermectin on microfilariae in the skin, it is difficult to explain its long-lasting effect followed by slow recovery or their very slow migration from the eye. The reappearance of microfilariae almost a year after treatment indicates that ivermectin is not macrofilaricidal. It is known from metabolism studies that almost all of the applied dose is excreted in a matter of days; therefore, residual effects cannot account for the long-term effect. In order to clarify the activity of ivermectin on adult worms, nodules were surgically removed from patients treated with ivermectin, and the worms were found to be very alive indeed. Nodules were removed at various intervals after treatment, and the following observations could be made. After 1 month, the microfilariae in the uterus were normal. After 2 months there was an increase in number; however, many showed abnormalities or were dead. After 3 and 6 months, almost all microfilariae were dead, but at 9 and 12 months, the uterus of the adult contained a mixture of abnormal and normal microfilariae. From this study it was concluded that although embryogenesis continued despite ivermectin treatment, no expulsion occurred, leading to death and resorption. With passage of time, the worms regained their ability to procreate. It was also established that even repeated treatment at short intervals did not kill the adult worm. Studies designed to measure the effect of ivermectin on the transmission of the disease have shown that the microfilariae uptake by flies, which spread the disease, is very much decreased when the whole community has been treated with ivermectin. It was concluded that in Central America, where the disease is transmitted by *Simulium ochraceum*, ivermectin may lead to eradication of the disease. In Africa, however, the vector is *Simulium yahense* and the geographical conditions are very different, making it less likely that total suppression of the disease can be achieved in the near future.

For a drug widely used in humans, safety is of paramount importance. Already in the very early trials it became evident that side reactions caused by ivermectin were less severe and less common than the ones observed after treatment with diethylcarbamazine (DEC), the drug commonly used in the past. These symptoms consisted or pruritus, fever, rashes, edema, ocular damage, and, in a few cases, postural hypotension. In one study, over 50,000 people were treated, and 9% had some adverse reactions, 2.4% had moderately adverse reactions, and 0.24% had severe reactions. These side effects were particularly evident with patients who carried a high load of microfilariae, indicating that the reaction could be due to the great number of dead microfilariae, not to the inherent toxicity of the drug. Repeated treatment with ivermectin illicited even milder side effects than the first treatment. Duke et al. [125] conducted a trial in which patients were given six doses at 0.15 mg/kg, over a period of 12 weeks with very few side effects. Despite this high dose, ivermectin had no effect on the adult worms, as shown by the surgical removal of nodules from treated patients. Collins et al. [126] treated residents of five hyperendemic communities in Guatemala with ivermectin every 6 months for 30 months. Significant reductions in the prevalence and intensity of skin infections were achieved. In one trial in which 92.7% of the population participated, prevalence was reduced from 74% at pretreatment, to 34.9% after four treatments. Further, the microfilarial skin density values were decreased to a level associated with low infectiousness for the vector, *Simulium ochraceum*. Side effects were mainly pruritus and facial edema. A large campaign was carried out in Cameroon by Prodhon et al. [127]. Over 20,000 people were treated at least once, and in some cases four times, at 6 months or yearly intervals. Clinical, parasitological, and ophthalmological examinations were done before and after treatment, side effects were noted, and arterial blood pressure monitored on a small sample. Vector transmission was evaluated by capture and dissection of blackflies. Microfilarial loads were reduced by 90% and 60%, 6 months and 1 year after treatment, respectively. Carriers of microfilariae in the anterior chamber of the eye were also reduced, and there also was a decline of chorioretinal lesion. The conclusion of the study was that the best therapy in a hyperendemic area consisted of two doses administered every six months. Whitworth [128] in a series of articles described ivermectin treatment in Sierra Leone. Four doses at six monthly intervals were given to 1745 patients [129]. After the first treatment, more adverse reactions were reported than after the second treatment. Adverse reactions were correlated with the size of the microfilarial load. More patients returned for retreatment after ivermectin than after receiving placebo, and adverse reactions were mild. In a different study [130], villagers, all suffering ocular onchocerciasis, were dosed at six monthly intervals with ivermectin or placebo. The treated group had significantly lower prevalences of microfilariae in the anterior chamber and cornea, and also lower punctate keratitis and iritis. Prothrombin ratios [131] were also recorded, and it was concluded—contrary to previous reports—that ivermectin has only a minimal effect on coagulation and presents no danger to the patients participating in large campaigns. Ivermectin did not reduce the degree of itching before and for 6 months after administration of the fourth dose of ivermectin. Although this was disappointing, it did not detract from the overall success of treating this dreadful disease [132]. Duke et al. [133], in contrast to other workers, reported mortality on adult worms caused by ivermectin. Patients were treated with 4, 8, or 11 doses of ivermectin (0.15 mg/kg) given once every 3 months. Four months after the last treatment, that is, 13, 25, or 34 months after the beginning of the trial, the nodules were excised. At the 25th (8th dose) and 34th (11th dose) month time period, mortality in female worms was in excess of 25.5% and 32.6%, respectively, as compared to the controls. None of the females were producing microfilariae, and embryo production was only 7.7% and 18.2% as compared to the controls. The number of live male worms in the nodules was also well below that of the controls.

By the end of 1991, 23 countries in Africa and 5 in the Americas had commenced campaigns to reach and treat as many of their people as feasible. Roughly, 2 million people have been treated with ivermectin to date, and at least another 6 million need treatment. Despite the generosity of Merck in supplying ivermectin for human use free of charge, large funds are required to reach and distribute the drug to persons in dire need of treatment.

VIII. ENVIRONMENTAL FACTORS

Due to their novel structure and relatively toxic nature, ivermectin, abamectin, and the milbemycin analogs were thoroughly investigated for their environmental impact [134]. Whatever the route of administration, ivermectin and its metabolites are excreted by most animals in the feces as shown by experiments with labeled compounds. Halley et al. [135] has reviewed the environmental impact of ivermectin and has calculated that the highest concentration of ivermectin in the soil would be about 0.2 ppb due to the spread of manure around commercial feedlots. Ivermectin and abamectin have low water solubility and are bound strongly to organic matter in soil as shown by various leaching experiments [136] when only ivermectin metabolites were detected in the effluent; these have a lower toxicity than ivermectin. Soil stability is dependent on temperature. In the summer, the half-life is about 2 weeks, whereas in the winter, it varies between 3 and 8 months. Both ivermectin and abamectin are rapidly degraded when exposed to sunlight as a thin film [137, 138]. No toxicity to soil microorganisms was found at the levels expected in the soil. Ivermectin is highly toxic to fish, such as trout and blue gill, with LC_{50} values of 3.0 and 4.8 ppb, respectively. This toxicity is greatly reduced, however, by the strong adsorption of ivermectin on suspended soil particles found in rivers and lakes [139].

One of the most controversial environmental aspects of ivermectin concerns its impact on cattle dung. As known from residue studies, ivermectin is excreted mainly unchanged by most animals, and due to its strong insecticidal activity, it affects both harmful and beneficial insects. Roncalli [140] has reviewed the literature up to about 1988. Various flies associated with the droppings of farm animals cause considerable damage by slowing food conversion and/or physical damage to animals. The residual ivermectin in the excreted dung prevents the development of the housefly and hornfly from larvae. When administered daily, at a low oral dose, it is highly effective on face flies, stable flies, and houseflies. Several trials were carried out on *Onthophagus gazella* (dung beetle) in Australia, with both ivermectin and abamectin. While at 0.3 mg/kg, the compound had no effect on the adult beetle; larvae failed to develop for up to 21 days after treatment. Somewhat similar results were observed with *Onthophagus binodis*. In 1987, Wall and Strong [141] published an article on the environmental consequences of treating cattle with ivermectin. Under field conditions, the feces of calves receiving 0.04 mg/kg/day did not degrade as fast as the feces of placebo-treated calves. This was attributed to the absence of feces-degrading insects, which, in turn, caused substantial fouling of the pasture land. After the appearance of this publication, some workers confirmed the detrimental effect of ivermectin on the degradation of cow dung, whereas others presented data arguing the opposite. Wardhaugh and Rodriguez [142] found that the dung from calves injected with ivermectin (0.2 mg/kg) killed the larvae of *Orthelia cornicina* for 32 days after treatment. *Copris hispanus* also suffered mortality, and in surviving beetles, feeding was suppressed, and the oviposition rate was reduced. Madsen et al. [143] in Denmark studied the residual effects of ivermectin in cattle dung under laboratory and field conditions. They reported marked inhibition of many beetles and larvae. Houseflies were affected for up to 2 months by ivermectin-treated dung. They concluded that decreased decomposition was due to beetle mortality, which was further aggravated by slow-release ivermectin treatments. Strong [144] reviewed the impact of avermectins on insects associated with cattle dung. He claimed very deleterious effects on adults, and larvae of *Diptera* and *Coleoptera*, at both high and low levels. Five-week-old dung of once-injected animals still showed harmful activity against insects, and when cattle were treated at 40mg/day/cow, their dung did not degrade for at least 3 months. He advocated further research on the consequences of nonlethal residues of the avermectins on the environment. Quite the opposite was claimed by many researchers, stating that ivermectin had no influence on the breakdown of cow dung. In a field study described by Jacobs et al. [145], 400 cattle were grazed for 5 months and treated with ivermectin at 21, 56, and 91 days. Ten months after the beginning of the trial, no bovine fecal residue was found, indicating a complete breakdown in less than 7 months. Schaper and Liebisch [146] in Germany also saw no delay in the decomposition of cattle dung under field conditions. To complicate matters further, Wardhaugh and Mahon [147] claimed that under laboratory conditions, dung from ivermectin-treated animals attracted more beetles than from nontreated animals. For cattle, this was still evident in feces produced 25 days after treatment, and these findings were also confirmed under field conditions; the authors also claimed that treated pats had a higher population of beetles than untreated ones. Mahon and Wardhaugh [148] in Australia showed that dung from treated sheep did not have a significant effect on *L. cuprina* populations. Ewert et al. [149] in a thorough study at the University of Illinois concluded that repeated ivermectin treatment in horses did not cause prolonged dung degradation, which, in turn, would have caused fouling of the pasture land. Although most of these investigations were based on bioassay, a physical procedure for the determination of ivermectin in cow dung has been developed by Sommer et al. [150].

IX. RESIDUE AND METABOLISM

Analytical procedures for determining ivermectin residues in tissue and plasma have been thoroughly described by Downing [151] and will not be repeated here. The following results which have appeared after publication of Campbell's book are of interest. Slanina et al. [152] investigated the effect of cooking on ivermectin residues in the edible tissues of swine and cattle and found that in minced beef boiling decreased the ivermectin residual by 45%, whereas frying led to a 50% decrease. The suggested withdrawal time for ivermectin in the edible tissue of swine is 21 days, and for cattle it is 28 days. Chiu and Lu [153] have reviewed the metabolism of ivermectin and the residual amount left in tissue. An interesting recent publication [154] gave further details on the absorption, tissue distribution, and excretion of tritium-labeled ivermectin in cattle, sheep, and rat. In a following article by Chiu et al. [155] the metabolic disposition of ivermectin in swine was described. Prabhu et al. [156] used high-performance liquid chromatography (HPLC) with fluorescence detection to determine ivermectin levels in swine at the parts per billion level. The recovery at levels between 6 and 150 ppb in the kidney, liver, fat, and muscle was 85%, with a quantitation limit of 5 ppb and a limit of detection of 1 ppb. The results obtained were similar to the ones obtained by the reverse isotope dilution method. The withdrawal time suggested was 5 days when ivermectin was given in feed at 2 ppm for 7 days. If treatment was by subcutaneous injection at 0.4 mg/kg, the withdrawal time suggested was 18 days. As previously mentioned, Sommer et al. [150] developed an analytical procedure suitable for determining fecally excreted ivermectin in cow dung.

A number of recent publications have appeared on the residue determination of abamectin on various crops. Feely and Wislocki [157] investigated the acetone unextractable residues in celery with the aid of tritiated and ^{14}C-labeled abamectin. About 16% in the leaves and about 8% in the stalks was unextractable. Some of the labeled nonextractable residue was incorporated into glucose. Using HPLC, Vuik [158] developed a rapid procedure for the determination of abamectin in lettuce and cucumber. Also using HPLC, Jongen et al.

[159] developed procedures for the determination of occupational exposure to abamectin and analyzed for residues on cotton gloves and on greenhouse foliage; air concentration could not be measured due to decomposition. On exposure to sunlight, the E (trans) 8,9 double bond isomerizes to the Z (cis) isomer. Prabhu et al. [160] developed a rapid and sensitive HPLC method for the quantitative estimation of these two isomers; the recovery was above 84%. Halley et al. [161] compared the metabolism of ivermectin and abamectin. In cattle, sheep, and rats, ivermectin was metabolized to the 24-hydroxymethyl compound (24-OHMe). In swine, the main component isolated was the 3″-O-desmethyl compound (3″-ODMe). In contrast, abamectin in cattle and goats yielded a mixture of 24-OHMe (major) and 3″-ODMe (minor) metabolites. In rats, 3″-ODMe was the major metabolite and 24-OHMe was the minor one. Sadakane et al. [162] administered to rats milbemycin α_3 and α_4, tritium labeled at the 5-position. Within 7 days almost all the drug was excreted. The residual was mainly in the fat and liver. Milbemycin α_3 was metabolized faster than milbemycin α_4. The main metabolites were the mono-hydroxylated and dihydroxylated compounds. The metabolism of abamectin was studied in various types of soil [137] and the half-life varied between 20 and 47 days. Extensive degradation took place; the major metabolite was shown to be the 8α-hydroxy compound. In sterile soil, no metabolism was observed.

X. RESISTANCE

As with all anthelmintics and insecticides, the development of resistance in the avermectin and milbemycin families was only a question of time. In 1988 in South Africa, van Wyk and Malan [163] first reported the development of resistance to ivermectin in four strains of *Haemonchus contortus*. In one strain, resistance developed rapidly, despite rotation with structurally unrelated anthelmintics. In follow-up studies, van Wyk and co-workers [164, 165] found four strains of *H. contortus* resistant to ivermectin, one them also being resistant to the benzimidazoles and the salicylanilides. In one case, resistance developed after as few as three treatments, suggesting cross-resistance between ivermectin and structurally unrelated anthelmintics. The spread of resistant *H. contortus* strains may be accelerated by the practice of transporting rams to many farms across South Africa in order to improve the quality of merino wool. Craig and Miller [166] in 1990 noted the development of resistance to *H. contortus* in goats treated with ivermectin for 5 years, with occasional rotation to the benzimidazoles. Resistance was apparent after both oral and subcutaneous administration. In a second publication, Devaney et al. [167] reported ivermectin-resistant strains of *H. contortus* in goats and calves that were treated with three to four times the recommended dose of ivermectin. Echevarria et al. [168] in Brazil also found ivermectin resistance of *H. contortus* in lambs artificially infected. Cross-resistance to albendazole but not levamisole was also apparent. In a second publication [169], two groups of lambs were infected with ivermectin-resistant or -susceptible strains of *H. contortus* and treated with ivermectin on various postinfection days. After necropsies, it was determined that although ivermectin was effective on susceptible strains, resistance to ivermectin by the resistant strain was present as early as the fourth larval stage. Pankavich et al. [170] claimed that moxidectin removed ivermectin-resistant worms from sheep. Shoop [171] questioned these findings and tested ivermectin-resistant isolates of *Ostertagia circumcincta* and *Trichostrongylus colubriformis* in sheep. Ivermectin needed 23 and 6 times more material to remove 95% of resistant *O. circumcincta* and *T. colubriformis*, respectively, as compared to nonresistant isolates; moxidectin needed 31 and 9 times more material to remove the resistant isolates. The findings of Conder et al. [172] also disagreed with the Pankavich [170] paper. Using standard microelectrode techniques, they showed that moxidectin induced loss of membrane resistance in the shore crab muscle preparation, in a pattern similar to that of ivermectin, indicating a common mode of action. Also, moxidectin, in jirds infected with ivermectin-resistant *H. contortus*, cleared only 47% of the worms at a dose that cleared 98% of the ivermectin-susceptible strain. These results strongly suggest co-resistance between moxidectin and ivermectin. These findings are not surprising considering the structural similarity between the two compounds. Resistance to abamectin by houseflies [173], German cockroaches [174], and Colorado potato beetles [175] was also observed.

XI. MARKET SIZE

In 1989, worldwide animal health sales were estimated at US$9 billion, with the United States accounting for $2.57 billion. These figures include feed additives, pharmaceuticals, biologics, and diagnostics. Only one animal health product—ivermectin—had sales that would place it among the top 50 human pharmaceutical compounds. During 1991, worldwide sales for ivermectin were US$565 million [176]; past sales figures and future estimates are presented in Table 1. Table 2 shows

Table 1 Worldwide Sales and Estimate for Ivermectin Dollars in Millions

1988	1989	1990	1991	1992E	1993E	1994E	1995E
$395	$455	$535	$565	$590	$615	$640	$665

Table 2 1991 U.S. Ivermectin Sales

Eqvalan (equine anthelmintic)	$8,466
Heartgard (canine heartworm)	67,206
Ivomec (rumen anthelmintic)	105,390
Ivomec F (rumen anthelmintic)	8,372
Ivomec Sheep (rumen anthelmintic)	842
Ivomec (swine anthelmintic)	10,480
Total ivermectin sales	$200,756

1991 U.S. sales figures for various formulations of ivermectin. Worldwide sales during 1991 for Abamectin (avermectin B_1) approached US$50 million [177]. Some Wall Street analysts expect moxidectin—an anthelmintic marketed by American Cyanamid—to reach 15–20% of the market share of ivermectin in the mid to late nineties; in Argentina it has already achieved a 10% market share.

The milbemycin derivative Interceptor marketed by Ciba-Geigy against canine heartworm had sales of about $18 million in the United States.

REFERENCES

1. H. Mishima, M. Kurabayashi, C. Tamura, S. Sato, H. Kuwano, and A. Saito, *Tetrahedron Lett.*, 10:711–714 (1975).
2. a, R.W. Burg, B.M. Miller, E.E., Baker, J. Birnbaum, S.A. Currie, R. Hartman, Y-L. Kong, R.L. Monaghan, G. Olson, I. Putter, J.B. Tunac, H. Wallick, E.O. Stapley, R. Oiwa, and S. Omura, *Antimicrob. Agents Chemother.*, 15:361–367 (1979); b, T.W. Miller, L. Chaiet, D.J. Cole, L.J. Cole, J.E. Flor, R.T. Goegelman, V.P. Gullo, H. Joshua, A.J. Kempf, W.R. Krellwitz, R.L. Monaghan, R.E. Ormond, K.E. Wilson, G. Albers-Schönberg, and I. Putter, *Antimicrob. Agents Chemother.*, 15:365–371 (1979); c, J.R. Egerton, D.A. Ostlind, L.S. Blair, C.H. Eary, D. Suhayda, S. Cifelli, R.F. Riek, and W.C. Campbell, *Antimicrob. Agents Chemother.*, 15:372–378 (1979).
3. (a) J.C. Chabala, M.H. Fisher, and H.M. Mrozik, U.S. Patents 4,171,314 and 4,173,571 (1979). (b) H. Mrozik, J.C. Chabala, P. Eskola, A. Matzuk, F. Waksmunski, M. Woods, and M.H. Fisher, *Tetrahedron Lett.*, 24:5333–5336 (1983).
4. W.C. Campbell (ed.), *Ivermectin and Abamectin*, Springer-Verlag, New York, 1989.
5. M.H. Fisher and H. Mrozik, *Macrolide Antibiotics* (S. Omura, ed.), Academic Press, New York, p. 553 (1984).

6. M.H. Fisher and H. Mrozik, *Annu. Rev. Pharmacol. Toxicol.*, 32:537–53 (1992).

7. H.G. Davies and R.H. Green, *Natural Product Reports*, 3:87–121 (1986).

8. H.G. Davies and R.H. Green, *Chem. Soc. Rev.*, 20:211–269 (1991).

9. H.G. Davies and R.H. Green, *Chem. Soc. Rev.*, 20:271–339 (1991).

10. M.V.J. Ramsay, S.M. Roberts, J.C. Russell, A.H. Shingler, A.M.Z. Slawin, D.R. Sutherland, E.P. Tiley, and D.J. Williams, *Tetrahedron Lett.*, 28:5353–5356 (1987).

11. J.D. Hood, R.M. Banks, M.D. Brewer, J.P. Fish, B.R. Manger, and M.E. Poulton, *J. Antibiotics*, 42:1593 (1989).

12. M.A. Haxell, H. Maeda, and J. Tone, European Patent EP 334,484 (1989).

13. C.J. Dutton, S.P. Gibson, A.C. Goudie, K.S. Holdom, M.S. Pacey, and J.C. Ruddock, *J. Antibiotics*, 44:357–365 (1991).

14. T.S. Chen, E.S. Inamine, O.D. Hensens, D. Zink, and D.A. Ostlind, *Arch. Biochem. Biophys.*, 269:544–547 (1989).

15. S. Omura, H. Ikeda, and H. Tanaka, *J. Antibiotics*, 44:560–563 (1991).

16. T. Rezanka, H. Míková, and M. Jurková, *FEMS Microbiol. Lett.*, 96:31–36 (1992).

17. K. Nakagawa, K. Sato, Y. Tsukamoto, and A. Torikata, *J. Antibiotics*, 45:802–805 (1992).

18. H. Mishima, *Pestic. Chem. Hum. Welfare Environ. Proc. Int. Congr. Pestic Chem.*, 2:129 (1983).

19. H. Mishima, J. Ide, S. Muramatsu, and M. Ono, *J. Antibiotics*, 36:980 (1983).

20. G. Albers-Schönberg, B.H. Arison, J.C. Chabala, A.W. Douglas, P. Eskola, M.H. Fisher, A. Lusi, H. Mrozik, J.L. Smith, and R.L. Tolman, *J. Am. Chem. Soc.*, 103:4216–4221 (1981).

21. J. P. Springer, B.H. Arison, J.M. Hirshfield, and K. Hoogsteen, *J. Am. Chem. Soc.*, 103:4221–4224 (1981).

22. A. Neszmélyi, D. Machytka, A. Kmety, P. Sándor, and G. Lukacs, *J. Antibiotics*, 42:1494 (1989).

23. D. Diez-Martin, P. Grice, H.C. Kolb, S.V. Ley, and A. Madin, *Tetrahedron Lett.*, 31:3445–3448 (1990).

24. G.H. Baker, R.J.J. Dorgan, J.R. Everett, J.D. Hood, and M.E. Poulton, *J. Antibiotics*, 43:1069 (1990).

25. G.T. Carter, J.A. Nietsche, and D.B. Borders, *J. Chem. Soc., Chem. Commun.*, 402 (1987).

26. S. Rajan and G. W. Stockton, *Magn. Reson. Chem.*, 27:437 (1989).

27. M.A. Haxell, B.F. Bishop, P. Bryce, K.A.F. Gration, H. Kara, R.A. Monday, M.S. Pacey, and D.A. Perry, *J. Antibiotics*, 45:659 (1992).

28. H. Mrozik, B.O. Linn, P. Eskola, A. Lusi, A. Matzuk, F.A. Preiser, D.A. Ostlind, J.M. Schaeffer, and M.H. Fisher, *J. Med. Chem.*, 32:375–381 (1989).

29. T.K. Jones, H. Mrozik, and M.H. Fisher, *J. Org. Chem.*, 57:3248–3250 (1992).

30. T.A. Blizzard, G.M. Margiatto, H. Mrozik, W.L. Shoop, R.A. Frankshun, and M.H. Fisher, *J. Med. Chem.*, 35:3873–3878 (1992).

31. B. Frei and H.B. Mereyala, European Patent Appl. EPO 235085; *Chem. Abstr.*, 109:110836d (1987).

32. T. Blizzard, L. Bostrom, G. Margiatto, H. Mrozik, and M.H. Fisher, *Tetrahedron Lett.*, 32:2723–2726 (1991).

33. S. Hanessian and P. Chemla, *Tetrahedron Lett.*, 32:2719–2722 (1991).

34. T.L. Shih, M.A. Holmes, H. Mrozik, and M.H. Fisher, *Tetrahedron Lett.*, 32:3663–3666 (1991).

35. P.T. Meinke, P. Sinclair, H. Mrozik, S. O'Connor, D.A. Ostlind, W.L. Shoop, B.H. Arison, and M.H. Fisher, *Bioorg. Medic. Chem. Lett.*, 2:537–540 (1992).

36. T. Blizzard, G. Margiatto, B. Linn, H. Mrozik, and M.H. Fisher, *Bioorg. Medic. Chem. Lett.*, 1:369–372 (1991).

37. B. Fraser-Reid, J.C. López, and R. Faghih, *Tetrahedron*, 48:6763–6768 (1992).

38. M.J. Hughes, E.J. Thomas, M.D. Turnbull, R.H. Jones, and R.E. Warner, *J. Chem. Soc., Chem. Commun.*, 755 (1985).

39. G.I. Kornis, M.F. Clothier, S.J. Nelson, F.E. Dutton, and S.A. Mizsak, *Synthesis and Chemistry of Agrochemicals* (D.R. Baker, J.G. Fenyes, W.M. Moberg, eds.), ACS Symposium Series, No. 443, American Chemical Society, Washington, DC, pp. 422–435 (1991).

40. S.R. Schow, M.E. Schnee, and J.J. Rauh, *Synthesis and Chemistry of Agrochemicals* (D. R. Baker, J.G. Fenyes, W.M. Moberg, eds.), ACS Symposium Series, No. 443, American Chemical Society, Washington, DC, pp. 437–445 (1991).

41. J. Scherkenbeck and U. Wachendorff-Neumann, *Tetrahedron Lett.* 32:1719–1722 (1991).

42. T. Blizzard, M.H. Fisher, H. Mrozik, and T.L. Shih, *Recent Prog. Chem. Synthesis Antibiotics* (G. Lukacs, ed.), Springer-Verlag p. 65 (1990).

43. S.V. Ley and A. Armstrong, *Strategies and Tactics in Organic Synthesis* (T. Lindberg, ed.), Academic Press, New York, Vol. 3, p. 237 (1991).

44. G.R. Lankas and L.R. Gordon, *Ivermectin and Abamectin* (W.C. Campbell, ed.), Springer-Verlag, New York, pp. 89–112 (1989).

45. N. Matsunuma, K. Hirano, K. Kimura, N. Miyakoshi, K. Yamashita, and H. Masuda, *Sankyo Kenkyusho Nempo*, 35:71–97 (1983).

46. G.A. Oakley, *Ivermectin—The Veterinary Handbook*, Anpar Books, Berkhamsted, U.K. (1990).

47. G.W. Benz, R.A. Roncalli, and S.J. Gross, *Ivermectin and Abamectin* (W.C. Campbell, ed.), Springer-Verlag, New York, pp. 215–229 (1989).

48. P.K. Wilkinson, D.G. Pope, and F.P. Baylis, *J. Pharm. Sci.*, 74:1105–1107 (1985).

49. J.S. Eagleson and G.R. Allerton, *Aust. Vet J.*, 69: 133–134 (1992).

50. M.D. Soll, J.A. Dassonville, and C.J.Z. Smith, *Parasitol. Res.*, 78:120–122 (1992).

51. S.A. Bisset, R.V. Brunsdon, and S. Forbes, *N.Z. Vet. J.*, 38:4–6 (1990).

52. D.G. Pope, P.K. Wilkinson, J.R. Egerton, and J. Conroy, *J. Pharm. Sci.*, 74:1108–1110 (1985).
53. S.M. Taylor and J. Kenny, *Med. Vet. Entomol.*, 4:147–150 (1990).
54. G. Keck, L. Combier, C. Boulard, J.P. Christophe, and G. Levasseur, *Point-Veterinaire*, 23:927–931 (1992).
55. M.D. Soll, I.H. Carmichael, H.R. Scherer, and S.J. Gross, *Vet. Parasitol.*, 42:67–71 (1992).
56. K. Wohlgemuth, J.J. Melancon, H. Hughes, and M. Biondini, *Bovine Practitioner*, 1989:61–62, 64–66 (1989).
57. S.W. Bauck, G.K. Jim, P.T. Guichon, K.M. Newcomb, J.L. Cox, and R.A. Barrick, *Can. Vet. J.*, 30:161–164 (1989).
58. M.D. Soll, I.H. Carmichael, and R.A. Barrick, *Prev. Vet. Med.*, 10:251–256 (1991).
59. S.W. Bauck, C.A. Piche, and K.M. Newcomb, *Can. Vet. J.*, 33:394–396 (1992).
60. R.L. Sharma and T.K. Bhat, *J. Vet. Parasitol.*, 4:5–11 (1990).
61. P.B. Hughes and G.W. Levot, *J. Aust. Entomol. Soc.*, 29:109–111 (1990).
62. M.D. Soll, I.H. Carmichael, G.E. Swan, and A. Abrey, *Vet. Rec.*, 130:572–574 (1992).
63. T.G. Easterly, K.J. Jenkins, and T.R. McCabe, *Wildl. Soc. Bull.*, 20:34–39 (1992).
64. C.R. Reinemeyer, B.W. Rohrbach, V.M. Grant, and G.L. Radde, *Vet. Parasitol.*, 42:111–122 (1992).
65. E. Palmer, *Svensk Veterinartidning*, 41:853–855 (1989).
66. T. Hiepe, H.F. Matthes, G. Werner, W. Zoephel, H. Stark, and H.G. Engel, *Monatsh Veterinaermed*, 44:501–503 (1989).
67. J.J. Arends, C.M. Stanislaw, and D. Gerdon, *J. Animal Sci.*, 68:1495–1499 (1990).
68. T.B. Stewart, D.L. Leon, M.C. Fox, L.L. Southern, and K.E. Bodak, *Vet. Parasitol.*, 39:253–266 (1991).
69. W.C. Campbell, W.H.D. Leaning, and R.L. Seward, *Ivermectin and Abamectin* (W.C. Campbell, ed.), Springer-Verlag, New York, pp. 234–244 (1989).
70. R.J. Bell and J.E. Holste, *Can. Vet. J.*, 31:519–521 (1990).
71. O. Nilsson, J. Hellander, and P. Jerneld, *Svensk Veterinartidning*, 42:167–171 (1990).
72. W.C. Campbell, *Ivermectin and Abamectin* (W.C. Campbell, ed.), Springer-Verlag, New York, pp. 245–259 (1989).
73. J.D. Pulliam and J.M. Preston, *Ivermectin and Abamectin* (W.C. Campbell, ed.), Springer-Verlag, New York, pp. 149–161 (1989).
74. S.P. Rohrer and D.V. Evans, *Vet. Res. Commun.*, 14:157–166 (1990).
75. A.J. Paul, W.J. Tranquilli, K.S. Todd, D.H. Wallace, and M. Soll, *Vet. Med.*, 86:623–625 (1991).
76. J.N. Clark, C.P. Daurio, B.J. Skelly, E.N. Cheung, and A.R. Jeffcoat, *J. Vet. Pharmacol. Therap.*, 15:217–220 (1992).

77. M. Tagawa, A. Takiyama, H. Ejima, and K. Kurokawa, *Bull. Nippon Veter. Zootech. College*, 33:97–99 (1984).
78. M. Horie and S. Noda, *J. Japan Veter. Med. Assoc.*, 39:422–426 (1986).
79. T. Sakamoto, I. Seki, K. Kikuchi, H. Nakahara, and H. Ogasawara, *J. Fac. Agric. Iwate Univ.*, 17:197–210 (1985).
80. T. T. Fukase, In, S. Chinone, S. Akihama, and H. Itagaki, *J. Vet. Med. Sci.*, 53:817–822 (1991).
81. K. Shiramizu and M. Abu, *J. Japan Vet. Med. Assoc.*, 38:353–356 (1985).
82. K. Shiramizu, M. Abu, K. Haida, and Y. Fukuda, *J. Vet. Med.*, 1986:33–36 (1986).
83. M. Horie, S. Noda, M. Akashi, and H. Mori, *J. Vet. Med.*, 1986:259–261 (1986).
84. M. Tagawa, A. Takiyama, and K. Kurokawa, *Japan J. Vet. Sci.*, 48:1267–1270 (1986).
85. Y. Sasaki, H. Kitagawa, H. Okachi, Y. Kajita, and K. Ishihara, *Nippon Juigaku Zasshi*, 48:579–586 (1986).
86. A. Takiyama, *Bull. Nippon Vet. Zootech. College*, 1987:150–152 (1987).
87. Y. Sasaki, H. Kitagawa, and K. Ishihara, *Japan J. Vet. Sci.*, 48:1207–1214 (1986).
88. H. Kitagawa, Y. Sasaki, and K. Ishihara, *Nippon Juigaku Zasshi*, 48:517–522 (1986).
89. Y. Sasaki, H. Kitagawa, Y. Kajita, H. Okachi, and K. Ishihara, *Japan J. Vet. Sci.*, 48:1253–1256 (1986).
90. H. Kitagawa, Y. Sasaki, K. Ishihara, and K. Ishizako, *Japan J. Vet. Sci.*, 50:1184–1191 (1988).
91. M.D. Soll, *Ivermectin and Abamectin*, (W.C. Campbell, ed.), Springer-Verlag, New York, pp. 260–286 (1989).
92. G.A. Conder, L.W. Jen, K.S. Marbury, S.S. Johnson, P.M. Guimond, E.M. Thomas, and B.L. Lee, *J. Parasitol.*, 76:168–170 (1990).
93. V. Richardson, *Vet. Rec.*, 130:432 (1992).
94. V.I. Dschobava and W. Tscherner, Erkrankungen der Zootiere, Verhandlungsbericht des 32, Internationalen Symposiums uber die Erkrankungen der Zoo und Wildtiere vom 23 Mai bis 27 Mai 1990 in Eskilstuna, 1990, 47–51 (1990).
95. B.E. Joseph, D.W. Wilson, R.V. Heinrickson, P.T. Robinson, and K. Benirschke, *Lab Anim. Sci.*, 34:360–364 (1984).
96. E. Kutzer, *Wiener Tierarztliche Monatsschrift*, 77:309–312 (1990).
97. M. Nordkvist, D. Christensson, and C. Rehbinder, *Rangifer*, 4:10–15 (1984).
98. B. Robin, K. Konig, and M.D. Anstey, *Rev. Scientif. Tech.*, 8:147–161 (1989).
99. M.E. Fowler, *Avian/Exotic Practice*, 3:22–25 (1986).
100. J. Frolka and J. Rostinska, *Veterinarstvi*, 35:415–416 (1985).
101. A. Kuntze and O. Kuntze, *Berliner Munchener Tierarztliche Wochenschrift*, 104:46–48 (1991).

102. N.O. Stanchi and C.S. Grisolia, *Veterinaria*, 3:578–581 (1986).
103. J. Letcher and M. Glade, *J. Am. Vet. Med. Assoc.*, 200:537–538 (1992).
104. N. Kummerfeld and N.C. Schafer, *Kleintierpraxis*, 32:293–296 (1987).
105. J.C. Webb, J.G. Burg, and F.W. Knapp, *J. Econ. Entomol.*, 84:1266–1269 (1991).
106. E.T. Lyons, S.C. Tolliver, J.H. Drudge, D.E. Granstrom, S.S. Collins, and S. Stamper, *Vet. Parasitol.*, 41:255–284 (1992).
107. S. Ranjan, C. Trudeau, R.K. Prichard, R. von Kutzleben, and D. Carrier, *Vet. Parasitol.*, 41:227–231 (1992).
108. P.J. Scholl, F.S. Guillot, and G.T. Wang, *Vet. Parasitol.*, 41:203–209 (1992).
109. M. Eysker and J.H. Boersema, *Vet. Q.*, 14:79–80 (1992).
110. G.L. Zimmerman, E.P. Hoberg, and J.A. Pankavich, *Am. J. Vet. Res.*, 53:1409–1410 (1992).
111. R.A. Dybas, *Ivermectin and Abamectin* (W.C. Campbell, ed.), Springer-Verlag, New York, pp. 287–310 (1989).
112. J.V. French and J. Villarreal, *J. Rio Grande Val Hortic. Soc.*, 43:9–16 (1990).
113. J.G. MacConnell, R.J. Demchak, F.A. Preiser, and R.A. Dybas, *J. Agric. Food Chem.*, 37:1498 (1989).
114. J.G. Morse, J.A. Immaraju, and O.L. Brawner, *Citrograph*, 73:112–115 (1988).
115. M.A. Harris, J.W. Begley, and D.L. Warkentin, *J. Econ. Entomol.*, 83:2380–2384 (1990).
116. J.A. Immaraju, T.D. Paine, J.A. Bethke, K.L. Robb, and J.P. Newman, *J. Econ. Entomol.*, 85:9–14 (1992).
117. T.S. Corbitt, A.S.J. Green, and D.J. Wright, *Crop Protect.*, 8:127–132 (1989).
118. D.C. Deecher, J. Brezner, and S.W. Tanenbaum, *J. Econ. Entomol.*, 82:1395–1398 (1989).
119. Z.Q. Zhang and J.P. Sanderson, *J. Econ. Entomol.*, 83:1783–1790 (1990).
120. A.H. Hara and T.Y. Hata, *Sci. Hortic.* (AMST), 51:155–163 (1992).
121. D.G. Cochran, *J. Econ. Entomol.*, 83:1243–1245 (1990).
122. B.M. Greene, *J. Infect. Diseases*, 166:15–21 (1992).
123. W.C. Campbell, *Annu. Rev. Microbiol.*, 45:445–474 (1991).
124. J.R. Egerton, E.S. Brokken, D. Suhayda, C.H. Eary, J.W. Wooden, and R.L. Kilgore, *Vet. Parasitol.*, 8:83–88 (1981).
125. B.O. Duke, M.C. Pacque, B. Munoz, B.M. Greene, and H.R. Taylor, *Bull. World Health Organ.*, 69:163–168 (1991).
126. R.C. Collins, C. Gonzalesperalta, J. Castro, G. Zeaflores, M.S. Cupp, F.O. Richards, and E.W. Cupp, *Am. J. Trop. Med. Hyg.*, 47:156–169 (1992).
127. J. Prodhon, M. Boussinesq, G. Fobi, J.M. Prudhom, P. Enyong, C. Lafleur, and D. Quillevere, *Bull. World Health Organ.*, 69:443–450 (1991).
128. J. Whitworth, *Parasitol. Today*, 8:138–140 (1992).
129. J.A. Whitworth, D. Morgan, G.H. Maude, M.D. Downham, and D.W. Taylor, *Trans. R. Soc. Trop. Med. Hyg.*, 85:501–505 (1991).
130. J.A.G. Whitworth, C.E. Gilbert, D.M. Mabey, G.H. Maude, D. Morgan, and D.W. Taylor, *Lancet*, 338:1100–1103 (1991).

131. J.A.G. Whitworth, C.R.M. Hay, A.M. McNicholas, D. Morgan, G.H. Maude, and D.W. Taylor, *Ann. Trop. Med. Parasitol.*, *86*:301–305 (1992).

132. J.A.G. Whitworth, A.J.F. Luty, G.H. Maude, D. Morgan, M.D. Downham, and D.W. Taylor, *Trans. R. Soc. Trop. Med. Hyg.*, *86*:281–283 (1992).

133. B.O.L. Duke, G. Zeaflores, J. Castro, E.W. Cupp, and B. Munoz, *Am. J. Trop. Med. Hyg.*, *46*:189–194 (1992).

134. B.A. Halley, R.J. Nessel, and A.Y.H. Lu, *Ivermectin and Abamectin* (W.C. Campbell, ed.), Springer-Verlag, New York, pp. 162–172 (1989).

135. B.A. Halley, T.A. Jacob, and A.Y.H. Lu, *Chemosphere*, *18*:1543–1563 (1989).

136. V.F. Gruber, B.A. Halley, S-C Hwang, and C.C. Ku, *J. Agric. Food Chem.*, *38*:886–890 (1990).

137. P.G. Wislocki, L.S. Grosso, and R.A. Dybas, *Ivermectin and Abamectin* (W.C. Campbell, ed.), Springer-Verlag, New York, pp. 182–200 (1989).

138. L.S. Crouch, W.F. Feely, B.H. Arison, W.J.A. Vandenheuvel, L.F. Colwell, R.A. Stearns, W.F. Kline, and P.G. Wislocki, *J. Agric. Food Chem.*, *39*:1310–1319 (1991).

139. B.A. Halley, R.J. Nessel, A.Y.H. Lu, and R.A. Roncalli, *Chemosphere*, *18*:1565–1572 (1989).

140. R.A. Roncalli, *Ivermectin and Abamectin* (W.C. Campbell, ed.), Springer-Verlag, New York, pp. 173–181 (1989).

141. R. Wall and L. Strong, *Nature*, *327*:418–421 (1987).

142. K.G. Wardhaugh and M.H. Rodriguez, *J. Appl. Entomol.*, *106*:381–389 (1988).

143. M. Madsen, B.O. Nielsen, P. Holter, O.C. Pedersen, J.B. Jespersen, K.M. VagnJensen, P. Nansen, and J. Gronvold, *J. Appl. Ecol.*, *27*:1–15 (1990).

144. L. Strong, *Bull. Entomol. Res.*, *82*:265–274 (1992).

145. D.E. Jacobs, J.G. Pilkington, M.A. Fisher, and M.T. Fox, *Vet. Rec.*, *123*:400 (1988).

146. R. Schaper and A. Liebisch, *Tierarztliche Umschau*, *46*:12–18 (1991).

147. K.G. Wardhaugh and R.J. Mahon, *Bull. Entomol. Res.*, *81*:333–340 (1991).

148. R.J. Mahon and K.G. Wardhaugh, *Austral. Vet. J.*, *68*:173–177 (1991).

149. K.M. Ewert, J.A. DiPietro, C.S. Danner, Jr., and L.M. Lawrence, *Vet. Rec.*, *129*:140–141 (1991).

150. C. Sommer, B. Steffansen, J. Springborg, and P. Nansen, *Acta Vet. Scand. Supp.*, *87*:391–393 (1991).

151. G.V. Downing, *Ivermectin and Abamectin* (W.C. Campbell, ed.), Springer-Verlag, New York, pp. 324–335 (1989).

152. P. Slanina, J. Kuivinen, C. Ohlsen, and L.G. Ekstrom, *Food Addit. Contam.*, *6*:475–481 (1989).

153. S.H.L. Chiu and A.Y.H. Lu, *Ivermectin and Abamectin* (W.C. Campbell, ed.), Springer-Verlag, New York, pp. 131–143 (1989).

154. S.H.L. Chiu, M.L. Green, F.P. Baylis, D. Eline, A. Rosegay, H. Meriwether, and T.A. Jacob, *J. Agric. Food Chem.*, *38*:2072 (1990).

155. S.H.L. Chiu, E. Sestokas, R. Taub, M.L. Green, F.P. Baylis, T.A. Jacob and A.Y.H. Lu, *J. Agric. Food Chem.*, *38*:2079 (1990).

156. S.V. Prabhu, T.A. Wehner, and P.C. Tway, *J. Agric. Food Chem.*, *39*:1468–1471 (1991).

157. W.F. Feely and P.G. Wislocki, *J. Agric. Food Chem.*, *39*:963–967 (1991).

158. J. Vuik, *J. Agric. Food Chem.*, *39*:303–305 (1991).

159. M.J. Jongen, R. Engel, and L.H. Leenheers, *Am. Ind. Hyg. Assoc. J.*, *52*:433–437 (1991).

160. S.V. Prabhu, R.J. Varsolona, T.A. Wehner, R.S. Egan, and P.C. Tway, *J. Agric. Food Chem.*, *40*:622–625 (1992).

161. B.A. Halley, N.I. Narasimhan, K. Venkataraman, R. Taub, M.L.G. Erwin, N.W. Andrew, and P.G. Wislocki, Fourth Chem. Congress of North America, Abstr. Am. Chem. Soc., #128 (1991).

162. S. Sadakane, K. Tanaka, E. Muraoka, and M. Ando, *J. Pestic. Sci.*, *17*:147–154 (1992).

163. J.A. Van Wyk and F.S. Malan, *Vet. Rec.*, *123*:226–228 (1988).

164. J.A. Van Wyk, F.S. Malan, H.M. Gerber, and R.M.R. Alves, *J. Vet. Res.*, *56*:41–49 (1989).

165. J.A. Van Wyk, P.C. Van Schalkwyk, G.F. Bath, H.M. Gerber, and R.M.R. Alves, *J. South African Vet. Assoc.*, *62*:171–175 (1991).

166. T.M. Craig and D.K. Miller, *Vet. Rec.*, *126*:560 (1990).

167. J.A. Devaney, T.M. Craig, and L.D. Rowe, *Int. J. Parasitol.*, *22*:369–376 (1992).

168. F.A.M. Echevarria, J. Armour, and J.L. Duncan, *Vet. Parasitol.*, *39*:279–284 (1991).

169. F.A.M. Echevarria, J. Armour, M.F. Borba and J.L. Duncan, *J. Parasitol.*, *78*:894–898 (1992).

170. J.A. Pankavich, H.Berger, and K.L. Simkins, *Vet. Rec.*, *130*:241–243 (1992).

171. W.L. Shoop, *Vet. Rec.*, *130*:563 (1992).

172. G.A. Conder, D.P. Thompson, and S.S. Johnson, *Vet. Rec.* *132*:651–652 (1993).

173. J.G. Scott, R.T. Roush, and N. Liu, *Experientia*, *47*:288–291 (1991).

174. J.G. Scott, *J. Agric. Entomol.*, *8*:77–82 (1991).

175. J.M. Clark, J.A. Argentine, and H. Lin, 204th American Chemical Society National Meeting, Abstr. Am. Chem Soc. #107 (1992).

176. Merck & Co., Inc. and Lehman Brothers Estimates.

177. *Decision Resources*, Burlington, MA pp. 28, 32, 1992.

5

Animal Health Products

David J. Wadsworth

Ciba-Geigy Limited, Basel, Switzerland

I. HISTORICAL OVERVIEW

Until the introduction of the sulphonamide antibiotics in the first half of the twentieth century the only treatments available to the practicing veterinarian were based on a great variety of herbal sources of dubious efficacy. The Romans treated parasitic mange in sheep by the application of a mixture of the juice of boiled lupins, dregs of old wine, and olive lees. Skin lesions were treated by scraping the skin and then applying hemlock juice that had been salted and stored in pots in a dung heap for over a year. These types of treatments have been aptly described [1] as being "esteemed according to whether they were rare, complex or unpleasant. A drug which combined all three qualities was irresistible." Even well into the current century dubious practices, such as the injection of turpentine into the windpipes of cattle in order to control lungworm, were commonly performed. The landmark discovery of the antibiotic effect of the penicillium mould by Sir Alexander Fleming in 1929 led to a scientific revolution in terms of the treatment and prevention of diseases in both the medical and veterinary areas. In 1992 the estimated value of the worldwide animal health (AH) market had grown to US$ 12,000 million with the major markets being in North America and western Europe (Table 1, Figure 1) and many of the world's largest pharmaceutical companies investing heavily in AH

Table 1 The AH World Market; Major Submarkets 1992

Submarket	Value (in millions US$)	Percentage of total
Pharmaceuticals	5900	48
Antibacterials	2000	18
Endoparasiticides	1500	12
Ectoparasiticides	770	6
Nutritionals	700	5
Others	1000	7
(vet specialities, disinfectants, etc.)		
Feed additives	4550	39
Biologicals	1550	13
Total market	12000	100

research and development [2]. The animal health market can be divided into four major areas: nutritional feed additives, medicinal feed additives, biologicals, and pharmaceuticals (Figure 2). Each area can be further subdivided and each of these subdivisions will be considered in turn, with specific reference to the role played by products containing compounds of natural origin, if any are present.

II. NUTRITIONAL FEED ADDITIVES: VITAMINS AND AMINO ACIDS

In 1992 the total value of the world market for vitamin and amino acid animal feed additives was in the order of US$ 1975 million. The ma-

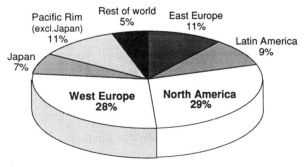

Figure 1 Animal health world market; market size by country 1992.

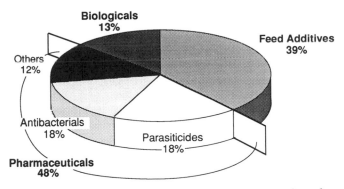

Figure 2 Animal health world market; major submarkets 1992.

jor vitamin products are vitamin A (*1*) (Hoffmann–LaRoche/BASF, US\$ 258 million) [3] and vitamin E (*2*) (Hoffmann–LaRoche/BASF/ Rhône-Poulenc, US\$ 222 million). The essential nutrients (*R, S*)-methionine (*3*) (Degussa and Rhône-Poulenc, US\$ 705 million), which is also indicated in the therapy of ketosis in cattle and sheep as well as urinary infections in domestic pets, and lysine (*4*) (Ajinomoto and Orsan, US\$ 205 million) are the major products in the amino acid field. A more detailed discussion of this area is beyond the scope of this chapter.

(*1*)

(*2*)

(*3*)

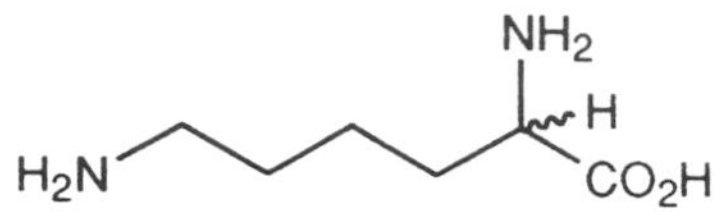

(4)

III. MEDICINAL FEED ADDITIVES

A. Antibiotics

Since the introduction of Prontosil, the first of the sulphonamides, in 1935, followed five years later by the penicillins, the therapy of human infectious diseases has shown dramatic improvements that have been mirrored in veterinary practice. Economically serious diseases such as bacterial mastitis, bacterial enteritis, pneumonia, and foot and leg infections in cattle, sheep, and horses can now be successfully controlled. The increased health and productivity of farm animals since the 1940s is illustrated by the observed doubling of the milk output per dairy cow and the reduction in the time needed to produce beef by two-thirds. Improvements in breeding techniques and nutrition have equally contributed to such dramatic increases in the efficiency of food production. The world AH antibiotic market amounted to US$ 840 million in 1992 with an estimated growth rate to 1997 of 1.19% annually. The major market area is North America with 43% of worldwide sales.

1. *Penicillins*

The serendipitous discovery by Fleming of the antibacterial effect shown by a penicillium mould contaminating an agar plate culture of staphylococci has been well documented [4]. The pioneering work of Florey and coworkers [5] led to the clinical introduction of Penicillin G for human use in 1941, which was rapidly followed by other semi-synthetic penicillin derivatives and the related cephalosporins [6]. The

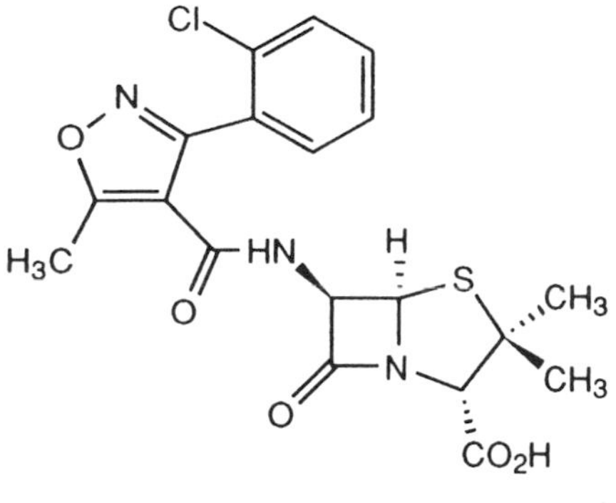

(5)

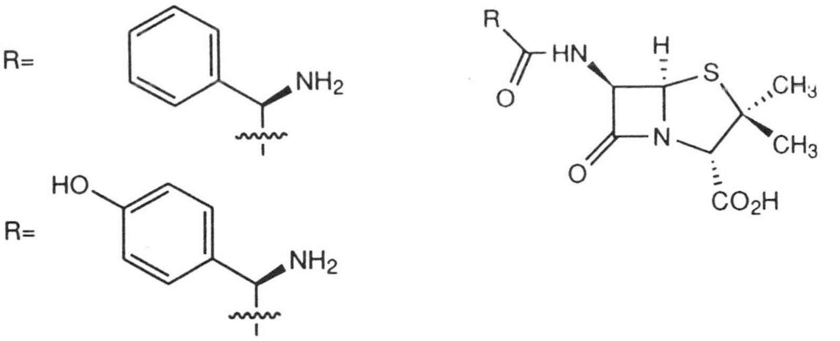

R=

(6)

R=

(7)

main penicillins in use for animal therapy are the semisynthetic cloxacillin (5), which is active against the range of gram-positive organisms including *Staphylococcus aureus* which causes mastitis in dairy cattle, ampicillin (6) and amoxycillin (7), which are available from a wide variety of manufacturers. Current sales of penicillins in the animal health market are in the order of US$ 130 million per year.

2. *Tetracyclines*

The first tetracycline, chlorotetracycline (8), was isolated from a culture of the actinomycete *Streptomyces aureofaciens* in 1948 by Duggar and was soon followed in 1950 by the discovery of oxytetracycline (9) by Finlay and co-workers [7]. This family of antibiotics show a broad spectrum of activity against gram-positive and gram-negative bacteria and have found widespread use in both human and animal therapies, although

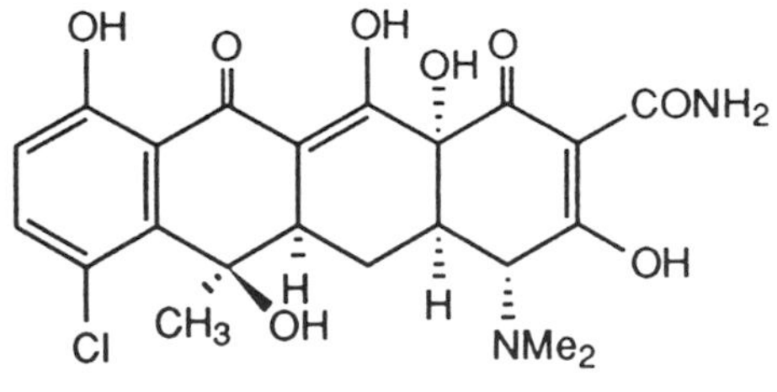

(8)

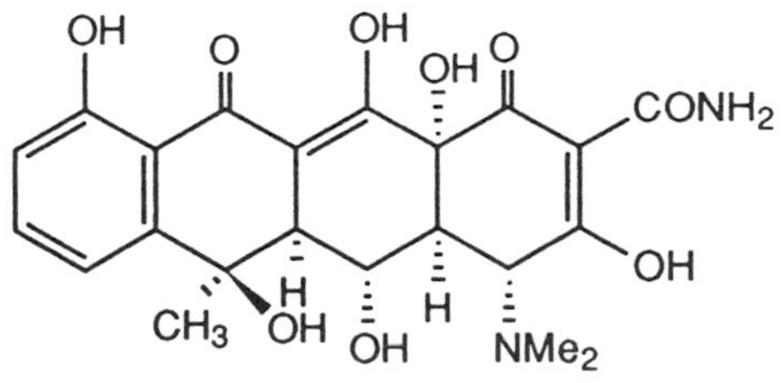

(9)

resistance to the tetracyclines is not uncommon. Both of these active ingredients feature prominently in the list of the best-selling animal health products (excluding nutritionals), with oxytetracycline (Terramycin, Pfizer) ranking in second place with annual sales of US$ 258 million in 1992 and the longer-acting chlorotetracycline (Aurofac/Aureomycin, American Cyanamid) in third position, with sales of US$ 166 million.

3. Bambermycin, Virginiamycin, Spiramycin, and Zinc Bacitracin

Bambermycin, virginiamycin, and spiramycin achieved total combined sales of US$ 172 million in 1992. These three products are also sold as growth promoters and are considered in more detail in Section IIIC1. The structure of the antibiotic zinc bacitracin, which is prepared by the action of zinc salts on bacitracin broth, has not yet been determined. Total sales in 1992 were US$ 74 million, with bacitracin gaining market share in the swine and poultry sectors from the tetracycline-based products.

4. Lincomycin

Lincomycin (10) was isolated in 1962 from a culture of *Streptomyces lincolnensis* that was found in a soil sample taken from Lincoln, Nebraska [8]. Lincomycin shows excellent antistaphylococcal and antistreptococcal activities and is especially valuable in the control of mycoplasmal infections in pigs and for the treatment of skin and lung infections as it accumulates in these organs. In 1992 lincomycin (Lincocin, Upjohn) was placed seventh in the ranking of the best-selling animal health products, with total sales of US$ 85 million.

(10)

5. Tylosin

a. ISOLATION AND STRUCTURAL STUDIES Tylosin (11) is a 16-membered macrolide that belongs to the family of macrolide antibiotics. It was first isolated by workers at Eli Lilly in 1960 from a culture of the actinomycete *Streptomyces fradiae* that was discovered in soil samples obtained from Nongkhai, Thailand [9].

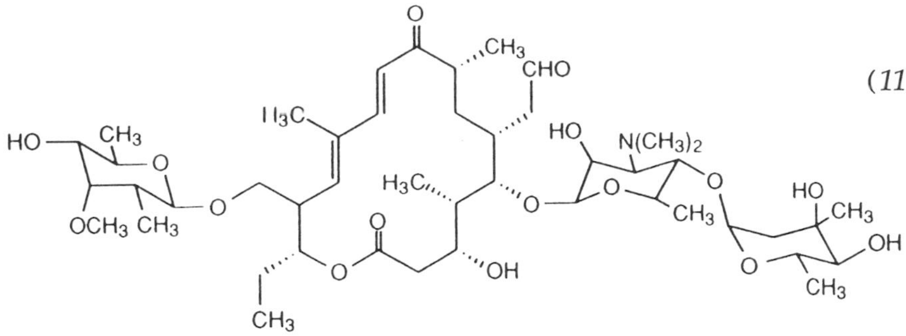

(11)

Degradation of tylosin under mildly acidic conditions affords the sugar mycarose (12) and a new antibiotic substance, desmycosin (13). Upon more vigorous treatment with acid, a second nitrogen-containing sugar, mycaminose (14), is obtained with concomitant destruction of the aglycone of the molecule [10]. In 1964 the third constituent sugar, mycinose (15), was identified using more controlled conditions for the decomposition of tylosin by acid. [11]. Further degradation studies, coupled with comparisons of the possible structures of tylosin with other known 16-membered macrolides, led to the conclusion that the new antibiotic had the structure 11, with the mycarosyl-mycaminosyl residue attached at either C-3 or C-5 [12]. Final confirmation of the structure of tylosin came in 1975 from detailed nuclear magnetic resonance (NMR) studies on desmycosin and its tetra-acetyl derivative [13]. The absolute configuration of tylosin was established in 1982 by an x-ray crystallographic study on 5-O-mycarosyl tylactone [14] with the total synthesis being announced in the same year [15].

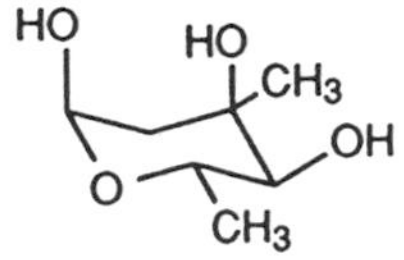

(12)

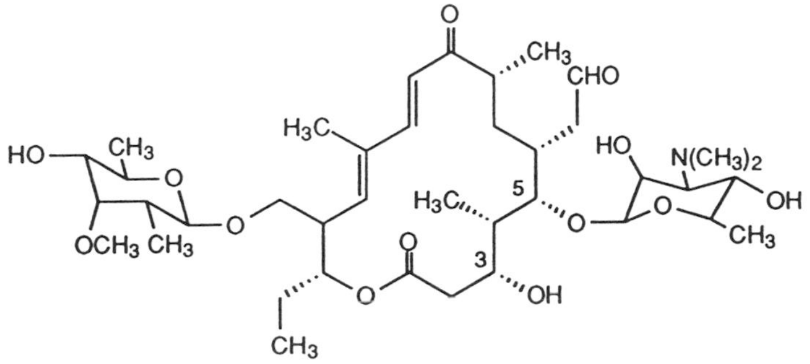

(13)

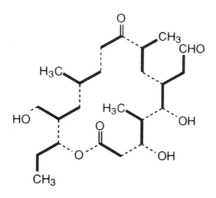

(14)

(15)

b. BIOSYNTHESIS The tylosin aglycone macrocycle is biosynthesized from two acetates, five propionates, and one butyrate unit (Figure 3) [17]. Following the construction of the carbon backbone and subsequent macrolide ring closure by a polyketide synthase (PKS) multienzyme complex (see Section III-B for further details), the sugars are attached and the 3-hydroxy group of the mycinose unit methylated.

c. BIOLOGICAL PROFILE Tylosin is active against gram-positive bacteria including the genera *Streptococcus*, *Bacillus*, and *Staphylococcus* and has been used successfully in animals for treating respiratory and

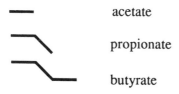

Figure 3 The biosynthetic origin of tylosin aglycone.

genito-urinary tract infections as well as otitis, cellulitis, and secondary bacterial infections associated with viral disease or postoperative infections. Tylosin also shows good activity against *Mycoplasma* and *Treponema hyodysenteriae* making it particularly valuable for use in cases of mycoplasmosis in chickens and turkeys and also in outbreaks of swine dysentery. Additionally, tylosin shows a growth-promoting effect when added at subtherapeutic levels to the feed of pigs and this effect is discussed further in Section IIIC-1. Tylosin, in common with other macrolide antibiotics, inhibits bacterial protein synthesis by binding to ribosomes and inhibiting peptidal transferases [18].

Tylosin is relatively nontoxic. The intravenous LD_{50} in mice is 400 mg/kg and oral doses up to 5 g/kg show no effects in rats. Long-term studies on rats fed 1% tylosin in their daily diet for 2 years showed no changes in growth rates and no observed pathological conditions. Dogs have tolerated daily doses of 100 mg/kg for more than 1 year without any untoward clinical signs. Even at a dose of 200 mg/kg in dogs, only minimal side effects of nausea and vomiting were observed [19]. Animals treated with tylosin therapeutically must not be slaughtered for human consumption during treatment or for at least 7 days after the last treatment. Milk for human consumption may only be taken from cows 4 days after the last treatment.

d. PRODUCT PROFILE Tylosin (Tylan, Eli Lilly and generics) is available in a variety of dosage forms. It is the single largest selling antibiotic administered in feed with estimated sales of US$ 150 million in 1992.

B. Anticoccidials

Coccidiosis is an infection of the intestinal epithelial and liver cells of vertebrates by intracellular protozoal parasites belonging to the phylum Apicomplexa [20]. These organisms were discovered by Leeuwenhoek in 1674 when he observed oöcysts of Eimera stiedae in the bile ducts of rabbits. Coccidia, belonging almost exclusively to the genus Eimeria, cause clinical or subclinical disease in chickens, turkeys, geese and ducks and are probably involved in 5-10% of deaths that occur in poultry flocks, inflicting economic losses in the order of hundreds of millions of dollars. Consequently, the prevention and control of coccidiosis in chickens is commercially important and several natural products are used against this type of protozoal infection. The coccidiostat market has an estimated size in 1992 of US$ 525 million and a growth rate of 2.2% p.a. with the largest markets being Western Europe (29%) and North America (26%).

POLYETHER IONOPHORES

The major anticoccidial products belong to the class of polyether ionophores which complex efficiently Group I or II metal cations [21]. The
cations coordinate to the heterocyclic ring oxygen atoms in a cavity
formed by the folding of the ionophore around the metal. The result-
ant transport of the chelated metal ions across the cell membranes by
the polyethers uncouples oxidative phosphorylation leading to the death
of the cell. Unfortunately, this effect is also observed with mammalian cells, and careful dosing is required to prevent the appearance of
toxic symptoms in animals. The polyether ionophores are
biosynthesized from simple fatty acid building blocks such as acetate,
propionate, and butyrate. A specific polyketide synthase (PKS) multienzyme complex sequentially assembles the carbon chain backbone of
each ionophore by a series of condensation, reduction, and dehydration events. It has been postulated, and observed in model systems,
that the heterocyclic ring systems commonly found in the polyether ionophores are subsequently constructed from a series of ring-opening
reactions of the polyepoxides derived from acyclic polyene precursors.
A comprehensive review of the chemical and biochemical aspects of
polyether ionophore antibiotic biosynthesis appeared in 1991 [22].

1. *Monensin*

a. ISOLATION AND STRUCTURE DETERMINATION Monensin
(*16*), an ionophore antibiotic, is the major component in a group of
closely related compounds first isolated from a strain of *Streptomyces
cinnamonensis* by workers at the Eli Lilly Research Laboratories in 1967
[23].

(*16*)

Monensin (formerly called monensinic acid), either as the free acid
or its salts, is sparingly soluble in water but completely soluble in most

organic solvents. The structure of monensin was elucidated by a combination of derivatisation, degradation, NMR studies and finally an x-ray crystallographic study of its silver salt [24]. In the crystal structure, the molecule is wrapped around the metal, with six oxygen atoms situated within 2.7 Å of the silver ion. The exterior surface of the molecule is lipophilic, which results in the low solubility of monensin salts in water. The total synthesis of monensin has been achieved by a number of groups, the first being that published by Fukuyama et al. in 1979 [25].

b. BIOSYNTHESIS Monesin is biosynthesized from five acetate, seven propionate, and one butyrate units with the O-methyl group being derived from S-adenosyl methionine [26] (Figure 4).

c. BIOLOGICAL PROFILE Monensin shows antibacterial, antifungal, and anticoccidial activity as well as growth-promoting effects in ruminants (see Section C3). It exhibits its anticoccidial action by the suppression of the development of the sporozoite into the schizont stage and has to be present before infection occurs. It has also been postulated that monensin and other ionophores may also act by inhibiting nutrient transport in the host cell, thus indirectly affecting coccidial de-

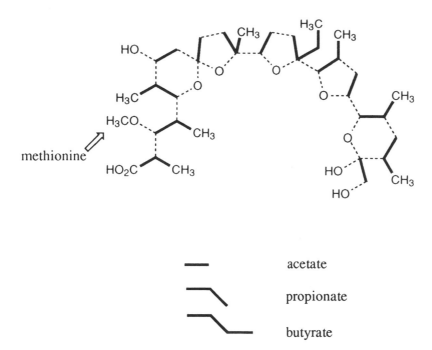

Figure 4 The biosynthetic origin of monensin.

velopment [27]. Some evidence of resistance has been documented and "shuttle" programs, in which coccidiostats are rotated, are now recommended by most manufacturers (for example Eli Lilly recommends rotation with its other coccidiostat narasin). The LD_{50} of monensin is 45 mg/kg in mice and 285 mg/kg in chicks [23]. Accidental ingestion of monensin and other ionophores by horses can be fatal and poultry treated with monensin must not be slaughtered for human consumption until at least 3 days after the end of treatment.

d.　PRODUCT PROFILE.　Monensin sodium salt (Monensin, Coban, Rumensin, Romensin, Eli Lilly and licensees) for poultry use is sold in a 5–20% premix and is used preventatively by inclusion in the feed at a rate of 100–120 g/ton for broiler chickens. Monensin is one of the leading selling coccidiostats with estimated sales in 1992 of US$ 135 million.

2.　*Lasalocid A*

a.　ISOLATION AND STRUCTURE DETERMINATION　Lasalocid A (*17*) (originally designated as X-537A) was isolated in 1951 by a group from Hoffmann–LaRoche from cultures of *Streptomyces lasaliensis* that was present in soil samples collected in Hyde Park, Massachusetts [28]. Butyl acetate extraction of the cells afforded the sodium salt of lasalocid A, which, in common with many of the ionophores, was soluble in benzene and hot petroleum ether but not in water. This characteristic greatly simplified the separation of lasalocid A from other metabolites.

(*17*)

In the original publication describing the isolation of lasalocid A the only structural information disclosed was that an aromatic ring substituted by a hydroxyl and a carbonyl group were present and that the compound was optically active. In 1970, two articles were published simultaneously in which the full structure was elucidated using a combination of both chemical and physical methods [29, 30]. The β-ketol system could be cleaved with 10% aqueous sodium hydroxide to afford the more easily identified acid (*18*) and ketone (*19*). The initial product of the retro-aldol reaction is presumably aldehyde (20) which, under the reaction conditions, cyclizes to (18). Complexation of two molecules of the mono-anion of lasalocid A with a Ba^{2+} ion formed crystals that were suitable for x-ray analysis that gave final proof of the structure. The total synthesis of lasalocid A has been achieved several times, with the first report appearing in 1978 from Nakata and co-workers [31].

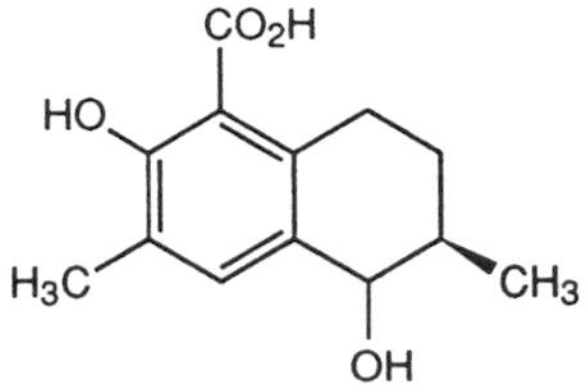

(18)

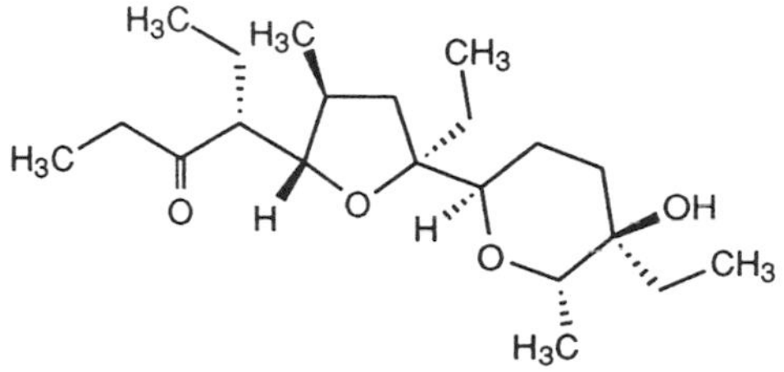

(19)

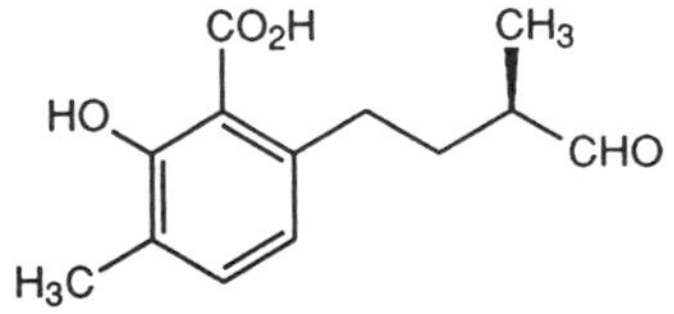

(20)

b. BIOSYNTHESIS   It was shown, using a variety of labeled precursors, that the carbon skeleton of lasalocid is made up from five acetate, four propionate, and three butyrate units [32] (Figure 5).

c. BIOLOGICAL PROFILE   Lasalocid A shows good activity against gram-positive bacteria and mycobacteria but is inactive against gram-negative bacteria and fungi. Importantly, anticoccidial activity as well as growth-promoting effects in ruminants have been observed (see Section IIIC). It forms a range of complexes with different metals than those complexed by monensin and is used mainly in the control of coccidiosis in broiler chickens.

Lasalocid A is relatively toxic with an LD_{50} in mice of 11 mg/kg when administered subcutaneously, 40 mg/kg for intraperiotoneal application, and 146 mg/kg when given orally. Feed containing lasalocid A must be withdrawn at least 5 days before the slaughter of chickens.

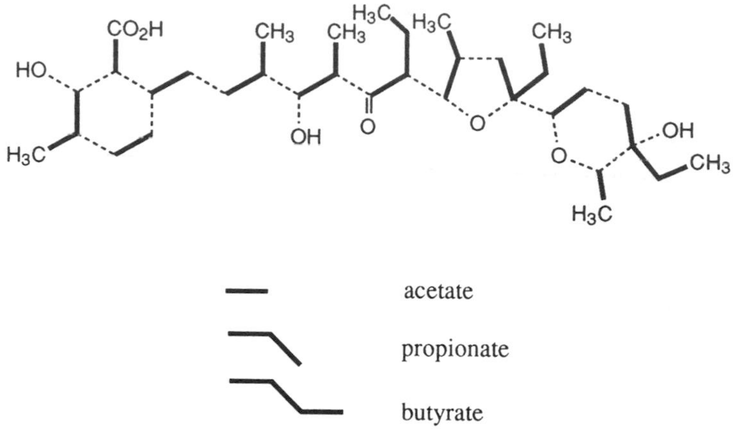

Figure 5 The biosynthetic origin of lasalocid A.

d. PRODUCT PROFILE Lasalocid A (Avatec, Bovatec, Hoffmann-LaRoche) is available as a 15% premix in a cereal carrier. It is used for the prevention of coccidiosis at a dose of 90 g/ton but is not recommended for use against already established outbreaks. Sales in 1992 were US$ 26 million, albeit with greater success as a cattle growth promoter (see Section IIIC3).

3. Salinomycin

a. ISOLATION AND STRUCTURE DETERMINATION Salinomycin (21) was first isolated in 1972 from a culture of *Streptomyces albus* by workers at the Kaken Chemical Co. Ltd. in Japan [33].

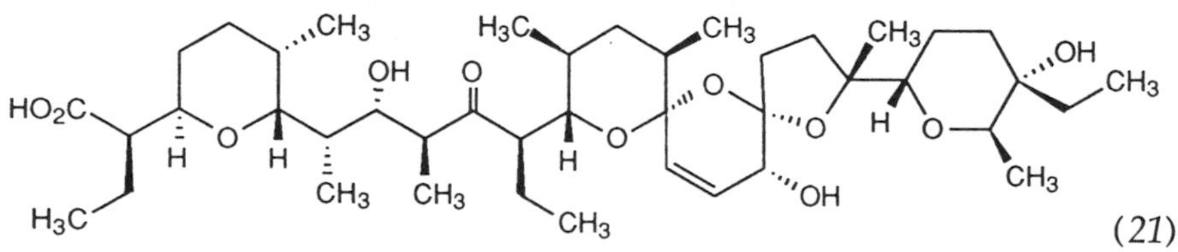

(21)

At the time of its isolation certain similarities between salinomycin and other polyethers such as monensin were noticed, and in 1973 the structure of the new antibiotic coccidiostat was published [34]. Unlike many of the other polyether antibiotics, salinomycin gave no crystallizable metal salts suitable for x-ray analysis; however, the derived *p*-iodophenacyl ester allowed the structure and the absolute configuration to be solved by the heavy-atom x-ray crystallographic method. The biosynthetic origin of salinomycin is similar to that shown for narasin (see Section IIIB4) with one propionate unit being substituted by an acetate. The total synthesis of salinomycin was first reported in 1981 by Kishi and co-workers [35].

b. BIOLOGICAL PROFILE   Salinomycin is active against a variety of gram-positive bacteria including mycobacteria and some filamentous fungi. No activity is observed against gram-negative bacteria or yeast. As a commercially important feature, salinomycin demonstrates good activity in chickens infected with *Eimeria tenella*. In comparison with monensin and lasalocid A, salinomycin is more effective against a broad spectrum of *Eimeria* species at a range of doses [36]. Salinomycin is also effective as a ruminant growth promoter (see Section IIIC). The LD_{50} of salinomycin in mice is 18 mg/kg when administered intraperitoneally and 50 mg/kg orally. Chickens intended for human consumption should not be slaughtered until at least 5 days after the last treatment. However, salinomycin is not suitable for the treatment of coccidiosis in turkeys.

c. PRODUCT PROFILE   Salinomycin (Coxistac, Salocin, Ovicox, Biocox, Kaken and licensees) is available for incorporation in broiler feed at a recommended level of 60 g/ton in the finished feed. In 1992 salinomycin was the leading coccidiostatic product, with estimated worldwide sales of US$ 145 million.

4. *Narasin*

a. ISOLATION AND STRUCTURAL STUDIES   The isolation of narasin (22) from the aerobic fermentation of *Streptomyces aureofaciens* was announced by workers at Eli Lilly in 1975 [37].

(22)

A comparison of the mass spectra of narasin and salinomycin made clear that the two products were structurally very similar [38]. Narasin possesses one extra methyl group, the position and configuration of which was elucidated using ^{13}C-NMR techniques [39]. The total synthesis of narasin, together with that of salinomycin, was first reported in 1981 by Kishi and co-workers [35].

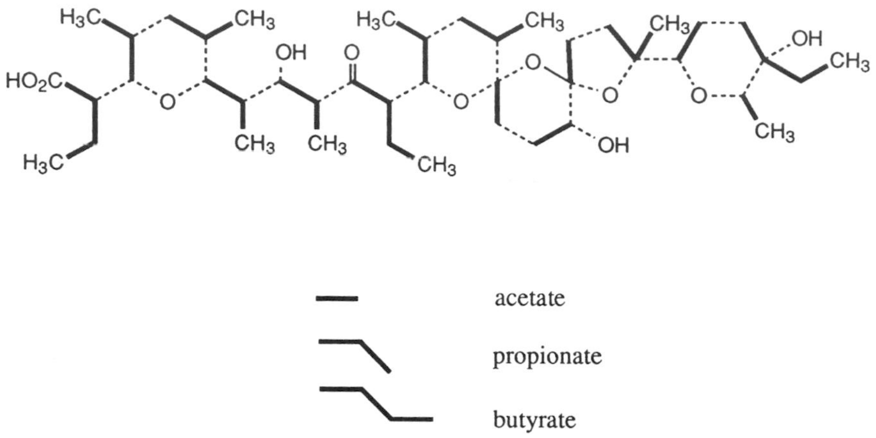

Figure 6 The biosynthetic origin or narasin.

b. BIOSYNTHESIS Narasin is biosynthesized from five acetate, seven propionate, and three butyrate units [40] (Figure 6).

c. BIOLOGICAL PROFILE Narasin is active against gram-positive bacteria, anaerobic bacteria, and the viruses that cause herpes, polio, transmissible gastroenteritis, Newcastle disease, and infectious bovine rhinotracheitis. Addition of narasin to the feed of chickens protects against coccidial infections equal to or slightly better than monensin. Additionally, the reduction in intestinal lesions is greater with narasin medication than with monensin [41]. Narasin also increases the efficiency of feed utilization in ruminants (see Section IIIC).

The intraperitoneal LD_{50} of narasin in mice is 7.15 mg/kg and poultry intended for human consumption must not be slaughtered while receiving narasin or for 5 days after discontinuation of treatment.

d. PRODUCT PROFILE Narasin (Monteban, Eli Lilly) is available as a 10% premix for the prevention of coccidiosis in broiler chickens and dose levels should be in the region of 70 g/ton of feed. In 1992, narasin achieved sales in this area of US$ 41 million.

5. *Maduramicin*

a. ISOLATION AND STRUCTURAL STUDIES Maduramicin (23) (formerly known as X-14868A and not to be confused with maduramycin which was isolated from *Actinomadura rubra*) was isolated as its sodium salt from a culture of a *Nocardia* species by workers at Hoffmann–LaRoche in 1981 [42] and as the free acid hydrate from *Actinomadura yumaense* by American Cyanamid [43].

(23)

The structure of maduramicin was elucidated by x-ray crystallography in 1982 [44] with the full ^{13}C-NMR spectrum being assigned in 1984 [45].

b. BIOSYNTHESIS The biosynthesis of maduramicin was studied using ^{13}C-, ^{14}C-, and ^{18}O-labeled precursors [45]. The 47 carbon, 7-ring system was shown to be built up from 8 acetate, 7 propionate, and 4 methionine molecules (Figure 7).

c. BIOLOGICAL PROFILE Maduramicin is active against gram-positive bacteria but not against gram-negative bacteria or fungi [46] and, in addition, displays antimalarial [47] and nematocidal activities [48]. Maduramicin is an active coccidiostat at doses an order of magnitude

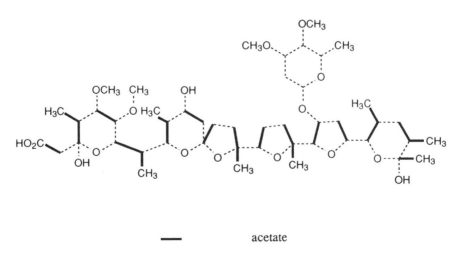

————— acetate

⟍⟋ propionate

Figure 7 The biosynthetic origin of maduramicin.

lower than those necessary for activity with monensin, lasalocid A, or salinomycin. Animals receiving maduramicin must not be slaughtered during treatment and broiler chickens may be only slaughtered for human consumption 5 days after the last dosing.

d. PRODUCT PROFILE Maduramicin ammonium salt (Cygro, American Cyanamid) is available as a 10% premix and should be administered at a medication level of 50 g/ton. In 1992 maduramicin recorded sales of US\$ 39 million, although it was withdrawn from the U.S. market due to regulatory problems.

C. Growth Promoters

Growth promoters in animals can be defined as substances that are not direct nutrients but, when administered regularly, lead to an increase in the rate of growth and an improved feed conversion efficiency. There are three major classes of growth promoters; the antibacterials, the hormones, and a class of compounds that act specifically by improving rumen activity and ruminal propionic acid levels.

1. *Antibiotic Growth Promoters*

It was first shown in 1955 that the addition of small quantities of penicillin to chick feed (10 g/ton) led to a marked increase in growth rate which could be reproduced when the chicks were kept in a germ-free environment [49]. Further studies have shown that this effect is even more pronounced in broilers and pigs, and most new antibiotics are now screened in these animals for growth-promoting activity. However, because of concerns about the potential increase in the rate of emergence of resistant strains of bacteria, it was decided to limit the types of antibiotics that could be used as growth promoters to those that had little value in human medicine. The three major products in this area are virginiamycin, the bambermycins, and spiramycin, which showed combined sales of US\$ 171 million in 1992. Virginiamycin (Smith Kline Beecham) was first isolated in 1955 from a *Streptomyces* species related to *Streptomyces virginiae* [50] and contains a number of components, with virginiamycin M_1 (24) predominating. The bambermycin antibiotic complex (Hoechst) was isolated from cultures of *Streptomyces bambergiensis* and consists of at least four active ingredients, with moenomycin A (25) being the major component [51]. Spiramycin (26) (Spira, Rhône-Poulenc) is produced by *Streptomyces ambofaciens* that was found originally in soil samples taken from northern France in 1952 [52].

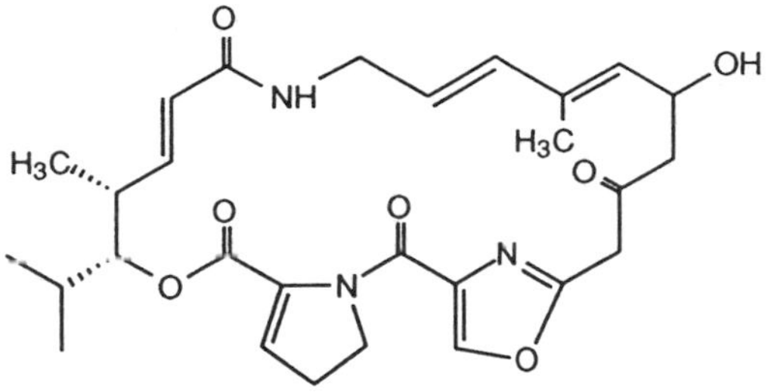

(24)

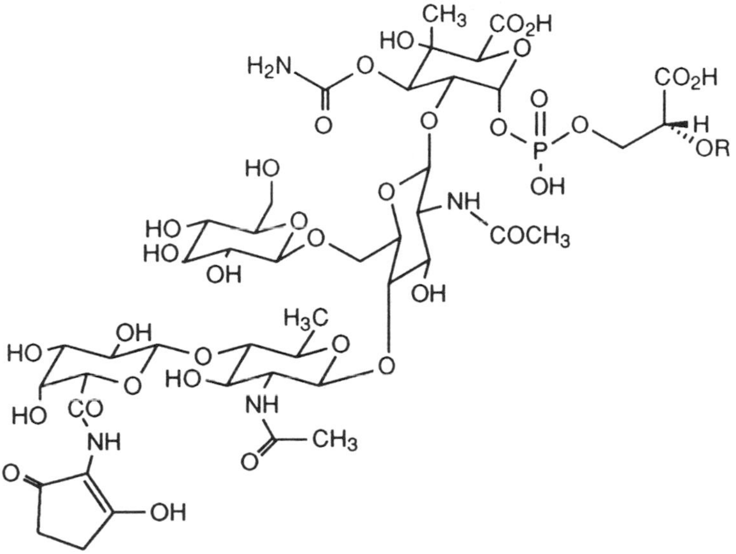

(25)

R=

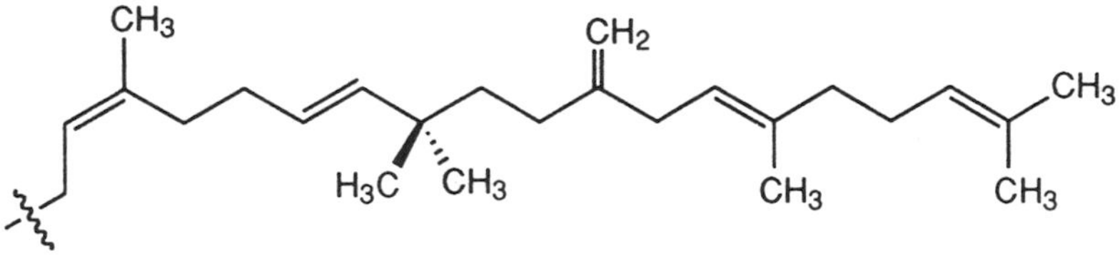

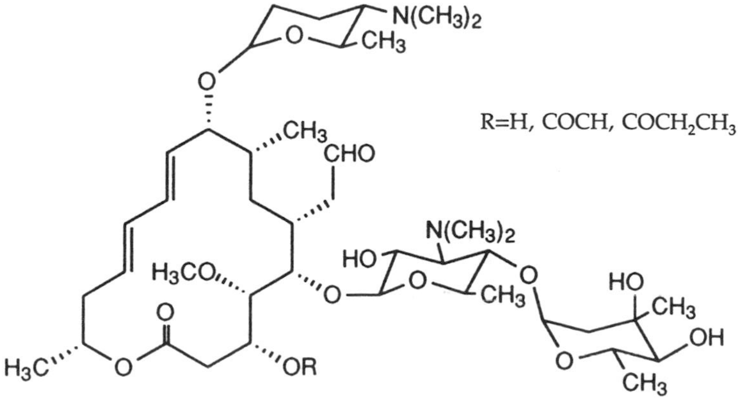

R=H, COCH, COCH$_2$CH$_3$

(26)

2. *Hormonal Growth Promoters*

The castration of male cattle in order to facilitate handling is a common practice throughout the world. This radical surgical measure, however, leads to reduced levels of several hormones, such as testosterone, which play important roles in the growth process and feed conversion. Only implants of the natural steroidal hormones oestradiol (27), progesterone (28), and testosterone (29), which have anabolic effects, have remained in use for growth-promoting purposes in a limited number of countries.

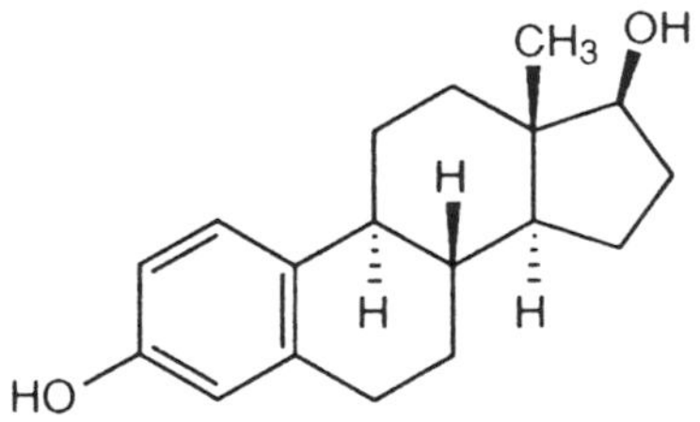

(27)

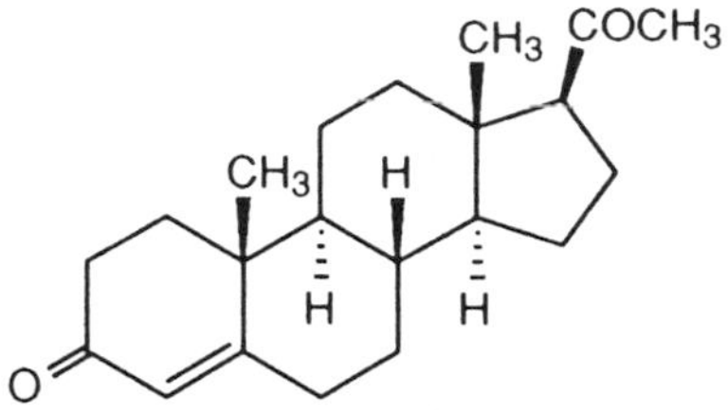

(28)

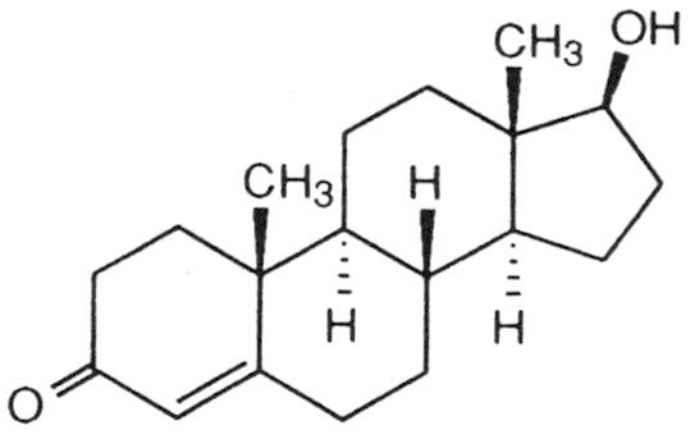

(29)

Melengesterol acetate (MGA) (30) is a pregestational steroid that is used to suppress oestrus behavior in feedlot heifers. It is also approved for use in the United States in combination with monensin and lasalocid A and achieved sales of US\$ 15 million in 1992.

3. *Compounds Affecting Ruminal Efficiency*

Cattle, sheep, and other ruminants are able to digest relatively coarse fibers such as straw, hay, and grass by means of fermentation in the

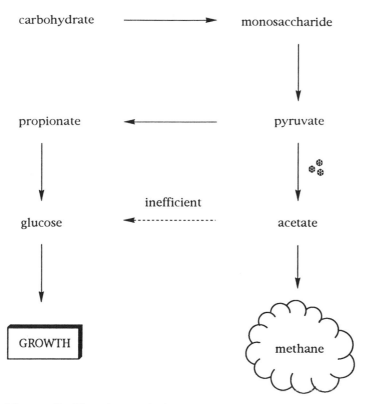

(30)

rumen (one of the four stomachs). One of the major inefficiencies of this fermentation process is the favored breakdown of pyruvate to acetate and methane (Figure 8). In the 1970s it was shown that the polyether antibiotic and coccidiostat monensin (Section IIIB1), when added to the feed of ruminants, exhibited a significant growth-promoting effect, and it was postulated that the compound acts on the organ-

Figure 8 The chemical degradation of dietary carbohydrate by rumen microflora.

isms that convert pyruvate to acetate, thus increasing the relative concentration of propionate in the rumen. This leads to a better utilization of the carbohydrate feedstock and an overall increase in the rate of growth.

In 1992 the estimated world market size for this type of growth promoter was US$ 170 million. The leading products in this area are monensin (Eli Lilly, US$ 65 million in 1992) and lasalocid A (Hoffmann–LaRoche, US$ 48 million in 1992). Other important natural products used as growth promoters are avoparcin, narasin, and salinomycin. All of these products, with the exception of avoparcin, have already been discussed by reference to their coccidiostatic properties, and, therefore, only avoparcin will be considered in detail here.

a.   AVOPARCIN

1.   *Isolation and Structural Studies*   Avoparcin (*31, 32*) (originally designated LL-AV290) was isolated from a culture of *Streptomyces candidus* by workers at American Cyanamid in 1967 [53].

Initial degradative studies on avoparcin indicated that it was a glycopeptide similar to the known vancomycin and ristocetins [54]. A partial structure based on degradation studies was published in 1974

α-avoparcin R=H   (*31*)

β-avoparcin R=Cl   (*32*)

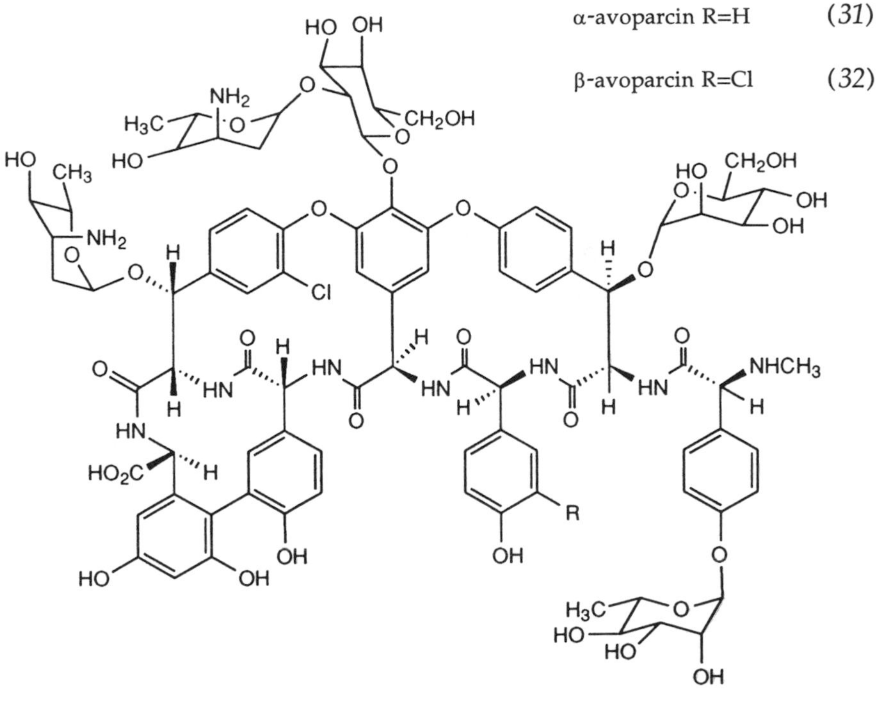

[55] which was extended by work reported in 1979 that demonstrated that avoparcin consisted mainly of two components (α and β) in the ratio of about 1:4 [56]. In 1980 a publication appeared in which the full structures of the avoparcins were established with the exceptions of the exact positions of the chlorine atom in β-avoparcin and the ristosamine and mannose sugar moieties [57]. These questions were finally answered in 1981 [58]. The absolute stereochemistry of avoparcin was disclosed in 1983 [59].

2. *Biological Profile* Avoparcin exhibits antibiotic activity against gram-positive and gram-negative as well as acting as a growth promoter in poultry and ruminants. The antibiotic effect of the glycopeptide class of antibiotics involves their binding the D-ala-D-ala terminal positions of the cross-linking pentapeptide, thus interrupting cell wall formation in bacteria. The net result is the proliferation of weak spots in the cell walls which results in cell lysis.

Avoparcin is relatively nontoxic with an oral LD_{50} in mice, rats, and chickens higher than of 10 g/kg [60]. The natural product is poorly absorbed from the gastrointestinal tract, leaves no detectable tissue residues and has a very wide margin of safety.

3. *Product Profile* Avoparcin (Avotan, American Cyanamid) is available as a feed additive for poultry and ruminants as a 2–10% premix in an oiled chalk carrier. The final dose level is in the region of 7.5–15 g/ton. In 1992 avoparcin showed sales of US$ 42 million.

IV. PHARMACEUTICALS

A. Ectoparasiticides

The range of products available to the veterinarian for the control of ectoparasites (ticks, flies, and lice) has largely reflected the array of compounds developed as insecticidal agents. Of particular note are the synthetic pyrethroids (for example, Deltamethrin, Roussel-Uclaf, *33*) that were developed to counteract the photoinstability of the natural pyrethrins (e.g., pyrethrin I, *34*) which were isolated from pyrethrum flowers. The ectoparasiticide market was estimated to be worth US$ 585 million in 1992. These compounds are discussed in more detail in Chapter 2 of this book.

(*33*)

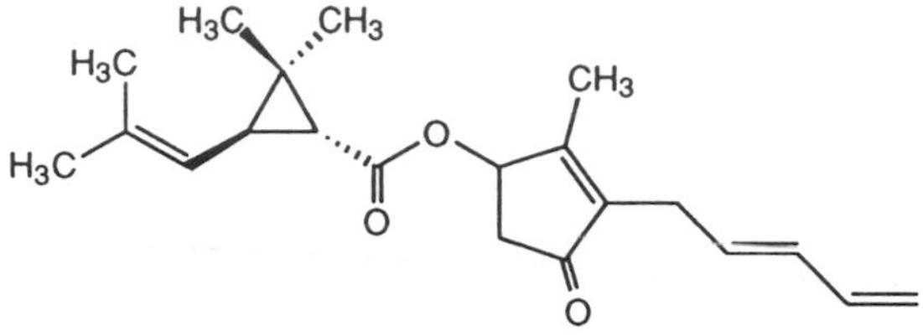

(34)

B. Endectocides

The enormous variety of helminths (tapeworms, flukes, and round-worms) found in companion and farm animals are important targets for chemotherapy. The great economic losses caused by these organisms has prompted a search for novel anthelmintics. The isolation of the family of avermectins from the actinomycete *Streptomyces avermitilis* and the subsequent preparation of the reduced 22,23-dihydroavermectin B_1 (ivermectin) (35) marked a major milestone in the treatment of nematode helminths and other pests [61]. Ivermectin (Ivomec, Merck) is currently the largest selling product in the animal health world market with estimated sales of US\$ 500 million in 1992. Ivermectin is a specific example of an endectocide (active against both endoparasites and ectoparasites), showing good activity against a variety of ectoparasites when administered to cattle, sheep, horses, and pigs [62]. The

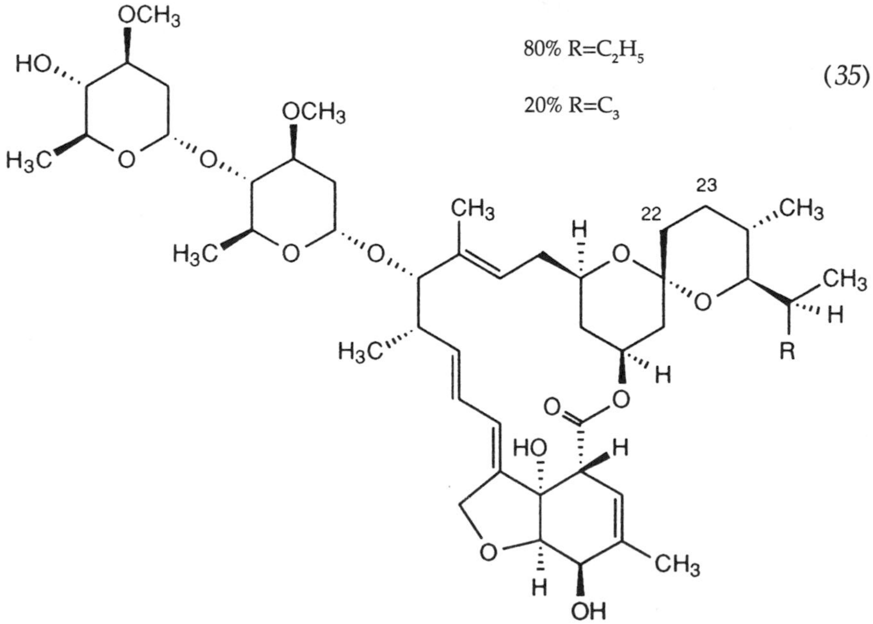

80% R=C_2H_5

20% R=C_3

(35)

avermectins, and the related milbemycins, are discussed in more detail in Chapter 4 of this book.

C. Hormones

Hormone therapy increases the profitability of breeding herds and flocks by improving the success rate of artificial insemination procedures (by the use of progestins and prostaglandins) and by the induction of birth at the desired time (corticosteroids and prostaglandins). Additionally, estrus can be synchronized and aberrant sexual behavior in cats and dogs avoided by progesterone derivatives called progestins. Major products in this area include the naturally occurring human and equine gonadotrophins (HCG, HMG, and PMSG), the progestins [for example, medroxyprogesterone acetate (MAP) (36)] and cloprostenol (37) (ICI) which is a synthetic analog of prostaglandin $F_{2\alpha}$ (38).

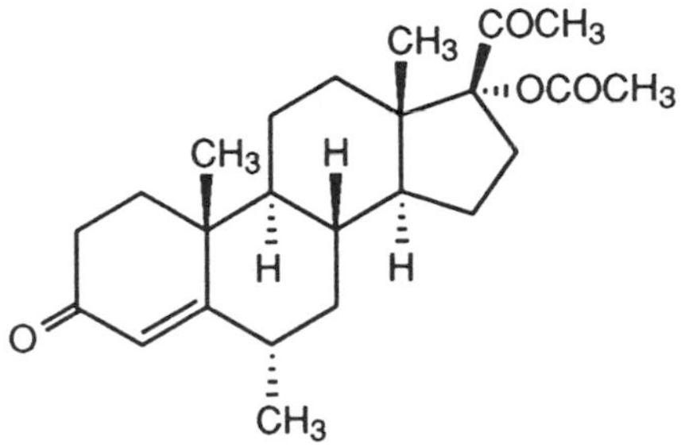

(36)

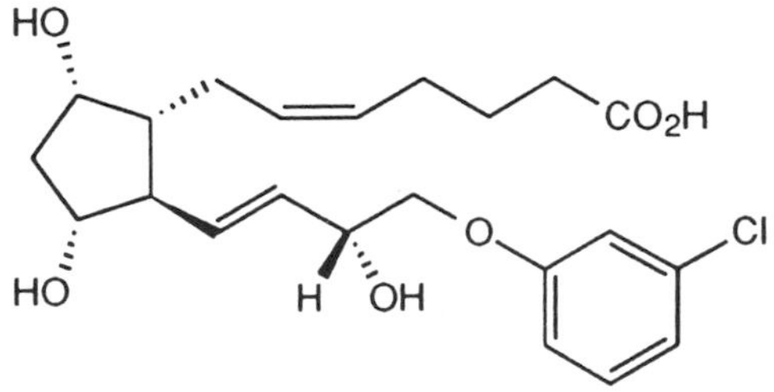

(37)

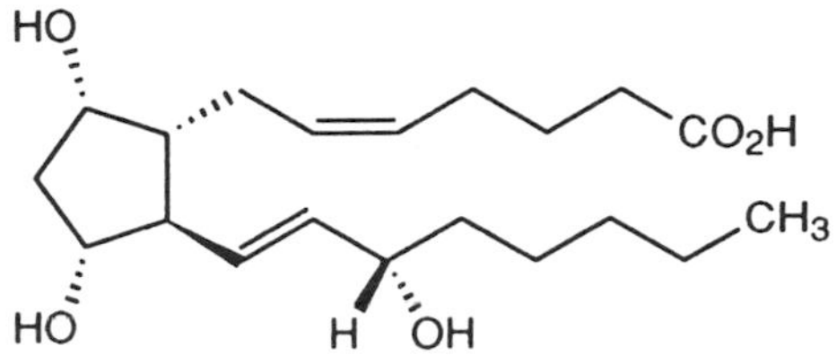

(38)

V. CONCLUSION

As in the field of human medicine, natural products and their derivatives have played a major role in the control of pathogens in the area of animal health. The improving health and productivity of farm animals contributes directly to the ever-increasing standards of nutrition today enjoyed (and demanded) by a large section of the world's population. Currently, many research groups are focusing on the discovery of new natural products as sources for medicinal and veterinary products, not only in the light of the supposedly "green" nature of such compounds but also as sources of unusual structural features and novel modes of action. The future health of our farm and companion animals, and ultimately ourselves, will depend on the success of this search.

REFERENCES

1. E.S. Turner, *Call the Doctor: A Social History of Medical Men*, Michael Joseph, London, p. 320 (1958).
2. Vivash Jones / County NatWest Woodmac, *1992 Summary Report*, London (1994).
3. W.H. Sebrill and R.S. Harris, *The Vitamins*, Vol. 1, 2nd ed., Academic Press, New York, (1967).
4. S.J. Selwyn, *J. Antimicrob. Chemother.*, 5:249 (1979).
5. H.W. Florey, E. Chain, N.G. Heatley, M.A. Jennings, A.G. Sanders, E.P. Abraham, and M.E. Florey, *Antibiotics, Vol. 1–2*, Oxford University Press, London (1949).
6. G.G.F. Newton and E.P. Abraham, *Biochem. J.*, 62:651 (1956).
7. For a comprehensive review: W. Dürckheimer, *Angew. Chem.*, 14:721 (1975).
8. D.J. Maso, A. Dietz, and C. DeBoer, *Antimicrob. Ag. Chemother.*, 1962:554 (1962).
9. R.L. Hamill, M.E. Haney, Jr., J.M. McGuire, and M.C. Stamper, *U.S. Patent* 3, 178, 341 (1967) to Eli Lilly.
10. R.B. Woodward, *Angew. Chem.*, 69:50 (1957).
11. R.B. Morin and M. Gorman, *Tetrahedron Lett.*, 4:2339 (1964).
12. R.B. Morin, M. Gorman, R.L. Hamill, and P.V. Demarco, *Tetrahedron Lett.*, 11:4737 (1970).
13. H. Achenbach, W. Regel, and W. Karl, *Chem. Ber.*, 108:2481 (1975).
14. N.D. Jones, M.O. Chaney, H.A. Kirst, G.M. Wild, R.H. Baltz, R.L. Hamill, and J.W. Paschal, *J. Antibiotics*, 35:420 (1982).
15. K. Tatsuta, Y. Amemiya, Y. Kanemura, H. Takahashi, and M. Kinoshita, *Tetrahedron Lett.*, 23:3375 (1982).
16. E.T. Seno, R.L. Pieper, and F.M. Huber, *Antimicrob. Ag. Chemother.*, 11:455 (1977).

17. S. Omura, A. Nakagawa, H. Takeshima, J. Miyazawa, C. Kitao, F. Piriou, and G. Lukacs, *Tetrahedron Lett.*, *16*:4503, (1975).
18. G.C. Brander, *Chemicals for Animal Health Control*, Taylor and Francis, London, p. 60 (1986).
19. R.C. Anderson and H.M. Worth, unpublished results.
20. R.F. Gordon and F.T.W. Jordan, *Poultry Diseases*, Baillière Tindall, London, p. 166 (1982).
21. J.W. Westley, *Polyether Antibiotics Naturally Occurring Acid Ionophores, Vol. 1 Biology* (J.W. Westley, ed.), Marcel Dekker, New York (1983).
22. J.A. Robinson, *Prog. Chem. Org. Nat. Prod.*, *58*:1 (1991).
23. M.E. Haney and M.M. Hoehn, *Antimicrob. Ag. Chemother.*, *1967*:349 (1967).
24. A. Agtarap, J.W. Chamberlin, M. Pinkerton, and L. Steinrauf, *J. Am. Chem. Soc.*, *89*:5737 (1967).
25. T. Fukuyama, K. Asaka, D.S. Karanewsky, C.L.J. Wang, G. Schmid, and Y. Kishi, *J. Am. Chem. Soc.*, *101*:262 (1979).
26. L.E. Day, J.W. Chamberlin, E.Z. Gordee, S. Chen. M. Gorman, R.L. Hamill, T. Ness, R.E. Weeks, and R. Stroshane, *Antimicrob. Ag. Chemother.*, *4*:410 (1973).
27. C.C. Wang, *The Biology of Coccidia* (P.L. Long, ed.), University Park Press, Baltimore, p. 167 (1982).
28. J. Berger, A.I. Rachlin, W.E. Scott, L.H. Sternbach, and M.W. Goldberg, *J. Am. Chem. Soc.*, *73*:5295 (1951).
29. J.W. Westley, R.H. Evans, Jr., T. Williams, and A. Stempel, *J. Chem. Soc., Chem. Commun.*, *1970*:71 (1970).
30. S.M. Johnson, J. Herrin, S.J. Lui, and I.C. Paul, *J. Chem. Soc., Chem. Commun.*, *1970*:72 (1970).
31. T. Nakata, G. Schmid, B. Vranesic, M. Okigawa, T. Smith-Palmer, and Y. Kishi, *J. Am. Chem. Soc.*, *100*:2933 (1978).
32. J.W. Westley, R.H. Evans, G. Harvey, R.G. Pitcher, D.L. Preuss, A. Stempel, and J. Berger, *J. Antibiotics*, *27*:288 (1974).
33. Y. Tanaka, H. Saito, Y. Miyazaki, H. Sugawara, J. Nagatsu, and M. Shibuya, *U.S. Patent 3, 357, 948* (1974) to the Kaken Chemical Co.
34. H. Kinashi, N. Otake, and H. Yonehara, *Tetrahedron Lett.*, *14*:4955 (1973).
35. Y. Kishi, S. Hatakeyama and M.D. Lewis, (Front. Chem. Plenary Keynote Lect. 28th IUPAC Congr.) (1981).
36. T.T. Migaki, L.R. Chappel, and W.E. Babcock, *Poultry Sci.*, *58*:1192 (1979).
37. L.D. Boeck, M.M. Hoehn, R.E. Kastner, R.W. Wetzel, N.E. Davies, and J.E. Westhead, *Dev. Ind. Microbiol.*, *18*:471 (1976).
38. J.L. Occolowitz, D.H. Berg, M. Dobono, and R.L. Hamill, *Biomed. Mass Spectros.*, *3*:272 (1976).
39. H. Sato, T. Yahagi, Y. Miyazaki, and N. Otake, *J. Antibiotics*, *30*:530 (1977).
40. D.E. Dorman, J.W. Paschal, W.M. Nakatsukasa, L.L. Huckstep, and N. Neuss, *Helv. Chim. Acta*, *59*:2625 (1976).
41. M.D. Ruff, W.M. Reid, A.P. Rahn, and L.R. McDougald, *Poultry Sci.*, *59*:2008 (1980).

42. C-M. Liu, B. Prosser, and J. Westley, U.S. Patent 4, 278, 663 (1981) to Hoffmann–LaRoche.

43. D.P. Labeda, J.H.E.J. Martin, and J.J. Goodman, U.S. Patent 4, 407, 946 (1983) to American Cyanamid.

44. J.W. Westley, C-M. Liu, J.F. Blount, R.H. Evans, L.H. Sello, N. Troupe, and P.A. Miller, *Trends in Antibiotic Research* (H. Umezawa, ed.), Japan Antibiotics Research Association, Tokyo (1982).

45. H.R. Tsou, S. Rajan, R. Fiala, P.C. Mowery, M.W. Bullock, D.B. Borders, J.C. James, J.H. Martin, and G.O. Morton, *J. Antibiotics*, 37:1651 (1984).

46. C-M. Liu, T.E. Hermann, A. Downey, B. La T. Prosser, E. Schildknecht, N.J. Palleroni, J.W. Westley, and P.A. Miller, *J. Antibiotics*, 36:344 (1983).

47. A.L. Oronsky, U.S. Patent 4, 496, 549 (1985) to American Cyanamid.

48. I.B. Wood, U.S. Patent 4, 510,134 (1985) to American Cyanamid.

49. M.E. Coates, M.K. Davies, and S.K. Kon, *Br. J. Nutr.*, 9:110 (1955).

50. P. de Somer and P.J. van Dijck, *Antibiot. Chemother.*, 5:632 (1955).

51. F. Lindner and K.H. Wallhäusser, U.S. Patent 3, 674, 866 (1966) to Hoescht.

52. L. Ninet and J. Verrier, U.S. Patent 2, 943, 023 (1961) to Rhône-Poulenc.

53. M.P. Kunstmann and J.N. Porter, U.S. Patent 3, 338, 786 (1967) to American Cyanamid.

54. N.P. Kunstmann, L.A. Mitscer, J.N. Porter, A.J. Shay, and M.A. Darkin, *Antimicrob. Ag. Chemother.*, 1968:248 (1968).

55. J.J. Hlvaka, P. Bitha, J.H. Boothe, and G. Moreton, *Tetrahedron Lett.*, 15:175 (1974).

56. W.J. McGahren, J.H. Martin, G.O. Morton, R.T. Hargreaves, R.A. Leese, F.M. Lovell, and G.A. Ellestad, *J. Am. Chem. Soc*, 101:2237 (1979).

57. W.J. McGahren, J.H. Martin, G.O. Morton, R.T. Hargreaves, R.A. Leese, F.M. Lovell, G.A. Ellestad, E. O'Brien, and J.S.E. Holker, *J. Am. Chem. Soc*, 102:1671 (1980).

58. W.J. McGahren, I.M. Armitage, F. Barbatschi, W.E. Gore, G.O. Morton, R.A. Leese, F.M. Lovell, and G.A. Ellestad, *J. Am. Chem. Soc.*, 103:6522 (1981).

59. G.A. Ellestad, W. Swenson, and W.J. McGahren, *J. Antibiotics*, 36:1683 (1983).

60. American Cyanamid, company literature.

61. W.C. Campbell (ed.), *Ivermectin and Avermectin*, Springer-Verlag, New York (1989).

62. W.C. Campbell, M.H. Fisher, E.O. Stapley, G. Albers-Schönberg, and T.A. Jacob, *Science*, 221:4613 (1983).

6

Herbicides and Plant Growth Regulators

Richard James Stonard

Monsanto Company, St. Louis, Missouri

Margaret A. Miller-Wideman

G. D. Searle, St. Louis, Missouri

I. INTRODUCTION

During the first five decades (1940s–1980s) of the use of synthetic organic chemicals in crop agriculture, chemicals of natural origin made a significant contribution to the improvement in yield and quality which resulted from the widespread adoption of this agronomic practice. This contribution is attributable principally to the development of the pyrethroids from the natural product, pyrethrin, and their impact on the control of crop-damaging insects. Other examples of natural products serving as prototypes for insecticide development include physostigmine, nereistoxin, and juvenile hormones. During this time period, several fungicides of microbial origin (e.g., polyoxin and validamycin) were utilized in controlling plant pathogens, but this practice was restricted mainly to Japan. Recently strobilurin, a fungal metabolite, has served as an important prototype for the development of the β-methoxyacrylate family of fungicides and pyrrolnitrin has served as the prototype for phenylpyrroles. During the 1960s the cost-effective production of gibberellic acid as an unpurified fermentation broth of the fungus *Gibberella fujikuroa* led to the widespread commercial utilization of this family of plant-growth-regulating chemicals. Although there were no natural herbicides of consequence in commercial use until the late 1980s, one of the first synthetic organic herbicides to be widely used (2,4-D) was the product of an auxin analog synthesis program. During the past decade, research aimed at identifying natural substances that might serve as commercial herbicides per se or as prototypes for the development of synthetic analogs has intensified. The intent of this chapter is to provide an overview of the methodologies employed in the search for these compounds and to present examples of the active secondary metabolites which have been discovered during the past several years [1].

II. WHY MICROBIAL SECONDARY METABOLITES AS A SOURCE OF BIOACTIVITY?

Serendipitous screening of microbial fermentation broths for activities of interest in the field of human pharmaceuticals has been a respected practice for several decades. This strategy has provided researchers with a wealth of valuable new drugs, most notably in the antibiotics area. This approach remains a core discovery strategy in most pharmaceutical companies today. The incorporation of this approach into the agrochemical discovery paradigm is a much more recent occurrence and remains a subject of much debate. What is not debated is the remarkable aptitude that microorganisms have for producing structurally diverse secondary metabolites.

The proponents of empirical screening for agricultural activities from fermentation broths further recognize that natural substances offer biodegradability, target specificity, and unique mechanisms of action, all of which are important characteristics of modern pesticides. Opponents cite the high cost-of-goods of the natural substance per se or that they are frequently too structurally complex to serve as models for analog synthesis programs. Furthermore, the positive attributes of natural products are in many cases also viewed as negatives; for example, insufficient environmental stability to be commercially efficacious and too narrow a spectrum of activity to be stand-alone products. Although significant hurdles to commercialization of a natural product in agriculture exist, the success of products based on microbial secondary metabolites such as avermectin, phosphinothricylalanylalanine (1), and strobilurin provide clear justification for the

R=H   (1)
R=L-leucine   (3)
R=L-alanine   (4)

search given the ever-decreasing rate of commercialization of novel synthetic agrochemicals having unique modes of action.

Herbicides of microbial origin possess one further advantage over compounds of purely synthetic origin. This is the ready access to a gene encoding for an enzyme or enzymes capable of inactivating the natural product. In the case of *Streptomyces hygroscopicus*, which produces phosphino-thricylalanylalanine, a gene (*bar*) encoding for an acetyl-transferase was found to be clustered with the genes encoding for the biosynthetic enzymes on an 18-kb gene [2]. Acetylated phosphino-thricin, the product of the action phosphinothricin acetyl-transferase on phosphinothricin, is devoid of herbicidal activity. Crops genetically engineered to express the *bar* gene represent an excellent vehicle for enhancing the utility of phosphinothricin-based herbicides as a result of the improvement in selectivity of this nonselective herbicide. This option should be considered when assessing the value of microbial herbicide discovery programs. Crops engineered to be resistant to environmentally benign postemergence herbicides such as glyphosate and phosphinothricin present numerous economic and environmental advantages for the farmer.

III. RECENTLY DISCOVERED HERBICIDAL SUBSTANCES OF MICROBIAL ORIGIN

The most significant development in the quest for herbicidal natural products from microbial fermentation broths occurred in 1971 when Bayer et al. [3] discovered the *S. hygroscopicus* metabolite phosphinothricylalanyl-alanine (*1*). Since that time, this compound has been developed into commercial herbicides by both Meiji Seika (Herbiace®) and Hoechst AG, i.c., Basta® [3]. Basta® is an attractive product having broad-spectrum, postemergent herbicidal activity, low rates of application (0.84–1.68 kg ai/ha) and low toxicity (oral LD_{50}: 2170 mg/kg; dermal LD_{50}: 1400 mg/kg versus male rats). The active ingredient in Basta® is synthetic phosphinothricin (*2*), whereas Meiji Seika employs the fermentation-produced tripeptide, phosphino-thricylalanylalanine or bialaphos [4a]. Since the discovery of phosphino-thricylalanylalanine, several related phosphinothricin-containing, naturally occurring peptides have been discovered, including phosalacine (*3*) [5] and trialaphos (*4*)

(2)

[6]. Although many structurally unique natural product herbicides have been discovered from microbial sources, only those based on the amino acid, phosphinothricin have been introduced into commerce.

It is noteworthy that the majority of the microbially derived phytotoxic compounds that have been discovered during the past decade are water-soluble peptides, amino acids, or nucleosides: for example, α-methylene-β-alanine (5) [7, 8], oxetin (6) [9], cis-2-amino-1-hydroxy-cyclo-butane-1-acetic acid (7) [10], isoxazole-4-carboxylic acid (8) [11], hydantocidin (9) [12], 5′-deoxy-guanosine (10) [13], coaristeromycin (11) [13], 5′-deoxytoyocamycin (12) [14], SF 2494 (13) [15], phthoxazolin (14) [16], homoalanosine (15) [17], cyanobacterin (16) [18], and altemicidin (17) [7, 19].

Metabolites possessing whole plant activity which were discovered recently and which are devoid of nitrogen include herboxidiene (18) [20a], kaimonolide A (19) [21] and B (20) [22], arabenoic acid (21) [23], and cornexistin (22) [24].

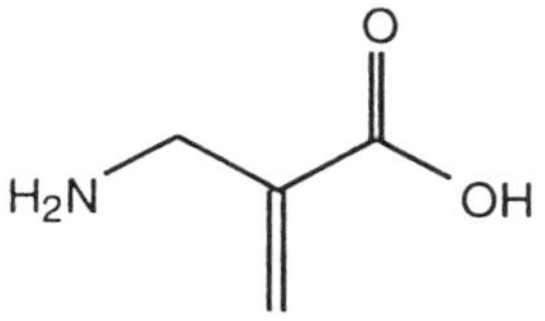

(5)

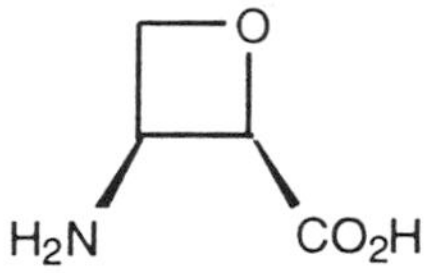

(6)

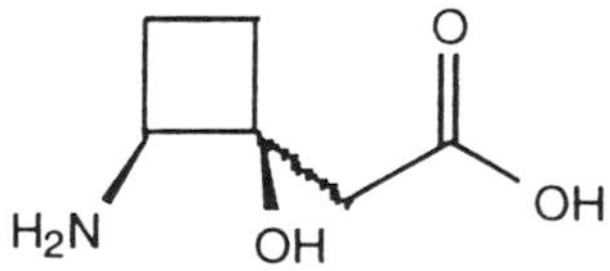

(7)

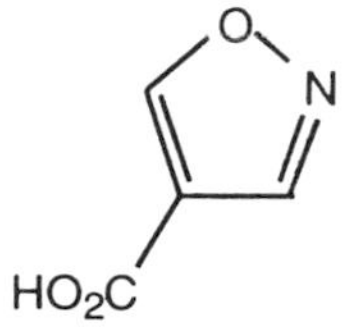

(8)

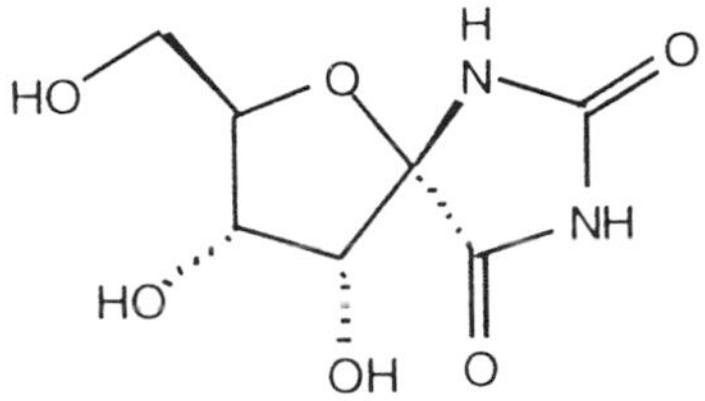
HO
O
H
N
O
NH
HO
OH
O

(*9*)

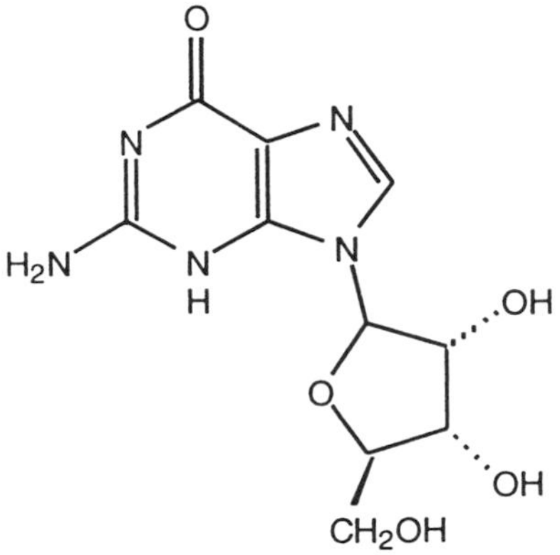
O
N
N
H2N
N
H
N
OH
O
OH
CH2OH

(*10*)

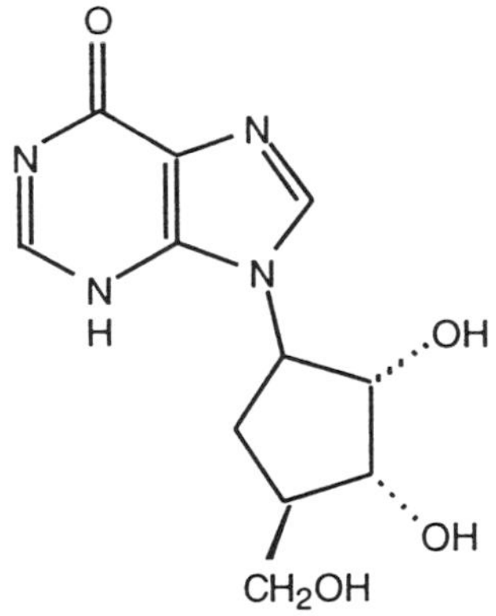
O
N
N
N
H
N
OH
OH
CH2OH

(*11*)

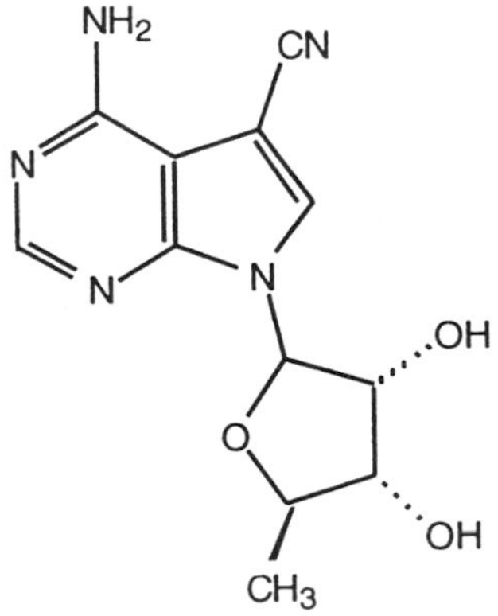
NH2
CN
N
N
N
OH
O
OH
CH3

(*12*)

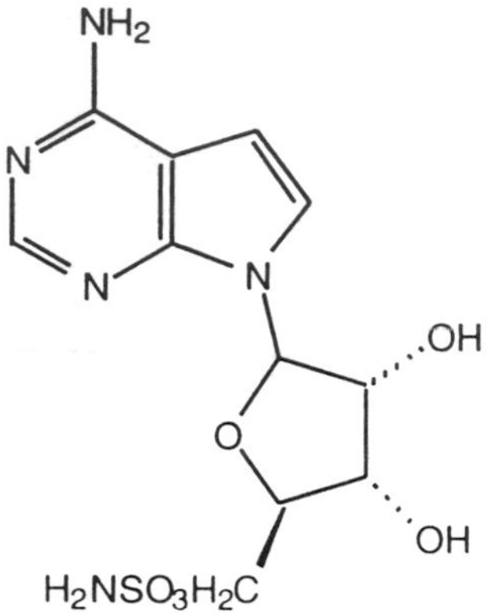

NH2
N
N
N
OH
O
OH
H2NSO3H2C

(*13*)

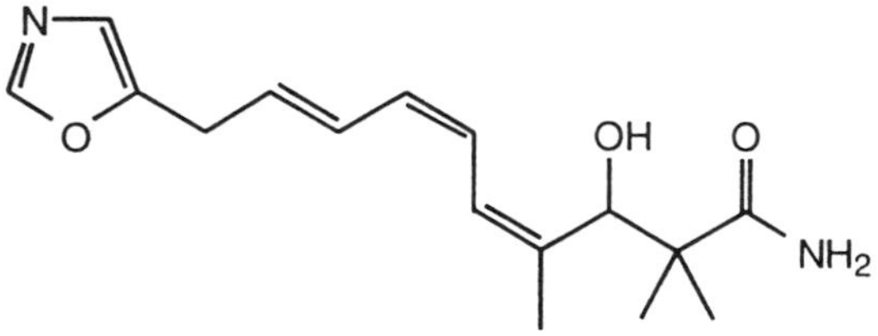

N
O
OH
O
NH2

(*14*)

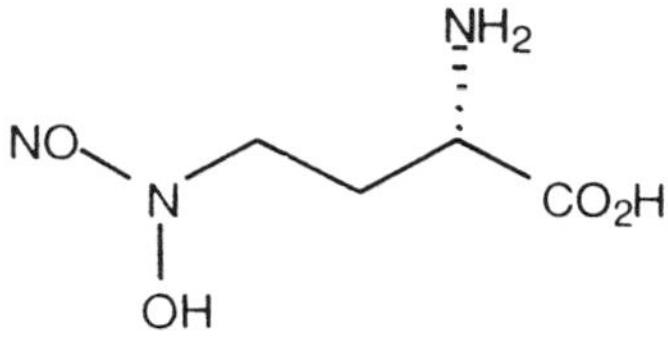

NH2
NO
N
CO2H
OH

(*15*)

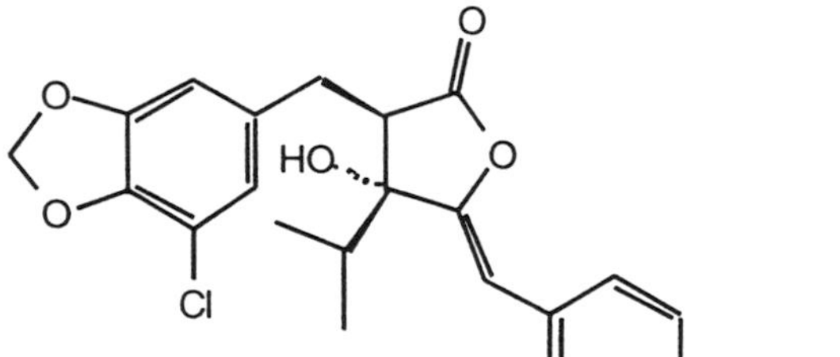

O
O
O
HO
O
Cl
OCH3

(*16*)

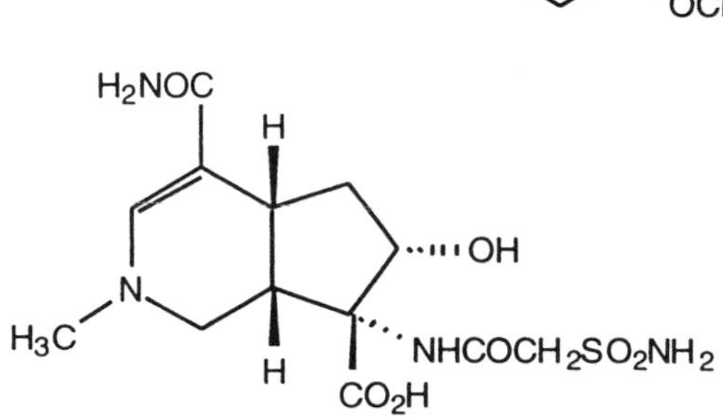

H2NOC
H
OH
N
H3C
NHCOCH2SO2NH2
H
CO2H

(*17*)

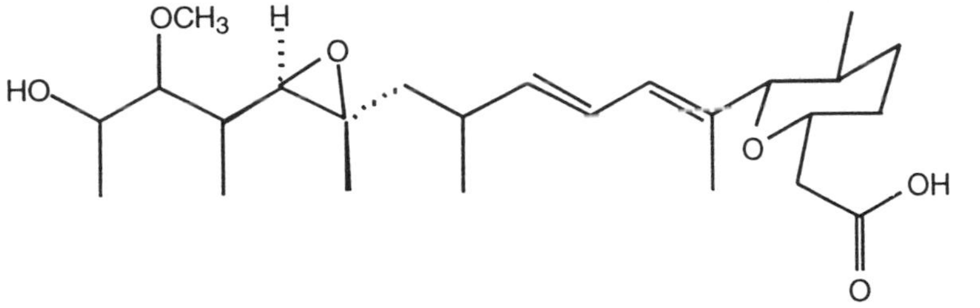
OCH3
H
O
HO
O
OH
O

(18)

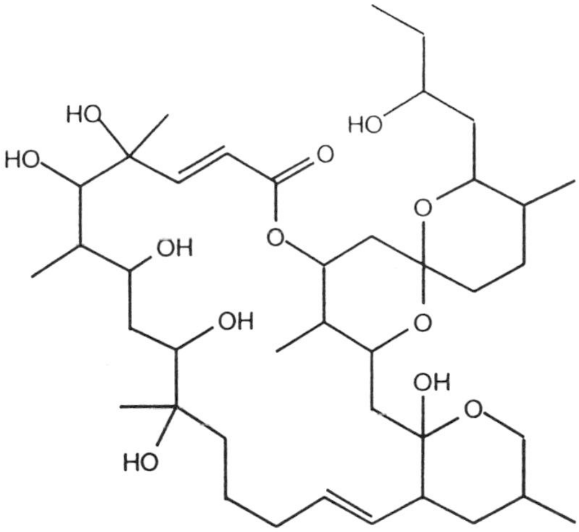
HO
HO
HO
O
O
OH
O
OH
O
OH
O
HO

(19)

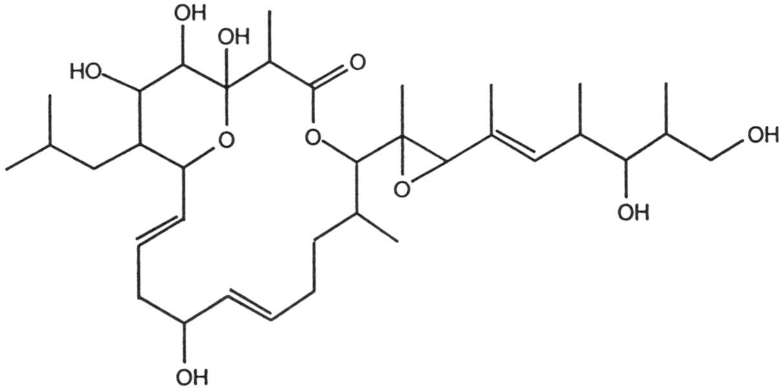
OH
OH
HO
O
O
O
OH
O
OH
OH

(20)

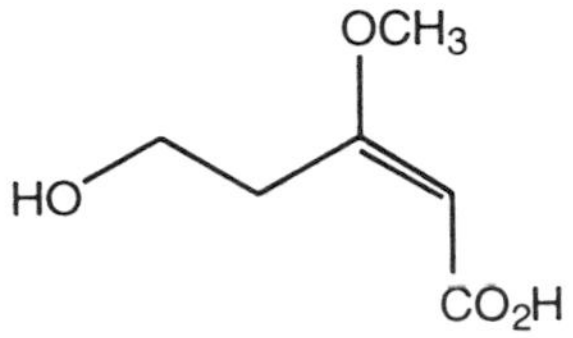
OCH3
HO
CO2H

(21)

(22)

As discussed in the next section, a myriad of assays have been utilized in the discovery of herbicidal natural products such as compounds 1–22 (see Table 1 for examples). Unfortunately, few of these compounds have proven to have activity on intact weeds that is sufficient to warrant either synthetic follow-up or advanced testing of the fermentation-produced metabolite. From the available information, the exceptions to this generalization appear to be homoalanosine, herboxidiene, isoxazole-4-carboxylic acid, altemicidin, and hydantocidin.

Synthesis work aimed at producing an analog of altemicidin (17) with improved unit activity was discontinued [7] following the report of its independent discovery [19] and the suggestion of a potential toxicological issue with this compound. Hydantocidin (9) is a broad-spectrum herbicide with very good activity against both narrow-leafed and broad-leafed weeds [12]. Isoxazole-4-carboxylic acid (8) appears to be a nonselective herbicide with a good spectrum of weed control and low use rate. Homoalanosine (15) is also an effective broad-spectrum herbicide and has shown good activity in the field at rates as low 10 g/acre [17]. Herboxidiene (18) is a complex polyketide showing highly selective activity against several annual weed species at rates as low as 7 g/acre in the greenhouse [20b]. Attempts to simplify the structure by a combination of synthesis and degradative reactions failed to improve upon spectrum or maintain the original biological activity. This lead was discontinued owing to the cost of production of the compound per se relative to its limited spectrum of weed control. None of the aforementioned compounds (3–22) are currently registered for commercial use.

IV. THE DETECTION OF HERBICIDAL SUBSTANCES PRODUCED BY MICROORGANISMS

Historically, the primary focus of the natural products researcher has been structure elucidation of novel compounds. Frequently, biological activity was viewed as a means to an end, and little attention was paid to the significance of the bioassay. More recently, as interest in secondary metabolites as sources of unique prototypes for synthesis, novel modes of action and as agricultural products per se has grown, interest in optimizing their detection at a primary screening level in a selective and cost-effective manner has also grown. Unlike the biologist who screens synthetic chemistry with full knowledge of the chemical structure and, more importantly, the concentration of the test material, the biologist participating in the screening of natural products is at a disadvantage as a consequence of the absence of this information. Not until later stages in the purification process does knowledge of the potency of the active component become available. This places great importance on having highly sensitive and reproducible assays when searching for phytotoxic activity from complex matrices such as fermentation broths. Although this is true for any test, it is imperative for natural product screens when the source of the activity is typically in limited supply.

The ideal natural product assay utilizes very little test material, is sensitive, is comprehensive in detecting potentially interesting modes of action (unless deliberately designed to be specific), is highly reproducible from test to test, has a quick turn around time, is easy to read, is relatively inexpensive to perform, and is adaptable to high-throughput screening. Many of these criteria are relevant to the academic researcher. All are relevant to the industrial researcher involved in high-throughput screening against multiple targets.

Early herbicidal natural products discovery efforts utilized very simple seed-germination-inhibition tests or whole plant observation assays following spraying with the test solution. More recently, assays targeting specific enzymes or metabolic pathways have been used (i.e., antimetabolite, chlorophyll synthesis inhibition, de novo starch synthesis inhibition, etc.). Occasionally, inhibitory activity of an enzyme is known as a result of nonplant-related studies as for mevinolin, a coenzyme A reductase inhibitor. The herbicidal activity of mevinolin was postulated and was later confirmed using a simple whole plant assay [25].

Brief descriptions of assays used in the discovery or confirmation of phytotoxic secondary metabolites are provided below. Table 1, although not intended to be comprehensive, provides an overview of the success of each type of assay in the discovery and purification of herbicidally active secondary metabolites. Also discussed below are a number of assays that have not been reported to have been used in natural product discovery but which satisfy many of the criteria necessary for a successful discovery tool.

Table 1 Producing Organism, Compound, and Type of Assay Used to Discover or Confirm Phytotoxic Activity

Organism	Compound	Activity	Ref.
Aspergillus terreus	Acetylaranotin	Whole plant	33
Alternaria eichorniae	Alteichin	Whole plant & protoplasts	28
Streptomyces sp.	Altemicidin	Matrix screen	7
Alternaria porri	Altersolanol A&B	Whole plant	36
Alternaria mali	AM-toxins	Excised leaf assay	47
Bacillus subtilis	Amicoumacin B	Matrix screen	7
Streptomyces sp.	Anisomycin	Whole plant	32
Penicillium sp.	Anthglutin	Matrix screen	7
Unidentified fungus	Arabenoic acid	Matrix screen	23
Actinoplane sp.	Arabinoside A	Matrix screen	7
Streptomyces sp.	Azaserine	Whole plant & de novo starch synthesis inhibition	35, 93
Baccharis megapotamica	Baccharinol	Wheat coleoptile	63
Streptomyces hygroscopicus	Bialaphos	Whole plant	4b
Actinomycete sp.	Borrelidin	Whole plant	43
Botryotinia squamosa	Botrydienol	Whole plant	34
Alternaria carthami	Brefeldin A	Excised leaf assay	50
Diplodia macrospora	Chaetoglobosin K	Wheat cleoptile	68
Cinnamomum cassia	Cinnamic acid	Whole plant	31
Streptomyces sp.	Cispentacin	Matrix screen	7
Penicillium charlesii	Citreoviridin	Wheat coleoptile	62
Stagonospora apocynii	Citrinin	Excised leaf assay	46
Cladosporium cladosporioides	Cladospolide A & B	Whole plant	29
Aspergillus repens	Cladosporin	Wheat coleoptile	61
Streptomyces sp.	Coaristeromycin	Matrix screen	13
Actinomycete sp.	Coformycin	Matrix screen	7
Colletotrichins nicotianae	Colletotrichin	Excised leaf assay	49
Monascus sp.	Compactin	Callus inhibition	84
Oryza sativa	*p*-Coumeric	Germination inhibition	57
Streptoverticillium sp.	Cyclocarbamide A & B	Germination inhibition	54
Penicillium cyclopium	Cyclopenin	Wheat coleoptile	75
Phomopsis sp.	Cytochalasins	Wheat coleoptile	64
Alternaria macrospora	αβ-Dehydrocurvularin	Excised leaf assay and protoplasts	48
Oryza sativa	S(+)-Dehydrovomifoliol	Germination inhibition	57
Thermoactinomycete sp.	5'-Deoxyguanosine	Matrix screen	13
Nodulisporium hinnuleum	Desmethoxyviridiol	Wheat coleoptile	72
Streptomyces sp.	7-Deoxy-D-glycero-D-gluco-heptose	Chlorophyll synthesis Inhibition	97
Fusarium oxysporum f. sp. *Carthami*	Diacetoxyscirpenol	Whole plant	40
Aspergillus ustus	Dihydropergillin	Wheat coleoptile	66
Streptomyces sp.	2,5-Dihytdrophenyl-alanine	Matrix screen	7
Streptomyces filipenensis	Filipin	Germination inhibition	55
Actinoplane sp.	Formycins A & B	Whole plant	42
Fusarium oxysporum f. sp. *carthami*	Fusaric acid	Whole plant	40

(*continued*)

Table 1 Continued

Organism	Compound	Activity	Ref.
Streptomyces oyocaensis	Gabaculine	γ-Amino-butyrate Aminotransferase inhibitor	99
Streptomyces sp.	Gougerotin	Whole plant	39
Penicillium sp.	Hadacidin	Whole plant & de novo starch synthesis inhibition	44, 93
Streptomyces saganonensis	Herbicidins	Germination inhibition	53
Streptomyces hygroscopicus	Herbimycin A & B	Whole plant	27
Streptomyces chromofuscus	Herboxidiene	Matrix screen	20b
Streptomyces galilaeus	Homoalanosine	Whole plant	17
Streptomyces hygroscopicus	Hydantocidin	Germination inhibition	12
Streptomyces rochei	*cis*-2-Amino-1-hydroxy-cyclobutane-1-acetic acid	Matrix screen	10
Streptomyces sp.	*trans*-4-Hydroxy-L-proline	Matrix screen	7
Aspergillus candidus	Hydroxyterphenyllin	Wheat coleoptile	70
Oryza sativa	Ineketone	Germination inhibition	57
Fusarium oxysporum f. sp. *carthami*	Lycomarasmin	Whole plant	40
Streptomyces sp.	α-Methylene-β-alanine	Matrix screen	7
Kitasatosporia sp.	β-Methyltryptophan	Matrix screen	7
Aspergillus terreus	Mevinolin	Whole plant	25
Oryza sativa	Momilactone C	Germination inhibition	57
Fusarium moniliforme	Moniliformin	Wheat coleoptile	74
Drechslera turcica	Monocerin	Whole plant	38
Streptomyces griseus	Naramycin B	Whole plant	41
Aspergillus niger	Nigerazine B	Whole plant	30
Chaetomium trilaterale	Oosporein	Wheat coleoptile	71
Helminthosporium oryzae	Ophiolbolin A	Oat coleoptile	73
Aspergillus niger	Orlandin	Wheat coleoptile	69
Streptomyces sp. OM-2317	Oxetin	Enzyme inhibition	9
Perenniporia medullaepanis	Pereniporin A	Amino acid antimetabolite	96
Aspergillus ustus	Pergillin	Wheat coleoptile	65
Kitasatosporia phosalacinea	Phosalacine	Enzyme inhibition	5
Dreschlera sorokiana	Prehelminthosporol	Wheat coleoptile	67
Penicillium sp.	Pyrrol-3-alanine	Amino acid antimetabolite	7
Alternaria helianthi	Radicinin	Whole plant	37
Baccharis megapotamica	Roridin	Wheat coleoptile	63
Anastrophyllum minutum	Sphenolobane	Germination inhibition	52
Stemphylium botryosum	Stemphylotoxin I	Excised leaf & cell suspension	87
Pseudomonas tabaci	Tabtoxinine	Algal inhibition	51
Alternaria alternata	Tentoxin	Whole plant	45
Phomopsis sp.	Terrein	Whole plant	33
Streptomyces sp.	Toyocamycin	Whole plant	32
Baccharis megapotamica	Trichoverrin B	Wheat coleoptile	63
Baccharis megapotamica	Verrucarin	Wheat coleoptile	63
Gliocladium virens	Viridiol	Germination inhibition	56
Alternaria tagetica	Zinniol	Excised leaf assay & callus inhibition	86

A. Whole Plant, Excised Leaf, and Algal Assays

The majority of the known phytotoxic natural products have been initially detected and subsequently purified to homogeneity using whole plant assays. Whole plant assays may take many forms. Inhibition of seedling development or root elongation, algal growth inhibition, symptoms on leaves excised from whole plants, or observable effects from postemergent (POE) application to mature greenhouse grown plants are the most commonly referenced. Although the POE assay requires relatively large amounts of test material, resources, and time, it is ideal for discerning those characteristics that make a natural product a potential commercial herbicide lead (i.e., uptake, mode of action, and unit activity). Bialaphos [phosphinothricylalanylalanine (1)], isolated from *Streptomyces hygroscopicus*, was originally discovered because of its activity against sheath bright disease of rice [26]. However, its herbicidal activity was demonstrated using a POE assay [4b]. Bialaphos is rapidly metabolized in plants to the active amino acid phosphinothricin (2).

Other herbicidally active natural products that have been found or their activity confirmed using seedling assays or POE spraying of whole plants include herbimycin [27], homoalanosine [17], alteichin [28], cladospolides A and B [29], nigerazine B [30], cinnamic acid [31], anisomycin, and toyocamycin [32], acetylaranotin and terrein [33], botrydienal [34], azaserine [35], altersolanol A and B [36], radicinin [37], monocerin [38], mevinolin [25], gougerotin [39], fusaric acid, lycomarasmin and diacetoxyscirpenol [40], naramycin B [41], formycins A and B [42], borrelidin [43], hadacidin [44], and tentoxin [45]. Homalanosine was also evaluated under flooded rice paddy conditions where it was found to be selectively active against a number of weed species, without damaging rice plants.

Excised leaf assays have frequently been used to determine phytotoxicity. Young, single leaves are removed (with or without the petiole) from the plant and placed in a moist chamber under suitable photoperiod conditions. The test sample may be applied in a variety of ways. Occasionally, the sample has been applied to the abaxial or underside of the leaf, where the leaf cuticle is usually thinner. More often, the sample has been applied to the adaxial leaf surface after a slight abrasion or pinpoint injury has been made. It has been recommended that the sample be applied with a wetting agent such as Tween particularly if the leaf is pubescent, or adsorbed onto silica gel which is then placed on the leaf surface for the duration of the experiment. Visual ratings in the form of chlorotic and necrotic regions are usually made between 18 hr to 1 week after sample application. The phytotoxic activities of citrinin [46], AM-toxin [47], α,β-dehydrocurvularin [48], the colletotrichins [49], and brefeldin A [50] were determined using excised leaf assays. The results of leaf excision assays may be compromised as a result of interference resulting from plant defense mechanisms triggered by wounding. This phenomenon may result in false positive readings. However, this assay retains its usefulness in the case of toxins similar to the AM-toxin. AM-toxin activity is host-specific to the woody perennial, apple (*Malus* sp.), and seedling or whole plant assays would be difficult to perform as a result of an extremely long turnaround time. Thus, the excised assay remains the assay of choice.

Algae may be considered whole plants from a physiological and biochemical perspective even though their gross morphology differs greatly from higher plants. Various algal species have been utilized, although usually not independently, in the discovery or confirmation of herbicidal activity. *Chlorella vulgaris* was used in the investigation of the mode of action of tabtoxinine or "wildfire" toxin [51]. The advantages of using an algal culture were its rapid, quantitatively measurable growth, its easy culturability, and high susceptibility to the toxin.

B. Seed Germination Inhibition

Sphenoblane [52], herbicidins A and B [53], hydantocidin [12], cyclocarbamides A and B [54], filipin [55], viridiol [56], ineketone, dehydrovomifoliol, momilactone, and ρ-coumaric acid [57] have been purified using seed-germination-inhibition assays. Small seeded species such as lettuce (*Lactuca* sp.) and Chinese cabbage (*Brassica chinensis*) are routinely used in germination assays based on the hypothesis that their greater surface to seed volume ratio makes them more sensitive to their surrounding environment, and because they are quick to germinate, requiring no additional manipulation other than imbibition.

Seeds are typically placed on a moist, absorbent surface such as filter paper or cotton to which a surfactant such as Tween and the test sample have been added. After incubation in the dark at 25–28°C for 3 days, germination inhibition can usually be rated. Minimum inhibitory concentration can also be determined based on the lowest sample concentration required for 100% inhibition. Seed-germination-inhibition tests are simple to perform, easy to read, relatively quick, and have reasonably good predictive capability with regard to whole plant activity. Both hydantocidin and the herbicidins, initially purified using seed germination inhibition, were found to have significant activity against both monocotyledonous and dicotyledonous species in whole plant assays. Hydantocidin was comparable to bialaphos and glyphosate in its activity at 500 and 250 ppm against such weeds as barnyard grass (*Echinochloa crus-galli*), giant foxtail (*Setaria faberi*), common cocklebur (*Xanthium strumarium*), and morning glory (*Ipomoea* sp.) [12].

Seed germination assays performed on agar lend themselves to direct testing of inhibition by microorganisms without the need for fermentations and/or purifications. DeFrank [58] used an agar-based medium (A-9) which supported rapid growth of many actinomycete species and was noninhibitory to test plants. In this assay, the test microorganism was streaked in a 1.0-cm band. The plate was incubated for 14 days at 28°C to allow the microorganism to grow, and metabolites produced by the microorganism to diffuse through the medium. When the actinomycetes were sufficiently well grown, surface sterilized seeds of barnyard grass (*E. crus-galli*) and cucumber (*Cucumis sativus*) were partially submerged in the agar in rows perpendicular to the microbial band. After a further 6 days of incubation, inhibition of germination and root/shoot elongation were measured. Heisey et al. [59] used essentially the same procedure with minor modifications. Although this method does detect phytotoxic secondary metabolites, it has several weaknesses. Both Heisey and DeFrank found that microorganisms shown to produce phytotoxins in agar do not always produce toxins when grown in submerged culture. Although solid media scale-up and subsequent active component purification has been performed in the past, it presents many difficulties to the researcher, and a submerged shake culture is much preferred. In addition, DeFrank was using activity in the seed rows as a toxicity gradient measurement which may not be accurate because it is based on the assumption that all active components will diffuse through the agar at essentially the same rate. Finally, this system works well with the slow-growing actinomycetes but is not recommended for use with bacteria and fungi which have a tendency to "overgrow" the entire agar plate.

Another agar-based assay developed and used by Dornbos [60] was performed in 24-well tissue culture plates. One milliliter of molten water agar was placed in a well and allowed to solidify. Six known phytotoxins, *trans*-cinnamaldehyde, coumarin, nigericin, juglone, plumbagin, and benzylisothiocyanate, were added in the appropriate organic solvent to the surface of the agar. The solvent was allowed to evaporate and surface sterilized seeds of imbibed alfalfa (*Medicago sativa*) and velvetleaf (*Abutilon theophrasti*) and unimbibed annual ryegrass (*Lolium multiflorum*) were placed in the wells. Following 3 days of incubation, seeds were rated for inhibition of germination and root/shoot elongation. These same phytotoxins were also tested in a more standard seed-germination-inhibition assay in a plate containing treated filter paper. Dornbos found that the agar-based bioassay gave the same results as the filter paper assay but required much less test material. This assay has been routinely used for natural product purification in the author's laboratories. It meets the necessary criteria of being relatively quick, easy to perform, with a fast turn-around, and requiring little test material.

C. Wheat Coleoptile Assay

Cladosporin [61], citreoviridin [62], roridin, baccharinol, verrucarin, trichoverrin B [63], cytochalasin [64], pergillin [65], dihydropergillin [66], prehelminthosporol [67], chaetoglobosin K [68], orlandin [69], hydroxy-terphenyllin [70], oosporein [71], desmethoxyviridiol [72], ophiobolin A [73], moniliformin [74], cyclopenin, and cyclopenol [75] have all been purified using a wheat (*Triticum aestivum*), coleoptile assay. In this assay, the final modification of which was published in 1964 [76], wheat seeds are germinated for 4 days in moist, coarse sand. After manual removal of the roots and caryopses, the first 2 mm of the shoot apices are cut off using a Van der Weij guillotine. The next 4 mm are cut and retained for bioassay. Usually tests are performed with 10 replicates placed in a test tube containing phosphate buffer supplemented with 2% sucrose and an aliquot of the compound being assayed. Test tubes are placed on a roller rack apparatus and slowly rotated for 18 hr at which time the coleoptile lengths are measured. Controls are bathed in buffer-sucrose solution. This test is capable of detecting pure compounds at 10^{-8} M [77].

A similar assay was developed by Nitsch and Nitsch [78] using the first internode of oats (*Avena sativa*). The oat seed was germinated and prepared in the same way as the wheat coleoptile except that the coleoptile was completely removed from the stems after the roots and caryopses had been cut away. The stem was then fed into the Van der Weij guillotine and, after removal of the first 2 mm, sections 4 mm in length were cut and assayed in the same way as the wheat coleoptiles.

Although the first internode assay has been claimed to be more sensitive than the wheat coleoptile assay in measuring the effect of auxin on shoot growth [78], the coleoptile assay is quicker and easier to perform. Recently, it has been found that the wheat coleoptile assay is also capable of detecting bacteriocidal and fungicidal activity [79, 80].

There is evidence to show that the wheat coleoptile assay is of some value in predicting activity in whole plant systems [80]. In a comparison of the wheat coleoptile assay with a greenhouse-grown whole plant assay including bean (*Phaseolus vulgaris*), corn (*Zea mays*), and tobacco (*Nicotiana tabacum*) it was shown that of 10 antibiotics tested, 8, azathioprine, cephalexin, cephaloglycin, cephalothin, gentamicin, griseofulvin, neomycin and novobiocin, were inhibitory to wheat coleoptile growth. Gentamicin, neomycin, and novobiocin were also found to be active in the whole plant assay. Of the two compounds found to be inactive in the wheat coleoptile assay, chloramphenicol and 5-fluorocytosine, chloramphenicol was phytotoxic to bean and corn plants. In addition to antibiotics, Cutler [80] claimed that this assay was capable of detecting mycotoxins.

D. Matrix Screening

In recent years, a number of assay matrix screening programs have been reported [7, 81, 82] in which samples are tested in a battery of different primary-level screens. The resulting biological "fingerprint" of information is collated and evaluated against a database of activities before proceeding with purification and/or follow-up testing. Assay matrices are difficult for the individual researcher. They require significant resources and are usually employed in a high-throughput screening program. They have been used by agricultural companies that have the necessary resources to implement such systems. Advances in automation, miniaturization, and computerized data manipulation enable screening matrices, if designed correctly, to provide a wealth of valuable information resulting in the efficient prioritization of leads. In the author's laboratories, a matrix screening program has resulted in the identification of several novel phytotoxic substances including α-methylene-β-alanine [7, 8], *cis*-2-amino-1-hydroxycyclobutane-1-acetic acid [10], herboxidiene [20b], arabenoic acid [23], and coaristeromycin and 5'-deoxy-guanosine [13]. In addition, several known natural products not previously reported to have herbicidal activity such as β-methyltryptophan, 2,5-dihydrophenylalanine, amicoumacin B, *trans*-4-hydroxy-L, anthglutin, altemicidin, bafilomycin C, hygromycin A, coformycin, arabinoside A, and cispentacin were identified [7, 83].

The aforementioned compounds were discovered using a screening matrix [7, 83] consisting of the following:

- A duckweed (*Lemna minor*) assay in which rosettes of two to three fronds were added to a 12-well tissue culture plate containing maintenance media and the test sample. After a 7-day incubation at appropriate temperature and photoperiod, the frond number per well and gross morphological and phenotypical effects such as chlorosis, frond cupping, reduced size, and failure of the daughter fronds to separate from the original rosette were recorded.

- An *Arabidopsis thaliana* assay in which seeds were sprinkled onto 1 ml of 1% water agar previously overlaid with test sample in solvent which was allowed to evaporate. After 4 days, under appropriate growth chamber conditions, the assay was evaluated for phytotoxicity, chlorosis, and germination inhibition.
- A root-elongation-inhibition assay using velvetleaf (*A. theophrasti*) and barnyard grass (*E. crus-galli*) in plastic germination pouches to which sterile water and test sample had been added. After 5 days, inhibition of elongation was measured as a percent reduction relative to controls.
- A whole plant greenhouse assay in which soil flats were seeded with soybean (*Glycine max*), velvetleaf (*A. theophrasti*), tomato (*Lycopersicon esculentum*), wheat (*T. aestivum*), barnyard grass (*E. crus-galli*), and amaranth (*Amaranthus caudatus*), and allowed to germinate and grow. At the first true leaf stage the plants were sprayed with the test sample and incubated another 7 days prior to a visual rating.

With the exception of the whole plant assay, these assays require relatively small amounts of test material and can be rated within a 7-day period. After a trial screening period, the whole plant assay was relegated to a secondary assay.

Other researchers have utilized matrix screens for the identification of synthetic herbicides. BASF [82] used a primary matrix screen consisting of wheat (*T. monococcum*) cell suspension, cress (*Lepidium sativum*) germination, and algal (*Scenedesmus acutus*) and duckweed (*L. paucicostata*) bioassays. Ricerca [81] has performed a duckweed (*L. minor*) assay, a small seeded species germination inhibition and root-elongation test, and a whole plant assay using tomato, all of which were performed in 24-well tissue culture plates. Low amounts of test material, fast turn-around time (1 week), flexibility in target species selection, and "weighting" of the individual assays for desired biases make matrix screening an attractive approach for discovering novel phytotoxins.

E. Callus and Cell Suspension Assays

The plant growth inhibitor compactin [84] was originally discovered as an inhibitor of 3-hydroxy-3-methylglutaryl coenzyme A reductase, the enzyme-catalyzing mevalonate synthesis from hydroxymethylglutaryl coenzyme A. It was postulated that compactin exerted growth regulatory activity as a result of the reduction of the level of mevalonate in plants. Growth regulators such as cytokinins, gibberellins, and ubiquinones are derived from mevalonate. Plant-growth-inhibitory activity of compactin was actually demonstrated using tobacco (*N. tabacum*) callus employing methods described by Linsmaier and Skoog [85] in which approximately 8 mg of tobacco pith callus were added to appropriate medium containing various concentrations of test material. After a 30-day incubation, the fresh weight of the callus was measured.

Zinniol [86] was isolated by the concurrent use of an excised leaf assay and a soybean callus assay. As above, the effect on callus tissue growth was determined by measuring the callus fresh weight.

The bioassay assisted purification of stemphyloxin I was performed using an excised tomato leaf assay [87]. However, exponentially growing tomato cells in suspension culture were used to determine the effect of stemphyloxin I on the incorporation of ^{14}C-labeled amino acids into proteins.

Although cell suspension cultures have not been widely used in the discovery of herbicidal natural products, they have been used in high-throughput screening systems for the discovery of synthetic herbicides [82]. The cell suspension assay used routinely by researchers at BASF was based on the inverse proportionality of the reduction in cell culture conductivity and the increase in growth. Thus, if a uniform cell suspension was placed in a test tube containing a test compound, and conductivity measured before and after an incubation period using a special microelectrode, increases in growth would be quantitatively measured. Grossmann [82] claimed the usefulness of the cell suspension assay was in identifying herbicidally active compounds that cannot penetrate into or translocate within the intact plant. In a study in which 2000 synthetic compounds having no activity in whole plant greenhouse screens were retested in the cell suspension assay, 10% were found to be active. In some cases, by changing the formulation of the compound (i.e., from a wettable powder to an emulsifiable or oil-miscible flowable concentrate), whole plant activity would be demonstrated. Cell suspension assays are attractive because they can be miniaturized, require small amounts of test material, minimal space and labor resources, and lend themselves to quantitative measurements. However, these assays should be used in combination with whole plant assays whenever possible, as they will tend to refer relatively high numbers of compounds with low correlation to whole plant activity.

Assays can be designed and implemented that take advantage of a specific mode of action (i.e., inhibition of photosynthesis, chlorophyll production, inhibition of amino acid synthesis, etc.). Several natural products have been identified using assays of this type.

F. Photosynthesis-Inhibition Assays

Analysis of photosynthesis in plants has depended on the measurement of the rate of O_2 exchange using a Warburg manometer or, more recently, an oxygen electrode. Although precise, these methods are time-consuming, require specialized equipment, and are not amenable to high-throughput screening efforts. Two photosynthesis inhibitors, 4-methylthio-1,2-dithiolane and 5-methylthio-1,2,3-trithiane, from the green alga *Chara globularis*, have been found [88] using a method involving ^{14}C and photosynthesis in the diatom *Nitzschia palea* [89]. Although successful in this case, this assay is laborious, time-consuming, and requires the use of radiolabeled materials.

Several alternative, straightforward photosynthesis inhibitor detection bioassays have been described and are listed below:

- Pumpkin (*Cucurbita pepo*) cotyledon leaf discs are floated on a phosphate-based medium under light. In the presence of a photosynthesis inhibitor, the leaf discs become oxygen depleted, lose buoyancy, and sink [90].
- Bean (*P. vulgaris,*) leaf discs are infiltrated with acidic sodium carbonate as a CO_2 source and floated in a solution. Because their intercellular spaces become filled with solution, they sink. Under light, the photosynthesizing, oxygen-generating discs will float to the surface, unless they are in the presence of a photosynthesis inhibitor [91].
- A slight modification of the bean leaf disc assay uses sodium hydrogen carbonate as the CO_2 source [92].
- Measurement of de novo starch synthesis; see below [93, 94].
- Measurement of chlorophyll production by photosynthesizing cultures of the green algae *C. pyrenoidosa*; see below [95].

The first three of the above assays can be referred to as leaf disc buoyancy assays. These have not been utilized in the purification of phytotoxic natural products. They have been validated using known, synthetic photosynthesis inhibitors such as diuron, linuron, and atrazine which were active in the 0.1–10-µg/ml range. Development of a precise, leaf infiltration system and a reduction in the number of leaf disc replicates (currently 50–60) would be necessary in order to make these assays attractive for primary natural product screening.

A novel method of detecting photosynthesis inhibitors was developed by Kida [94] wherein de novo starch synthesis is detected. Second leaf stage plants of barnyard millet (*Panicum crus-galli*) and Italian ryegrass (*L. multi-florum*) were transferred from a greenhouse to a dark chamber for 12 hr. Five-millimeter sections of the second leaf were assayed using iodine and found to be negative for starch. Three 5-mm segments from the same leaf were then transferred to petri plates containing potassium phosphate buffer, a surfactant, and the test sample. The plates were illuminated at 14 klux for 16 hr. Segments were removed from the plates, immersed in hot methanol for pigment extraction, and stained with iodine. Known photosynthesis inhibitors such as linuron and propanil were active in this assay at 1 µg/ml. Although most nonphotosynthetic inhibitors were inactive, it was shown that some respiratory inhibitors, such as sodium azide and 2,4-dinitrophenol, and some protein synthesis inhibitors, such as cycloheximide and streptothricin, also inhibit de novo starch synthesis. Using this assay, Kida [93] has determined azaserine and hadacidin, previously identified herbicidal in whole plant assays, to be inhibitors of de novo starch synthesis. More recently, this assay has been used by Kida [96] to isolate streptothricin-like antibiotics from a *Streptomyces sp.* that were also shown to have whole plant activity against barnyard millet. However, as these compounds had no effect on oxygen evolution in an algal system, it is suspected that they are acting as protein synthesis inhibitors. This, coupled with the significant amount of manipulation required in performing this test, tends to relegate it for use as a secondary or mode-of-action-type assay of utility in characterizing identified leads.

A method for the detection of photosynthetic and respiratory inhibitors was developed using the green algae *C. pyrenoidosa* [95]. When added to an appropriate growth medium under slightly elevated temperature and high light intensity, chlorophyll production could be measured after 18–36 hr following an extraction with methanol. Comparison of chlorophyll production with oxygen evolution measured at 1–2 hr from the beginning of the assay using a Warburg apparatus was favorable. This assay proved to be very sensitive to both photosynthetic inhibitors (atrazine, bromacil, fenuron, linuron, propanil) and respiratory inhibitors (dinoseb, sodium azide) with I_{50} values (concentration required to give 50% inhibition of either chlorophyll production or oxygen evolution) in the range 0.01–1.0 µg/ml. This assay was developed to monitor field residues of atrazine and has not been used in the identification phytotoxic natural products. Its principle weakness would appear to be its lack of selectivity. Any number of compounds that are not photosynthesis or respiratory inhibitors can reduce the growth of *Chlorella* and, subsequently, reduce chlorophyll production and oxygen evolution.

Closely related to photosynthesis inhibition is the inhibition of chlorophyll formation. A method developed by Kida [96, 97] using dark grown cells of the blue-green algae *Scenedesmus obliquus* has been used to identify 7-deoxy-D-glycero-D-glucoheptose from *Streptomyces purpeofuscus* which is active at 12.5 μg/ml. The alga was grown in the dark for 6 days at which time it was centrifuged, and the pellet resuspended in the basal medium containing the test solution. After 14 hr under 14 klux lighting, newly synthesized pigments were extracted with hot methanol and chlorophyll content measured spectrophotometrically at 665 nm. This assay is specific for inhibition of chlorophyll production, is amenable to miniaturization, requires little test material, and is relatively quick but requires a significant amount of handling.

G. Antimetabolite and Enzyme-Inhibitor Assays

Pyrrolyl-3-alanine [7], pereniporin A [96], phosalacine [5], and oxetin [9] have been isolated as a result of the use of amino acid antimetabolite assays. Phytotoxic activity was later confirmed in whole plant assays. Amino acid antimetabolite assays typically utilize bacterial and fungal systems, although occasionally algae are used. An antimetabolite is defined as a "structural analog of an essential metabolite (vitamin, hormone or amino acid, etc.) which is able to cause signs of deficiency of the essential metabolite in some living thing or in some biological reaction" [98]. Inhibition in minimal media which disappears in complex media or in the presence of supplemented amino acids, vitamins, hormones, etc. is indicative of an antimetabolite. The antimicrobial activity of pyrrol-3-alanine was reversed by the addition of histidine, whereas phosalacine and oxetin activities were reversed by the addition of L-glutamine. The activity of pereniporin A toward *Bacillus subtilis* was reversed by the addition of casamino acids.

Gabaculine [99] was discovered through an enzyme inhibitor screen (γ-aminobutyrate aminotransferase inhibitor). Herbicidal activity was later confirmed in a whole plant assay.

H. Translocation Assay

A translocation assay, in which test solution was applied to castor bean (*Ricinus communis*) cotyledons having intact endosperm and in which exudate from the base of the cotyledon was collected and analyzed for the presence of the test compound, was utilized by Lavrik [7] as a secondary assay for determining phloem-transport capabilities. Systemic, phloem-translocated herbicides such as glyphosate are rare. Thus, herbicidally active natural products possessing phloem mobility would have a decided competitive edge for prioritization in a screening program. This is a time-consuming assay that would be very difficult to perform en masse. It requires either an analytical method for determining whether the test material has been translocated, or the collected exudate has to be tested in a second bioassay.

V. CONCLUSIONS

The measure of any assay as a research tool is directly proportional to the validity of the results obtained. Using that criterion, whole plant, seed-germination-inhibition, and wheat coleoptile assays have proven their effectiveness in the discovery and/or confirmation of phytotoxic secondary metabolites. Many of the more specific mode-of-action assays have been utilized for an insufficient period of time to make a fair comparison. We speculate that antimetabolite screening holds great promise. Based on the author's experience, and the results tabulated by Ayer et al. [83], matrix screening has been particularly successful in the identification of novel phytotoxins and in determining new activities for known natural products.

The trend in herbicidal natural product discovery is definitely toward assay miniaturization and automation, mode-of-action assay development, and the use of sophisticated databases which both enhance communication between the screeners and the natural product chemists and make the dereplication of known substances and the discovery and purification of new ones more facile. Novel compounds found using traditional assays may point to unexploited pathways for control of plant growth (i.e., enzyme inhibitors, inhibitors of detoxification and protective systems, photosensitizers, etc.), thereby leading to the development of new screening procedures. The path from discovery to commercialization of a microbial secondary metabolite in agriculture is long and expensive. With the continued development of new assays and sophisticated screening tools, novel substances with potential to impact agrichemical discovery will continue to be found.

REFERENCES

1. For recent reviews in this area see: a. S. Omura, Y. Tanaka, *Proceedings of the 7th International Congress of Pesticide Chemistry*, 1st ed., VCH Verlagsgesellschaft, Weinheim, 1991, pp. 87–96. b. P. Solymosi, *Novenyvedelem (Budapest)*, 25(6):258–63 (1989). c. H.G. Cutler, *Weed*

Technol, 2(4):525–532 (1988). d. B.S. Deshpande, S.S. Ambedkar, and J.G. Shewale, *Enzyme Microb. Technol.,* 10(8):455–473 (1988). e. H.G. Cutler, *Handbook of Natural Toxins,* Vol. 6, Marcel Dekker, New York, 1991, pp. 411–438.

2. a. M. De Block, J. Botterman, M. Vandewiele, J. Dockx, C. Thoen, V. Gossele, N. Rao Movva, C. Thompson, M. Van Montagu, J. Leemans, *Embo,* 6(9):2513–2518 (1987). b. C. Thompson, N. Rao Movva, R. Tizard, R. Crameri, J. Davies, M. Lauwereys, and J. Botterman, *Embo,* 6(9):2519–2523 (1987).

3. E. Bayer, K.H. Gugel, K. Hagele, H. Hagenmaier, S. Jessipow, W.A. Konig, and H. Zahner, *Helv. Chim. Acta,* 224–239 (1972).

4. K. Kaneko and K. Tachibana, *Bio. Ind.,* 5(1):87–93 (1988). b. K. Kaneko and K. Tachibana, *J. Pest. Sci.,* 11(2):297–304 (1986).

5. S. Omura, M. Murata, H. Hanaki, K. Hinotozawa, R. Oiwa, and H.J. Tanaka, *Antibiotics,* 37(8):829–835 (1984).

6. H. Kato, K. Nagayama, H. Abe, R. Kobayashi, and E. Ishihara, *Agric. Biol. Chem.,* 55(4):1133–1134 (1991).

7. P. Lavrik, B. Isaac, S. Ayer, and R. Stonard, *Dev. Ind. Microbiol.,* 32(1):79–91 (1991).

8. B.G. Isaac, S.W. Ayer, and R.J. Stonard, *J. Antibiotics,* 44(7):795–796 (1991).

9. S. Omura, M. Murata, N. Imamura, Y. Iwai, and H.J. Tanaka, *J. Antibiotics,* 37(11):1324–1332 (1984).

10. S. Ayer, B. Isaac, K. Luchsinger, N. Makkar, M. Tran, and R.J. Stonard, *Antibiotics,* 44(12):1460–1462 (1991).

11. M. Uramoto, M. Ubukato, H. Osada, I. Yamaguchi, and K. Isono, *Agric. Biol. Chem.,* 55(5):1415–1416 (1991).

12. M. Nakajima, K. Itoi, Y. Takamatsu, T. Kinoshita, T. Okazaki, K. Kawakubo, M. Shindo, T. Honma, M. Tohjigamori, and T. Haneishi, *J. Antibiotics,* 44(3):293–300 (1991).

13. B. Isaac, S. Ayer, L. Letendre, and R.J. Stonard, *J. Antibiotics,* 44(7):729–732 (1991).

14. Y. Wang, H.P.C. Hogenkamp, R.A. Long, G.R. Revankar, and R.K. Robibs, *Carbohydr. Res.* 59:449–457 (1977).

15. M. Iwata, T. Sasaki, H. Iwamatsu, S. Miyadoh, K. Tachibana, K. Matsumoto, T. Shormura, M. Sezaki, and T. Watanabe, *Meiji Seika Kenkyu Nenpo,* 26:17–22 (1987).

16. S. Omura, Y. Tanaka, and Y. Takahashi, *Japan Kokai Tokkyo Koho,* JP 0301-0692 (1991).

17. S. Fushimi, S. Nishikawa, N. Mito, M. Ikemoto, M. Sasaki, and H. Seto, *J. Antibiotics,* 42(9):1370–1378 (1989).

18. F.K. Gleason, *Bioact. Compd. Mar. Org. Indo-U.S. Symp.,* Balkema, Rotterdam, 1991, pp. 37–41.

19. A. Takahashi, S. Kurasawa, D. Ikeda, Y. Okami, and T. Takeuchi, *J. Antibiotics,* 42:1556–1566 (1989).

20. a. B.G. Isaac, S.W. Ayer, R.C. Elliott, and R.J. Stonard, *J. Org. Chem.,* 57(26), 7220–7226 (1992). b. M. Miller-Wideman, N. Makkar, M. Tran, B. Isaac, N. Biest, and R. Stonard, *J. Antibiotics,* 45(6):914–921 (1992).

21. A. Hirota, T. Kanza, H. Okada, A. Iogai, and M. Nakayama, *Japan Kokai Tokkyo Koho,* JP 02 92, 291 (1990).
22. A. Hirota, H. Okada, T. Kanza, A. Iogai, and H. Hirota, *Agric. Biol. Chem.,* 54(9): 2489–2490 (1990).
23. B. Isaac, S. Ayer, and R. Stonard, *J. Antibiotics,* 44(7):793–794 (1991).
24. T. Haneishi, M. Nakajima, K. Koi, K. Furuya, S. Iwado, and S. Sato, European Patent, EP 88-303813 (1988).
25. T. Bach and H. Lichtenthaler, *Physiol. Plant,* 59:50–60 (1983).
26. Y. Kondo, T. Shomura, Y. Ogawa, T. Tsuruoka, H. Watanabe, K. Totsukawa, T. Suzuki, C. Moriyama, J. Yoshida, S. Inouye, and T. Niida, *Sci. Rept. Meiji Seika Kaisha,* 13:34–41 (1973).
27. S. Omura, Y. Iwai, Y. Takahashi, N. Sadakane, and A. Nakagawa, *J. Antibiotics,* 32(4):255–261 (1979).
28. D. Robeson, G. Strobel, G. Matusumoto, E. Fisher, M. Chen, and J. Clardy, *Experientia,* 40:1248–1250 (1984).
29. A. Hirota, H. Sakai, and A. Isogai, *Agric. Biol. Chem.,* 49(3), 731–735 (1985).
30. T. Iwamoto, S. Shima, A. Hirota, A. Isogai, and H. Sakai, *Agric. Biol. Chem.,* 47(4):739–743 (1983).
31. G. Pagani, G. Caccialanza, and L. Montanari, *Il Farmaco (Sci. Ed.),* 28(3):214–230 (1973).
32. O. Yamada, Y. Kaise, F. Futatsuya, S. Ishida, K. Ito, H. Yamamoto, and K. Munakata, *Agric. Biol. Chem.,* 36(11):2013–2015 (1972).
33. S. Kamata, H. Sakai, and A. Hirota, *Agric. Biol. Chem.,* 47(11):2637–2638 (1983).
34. T. Kimata, M. Natsume, and S. Marumo, *Tetrahedron Lett.,* 26(17):2097–2100 (1985).
35. A. Norman, *Science,* 121:213–214 (1955).
36. R. Suemitsu, Y. Yamada, T. Sano, and K. Yamashita, *Agric. Biol. Chem.,* 48(9):2383–2384 (1984).
37. B. Tal, J. Robeson, B. Burke, and A. Aasen, *Phytochemistry,* 24(4):729–731 (1985).
38. D. Robeson and G. Strobel, *Agric. Biol. Chem.,* 46(11):2681–2683 (1982).
39. S. Murao and H. Hayashi, *Agric. Biol. Chem.,* 47(5):1135–1136 (1983).
40. D. Chakrabarti, K. Chaudhury, and S. Ghosal, *Experientia,* 32(5):608–609 (1976).
41. D. Berg, M. Schedel, R. Schmidt, K. Ditgens, and H.Z. Weyland, *Naturforsch,* 37c:1100–1106 (1982).
42. E. Bischoff, H. von Hasso, U. Klein, P. Reinecke, M. Schedel, R. Schmidt, and G. Zoebelein, Bayer AG Patent DE 3143972 (1983).
43. M. Dorgerloh, A. Kretschmer, R. Steffens, G. Zoebelein, K. Tietjen, W. Roeben, W. Stendel, and O. Salcher, Bayer AG Patent DE 3607287 (1988).
44. R. Gray, G. Gauger, E. Dulaney, E. Kaczka, and H. Woodruff, *Plant Physiol.,* 39:204–207 (1964).
45. A. Lax, H. Shepherd, and J. Edwards, *J. Weed Technol.,* 2:540–544 (1988).
46. P. Venkatasubbaiah, A. Baudoin, and W.J. Chilton, *J. Phytopathol.,* 135:309–316 (1992).

47. T. Ueno, Y. Hayashi, T. Nakashima, H. Fukami, S. Nishimura, K. Kohmoto, and A. Sekiguchi, *Phytopathology,* 65:82–83 (1975).
48. D. Robeson and G. Strobel, *J. Nat. Prod.,* 48(1):139–141 (1985).
49. M. Gohbara, Y. Kosuge, S. Yamasaki, Y. Kimura, and S. Tamura, *Agric. Biol. Chem.,* 42(5):1037–1043 (1978).
50. K. Tietjen, E. Schaller, and U. Matern, *Physiol. Plant Pathol.,* 23:387–400 (1983).
51. A. Braun, *Phytopathology,* 45:659–664 (1955).
52. J. Beyer, H. Becker, M. Toyota, and Y. Asakawa, *Phytochemistry,* 26(4):1085–1089 (1987).
53. M. Arai, T. Haneishi, N. Kitahara, R. Enokita, K. Kawakubo, and Y. Kondo, *J. Antibiotics,* 29(9):863–869 (1976).
54. A. Isogai, S. Sakuda, K. Shindo, S. Watanabe, and A. Suzuki, *Tetrahedron Lett.,* 27(10):1161–1164 (1986).
55. V. Wallen and W. Bell, *Plant Dis. Rept.,* 40(2):129–132 (1956).
56. C. Howell, and R. Stipanovic, *Phytopathology,* 74:1346–1349 (1984).
57. T. Kato, M. Tsunakawa, N. Sasaki, H. Aizawa, K. Fujita, Y. Kitahara, and N. Takahashi, *Phytochemistry,* 16:45–48 (1977).
58. J. DeFrank and A. Putnam, *Weed Sci.,* 33:271–274 (1985).
59. R. Heisey, J. DeFrank, and A. Putnam, *Am. Chem. Soc. Symp.,* 268:337–349 (1985).
60. D. Dornbos and G.J. Spencer, *J. Chem. Ecol.,* 16(2):339–352 (1990).
61. J. Springer, H. Cutler, F. Crumley, R. Cox, E. Davis, and J. Thean, *J. Agric. Food Chem.,* 29:853–855 (1981).
62. R. Cole, J. Dorner, R. Cox, R. Hill, H. Cutler, and J. Wells, *J. Appl. Environ. Microbiol.,* 42(4):677–681 (1981).
63. H. Cutler and B. Jarvis, *Environ. Exp. Bot.,* 25(2):115–128 (1985).
64. R. Cole, J. Wells, R. Cox, and H. Cutler, *J. Agric. Food Chem.,* 29:205–206 (1981).
65. H. Cutler, F. Crumley, J. Springer, R. Cox, R. Cole, J. Dorner, and J. Thean, *J. Agric. Food Chem.,* 28:989–991 (1980).
66. H. Cutler, F. Crumley, J. Springer, and R. Cox, *J. Agric. Food Chem.,* 29:981–983 (1981).
67. H. Cutler, F. Crumley, R. Cox, E. Davis, J. Harper, R. Cole, and D. Sumner, *J. Agric. Food Chem.,* 30(4):658–662 (1982).
68. H. Cutler, F. Crumley, R. Cox, R. Cole, J. Dorner, J. Springer, F. Latterell, J. Thean, and A. Rossi, *J. Agric. Food Chem.,* 28:139–142 (1980).
69. H. Cutler, F. Crumley, R. Cox, O. Hernandez, R. Cole, and J. Dorner, *J. Agric. Food Chem.,* 27(3):592–595 (1979).
70. H. Cutler, J. LeFiles, F. Crumley, and R. Cox, *J. Agric. Food Chem.,* 26(3):632–635 (1978).
71. R. Cole, J. Kirksey, H. Cutler, and E. Davis, *J. Agric. Food Chem.,* 22(3):517–520 (1974).
72. R. Cole, J. Kirksey, J. Springer, J. Clardy, H. Cutler, and K. Garren, *Phytochemistry,* 14:1429–1432 (1978).
73. M. Orsenigo, *Phytopathol. Z.,* 29:189–196 (1957).

74. R. Cole, J. Kirksey, H. Cutler, B. Doupnik, and J. Peckham, *Science,* 179:1324–1326 (1973).
75. H. Cutler, F. Crumley, R. Cox, J. Wells, and R. Cole, *Plant Cell Physiol.,* 25(2):257–263 (1984).
76. C. Hancock, H. Barlow, and H. Lacey, *J. Exp. Bot.,* 15(43):166–176 (1964).
77. H. Cutler, *The Science of Allelopathy* (A.R. Putnam and C.S. Tang, eds.), Wiley-Interscience, New York, 1986, pp. 147–170.
78. J. Nitsch and C. Nitsch, *Plant Physiol.,* 31:94–111 (1956).
79. H. Cutler, *Bioregulators: Chemistry and Uses* (R. Ory and F. Ritting, eds.), American Chemical Society, Washington, D.C., 1984; pp. 153–170.
80. H. Cutler, *Proc. Plant Growth Reg. Soc. of America,* (1984), pp. 1–9.
81. S. Woodhead, A. O'Leary, S. Rabatin and K. Crosby, *Brighton Crop Prot. Conf.,* Vols. 1,2, and 3, British Crop. Protection Council, Surrey, 1990, pp. 873–878.
82. K. Grossman, R. Berghaus and G. Retzlaff, *Pest. Sci.,* 35:283–289 (1992).
83. S. Ayer, B. Isaac, D. Krupa, K. Crosby, L. Letendre and R. Stonard, *Pest. Sci.,* 27:221–223 (1989).
84. T. Hashizume, *Agric. Biol. Chem.,* 47(6):1401–1403 (1983).
85. E. Linsmaier and F. Skoog, *Physiol. Plantarum,* 18:100–127 (1965).
86. D. Robeson and G. Strobel, *Phytochemistry,* 23(8):1597–1599 (1984).
87. I. Barash, G. Pupkin, D. Netzer and Y. Kashman, *Plant Physiol.,* 69:23–27 (1982).
88. U. Anthoni, C. Christophersen, J. Madsen, S. Wium-Andersen, and N. Jacobsen, *Phytochemistry,* 19:1228–1229 (1980).
89. E. Steermann-Nielsen, *J. Cons. Perm. Int. Explorer Mer.,* 18:117–140 (1952).
90. B. Truelove, D. Davis, and L. Jones, *Weed Sci.,* 22(1):15–17 (1974).
91. K. Bielecki, H. Doroszewicz, E. Grzys, and Z. Szuwalska, *Acta. Agrobotanica,* 35(1):123–131 (1982).
92. S. Saltzman and B. Heuer, *Pest. Sci.,* 16:457–462 (1985).
93. T. Kida and H. Shibai, *Agric. Biol. Chem.,* 49(11):3231–3237 (1985).
94. T. Kida, S. Takano, T. Ishikawa, and H. Shibai, *Agric. Biol. Chem.,* 49(5):1299–1303 (985).
95. B. Kratky and G. Warren, *Weed Sci.,* 19(6):658–661 (1971).
96. T. Kida, *Novel Microbial Products for Medicine and Agriculture,* (A. Demain, G. Somkuti, J. Hunter-Cevera, and H. Rossmoore, eds.), Soc. for Indus. Micro. 1989, pp. 195–202.
97. T. Kida and H. Shibai, *Agric. Biol. Chem.,* 50(2):483–484 (1986).
98. D. Pruess and J. Scannell, *Adv. Appl. Microbiol.,* 17:17–62 (1974).
99. K. Kobayashi, S. Miyazawa, A. Terahara, H. Mishima, and H. Kurihara, *Tetrahedron Lett.,* 7:537–540 (1976).

7

Fungicides and Bactericides

C.R.A. Godfrey

Zeneca Agrochemicals, Bracknell, Berkshire, England

I. INTRODUCTION

The widespread use of fungicides and bactericides in agriculture is a relatively recent phenomenon, with most of the major developments taking place during the last 100 years. In earlier times, crop losses due to fungal pathogens and bacteria were scarcely recognized or at least poorly understood. More recently, with the increasing emphasis on intensive monoculture and the economic drive to maximize farmers' profits, the detrimental effects of fungi and bacteria on crops have gained in significance. Yield and quality losses now represent an unacceptable penalty to farmers, who have turned to the use of chemicals in order to control diseases and guarantee a satisfactory return on their investments. Fungicides and bactericides now form an important component of the total agrochemical business, with worldwide sales in 1993 of approximately $4.8 billion, equivalent to 19% of the total agrochemical market.

Natural products have played a significant role in the discovery of new fungicides and bactericides, either by their direct application to diseased or susceptible plants or through their exploitation in the design of analogs with optimized biological and physical properties. The earliest recorded use of naturally derived fungicidal remedies has been attributed to agriculturists of ancient Rome, who used various plant

311

extracts to ward off fungal infestations in their crops. In a lecture entitled "One Hundred Years of Fungicide Use," delivered at the Bordeaux Mixture Centenary Meeting in 1985 [1], Brent referred to one William Forsyth, Gardener to King George III, who used a decoction of sulfur, lime, tobacco, and elderbuds as an eradicant treatment for mildew on fruit trees. It is unclear just how useful these mixtures were in practice and which, if any, of the organic constituents were responsible for the fungicidal effect. Nevertheless, it is tempting to speculate that these early examples illustrate the first recorded applications of fungicidal natural products in agriculture. The use of discrete, characterized natural products (i.e., single, purified compounds as opposed to mixtures and crude extracts) on a commercial scale, however, is a modern phenomenon, dating from the 1960s with the advent of compounds such as the antibiotic streptomycin (1).

Natural products can be exploited in several ways in the discovery of new agricultural fungicides. First, as outlined above, they can be used as a crude extract for direct application to the crop. Alternatively, once purified, they can constitute the active ingredient of a formulated mixture or be used in admixture with a synthetic product. When a natural product is insufficiently active in its own right, semisynthesis, or chemical modification, can, in principle, lead to derivatives with optimized properties, such as improved level and spectrum of activity, reduced phytotoxicity, and increased photostability. Finally, as an extrapolation of the latter approach, the structure of a natural product can act as the source of inspiration for the pesticide chemist in the design of new totally synthetic products which, to the untutored eye, may bear little resemblance to the original lead compound. This chapter will focus both on the principal natural products that have been developed commercially for the treatment of plant diseases and those that have been used as leads for the design of new synthetic fungicides. A comprehensive review of natural products with antifungaland antibacterial activity was published by Worthington in 1988 [2].

II. NATURAL PRODUCTS IN COMMERCIAL USE

During the last 30 years or so a great deal of effort has been put into finding microbial products for use as commercial fungicides and bactericides. This has proved to be a difficult task for a number of reasons. First, natural products possessing marketable levels of activity against commercially important diseases are extremely hard to find, and their discovery requires the use of elaborate high-throughput screening methods (or at least a great deal of luck). Many natural products derived from fermentation broths are complex molecules, present in minute concentrations, that are difficult to purify on a large scale. They are often inherently unstable (for example, to sunlight) and, consequently, are insufficiently persistent in the field to deliver a useful effect. Some lack selectivity of action and this can manifest itself in the form of toxicity to plants or mammals. As a result, those which have succeeded have tended to be nonphytotoxic and safe to humans, livestock, and wildlife at their recommended doses.

This section will review the properties of those natural products that have been exploited commercially as fungicides and bactericides. These can be classified conveniently into two distinct groups: the aminoglycosides [e.g., streptomycin (1), validamycin A (2), and kasugamycin (3) (Figure 1)], and the nucleosides [e.g., polyoxins B (4) and D (5) and blasticidin S (6) (Figure 2)]. All of these compounds are manufactured in Japan by large-scale fermentation of different species of *Streptomyces*. Together they command less than 1% of the total fungicide market and are sold in only a handful of countries around the world. A further example of a nucleoside, mildiomycin (7), was registered for use in Japan in 1982 as a treatment for powdery mildews but has not yet made a significant impact in the marketplace. Other structurally unrelated antibiotics, cycloheximide (8) (actidione) [2, 3], cellocidin (9) [2, 4], and ezomycin A (10) [2, 5], have also been used as fungicides in Japan but are very minor products. None of these compounds will be considered in detail here.

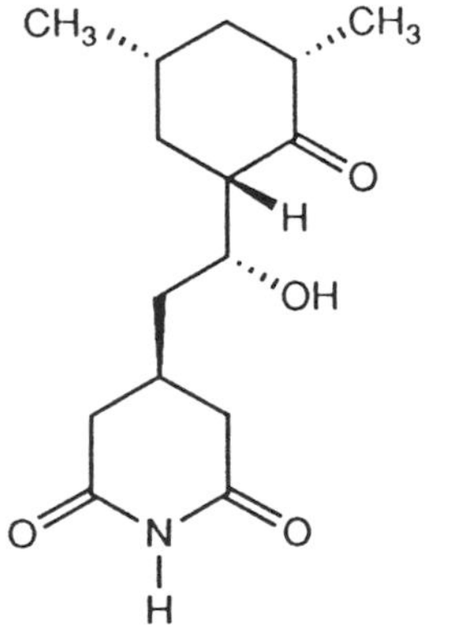

(8)

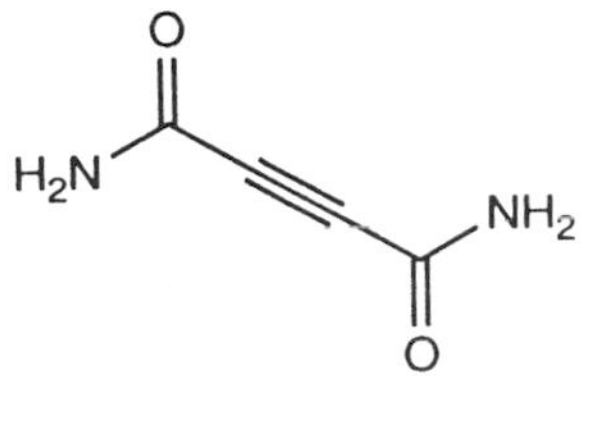

(9)

$$[N^2\text{-L-cystathiono}]$$

H₂NCOHN

β–cytosinyl

CO₂H   OH

H₂N

OH

(10)

A. Streptomycin

The first natural product to be widely used in agriculture was streptomycin (1). This compound is an antibacterial antibiotic that is still
used in medicine for the treatment of tuberculosis and, less commonly,
urinary and intestinal infections, mainly in combination with other antibiotics, such as penicillin. The compound was first reported by Schatz
et al. in 1944 [6] and its structure was elucidated by Brink and Folkers
3 years later [7]. It is a complex aminoglycoside which has been the
subject of a number of synthetic studies [8].

Streptomycin (1) is manufactured by Meiji Seika Kaisha Ltd. and
registered for agrochemical use in Japan, Bulgaria, Greece, and the
Netherlands, under a variety of trade names. It is produced by the fermentation of *Streptomyces griseus* (from which it is isolated as its
sesquisulfate), and it has been reported that 175 tons of the active ingredient were sold worldwide in 1986 [9]. It was first introduced as a
commercial product in 1962 and has found applications in crop protec-

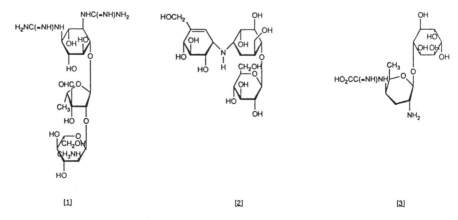

Figure 1 The aminoglycosides.

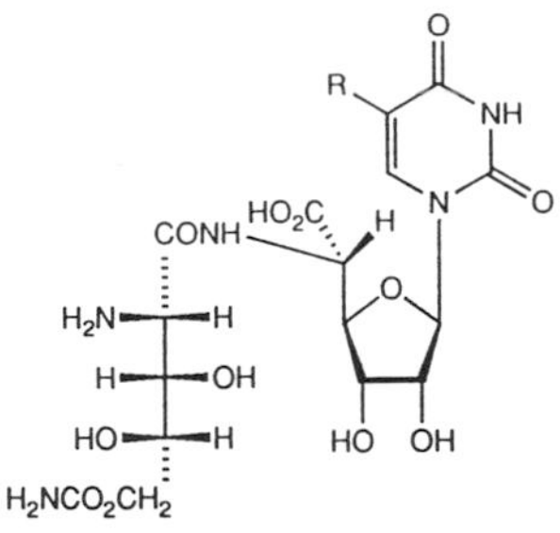

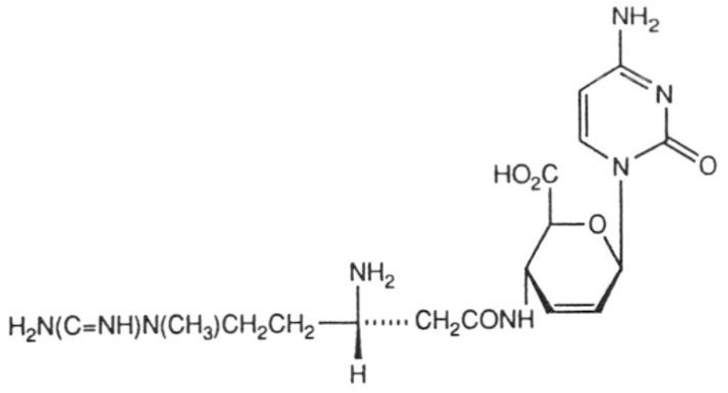

Figure 2 The nucleosides.

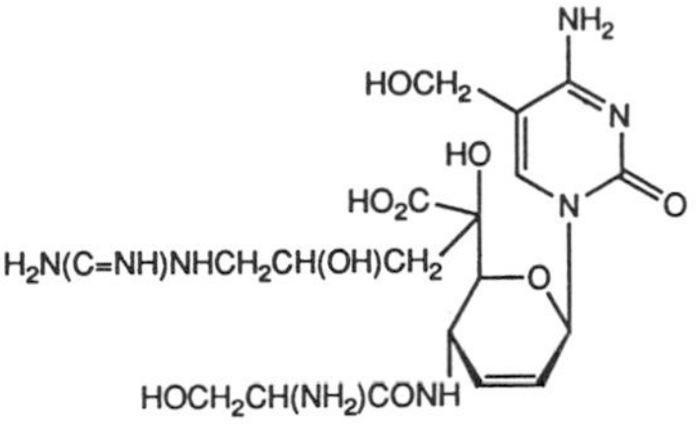

tion and horticulture as a systemic bactericide for the control of apple and pear fireblight and a wide variety of other bacterial rots, cankers, and wilts on stone fruit and vegetables. The mode of action of streptomycin is thought to be inhibition of aminoacyl-tRNA binding in the 70S ribosome complex [10]. It is much more effective against Gram-positive species of bacteria than Gram-negative species.

As is the case for the other aminoglycoside antibiotics, streptomycin has very low acute oral toxicity (LD_{50} >10,000 mg/kg bodyweight in mice) [11, 12]. The acute subcutaneous and intraperitoneal LD_{50}'s are 400 mg/kg and 340 mg/kg, respectively, for male mice and 325 mg/kg and 305 mg/kg, respectively, for female mice. Chronic toxicity testing on rats gave a maximum no-effect level (NOEL) of 125 ppm, and no effects were observed in skin irritation studies with rats.

B. Validamycin A

The second most important of the aminoglycosides in terms of tonnage sold (172 tons in 1986 [9]) is the antifungal antibiotic validamycin A

(2), which was first described by Iwasa et al. in 1971 [13]. It occurs as the major and most active component of a mixture of seven closely related compounds known as validamycins A to G, which are produced in the fermentation broth of *S. hygroscopicus* var. *limoneus*. Validamycin A (2) was assigned the structure (2) in 1980 by Suami et al. [14]. Synthetic studies on the validamycins and related compounds carried out by Ogawa and co-workers have led to a series of publications, culminating in the total synthesis of validamycins A and E in 1989 [15]. However, no biosynthetic studies on validamycin A (2) and its congeners appeared until 1987, when Rhinehart et al. published the results of feeding ^{13}C-labeled precursors to *S. hygroscopicus* var. *limoneus* [16].

Several Japanese groups have explored the effect on biological activity of chemical modifications to validamycin A (2) [17–19]. For example, it has been demonstrated that validoxylamine A (*11*), which lacks the D-glucose moiety of validamycin A (2), still retains one-tenth of its activity [17]. In another study, Asano and co-workers prepared all eight possible mono-β-D-glucosides of validoxylamine A and compared the activity of each isomer with validamycin A (2) against trehalase derived from *Rhizoctonia solani* [18]. None was more active than validamycin A itself. Other workers have shown that chemical derivatization of the D-glucose moiety in general leads to a decrease in biological activity [19].

(*11*)

Validamycin A is produced by the Japanese company Takeda Chemical Industries Ltd. and is sold in Japan, Bulgaria, and the Netherlands under the trade names Validicin and Solacol. It is manufactured on a large scale by the fermentation of *S. hygroscopicus* var. *limoneus*. As a commercial antibiotic, validamycin A is used as a treatment for the control of *Rhizoctonia* diseases such as rice sheath blight, black scurf on potatoes, and damping off in a variety of vegetables. It is formulated as a soluble concentrate at 30 g/L and as a dust for seed treatment at 3 g/kg. Its mode of action is still unclear, but it is known to decrease the maximum rate of hyphal extension and to increase hyphal branching without affecting the organism's specific growth rate [20].

In common with streptomycin (1), validamycin A displays an exceptionally low acute oral toxicity to mammals (LD$_{50}$ >20,000 mg/kg for both rats and mice) and is nonphytotoxic [11, 12]. Furthermore, in skin irritation studies on rabbits, a solution containing 500 g/L validamycin A produced no irritation, and no ill effects were observed when rats and mice were given 100 mg/kg and 2000 mg/kg, respectively, in 90-day feeding studies.

C. Kasugamycin

The antibiotic kasugamycin (3) was first isolated from *S. kasugaensis* in 1965 by Umezawa et al. [21] and its structure established a year later using a combination of chemical and x-ray studies [22]. It is an aminoglycoside comprising a 1D-*chiro*-inositol moiety linked to the oxalamidine derivative of kasugamine. Umezawa's group reported the formal total synthesis of kasugamycin in 1968 and has taken an active interest in elucidating its biosynthesis [23]. In 1969, some work was published by Cron and co-workers on the preparation of semisynthetic kasugamycin derivatives [24]. They were able to show that replacement of the oxalamidine group with aliphatic amidino groups afforded analogs of kasugamycin with improved activity against *Pseudomonas* organisms *in vitro*.

Kasugamycin is obtained on a large scale by fermentation of *S. kasugaensis* by Hokko Chemical Industry Co. Ltd. It is sold under the trade names Kasumin and Kasugamin as a systemic bactericide/fungicide for the control of diseases on rice [especially rice blast (*Pyricularia oryzae*)], fruit, and vegetables. However, it is not used on the same scale as streptomycin (1) and validamycin A, and only 91 tons were sold in 1986 [9]. Kasugamycin has low toxicity to plants and animals, and for this reason it has superseded blasticidin S for the control of rice blast. It is thought to prevent protein synthesis by inhibiting initiation complex formation on both 30S and 70S ribosomal subunits [31, 36]. Acute oral and percutaneous LD_{50}'s in the rat have been measured at 22,000 mg/kg and >4000 mg/kg, respectively [11, 12]. The same order of toxicity is also seen for mice and a value of >4000 mg/kg was obtained for male Japanese quail. A NOEL of 100 mg/kg daily was recorded in rats in 90-day feeding studies and no mutagenic or teratogenic effects were observed. The compound is not a skin or eye irritant in rabbits. In fish, the LC_{50} (48hr) is >40 mg/L (carp and goldfish) and for *Daphnia* >40 mg/L.

D. Polyoxins B and D

The polyoxins are a family of closely related nucleosides bearing peptidic chain substituents at C-4 of the furanose ring. A total of 13 polyoxins (polyoxins A–M) have been isolated and characterized from fermentation broths of *S. cacaoi* and all but two (C and I) are fungicidal [25]. The most important from a commercial point of view are polyoxins B (4) and D (5), which differ from each other solely in the oxidation state of the uracil component. The biosynthesis of the polyoxins has been shown to involve incorporation of uracil and C-3 of serine [26].

Although no total syntheses of polyoxins B and D have been completed, the synthesis of polyoxins C and J were published in the early 1970s [27] and a total synthesis of polyoxin J from *myo*-inositol has been reported recently [28]. Several routes to polyoxamic acid, a constituent amino acid of the polyoxins, have appeared in the literature [29, 30]. The closely related nikkomycins, which are of interest as potential drugs for the treatment of mycotic diseases in humans, have also attracted attention from synthetic chemists.

Both polyoxins B and D are produced in bulk by the fermentation of *S. cacaoi*, and in 1986 a total 76 tons of active ingredient were sold [9]. Polyoxin B is used for the control of a variety of fungal infections of fruit and vegetables, including *Alternaria* spp. on apples and pears and gray mould on vines and tomatoes. Polyoxin D (as its zinc salt) is mainly used on rice for the control of sheath blight. Both compounds are thought to disrupt cell wall biosynthesis by mimicking UDP-*N*-acetylglucosamine, the natural substrate for the enzyme chitin synthase [31]. Neither of the two compounds show high toxicity to mammals or fish. The acute oral LD_{50}'s for polyoxins B and D are relatively high—of the order 21,000 mg/kg (male and female rats) and 9,600 mg/kg, respectively [11, 12]. Polyoxin B gave an acute percutaneous LD_{50} of >20,000 mg/kg in rats with no signs of irritancy. No-effect levels in 2-year feeding studies in rats were >48,000 mg/kg diet (polyoxin B) and >50,000 mg/kg diet (polyoxin D).

E.   Blasticidin S

Blasticidin S (6) was first isolated from *S. griseochromogenes* in 1958 by Takeuchi et al. [32]. The structure and absolute stereochemistry of the compound was established by chemical means and final confirmation was provided by x-ray crystallography [33]. The compound contains two novel structural features: the blastidic acid and the cytosinine moieties. Although both of these fragments have been the subject of some synthetic interest, no reports on the total synthesis of blasticidin S have appeared. However, some biosynthetic work has been reported. For example, in 1968, Seto and co-workers proved that the compound is biosynthesised from D-glucose, L-α-arginine and L-methionine [34].

More recently, Guo and Gould have shown by radiolabeled feeding studies that cytosylglucuronic acid is incorporated intact into the cytosinine portion of blasticidin S [35].

Blasticidin S is manufactured by the Kaken Pharmaceutical Company in Japan. It is marketed under the tradename Bla-S for use as a contact fungicide with protectant and curative properties. Its main outlet is for the control of *Pyricularia oryzae* (rice blast) by foliar application, and it is normally used as the benzylaminobenzenesulfonate salt in order to reduce phytotoxicity. The fact that only 11 tons of blasticidin S were sold in 1986 [9] reflects its declining usage in favor of other products, such as kasugamycin, which are more effective and less phytotoxic in the treatment of rice blast. Blasticidin S acts by binding to the 50S ribosome in prokaryotes at the same site as gougerotin, thereby leading to inhibition of chain elongation [36]. Toxicology studies have demonstrated the higher toxicity of blasticidin S toward mammals when compared to the compounds discussed earlier. Published figures for acute oral LD_{50}'s are 39.5 mg AI (active ingredient)/kg and 53.3 mg benzylaminobenzenesulfonate/kg for mice and rats, respectively [11, 12]. Acute percutaneous LD_{50}'s are recorded at 220 mg/kg (mice) and 3100 mg/kg (rats) and the compound is an eye irritant. Toxicity toward carp is 8700 mg/L in water.

F. Mildiomycin

Mildiomycin (7) is a water-soluble, basic antibiotic produced by *Streptoverticillium rimofaciens* B-98891. Details of its initial isolation by ion exchange and absorption chromatography were reported in 1978 by Harada, Mizuta, and Kishi of Takeda Chemical Industries Ltd. [37] and its fungicidal properties were described shortly thereafter [38]. It was later established by a combination of degradative and spectroscopic studies that mildiomycin is a novel 5-hydroxymethylcytosine nucleoside with the structure (7) [39]. Biosynthetic studies showed that the 5-hydroxymethylcytosine moiety is produced by hydrolysis of 5-hydroxymethylcytidinemonophosphate, which, in turn, is derived from hydroxymethylation of cytidinemonophosphate [40].

Mildiomycin (7) shows promising eradicant (and some systemic) activity (>90% disease control at 25–50 ppm) against a variety of powdery mildews (e.g., *Sphaerotheca* spp., *Erysiphe* spp., *Podosphaera leucotricha*, and *Uncinula necator*) [2, 41]. However, although it was registered for use in Japan in 1982, it has not yet been developed into a commercially important product. Mildiomycin is claimed to have low toxicity toward plants, fish [LC_{50} > 40 mg/L (72 hr) for carp; >40 mg/

L (168 hr) for Japanese killifish], and mammals [$LD_{50} = 599$ mg/kg I.V., 5,250 mg/kg oral (mouse); 700 mg/kg I.V., 4,120 mg/kg oral (rat)] [42] and is thought to inhibit protein synthesis in fungi by blocking peptidyl-transferase [43].

III. NATURAL PRODUCTS AS LEADS FOR SYNTHESIS

It is clear from the previous section that complex natural products per se play a relatively minor role in comparison with their synthetic counterparts in the present-day market. Although agrochemical companies throughout the world continue to search for natural products to replace existing compounds, it is apparent that there is an increasing emphasis on the use of natural products as the starting point for the synthesis of simpler synthetic or semisynthetic compounds with optimized biological, physical, and environmental properties, rather than for use in their own right. Because the semisynthetic approach relies heavily on the ready availability of sufficient quantities of the natural starting materials, as well as on the development of appropriate synthetic methodology, it is less attractive than the total synthesis of simplified analogs, provided that the latter is a feasible option.

It is well established that the properties of a biologically active natural product can be changed by making appropriate modifications to its structure. Many pharmaceutical products have arisen in this way and the successful application of this approach to the discovery of new agrochemicals is illustrated throughout this book. It is possible for the chemist to alter the structure of molecules and to find those that bind most effectively to a target site. Structural changes can also be made to bring about beneficial changes in certain physical properties that might be necessary for a commercial product (such as water solubility, partition coefficient, or volatility, or to increase its stability toward metabolism or light). As is the case for the optimization of any lead molecule, structural modifications to natural products may be quite subtle (involving, for instance, the introduction or removal of a simple substituent on an aromatic ring) but nevertheless may produce profound effects. Alternatively, drastic changes to the overall structure of the natural product may be necessary in order to produce a viable end product, and this may require the synthesis of analogs that bear little resemblance to the starting molecule.

Recent examples of natural products that have been considered as leads for the discovery of new fungicides and that will be discussed in this section are pyrrolnitrin (*12*), the β-methoxyacrylates strobilurin A (*13*) and oudemansin A (*14*), hadacidin (*15*), thiolutin (*16*), griseofulvin (*17*), and pisiferic acid (*18*). Work on the first of these natural prod-

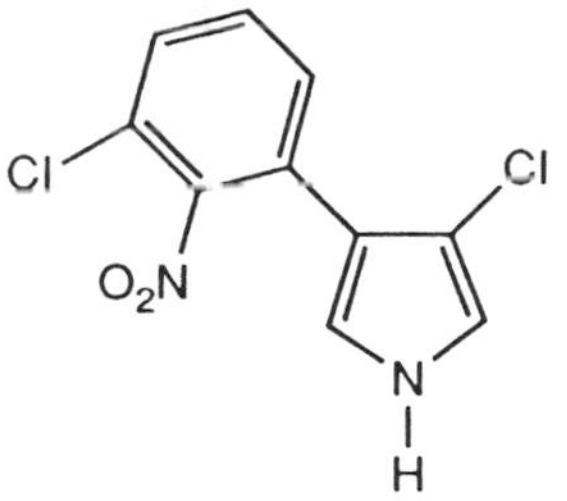

(12)

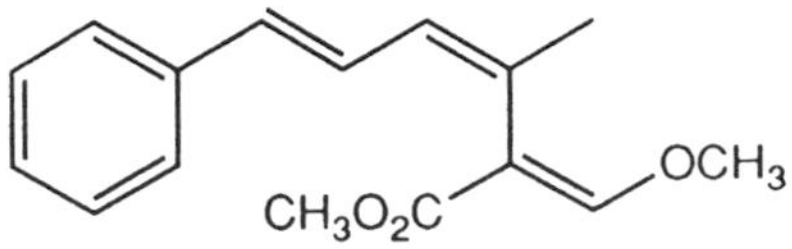

(13)

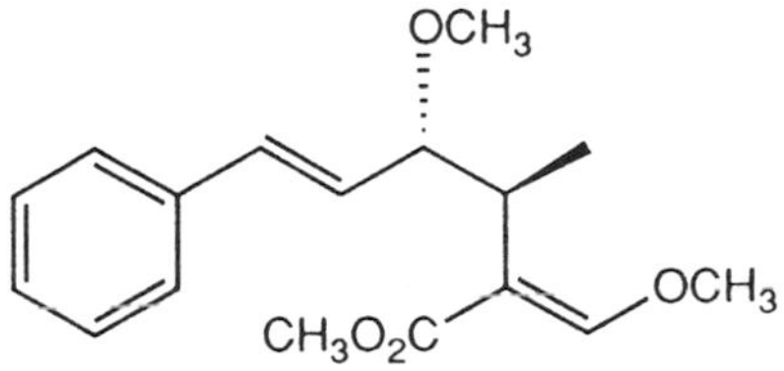

(14)

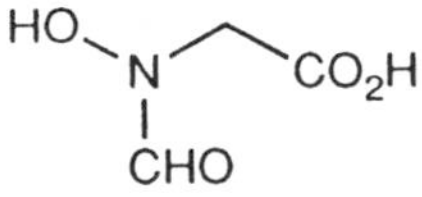

(15)

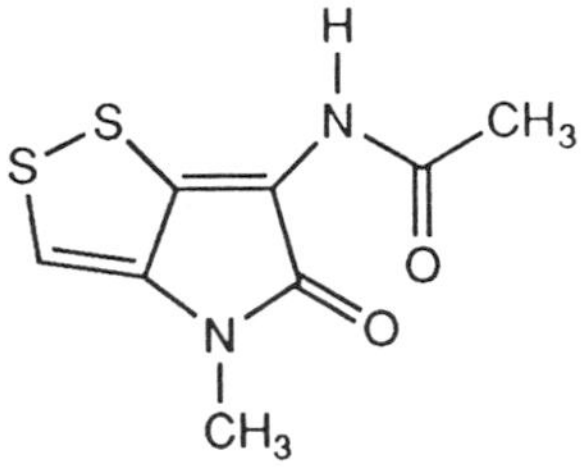

(16)

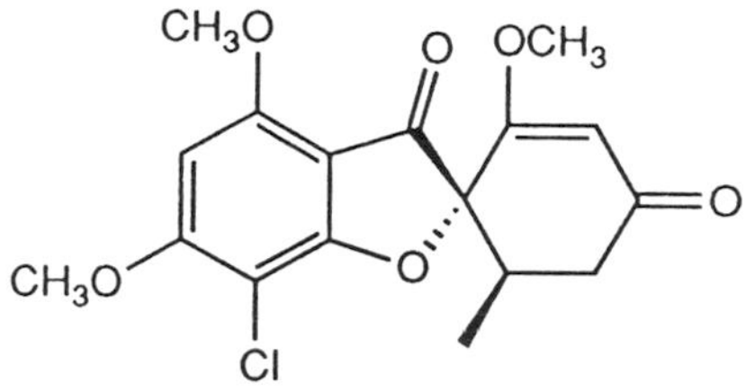

(17)

(18)

ucts has already led to to the development of fenpiclonil and CGA173506 as fungicidal seed treatments, and the synthetic β-methoxyacrylates are showing great potential as broad-spectrum fungicides.

A. Pyrrolnitrin

Pyrrolnitrin (12) was isolated from *Pseudomonas pyrrocinia* in 1964 by Arima and co-workers [44, 45]. This simple secondary metabolite has been of interest as a topical antifungal treatment for some years and is sold under the trade name Pyroace. The first indications of the potential of the natural product itself in agriculture appeared in a patent application filed in 1969 [9]. However, it soon became apparent that pyrrolnitrin was not sufficiently stable in sunlight to give an acceptable effect in the field, and much of the recent work in this area has focused on the synthesis of analogs with greater photostability. Ciba has led the way and its work has so far produced two promising compounds: fenpiclonil (formerly CGA 142705) (19) and CGA 173506 (20), which is still in development [9]. Several routes to these 3-cyano-4-aryl-pyrroles have been disclosed. Figure 3 illustrates two of these [9]. Substitution of the 3-chloro substituent of the pyrrole ring of the natural product with a cyano group dramatically improves the photostability. A group from Bayer AG has also been active in this area. They have published an article containing details of the synthesis and biological activity of a series of stabilized pyrrolnitrin analogs [46] and filed two patent applications [47].

Fenpiclonil (19) was announced at the Brighton Crop Protection Conference in 1988 [48] and CGA 173506 in 1990 [49]. The former compound has recently been introduced in Europe under the trade name Beret as a seed treatment for the control of snow mould, bunt, and glume blotch on cereals. It is claimed to have low toxicity toward mam-

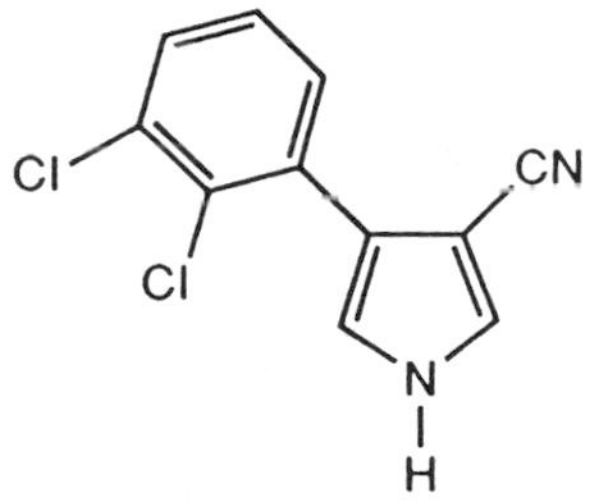

(19)

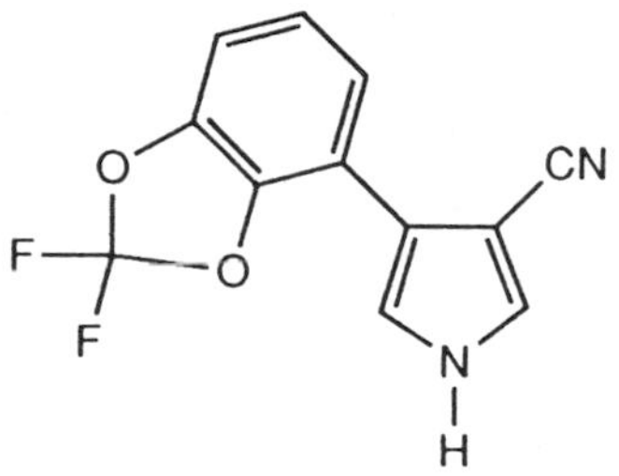

(20)

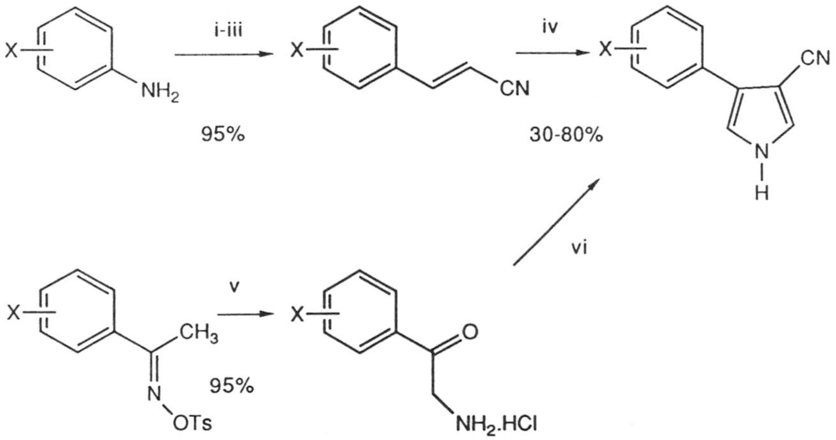

Figure 3 Synthesis of 3-cyanopyrrolnitrin analogs. *Reagents*: i, NaNO$_2$; ii, acrylonitrile, iii, Base; iv, TosMIC; v, NaOEt; vi, (CH$_3$)$_2$NCH:CHCN.

mals, birds, and honeybees and low mobility in the soil. It has a half-life in the soil of 150–250 days. Ciba is developing the more active analogs CGA 173506 (20) for use in the United States under the trade name Saphire.

Several studies on the mode of action of fenpiclonil have been published [50, 51]. In 1992, workers at INRA (France) drew attention to similarities in the activity of fenpiclonil, iprodione, and tolclofos-methyl and concluded that these compounds have the same mode of action [50]. More recently, Jespers et al. have described the biochemical effects of fenpiclonil on the fungus *Fusarium sulphureum*, one of the pathogens responsible for damping-off in cereals [51]. They concluded that the mode of action of fenpiclonil was novel and may be related to membrane-dependent transport processes. In contrast, pyrrolnitrin itself is known to be a respiration inhibitor.

B. The β-Methoxyacrylates

A second and potentially more important class of natural product-derived fungicides to emerge in the last decade are the β-methoxy-acrylates, which have also been termed "strobilurin analogs" [52]. These synthetic compounds are formally derived from a family of fungicidal natural products related to strobilurin A (13) and oudemansin A (14). Strobilurin A shows significant levels of fungicidal activity toward a range of plant pathogenic fungi *in vitro* but not *in vivo*, owing to its photochemical instability and its relatively high volatility. The β-methoxyacrylates are to be found in the mycelia of several genera of small basidiomycete fungi (*Strobilurus tenacellus* and *Oudemansiella mucida*, for example) which grow on decaying wood, and 14 related compounds (11 strobilurins and 3 oudemansins) have been characterized to date (Figures 4 and 5). A comprehensive review describing these fascinating natural products has recently been published by Clough [53].

Apart from its interesting fungicidal properties, one of the obvious attractions of strobilurin A as a starting point for further optimization is its structural simplicity when compared to the complex nucleosides and aminosugars described earlier in this chapter. Another important factor is the knowledge that these compounds are fungicidal by virtue of a novel mode of action. (They inhibit mitochondrial respiration in fungi by blocking the electron transfer at the cytochrome bc_1 complex [54].) Because no other commercial fungicides are active in this way (although many inhibit respiration at other sites), work in this area could lead to the development of synthetic fungicides that do

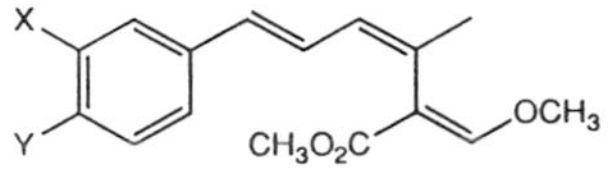

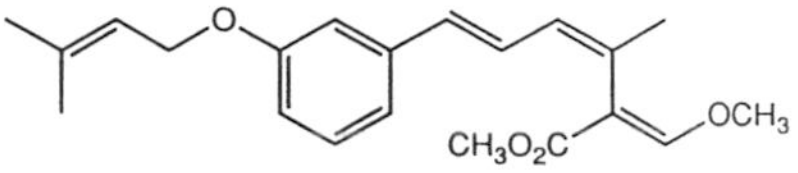

X = Y =H   Strobilurin A [13]
X = CH₃O, Y = Cl Strobilurin B
X = H, Y = CH3O   Strobilurin X
X = HO, Y = H   Strobilurin F*
X = CH₃O, Y = H   Strobilurin H

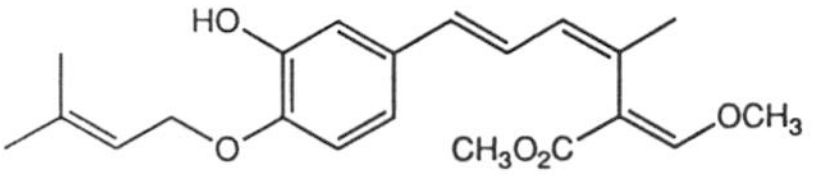

Strobilurin C

Strobilurin F*

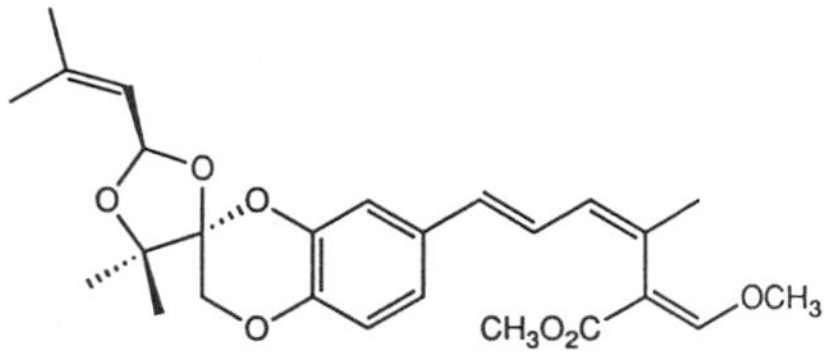

Strobilurin E

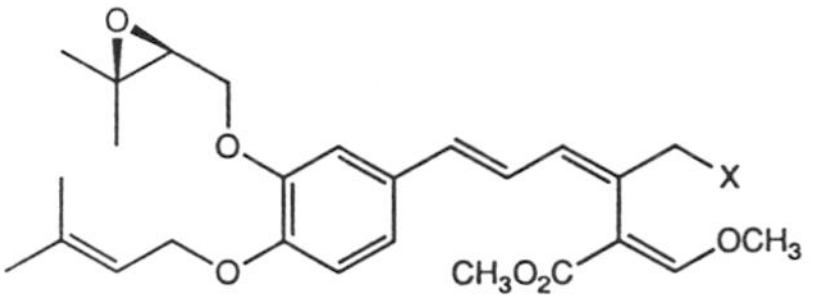

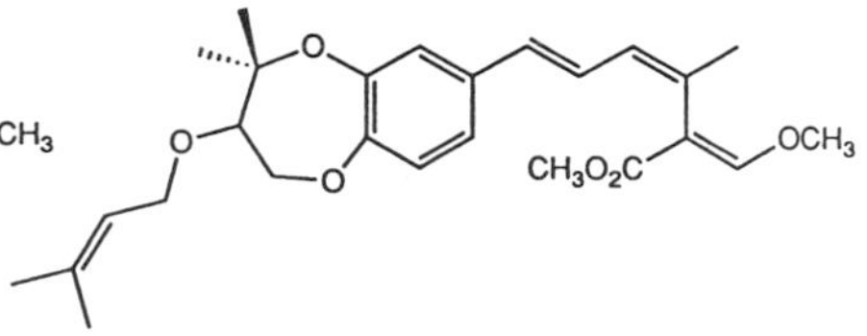

X = H   Strobilurin D
X = OH   Hydroxystrobilurin D

Strobilurin G

Figure 4 The strobilurins (two different compounds [marked *] have been designated as strobilurin F).

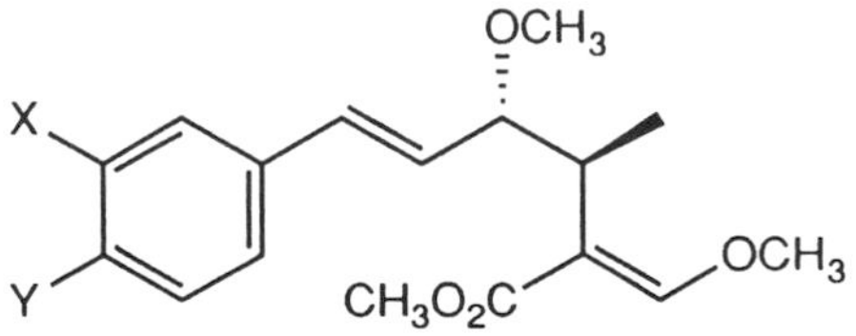

X = Y = H   Oudemansin A   [14]
X = CH₃O, Y = Cl   Oudemansin B
X = H, Y = CH₃O   Oudemansin X

Figure 5 The oudemansins.

not show cross-resistance to fungicides already in the marketplace. The inherent potential for mammalian toxicity with this mode of action has not materialized: Both strobilurin A and oudemansin A have low acute toxicity in mammals and no untoward toxic effects have been reported for any of the synthetic analogs currently in commercial development [52, 55].

The β-methoxyacrylates have attracted a lot of attention from industry since the early 1980s when chemists from Zeneca (formerly ICI) Agrochemicals and BASF (in collaboration with Professor W. Steglich, then at the University of Bonn) independently began work around strobilurin A, the simplest member of the series. Several detailed accounts of this research have been published [56]. Both groups adopted similar strategies in their attempts to improve on the fungicidal activity of the natural product and to overcome its lack of persistence. A common feature of both approaches was the replacement of the labile Z-double bond of strobilurin A with an *ortho*-disubstituted benzene ring to afford the stilbene (21), which conferred greater robustness to the molecule and a corresponding boost to its activity *in vivo*. In the course of this work it became apparent that it was possible to replace the 2-styryl side chain of the stilbene with a variety of substituents and retain activity against a broad spectrum of commercially important fungi. Similarly, replacing the phenyl ring bearing the β-methoxyacrylate moiety with other ring systems was found to be possible without deleterious effects on the fungicidal activity. By contrast, most modifications to the β-methoxyacrylate moiety (with the exception of the corresponding O-methyl oxime ethers and the β-methylthio analogs) lead to a dramatic fall in fungicidal activity.

(21)

This work has led to a flood of patent applications from many of the world's major agrochemical companies. Indeed, by the beginning of 1994, 14 companies had published a total of more than 160 patent applications, claiming both fungicidal and insecticidal activity. Research in this area of chemistry has so far yielded at least two development compounds. The first of these, ICIA5504 (22) and BAS 490 F (23), were both announced at the Brighton Crop Protection Conference in 1992 [52,

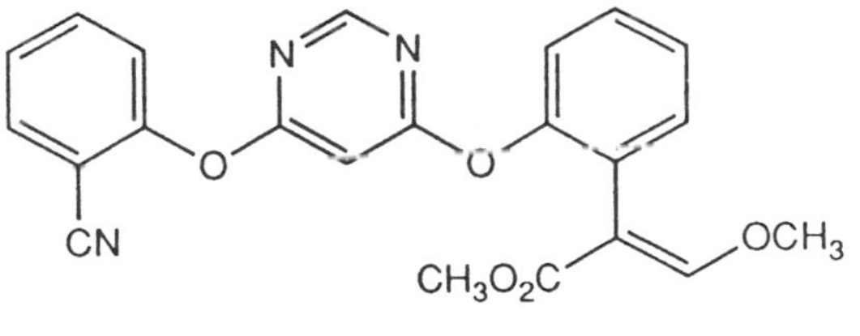

(22)

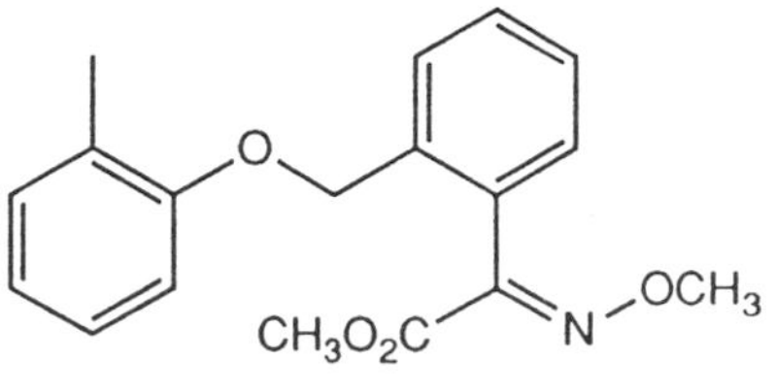

(23)

55]. ICIA5504 was discovered in 1988 at the end of a 5-year period in which many analogs were synthesized. Its laboratory preparation is illustrated in Figure 6.

ICIA5504 has eradicant, protectant, translaminar, and systemic properties, giving it the potential for use as a foliar, seed, or paddy water treatment and it is active against ascomycetes, basidiomycetes, deuteromycetes, and oomycetes. As a foliar treatment on wheat and barley, ICIA5504 has shown good control of a range of diseases that attack the stem base, leaves, and ears, and this leads to significant benefits in yield. As a seed treatment on barley, it has given good control of the important foliar pathogen powdery mildew. It controls both of the important fungal pathogens of rice, namely, rice blast and sheath blight, and may be applied either as a granule to paddy water or as a more traditional foliar spray. Furthermore, it has shown control of a wide range of fungal diseases of vines and other fruits. Both ICIA5504 and BAS 490 F showed low acute oral toxicity in rats and were negative in the Ames mutagenicity test [52, 55].

C. Hadacidin

In 1985, at the Bordeaux Mixture Centenary Meeting, workers from Shell presented two papers on the use of a novel experimental compound (code named WL87353) as a postharvest treatment for *Plasmopara viticola* (vine downy mildew) [57, 58]. WL87353 is the sodium salt of N-hydroxy-N-formylalanine (24) and has also been referred to in the literature by its trade name Hyformal [59]. It can be prepared from 2-bromopropionic acid according to Figure 7 [60]. Although WL87353 is formally derived from the natural product

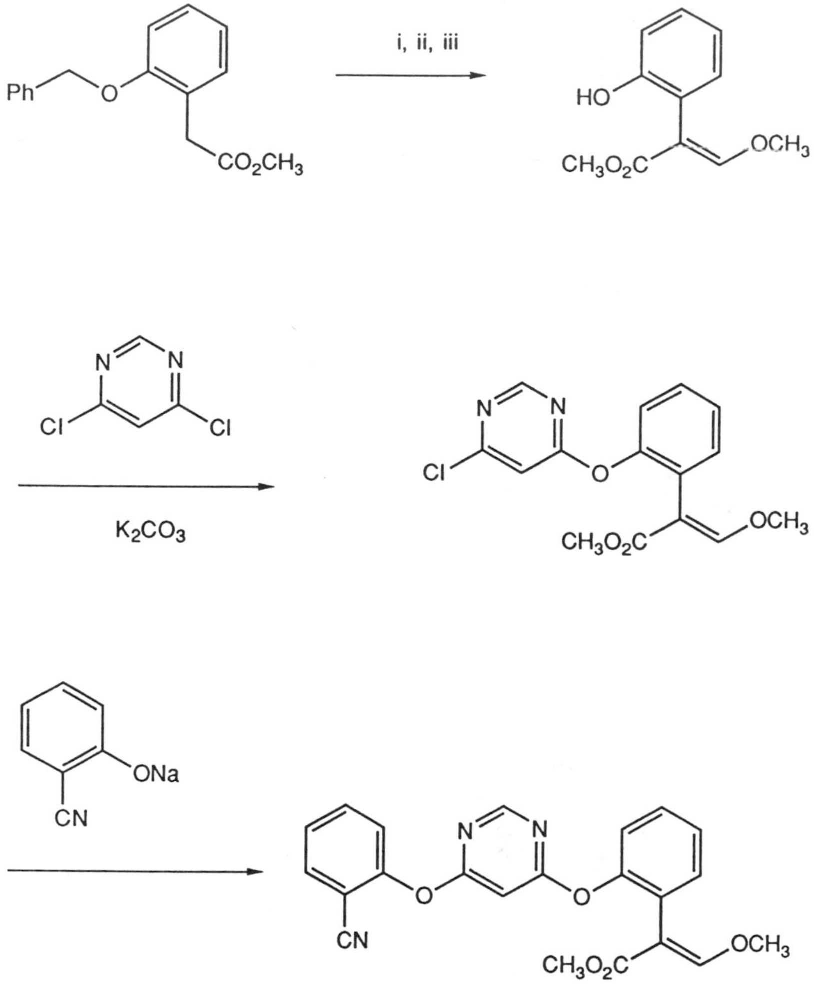

Figure 6 Laboratory synthesis of ICIA5504. *Reagents*: i, NaH/HCO$_2$CH$_3$; ii, K$_2$CO$_3$/(CH$_3$)$_2$SO$_4$; iii, H$_2$/Pd catalyst.

hadacidin (*N*-hydroxy-*N*-formylglycine) (*15*), a potent inhibitor of adenylosuccinate synthetase, it is unclear from the available literature whether WL87353 has the same mode of action or, indeed, if hadacidin actually served as the starting point for the discovery of this new compound.

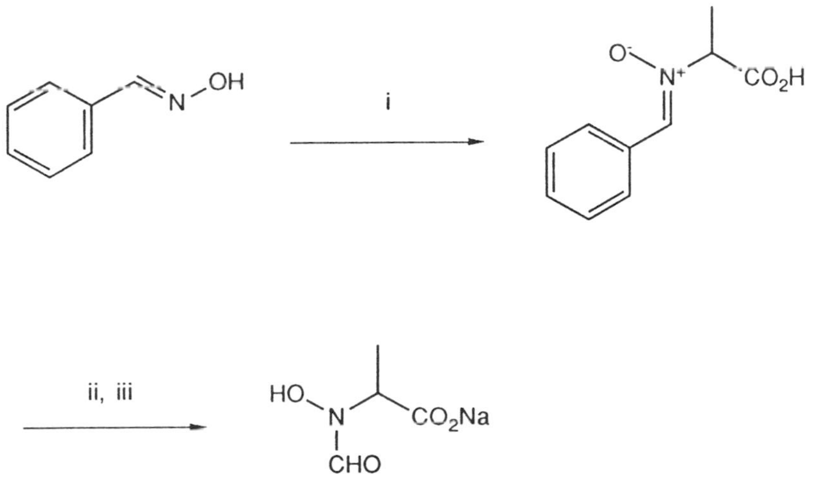

[24]

Figure 7 Preparation of WL87353 (24). *Reagents*: i, 2-Bromopropionic acid/ NaOEt; ii, HCO$_2$H/Ac$_2$O; iii, NaOH.

WL87353 is claimed to be a phloem mobile fungicide with excellent translocation properties and a long persistence of effect. Although it occasionally gave phytotoxicity when applied to vines in a routine summer schedule, its physical properties suggested a possible application as a postharvest treatment. Extensive experiments in the field demonstrated that a single treatment of vines with WL87353 as a late spray during the autumn brought about a prolonged suppression of vine downy mildew symptoms at least up until the end of flowering in the following season. However, although effective under optimum conditions at field rates of 2 kg/ha, WL87353 was not considered to be consistent enough in its performance at economic rates to justify further development. Several patent applications have appeared which claim fungicidal activity for a range of related compounds and mixtures with other known fungicides [60].

D. Thiolutin

The natural product thiolutin (*16*) and its homolog aureothricin (*25*) were the first members of the pyrrothine family of antibiotics to be isolated from strains of *Streptomyces* over 40 years ago [61]. Since then, nine further closely related analogs have been discovered (Figure 8)

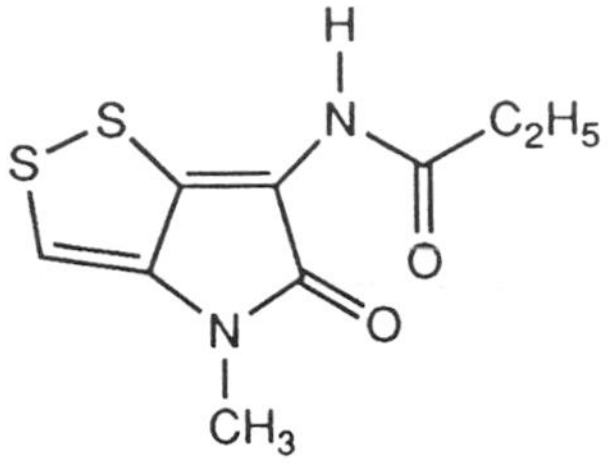

(25)

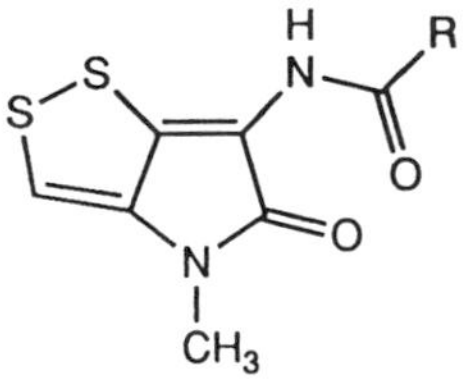

R = Methyl   Thiolutin [16]
R = Ethyl   Aureothricin [25]
R = iso-Propyl   Isobutyropyrrothin
R = n-Pentyl   Xenorhabdin IV
R = 4-Methylpentyl Xenorhabdin V

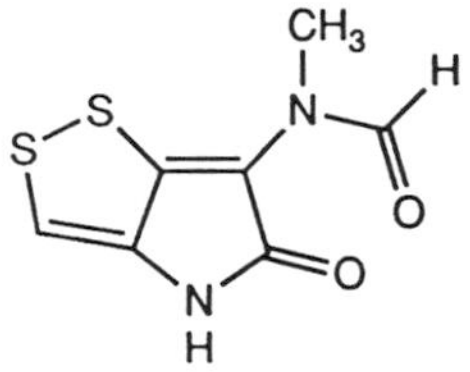

VD844

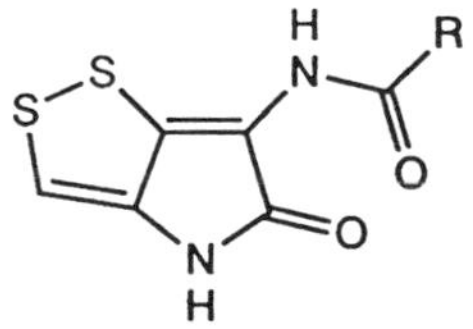

R = Methyl   Holomycin
R = H   VD846
R = n-Pentyl   Xenorhabdin I
R = 4-Methylpentyl Xenorhabdin II
R = n-Heptyl   Xenorhabdin III

Figure 8 The pyrrothines.

[65]. These compounds have shown interesting biological activity against a wide range of both gram-positive and gram-negative bacteria, as well as amoeboid parasites and pathogenic fungi. In particular, thiolutin (*16*) has been effective against a number of agricultural pathogens, including black rot and fire blight on apples, tobacco blue mould, and wilt on tomatoes. Thiolutin (*16*) is thought to act on nucleic acid metabolism at the RNA polymerase step. It has been reported to be a potent inhibitor of RNA chain elongation ($I_{50} = 2 \times 10^{-5}$ M) and its action is freely reversible [62].

The Pfizer company carried out some semisynthetic work based on pyrrothine itself (derived from thiolutin by hydrolysis) in the 1950s and reported their findings in the patent literature [63, 64]. More recently, in the early 1980s, a team of chemists at Zeneca (formerly ICI) Agrochemicals became interested in the pyrrothines as potential agricultural fungicides when in-house testing of samples of both thiolutin and holomycin highlighted protectant activity at 25 ppm against the commercially important fungal pathogens *Plasmopara viticola* (vine downy mildew) on vines and *Pythium ultimum* (damping-off) on potatoes.

A versatile synthesis of thiolutin, holomycin, and analogs was devised in order to determine the structural requirements for (and hence optimize) fungicidal activity. The synthetic route is outlined in Figure 9. Over 50 analogs of type *26* were prepared (including the natural products aureothricin, isobutyropyrrothin, and xenorhabdin IV), representing a wide range of substituents (R^1, R^2, R^3 = H, alkyl, alkenyl, aryl, aralkyl, etc.) attached to the pyrrothine nucleus. Many of these compounds showed fungicidal activity *in vivo* against a broad spectrum of commercially important fungal pathogens (*Plasmopara viticola, Phytophthora infestans, Cercospora arachidicola, Venturia inaequalis, Pyricularia oryzae,* and *Puccinia recondita*) at concentrations below 100 ppm. The best compounds showed good protectant activity against *Plasmopara viticola* at rates as low as 3 ppm.

(26)

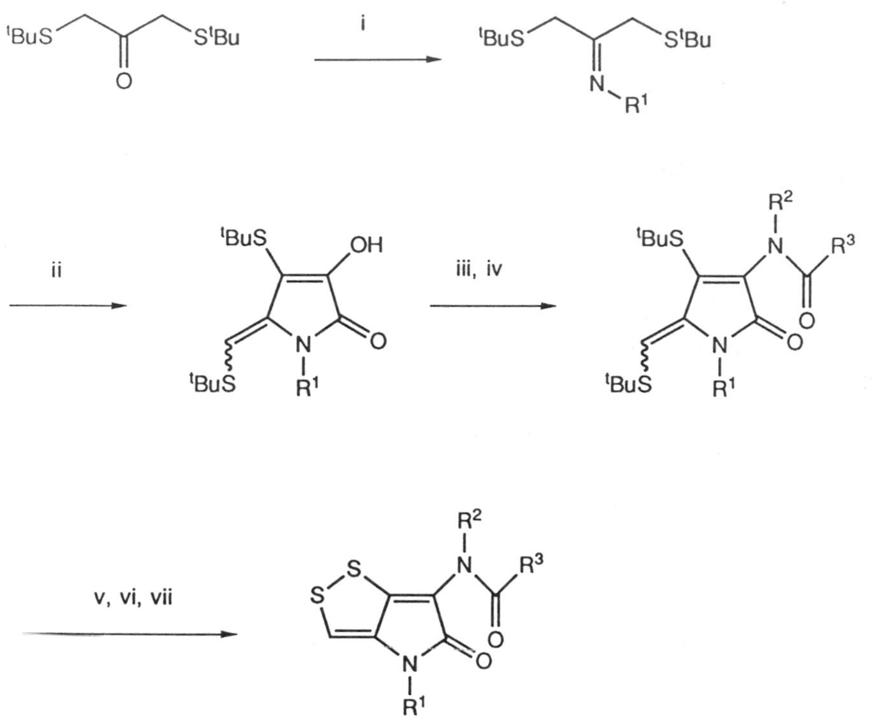

Figure 9 Synthesis of thiolutin analogs (26). *Reagents*: i, $R^1NH_2/TiCl_4$; ii, $(COCl)_2$; iii, R^2NH_2.HOAc (melt); iv, R^3COCl; v, $Hg(OAc)_2/TFA$; vi, H_2S; vii, I_2.

However, although significant improvements over the natural product leads were obtained, it was difficult to establish clear structure-activity relationships in the area. Furthermore, the synthetic analogs displayed predominantly protectant activity and little mobility within the plant. As a consequence, none of the compounds synthesized was of sufficient interest for further evaluation. A fuller account of work carried out in this area has been published recently [65].

E. Griseofulvin

Griseofulvin (17) was first isolated from *Penicillium griseofulvum* by Oxford et al. in 1939 [66] and its structure was later elucidated by Grove et al. [67]. It is produced industrially by fermentation of *P. griseofulvum*.

Griseofulvin has been the subject of several synthetic studies, most notably those of Stork and Tomasz [68] and Danishefsky and Walker [69]. Its fungicidal properties stem from its effect on the morphogenesis of fungi with chitinous cell walls [70]. Although originally introduced by Glaxo Laboratories Ltd. as an orally active antifungal antibiotic for the treatment of infections of the scalp, skin, and nails, griseofulvin is also used for a variety of veterinary and anthelmintic applications.

As an agricultural fungicide, griseofulvin has been less successful, despite the fact that it is reportedly systemic and nonphytotoxic. It has been examined in the past as a treatment for the control of certain plant diseases, including early blight of tomato, blossom blight of apples, and canker of melons, but no further developments in the use of the natural product itself in agriculture have been reported.

Although many semisynthetic analogs of griseofulvin have been prepared since the 1960s, the bulk of this work has inevitably been with pharmaceutical applications in mind and no analogs worthy of development as agrochemical fungicides have been disclosed. The main approach employed has been to chemically modify the natural product, which has been available in sufficient quantities from the fermentation process. As a result, analogs have been limited to grisane derivatives bearing a hydroxy or an alkoxy substituent on the aromatic ring.

Two recent approaches to analogs of griseofulvin, however, have relied on total synthesis [71, 72]. The first of these by Yamoto and co-workers involved the ring contraction of 6-ethoxycarbonyl-4,5,7-trimethoxycoumarin (27) to a 3(2H)-benzofuranone (28), which was subsequently elaborated to a racemic griseofulvin analog (29) bearing an ethoxycarbonyl substituent at the 5-position on the aromatic ring (Fig. 10) [71]. The product was unfortunately devoid of fungicidal activity. In 1990, a second Japanese group described the synthesis and biological activity of a series of six racemic analogs of griseofulvin (30–35) [72]. The chemistry was based on the total synthesis of griseofulvin by Danishefsky and Walker [69] and involved as a key step the [4+2] cycloaddition of the appropriate enones with the corresponding substi-

$R^1=R^2=H$ (30)

$R^1=Et, R^2=H$ (31)

$R^1=H, R^2=Et$ (32)

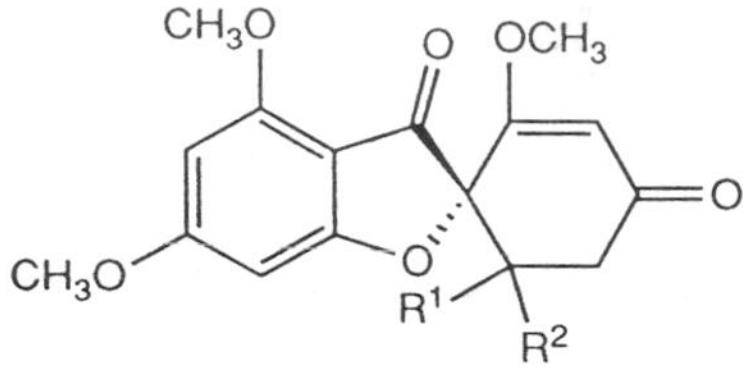

(33)

R¹=H, R²=Et   (34)

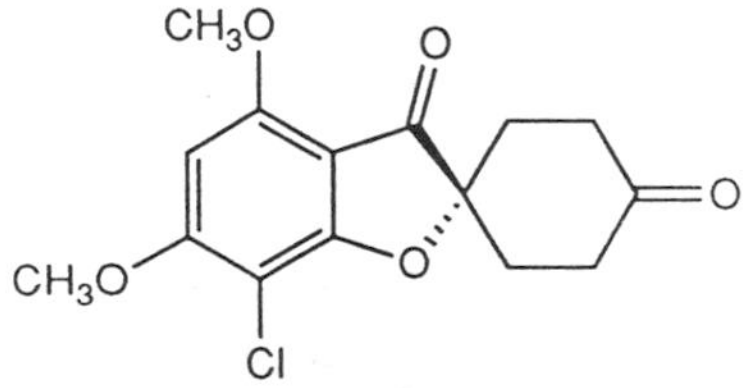

(35)

tuted butadienes. All of the analogs were less active than griseofulvin when assayed against *Botrytis cinerea* and *B. allii.*

F. Pisiferic Acid

In 1988, Kobayashi and co-workers described the synthesis and structure-activity relationships for 28 derivatives of pisiferic acid (*18*), a constituent of *Chamaecyparis pisifera* var. *plumosa* [73]. This natural product, an aromatic diterpene carboxylic acid, was isolated in the late 1970s and shown to have weak activity against *Pyricularia oryzae* (rice blast) [74].

In comparative studies, both pisiferic acid (*18*) and the commercial fungicide probenazole (*36*) were shown to suppress the intracellular hyphal growth of *P. oryzae* and to induce cytoplasmic denaturing of infected cells [73]. In view of the evident similarities between these fungicides in terms of their symptomology, together with the structural resemblances highlighted by computergraphic studies, it was proposed that both compounds possess the same mode of action.

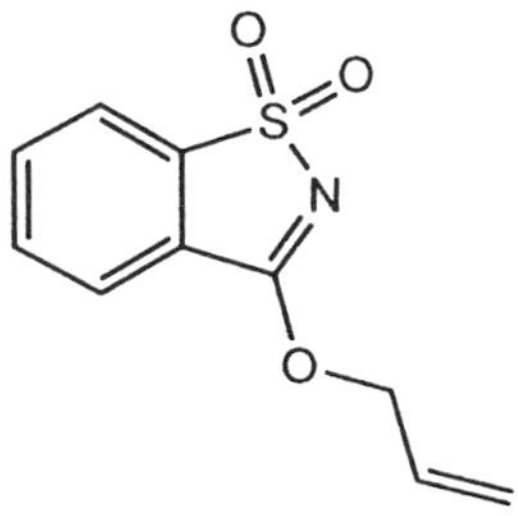

(36)

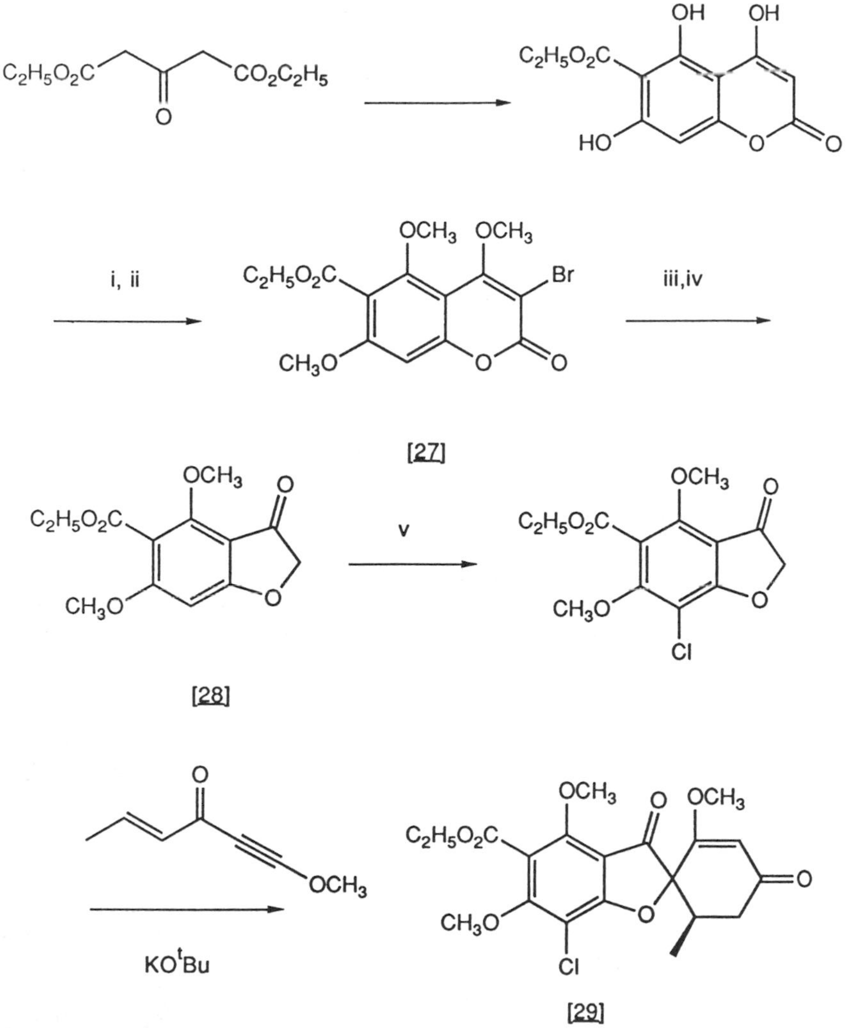

Figure 10 Preparation of griseofulvin analog (29). *Reagents*: i, $(CH_3)_2SO_4$; ii, Br_2; iii, KOH; iv, AcOH; v, Cl_2.

IV. CONCLUSIONS

Despite the obvious attractions of using natural products in agriculture, particularly at a time of great public concern about man-made pesticides, they still make up a vanishingly small proportion (<1%) of the global market for fungicides and bactericides. Furthermore, there are concerns, particularly among the more advanced countries in the world, that the widespread use of antibiotics in agriculture might lead to the creation of resistant strains, which could limit the efficacy of medical treatments for humans [75]. In contrast, commercial products derived from natural products (such as pyrrolnitrin and the β-methoxyacrylates) appear to have a much more promising future. In the absence of new outlets for existing products, it seems likely that the natural products currently in the marketplace will be progressively replaced by more effective natural and synthetic compounds that can offer advantages in terms of cost-efficacy, resistance-breaking potential, and environmental safety [76–78].

REFERENCES

1. K. Brent, *Fungicides for Crop Protection—100 years of Progress*, Monograph No. 31, British Crop Protection Council, Farnham, 1985, p. 11. For an excellent account of the early development of fungicides, see E.C. Large, *The Advance of the Fungi*, Jonathan Cape, London, 1940.
2. P.A. Worthington, *Natural Product Reports*, 47 (1988).
3. T. Misato, K. Ko, and I. Yamaguchi, *Adv. Appl. Microbiol.*, 21, 53 (1977); Y. Takaoka, *Ann. Phytopathol. Soc. Jpn.*, 34, 369 (1968); E.C. Kornfeld. R.G. Jones, and T.V. Parke, *J. Am. Chem. Soc.*, 71, 150 (1949).
4. Y. Okimoto and T. Misato, *Ann. Phytopathol. Soc. Jpn.*, 28, 209 (1963); S. Suzuki, G. Nakamura, K. Okama, and Y. Tomiyama, *J. Antibiotics*, 11, 81 (1958); S. Suzuki and K. Okuma, *J. Antibiotics*, 11, 84 (1958).
5. K. Sakata, A. Sakurai, and S. Tamura, *Agric. Biol. Chem.*, 38, 1883 (1974).
6. A. Schatz, *Proc. Soc. Exp. Biol. Med.*, 55, 66 (1944).
7. F.A. Kuehl, Jr., R.L. Peck, C.E. Hoffhine, Jr., E.W. Peel, and K. Folkers, *J. Am. Chem. Soc.*, 69, 1234 (1947).
8. H. Umezawa, *J. Antibiotics*, 27, 997 (1974).
9. R. Nyfeler and P. Ackermann, *Synthesis and Chemistry of Agrochemicals III* (D.R. Baker, J.G. Fenyes, and J.J. Steffens, eds.), ACS Symposium Series No. 504, American Chemical Society, Washington, D.C., 1992, p. 395.
10. T.J. Franklin and G.A. Snow, *Biochemistry of Antimicrobial Action*, 4th ed., Chapman and Hall, London, 1989, pp. 120–123 and references therein.
11. *The Agrochemicals Handbook*, 3rd edn., Royal Society of Chemistry, London, 1991.
12. *Japanese Pesticides Guide*, (K. Fukunaga, ed.) Japan Plant Protection Association, Tokyo, 1987.
13. T. Iwasa, E. Higashide, H. Yamamoto, and M. Shibata, *J. Antibiotics*, 24, 107 (1971); T. Iwasa, Y. Kameda, M. Asai, S. Horii, and K. Mizuno, *ibid.*, p. 119.
14. T. Suami, S. Ogawa, and N. Chida, *J. Antibiotics*, 33, 98 (1980).
15. Y. Miyamoto and S. Ogawa, *J. Chem. Soc. Perkin* I, 1013 (1989).

16. W-Z. Jin, K.L. Rinehart, Jr., and T. Toyokuni, *J. Antibiotics,* 40, 329 (1987); *J. Am. Chem. Soc.,* 109, 3481 (1987).
17. K. Yamamoto, *Noyaku Kagaku,* 3, 185 (1976).
18. N. Asano, Y. Kameda, and K. Matsui, *J. Antibiotics,* 44, 1406 and 1417 (1991).
19. A. Hasegawa, T. Kobayashi, H. Hibino, and M. Kiso, *Agric. Biol. Chem.,* 44, 143 (1980).
20. A.P.J. Trinci, *Exp. Mycol.,* 9, 20 (1985).
21. H. Umezawa, Y. Okami, T. Hashimoto, Y. Suhara, M. Hamada, and T. Takeuchi, *J. Antibiotics,* 18A, 101 (1965).
22. T. Ikekawa, H. Umezawa, and Y. Iitaka, *J. Antibiotics,* 19A, 49 (1966); Y. Suhara, K. Maeda, H. Umezawa, and M. Ohno, *Tetrahedron Lett.,* 1239 (1966).
23. Y. Suhara, F. Sasaki, K. Maeda, H. Umezawa, and M. Ohno, *J. Am. Chem.,* *Soc.,* 90, 6559 (1968).
24. M.J. Cron, R.E. Smith, I.R. Hooper, J.G. Keil, E.A. Ragan, R.H. Schreiber, G. Schab, and J.C. Godfrey, *Antimicrob. Ag. Chemother.,* 219 (1969).
25. K. Isono and S. Suzuki, *Heterocycles,* 13, 333 (1979); K. Isono, K. Asahi, and S. Suzuki, *J. Am. Chem. Soc.,* 91, 7490 (1969).
26. K. Isono and R.J. Suhadolnik, *Fed. Proc. Fed. Am. Soc. Exp. Biol.,* 32, 1966 (1973); K. Isono, S. Funayama, and R.J. Suhadolnik, *Biochemistry,* 14, 2992 (1975).
27. H. Kuzuhara, H. Ohrui, and S. Emoto, *Tetrahedron Lett.,* 5055 (1973); H. Ohrui, H. Kuzuhara, and S. Emoto, *ibid.* 4267 (1971).
28. N. Chida, K. Koizumi, Y. Kitada, C. Yokoyama, and S. Ogawa, *J. Chem. Soc. Chem. Commun.,* 111 (1994).
29. F. Tabusa, T. Yamada, K. Suzuki, and T. Mukaiyama, *Chem. Lett.,* 405 (1984); P. Garner and J.M. Park, *J. Organ. Chem.,* 53, 2979 (1988).
30. I. Savage and E.J. Thomas, *J. Chem. Soc. Chem. Commun.,* 717 (1989).
31. W. Koller, *Target Sites for Fungicide Action,* CRC Press, Boca Raton, FL, 1991, p. 255.
32. S. Takeuchi, K. Hirayama, K. Ueda, H. Sakai, and H. Yonehara, *J. Antibiotics,* 11, 1 (1958).
33. N. Otake, S. Takeuchi, T. Endo, and H. Yonehara, *Tetrahedron Lett.,* 1405 and 1411 (1965); J.J. Fox and K.A. Watanabe, *ibid.* 897 (1966); H. Yonehara and N. Otake, *ibid.* 3785 (1966); S. Onuma, Y. Nawata, and Y. Saito, *Bull. Chem. Soc. Jpn.,* 39, 1091 (1966); V. Swaminathan, J.L. Smith, M. Sundaralingam, C. Coutsogeorgopoulos, and G. Kartha, *Biochem. Biophys. Acta,* 655, 335 (1981).
34. H. Seto, I. Yamaguchi, N. Otake, and H. Yonehara, *Agric. Biol. Chem.,* 32, 1292 (1968).
35. J. Guo and S.J. Gould, *J. Am. Chem. Soc.,* 113, 5898 (1991).
36. P. Langcake, P.J. Kuhn, and M. Wade, *Progress in Pesticide Biochemistry,* Vol 3 (D. H. Hutson and T.R. Roberts, eds.), John Wiley and Sons, Chichester, 1983, p.1.

37. S. Harada, E. Mizuta, and T. Kishi, *J. Am. Chem. Soc,* *100,* 4895 (1978).

38. T. Iwasa, K. Suetomi, and T. Kusaka, *J. Antibiotics, 31,* 511 (1978).

39. S. Harada, E. Mizuta, and T. Kishi, *Tetrahedron, 37,* 1317 (1981).

40. H. Sawada, T. Suzuki, S. Akiyama, and Y. Nakao, *J. Ferment. Technol, 63,* 17 (1985).

41. T Misato, *Brighton Crop Prot. Conf.: Pests and Diseases—1984,* Vol. 3, British Crop Protection Council, Farnham, England, 1984, p. 929.

42. T. Iwasa, *Pestic. Chem.: Hum. Welfare Environ., Proc. Int. Congr. Pestic. Chem., Vol. 2,* 57 (J. Miyamoto and P.C. Kearney, eds.), Pergamon Press, Oxford, (1982).

43. E. Feduchi, M. Cosin, and L. Carrasco, *J. Antibiotics, 38,* 415 (1985).

44. K. Arima, H. Imanaka, M. Kousaka, A. Fukuda, and G. Tamura, *Agric. Biol. Chem., 28,* 575 (1964).

45. K. Arima, H. Imanaka, M. Kousaka, A. Fukuda, and G. Tamura, *J. Antibiotics, 18,* 211 (1965).

46. P.C. Knuppel, R. Lantzsch, and D. Wollweber, *Synthesis and Chemistry of Agrochemicals III* (D.R. Baker, J.G. Fenyes, and J.J. Steffens, eds.) ACS Symposium Series No. 504, American Chemical Society, Washington, D.C., 1992, p. 405.

47. Bayer AG, German Patents 3718375 (1988) and 3800387 (1989).

48. D. Nevill, R. Nyfeler, and D. Sozzi, *Brighton Crop Prot. Conf.: Pests and Diseases—1988, Vol. 1,* British Crop Protection Council, Farnham, 1988, p. 65.

49. K. Gehmann, R. Nyfeler, A.J. Leadbeater, D. Nevill, and D. Sozzi, *Brighton Crop Prot. Conf.: Pests and Diseases—1990,* Vol. 1, British Crop Protection Council, Farnham, 1990, p. 399.

50. P. Leroux, C. Lanen, and R. Fritz, *Pestic. Sci., 36,* 255 (1992).

51. A.B.K. Jespers, L.C. Davidse, and M.A. De Waard, *Pest. Biochem. Physiol., 45,* 116 (1993); *Pesticide Science, 40,* 133 (1994).

52. E. Ammermann, G. Lorenz, K. Schelberger, B. Wenderoth, H. Sauter, and C. Rentzea, *Brighton Crop Prot. Conf.: Pests and Diseases—1992,* Vol. 1, British Crop Protection Council, Farnham, England, 1992, p. 403.

53. J.M. Clough, *Natural Product Reports, 565* (1993).

54. W.F. Becker, G. von Jagow, T. Anke, and W. Steglich, *FEBS Letts., 132,* 329 (1981); R.W. Mansfield and T.E. Wiggins, *Biochim. Biophys. Acta, 1015,* 109 (1990); P.R. Rich, S.A. Madgwick, and T.E. Wiggins, *Biochem. Soc. Trans. 22,* 217 (1994); T.E. Wiggins, *ibid.,* 22, 221 (1994).

55. K. Beautement, J.M. Clough, P.J. de Fraine, and C.R.A. Godfrey, *Pestic. Sci., 31,* 499 (1991); J.M. Clough, P.J. de Fraine, T.E.M. Fraser, and C.R.A. Godfrey, *Synthesis and Chemistry of Agrochemicals, III* (D.R. Baker, J.G. Fenyes, and J.J. Steffens, eds.), ACS Symposium Series No. 504, American Chemical Society, Washington, D.C., 1992, p. 372; J.M. Clough, D.A. Evans, P.J. de Fraine, T.E.M. Fraser, C.R.A. Godfrey, and D. Youle, *Natural and Derived Pest Management Agents* (P. Hedin, R. Hollingworth, and J. Menn, eds.), American Chemical Society, Washington, D.C., 1993, p. 37.

56. J.R. Godwin, V.M. Anthony, J.M. Clough and C.R.A. Godfrey, *Brighton Crop Prot. Conf.: Pests and Diseases—1992*, Vol. 1, British Crop Protection Council, Farnham, 1992, p. 435.
57. C.L. Dunn and S.P. Klein, *Fungicides for Crop Protection—100 Years of Progress*, British Crop Protection Council, Farnham, 1985, 407.
58. M. Wade, D.P. Highwood, C.L. Dunn, J-M. Moncorge, and G. Perugia, *ibid.*, p. 455.
59. J.A. Page, *Pestic. Sci.*, *18*, 291 (1987).
60. Shell, European Patent 152129 (1985), European Patent 82577 (1983), European Patent 57027 (1982).
61. W.D. Celmer and I.A. Solomons, *J. Am. Chem. Soc.*, *77*, 2861 (1955), and references therein.
62. D.J. Tipper, *J. Bact.*, *119*, 795 (1974), and references therein.
63. Pfizer, U.S. Patent 2,689,854 (1954).
64. Pfizer, British Patent 692,066 (1953).
65. I. Dell, C.R.A. Godfrey, and D.J. Wadsworth, *Synthesis and Chemistry of Agrochemicals III*, (D.R. Baker, J.G. Fenyes, and J.J. Steffens, eds.), ACS Symposium Series No. 504, American Chemical Society, Washington, D.C., 1992, p. 384.
66. A.E. Oxford, H Raistrick, and P Simonart, *Biochem. J.*, *33*, 240 (1939).
67. J.F. Grove, J. Macmillan, T.C. Mulholland, and M.A.T. Rogers, *J. Chem. Soc.*, 3949 (1952).
68. G. Stork and M. Tomasz, *J. Am. Chem. Soc.*, *84*, 310 (1962), *86*, 471 (1964).
69. S. Danishefsky and F. Walker, *J. Am. Chem. Soc.*, *101*, 7018 (1979).
70. P.W. Brian, *Ann. Bot. Lond.*, *13*, 59 (1949).
71. M. Yamato, H. Yoshida, K. Ikezawa, and Y. Kohashi, *Chem. Pharm. Bull.*, *34*, 71 (1986).
72. B-S. Ko, T. Oritani, and K. Yamashita, *Agric. Biol. Chem.*, *54*, 2199 (1990).
73. K. Kobayashi, C. Nishino, H. Tomita, and M. Fukushima, *Phytochemistry*, *26*, 3175 (1987).
74. J.W. Ahn, K. Wada, S. Marumo, H. Tanaka, and Y. Osaka, *Agric. Biol. Chem.*, *48*, 2167 (1984), and references therein.
75. I. Yamaguchi, *Pestic. Sci.*, *35*, 391 (1992).
76. D.A. Evans and K.R. Lawson, *Pestic. Outlook*, *3*, 10 (1992); D.A. Evans and K.R. Lawson, *Milestones in 150 Years of the Chemical Industry* (P.J.T. Morris, W.A. Campbell, and H.L. Roberts, eds.), Royal Society of Chemistry, London, p. 68.
77. P. Hedin, *Naturally Occurring Pest Bioregulators* (P. Hedin, ed.), ACS Symposium Series No. 449, American Chemical Society, Washington, D.C., 1991, p. 1.
78. A.L. Young, *Pesticides—Minimising the Risks*, (N.N. Ragsdale and R. Kuhr (eds.), ACS Symposium Series No. 336, American Chemical Society, Washington, D.C., 1987, p. 1.

8

Rodenticides

A. Clare Elliott

Zeneca Agrochemicals, Bracknell, Berkshire, England

I. INTRODUCTION

A. The Need for Rodenticides

Of the total of approximately 5000 species of mammal which exist today, about 2000 are rodents, representing about 35% of the total mammal population [1].

Commensal rodents are those which inhabit areas of human population and activity, the word "commensal" meaning literally "sharing one's table." The three most important species in Europe are *Rattus norvegicus*, the Norway or brown rat, *Rattus rattus*, the black or roof rat, and *Mus musculus*, the house mouse. Several subspecies of *M. musculus* have been described [2]; although several forms of *R. norvegicus* and *R. rattus* are said to exist [1, 3], it is not clear whether these are true subspecies or simply color or geographical variants. In Britain, *R. norvegicus* is overwhelmingly the most important rodent pest on farms, whereas *M. musculus* and *R. norvegicus* appear to be of approximately equal importance on nonagricultural premises [4]. *R. rattus* is now rare in the United Kingdom.

Apart from the commensal rodents, many other species of rodent can be considered as pests on occasion; in Britain, the most important of these are the gray squirrel (*Sciurus carolinensis*) and the coypu (*Myocastor coypus*) [5].

341

Commensal rodents are regarded as pests for the following main reasons:

1. They damage and destroy growing crops and stored food and cause mechanical damage to other materials; the result is economic loss to farmers and food manufacturers. It is estimated that the world-wide postharvest loss due to rodent and insect damage might amount to 20% of its total value [6]. In Britain, it has been estimated that the total value of farm losses due to rodent damage is of the order of £10–20 millions per annum [4].
2. Contamination by rodents with their feces, urine, and hair renders human and animal foodstuffs unfit for consumption, causing further economic losses.
3. They spread disease. Although, in the United Kingdom and Europe, rats and mice do constitute a health risk wherever they live in close proximity to man, the spread of disease by rodents is not in most cases the primary reason for the adoption of control measures. However, in some countries, particularly in the tropics, rodent-borne diseases are of serious public health importance. Examples of the diseases transmitted from rats and mice to man include salmonellosis (food poisoning), leptospirosis (Weil's disease), and plague (the "Black Death" of the Middle Ages), all of which are caused by bacteria; typhus, caused by various species of *Rickettsia* organisms; and toxoplasmosis, a protozoal disease more usually transmitted to humans from cats and dogs [7]. In addition to spreading diseases to man, rodents may also cause their transmission to livestock and domestic pets.

Thus, for economic and health reasons, there is a clear need for the control of commensal rodent pests.

B. Historical Development of Rodent Control Measures

The earliest attempts to control commensal rodents used ancient poisons such as arsenic, crudely designed traps, and domestic predators, such as dogs, ferrets, and falcons [7]. These methods changed little until the early years of the twentieth century, when recognition of the role of rats and mice in the transmission of disease resulted in a search for more effective rodent control measures.

Natural products have played an important part in the subsequent development of effective methods in two major ways: first, some natural products have been used as rodenticides in their own right. The most important of these are scilliroside, reserpine, calciferol and cholecalciferol, and strychnine (Section II). Second, the highly effective and economically important anticoagulant rodenticides were developed from the natural product dicoumarin (Section III).

The natural products used as rodenticides per se are all acute poisons, as were all the rodenticides in common use until the advent of the anticoagulant rodenticides in the early 1950s; other acute rodenticides include fluoroacetamide [8, 9], sodium fluoroacetate [10], crimidine (1) [11], alphachloralose (2), norbormide (3) [12, 13], and zinc phosphide [13, 14]. The LD$_{50}$ values for some of these compounds and the concentrations of ai. in the bait are summarized in Table 1 [7].

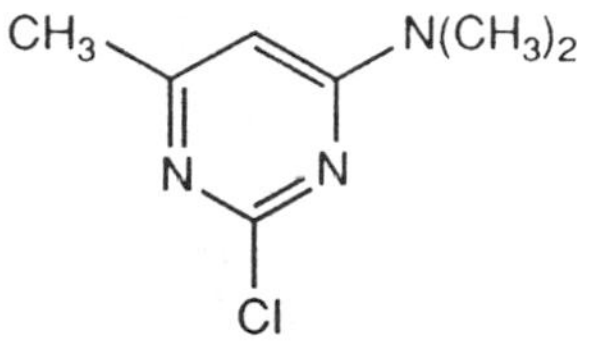

(1)

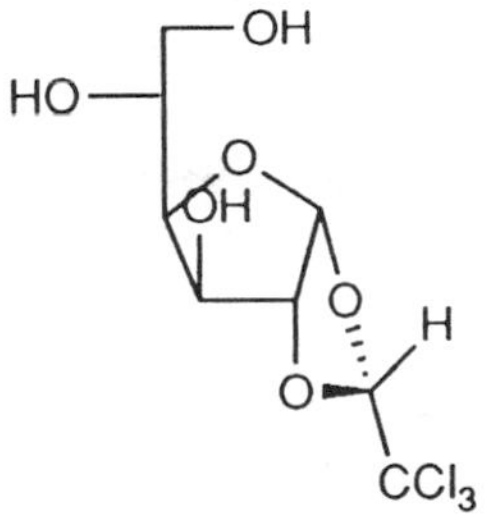

(2)

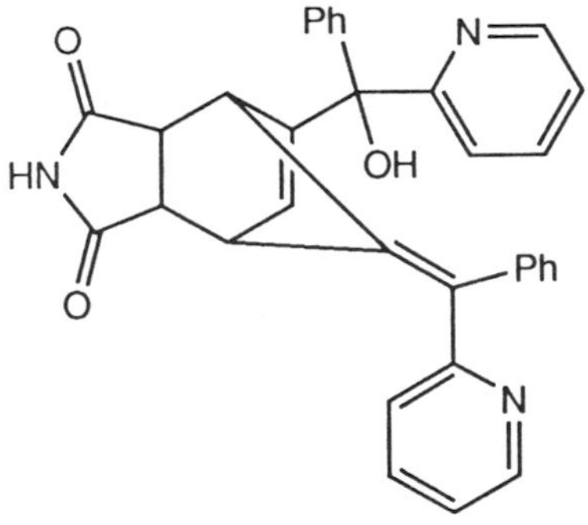

(3)

Although acute rodenticides are still used on occasion, they suffer from some serious disadvantages:

1. Because these compounds act relatively rapidly, it is possible for the rodent to ingest sublethal doses of the acute poison and then to

Table 1 Toxicity of Some Acute Rodenticides

Compound	% a.i. used in baits	LD$_{50}$ (R. norvegicus) (mg/kg)
Fluoroacetamide	2.0	13–16
Sodium fluoroacetate	0.25	5–10
Red squill	10.0	500
Scilliroside	0.5	1–5
Crimidine	0.5	1–5
Alphachloralose	4.0	300
Norbromide	1.0	12
Zinc phosphide	1.0	40

a.i. = active ingredient.

refuse further bait; this is known as "bait shyness." Laying of plain bait for several days before administering the poison ("prebaiting") can overcome this problem to some extent, but this practice is not always adopted, often for economic reasons.

2. The acute rodenticides show little selectivity, and there is, thus, a high risk of poisoning nontarget birds and animals, including rodent predators.

3. The acute compounds are generally less effective than the anticoagulants which superseded them, having LD$_{50}$ values (on R. norvegicus) of 1–500 mg/kg [7]; often 25% or more of the original infestation remains after treatment [15].

The introduction, in the 1950s, of the anticoagulant rodenticides dramatically improved the efficacy and safety of rodenticide use (see Section III); the first of these compounds to be used, from which the others were derived, was the natural product dicoumarin. The "first-generation" compounds, of which warfarin was the most used, were introduced in the 1950s and 1960s. Since the mid 1970s, a number of more potent "second-generation" anticoagulants have been developed and marketed for rodent control. These compounds possess several important advantages over the acute rodenticides:

1. Having generally lower LD$_{50}$ values (0.2–100 mg/kg), a single dose, particularly of one of the second-generation products, is usually sufficient to achieve a lethal effect. Thus, sublethal dosing and incomplete control are less likely.

2. Anticoagulants do not produce immediate symptoms, reducing the likelihood of bait shyness.

3. Although poisoning of nontarget animals may still occur, it has historically proved to be relatively rare, at least for the first-generation products; these are generally administered in several doses and rely on their cumulative effect to achieve control. Should accidental poisoning occur, an effective antidote, vitamin K_1, is available.

Anticoagulant rodenticides are currently used for the overwhelming majority of rat control treatments on premises of all types. Since 1979, when warfarin still accounted for 70–80% of all treatments [16], the use of the second-generation compounds has steadily increased.

Acute rodenticides are still used today in some situations, for example, when a rapid kill in an enclosed area is required or when there is extensive anticoagulant resistance. Two acute rodenticides, calciferol and alphachloralose, are used for a significant proportion of *M. musculus* treatments [4].

Although rodenticidal compounds are currently the most effective means available for the control of commensal rodent pests, other methods of rodent control are used in conjunction with rodenticides, or in situations in which the use of conventional rodenticides is not possible. These methods include traps, fumigants such as methyl bromide and calcium cyanide, chemical repellents (e.g., thiram and *N,N*-dimethyl-*S*-*tert*-butyl-sulphenyl-dithiocarbamate), and physical methods of rodent exclusion. The maintenance of high standards of hygiene and good housekeeping practices are also important in controlling the populations of commensal rodents.

II. NATURAL PRODUCTS USED AS RODENTICIDES

A. Scilliroside

Red squill, *Urginea maritima*, a member of the Liliaceae, is a large onion-like plant that grows wild in the coastal areas around the Mediterranean Sea [17]. The bulb extracts and dried powder from the red squill have been used for rodent control since the thirteenth century [18].

All plant parts, including the leaves, flower stalk, scales, and especially the roots and core of red squill contain scilliroside (*4*), a butadienolide glycoside, as the major toxicant; other scilla glycosides, including scillaren A (*5*), scillarenin (*6*), scilliglaucoside (*7*), and scillirubroside (*8*) are present in smaller amounts [19, 20]. The toxicity of the dried bulb varies substantially due to genetic variation in the

R=glucose   (4)

R=H   (9)

R=glucosylrhamnose   (5)

R=H   (6)

R=CHO R'=H   (7)

R=H, R'=OH   (8)

wild, seed-propagated plants differences in harvest time, and decomposition on storage [21, 22].

Scilliroside has a digitalis-like action and affects the cardiovascular and central nervous systems, causing convulsions and death. It is a relatively slow-acting poison, causing death after 24–48 hr. Because it causes convulsions, the use of red squill in the United Kingdom has been banned by the Animal (Cruel Poisons) Act 1963, although it is still used in many countries.

Until the 1940s, the toxicity of red squill was difficult to establish, because extracts vary so greatly in scilliroside content; it is estimated to be about 500 mg/kg [7]. The availability of samples of purified scilliroside enabled measurements of its toxicity to be made [23]. The oral LD_{50} was found to be 0.43 mg/kg for female and 1.35 mg/kg for male albino rats; the greater susceptibility of females had also been reported in studies using the crude extracts [24]. A more recent study using Charles River CD rats quotes LD_{50} values of 1.4 (female) and 5.3 (male) mg/kg [22]. The toxicity of scilliroside to 11 other species was also measured; these included the house mouse (0.17 mg/kg), cat (6 mg/kg), and field mouse (37 mg/kg). In practice, red squill appears to perform as well as most other acute poisons against *R. norvegicus* but not to be as efficient against *R. rattus* or *M. musculus*.

When administered to nontarget animals, red squill and scilliroside usually act as emetics and are, thus, considered to be relatively safe; however, a number of incidents of poisoning of domestic species have been reported [25].

One of the major limitations to the use of extracts of red squill and scilliroside as rodenticides is their poor palatability: Scilla glycosides are extremely bitter, and rats quickly learn to avoid the baits [25]. As a result, red squill has been relatively little used recently. However, the increasing resistance problems encountered with the anticoagulant rodenticides and the search for compounds with different modes of action has led to a reexamination of the potential of red squill as a source of rodenticidal compounds and a more detailed analysis of the components present [21, 22, 26].

Although scilliroside, as the glycoside, is bitter, its aglycone, scillirosidin (9), is tasteless [26], and, therefore, the aglycone might be more palatable. Earlier studies had shown that, in rats, scilliroside and scillirosidin are of similar toxicity [23]. Moreover, scillirosidin administered intravenously to cats is more toxic (0.12 mg/kg) than scilliroside (0.2 mg/kg). It has been suggested that the active toxicant is the aglycone and that this is produced on metabolism of the scilliroside by β-glycosidases in the gastrointestinal tract; scillirosidin would be more easily absorbed and be better able to penetrate the blood-brain barrier [21, 22]. Therefore, methods of producing scillirosidin from the red squill bulb have been investigated; however, chemical degradation methods also lead to hydrolysis of the lactone and acetyl groups [19], and the desacetyl compound is known to be much less active than the parent scillirosidin [21].

Enzymic approaches to glycoside cleavage have proved to be more successful: The commercially available enzyme naringinase was found to cleave scilliroside [22]. Several strains of *Aspergillus niger* also cleave the glycoside, both in its purified form and on treatment of the crude red squill extract [26]. Different strains were found to vary in the amount of desacetylscillirosidin also produced and, thus, to give products of varying toxicity. Assays of the activity of the aglycone preparation (containing scillirosidin, desacetylscillirosidin, and residual scilliroside) gave an LD_{50} of 8.0 mg/kg (on male rats), compared to 4.2 mg/kg for purified scilliroside.

It has been claimed that, because scillirosidin (9) is less bitter and, thus, more palatable than scilliroside (4), it should be more effective as a rodenticide [26], and that red squill might be a potential new economic crop for the southwest United States, where climatic conditions are similar to those in its natural habitat [26]. However, in common with all acute poisons, bait shyness could still occur; in addition, the nature of the poisoning symptoms induced by scilliroside and related compounds is likely to limit its use, even in countries where it is still permitted.

B. Reserpine

Reserpine (10) is an alkaloid extracted from the roots of various species of *Rauwolfia*, notably *R. serpentina* and *R. vomitoria* [27]. It can be produced synthetically [28] and has been used occasionally as a sedative and in the treatment of hypertension. A number of analogs of reserpine, including yohimbine (11), deserpidine (12), syrosingopine, and rescinamine, can also be extracted from *Rauwolfia* species; these compounds have a similar, although usually less potent, pharmacological effect [27].

(10)

The main effect of reserpine is on the central nervous system [29]; it causes the release of catecholamines from storage sites in the brain and nervous tissue, leading to a depletion of the brain reserves of cat-

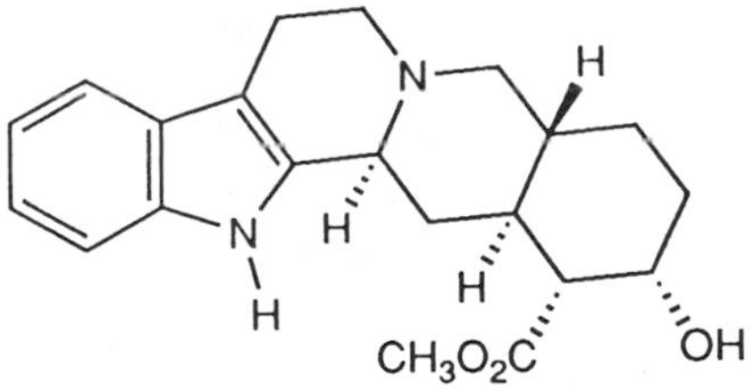

(11)

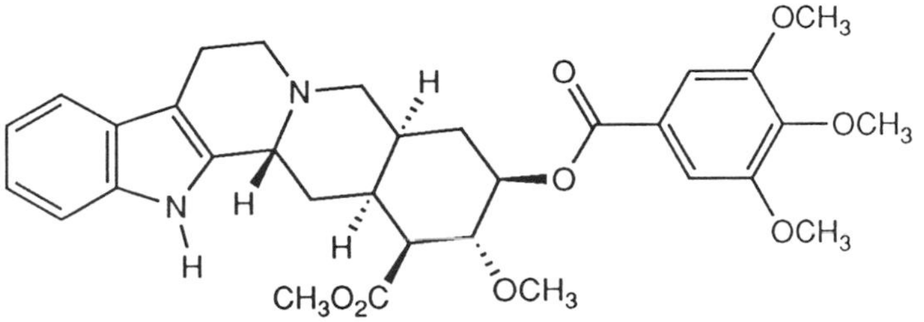

(12)

echolamines; one of these catecholamines is responsible for regulating body temperature, and its depletion results in a drop in body temperature and, eventually, hypothermia. A decrease in the amount of a second catecholamine, responsible for the conduction of nerve impulses, leads to central respiratory depression and deep sedation. The resulting inability to feed and consequent weight loss, together with the reduced body temperature, eventually results in death.

Reserpine is generally much more effective in controlling mice than rats; although the acute oral LD_{50} to mice is about 200 mg/kg, the toxicity decreases to as little as 2.5 mg/kg on accumulative administration for 4 days [1]. The acute oral LD_{50} to rats is about 420 mg/kg. The cumulative effect is much less marked in rats, thought to be the result of larger body size [27]: The smaller surface area:volume ratio reduces the rate of heat loss, and the greater food reserves enable the animal to survive until the effects of reserpine wear off. Larger body weight and high ambient temperature have been observed to decrease the toxicity to mice [30].

The toxicity of reserpine to nontarget animals has been studied in connection with its medical and veterinary uses; dogs seem to be most sensitive, although it is difficult to define the acute LD_{50} because few animals die [1]. However, most larger animals can survive the period of sedation and recover when the effects of reserpine wear off.

The use of reserpine in rodenticidal baits, at concentrations of 0.01–0.05% active ingredient, was patented by Rentokil in 1971 [31]. These were claimed to be effective against *M. musculus*. The major motivation for the introduction of reserpine as a rodenticide was the increasing level of resistance to warfarin. However, field trials showed that reserpine did not control mouse populations satisfactorily. In an attempt to improve the activity, a number of synthetic analogs prepared initially as hypotensive agents were tested [27]. Two of these, (*13*) and (*14*), were effective more rapidly than reserpine and were subsequently patented [32]. However, at about this time, the first of the highly active second-generation anticoagulants, difenacoum, was developed, reducing the need for an effective reserpine analog [30].

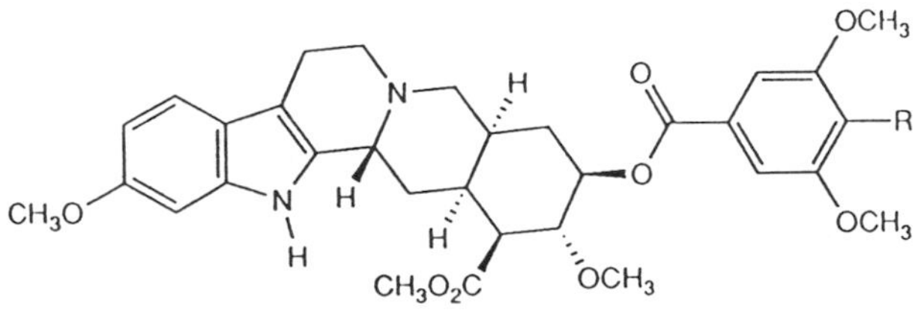

$R=CH_2CH_2CH_3$ (*13*)

$R=CH(CH_3)CH_2CH_3$ (*14*)

C. Ergocalciferol and Cholecalciferol

The vitamin D-related compounds, ergocalciferol (*15*) (often simply called calciferol) and cholecalciferol (*16*), have been introduced relatively recently as rodenticides [1]. Although ergocalciferol is also known as vitamin D_2 and cholecalciferol as vitamin D_3, the mention of the word vitamin on rodenticide labels is prohibited, for safety reasons, in some countries.

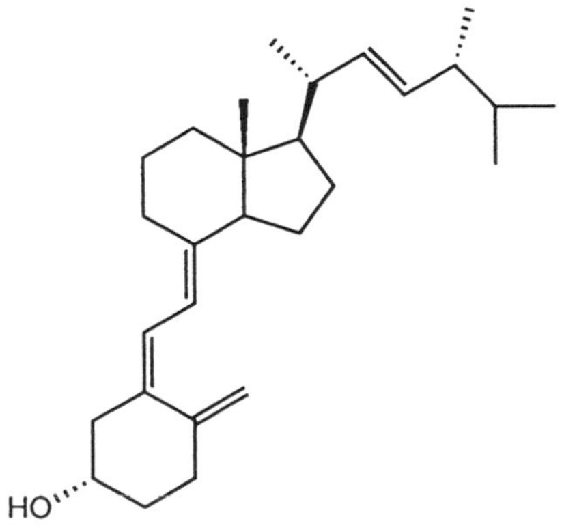

(*15*)

Ergocalciferol and cholecalciferol are naturally occurring vitamins, small quantities of which are essential for good health: A deficiency can

(16)

cause rickets in children and osteomalacia in adults. They are present in some foods, such as dairy products and some fish oils, and can also be manufactured in the body in the presence of sunlight.

In large amounts, ergocalciferol and cholecalciferol can cause toxic symptoms [33]. Under these circumstances, they stimulate the absorption of calcium from the intestine and mobilize skeletal reserves, leading to an increase in calcium levels in the blood [34]. The resulting hypercalcemia gives rise to calcification and degeneration of various soft tissues, particularly the kidneys, lungs, and heart.

The emergence of resistance to warfarin, and the resulting search for compounds with alternative modes of action, led to investigation of the potential of ergocalciferol [34] and, later, of cholecalciferol [35, 36] as rodenticides.

The acute LD_{50} of ergocalciferol to rats and mice is generally considered to be 30–100 mg/kg, although when administered for 5 consecutive days, the LD_{50} falls to 7 mg/kg. For cholecalciferol, values of 43.6 mg/kg on *R. norvegicus* and 42.5 mg/kg on *M. musculus* have been determined [35]. Laboratory tests and field trials established a concentration of ergocalciferol in the bait of 0.1% to be optimum; at this concentration, nearly all animals fed in the laboratory for 2 days died. In the field, control of *R. norvegicus* of 80% was achieved in one trial [1], and 12 out of 14 farms completely cleared in another case [37]. Against *M. musculus*, ergocalciferol was even more effective, achieving 79–100% control over seven sites [38].

When it was first introduced, it was claimed that warfarin and ergocalciferol had a synergistic effect on each other and that they should therefore be used in combination [37]. Further investigation showed no evidence for synergy, although the effects were additive [34]. A mixture of warfarin and ergocalciferol was introduced commercially by Sorex in 1974.

Laboratory tests, and, later, field trials [36] established the efficacy of the closely related compound, cholecalciferol (16), as a rodenticide, and it was introduced commercially in 1984 by Bell Laboratories in Wisconsin as Quintox. Bait containing 0.075% of the active ingredient was found to give excellent control: In laboratory tests, 100% mortality was obtained after 3 days' exposure to 0.075% bait [35]. In subsequent field trials, in both indoor and outdoor situations, 70–100% control of R. norvegicus, and almost complete control of M. musculus, were reported [36].

In addition to ergocalciferol and cholecalciferol, a series of synthetic derivatives of vitamin D have been patented by the Wisconsin Alumni Research Foundation as effective rodenticides. One of these, 1-α-hydroxycholecalciferol, has been registered in the United States by Bell Laboratories [1].

The two forms of calciferol are, generally, equally toxic to most mammals: Acute oral LD_{50} values for dogs have been quoted as 12–20 mg/kg [1] and 88 mg/kg [35]; cats are claimed to be more susceptible. The effect of ergocalciferol on a number of species has been documented [39]. Although there are no reports of ergocalciferol or cholecalciferol, when used as a rodenticide, having caused human deaths, there are several cases of poisoning, all the result of accidental or inadvertent overdosing, and the subacute symptoms are well documented [33].

A study of the secondary toxicity of cholecalciferol to dogs has been reported [35]: All six dogs survived feeding for 14 days on a diet consisting of rats killed by bait containing 0.075% cholecalciferol. No pathological abnormalities were observed, and it was thus concluded that cholecalciferol does not pose a secondary poisoning hazard to dogs.

D. Strychnine

Strychnine (17) is an alkaloid which, together with the dimethoxy analog, brucine (18), occurs in the seeds of species of the genus Strychnos, particularly Strychnos nux-vomica and S. wallichiana. Despite the fact that strychnine was first isolated in 1818 [40], it was not until 1931 that Robinson elucidated its structure [41]. This highly complex molecule was a challenging synthetic target; the first total synthesis was finally completed by Woodward in 1963 [42]. This extraordinary achievement remains a milestone in synthetic organic chemistry. The synthesis of strychnine and related Strychnos alkaloids are still of interest to chemists today [43].

The biosynthesis of strychnine and its analogs have also been of immense interest over many years; however, the detailed biosynthetic pathway is not yet known. These alkaloids are clearly of mixed

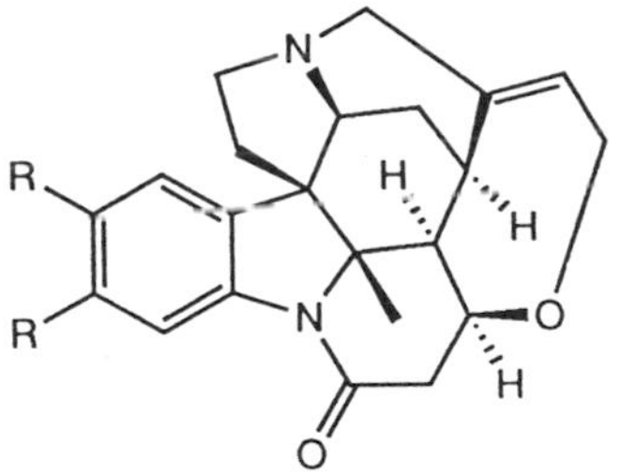

R=H (17)

R=OCH₃ (18)

biosynthetic origin; tryptophan and geraniol are both incorporated into strychnine, and the additional C_2 unit is derived from acetate. By analogy with the known routes to the *Vinca* alkaloids, a plausible pathway has been outlined [44].

Strychnine and its salts are very highly toxic to all mammals. For wild *R. norvegicus,* values of 4.8 mg/kg [45] and 6–8 mg/kg [7] have been reported. Strychnine produces violent muscular spasms, often within a few minutes, and death, due to paralysis of the central nervous system, usually follows within 30 min [7]. Because of the inhumane nature of the poisoning, most uses of strychnine as an animal poison have been banned in the United Kingdom since 1935 [46], but it has found limited use in other countries.

Strychnine is a highly bitter compound, one part of strychnine imparting a bitter taste to 500,000 parts of water; for this reason it is not effective against rats. It does appear to be more successful against mice and other rodents, such as gophers and squirrels [47].

The extremely poisonous nature of strychnine, together with its chemical stability, means that the poisoning of nontarget animals is a serious hazard, and there are many reports of such incidents, including the poisoning of domestic pets and children in the United States [47, 48]. Despite contentions that the responsible use of strychnine as a rodenticide, for example, by restricting its use to underground burrows and strict adherence to labeling instructions, should minimize the risk [49], it is clear that strychnine is too toxic to be of general use as a rodenticide.

III. NATURAL PRODUCTS AS LEADS FOR ANTICOAGULANT RODENTICIDES

A. Development of Anticoagulant Rodenticides from Dicoumarin

1. History

The anticoagulant properties of the natural product dicoumarin, 3,3'-methylenebis(4-hydroxycoumarin) (19), were discovered after an in-

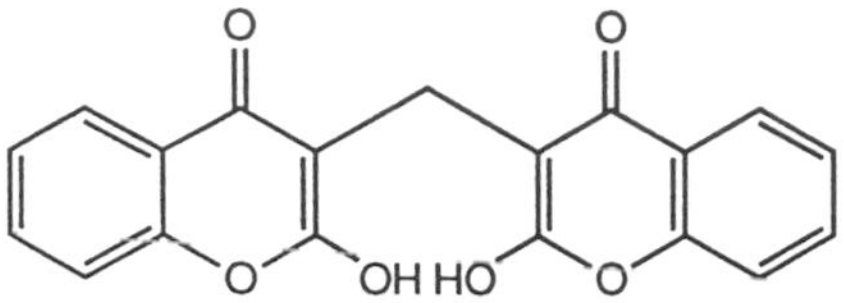

(19)

vestigation of the causes of a disease suffered by cattle which had been feeding on "spoiled" sweet clover [1]. The disease involved hemorrhaging; by 1929, this was attributed to a deficiency of thrombin in the blood [50], but it was not until 1939 that the chemical in the clover which had caused the hemorrhaging was isolated and identified as dicoumarin [51].

Following the identification of dicoumarin as an anticoagulant, synthetic material was prepared and patented in 1941 by the Wisconsin Alumni Research Foundation, although a patent for its use as a rodenticide was not obtained until 1952. Meanwhile, using dicoumarin as a lead, a program of synthesis was begun with the aim of improving the anticoagulant activity of the natural product. The 42nd compound in the series, 4-hydroxy-3-(3-oxo-1-phenylbutyl)coumarin (20) was synthesized in 1942 and found to be more effective than any of the others in causing hemorrhaging in rodents, being 50 times more toxic than the naturally occurring dicoumarin. A patent was granted in 1947, and the compound, named "warfarin" (after the Wisconsin Alumni Research Foundation), was registered for sale as a rodenticide in 1950. Over the next 20 years, warfarin became the best known and most widely used rodenticide: in the first 15 months of use alone, more than a million pounds of the concentrate and ready-to-use product were sold under 300 different trade names [1].

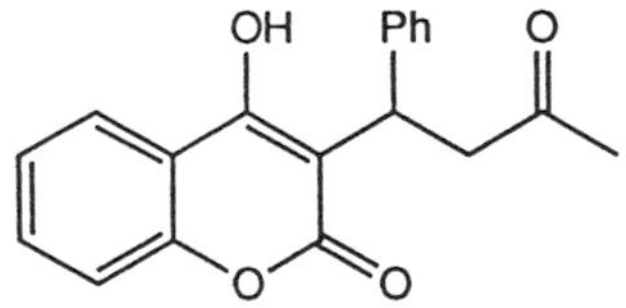

(20)

The advent of anticoagulant rodenticides dramatically improved the efficacy and safety of rodenticide use, and during the 1950s and 1960s a number of other anticoagulants were introduced. Some of these were, like warfarin, 4-hydroxycoumarins, for example, coumachlor (21) (marketed as Tomorin), coumafuryl (22), and coumatetralyl (marketed as Racumin) (23). The rodenticidal activity of the 1,3-indanediones was also recognized and exploited, leading to the development of, for example, pentolactin (24), diphacinone (25), and chlorophacinone (26).

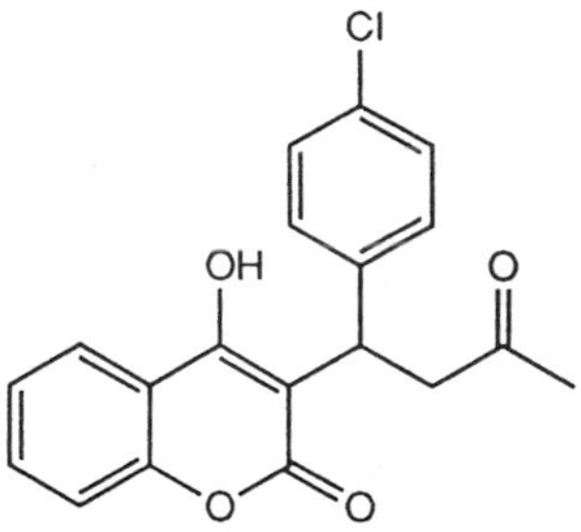

(21)

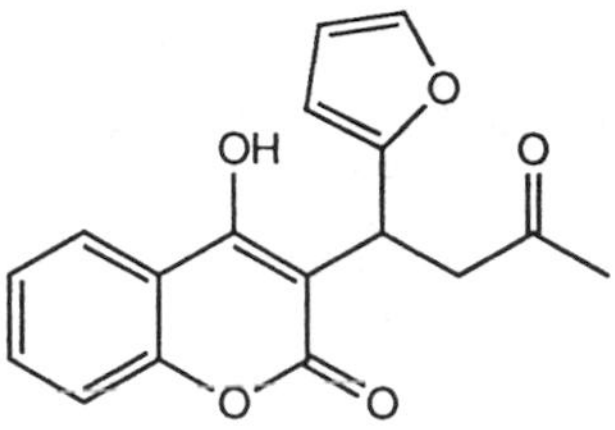

(22)

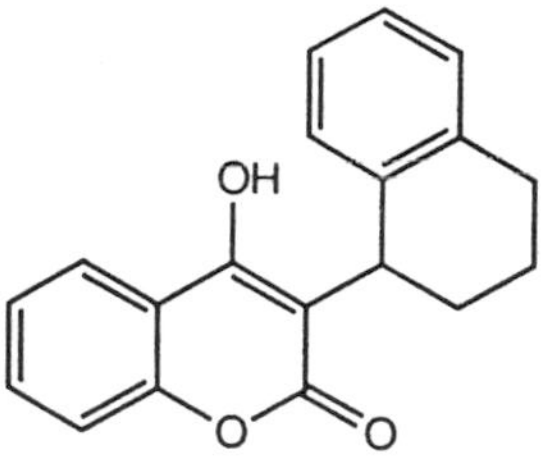

(23)

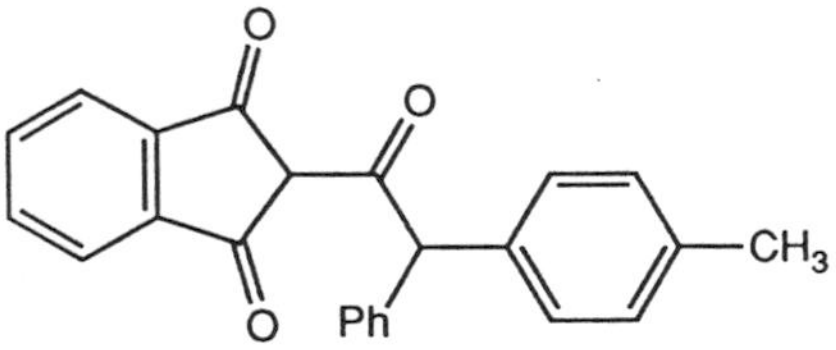

(24)

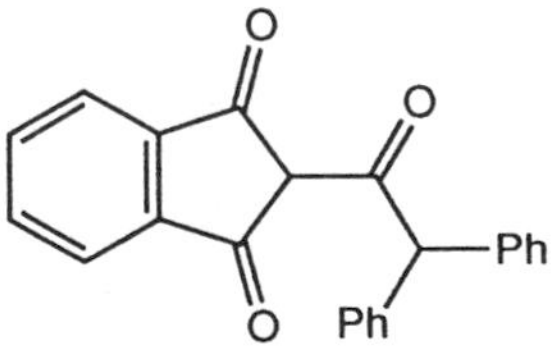

(25)

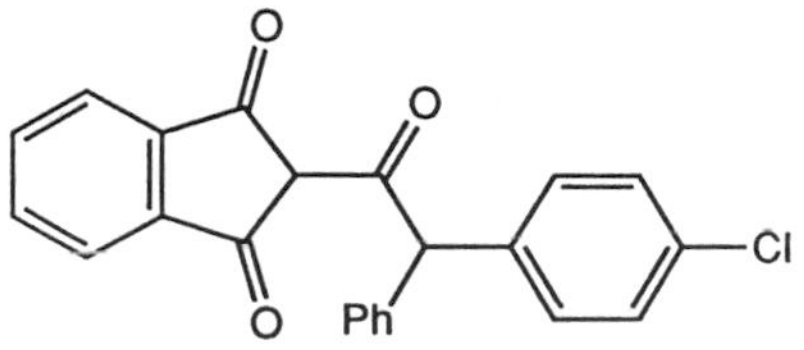

(26)

These early coumarins and indanediones were collectively known as the "first-generation anticoagulant rodenticides." Reviews of the early history of anticoagulants have been written by Link [52] and Mills [53].

Because the anticoagulants were so much more effective than the earlier acute poisons, they were soon used to the almost complete exclusion of other rodenticides. As a result of the consequent selection pressure, resistance to the first-generation rodenticides began to develop during the 1960s and soon became a serious problem.

The development of extensive resistance to the first-generation anticoagulants prompted a search for rodenticides that were active against resistant rodents. Two lines of research were pursued: the search for novel nonanticoagulant rodenticides and the development of anticoagulants that were, nevertheless, active against the resistant species. As a result, a number of nonanticoagulants were introduced, such as alphachloralose, reserpine, and calciferol (Section II), but none of these was very effective.

The advantages of anticoagulants over acute rodenticides were widely recognized and the search for resistance-breaking anticoagulants led to the development, in the 1970s, of a new series of anticoagulant rodenticides, the "second-generation anticoagulants." The first of these to be introduced, by Ward-Blenkinsop in 1974 [54], was difenacoum (27); the closely related but more toxic brodifacoum (28) was first marketed in 1978 (55). Both difenacoum and brodifacoum are now sold by ZENECA (formerly part of the ICI Group) and, mainly in the United Kingdom, by Sorex.

R=H   (27)

R=Br   (28)

Another anticoagulant, bromadiolone (29), was introduced at about the same time as brodifacoum, although it had actually been patented as a rodenticide, by Lipha, in 1967 [56–58]. Two further second-generation anticoagulants have subsequently been discovered: flocoumafen (30), which was introduced by Shell in 1986 [59], and Lipha's difethialone (31). Not only are all these second-generation an-

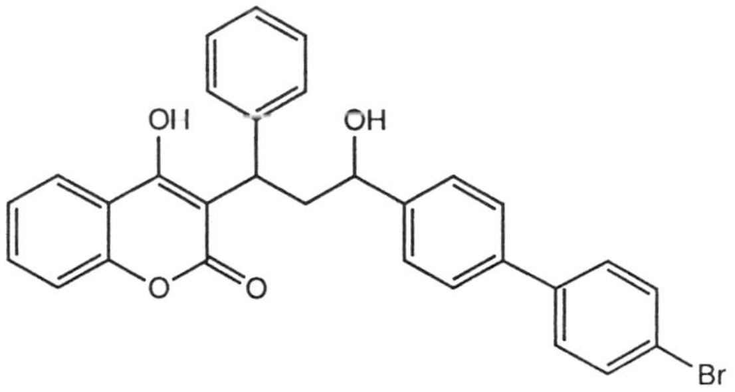

(29)

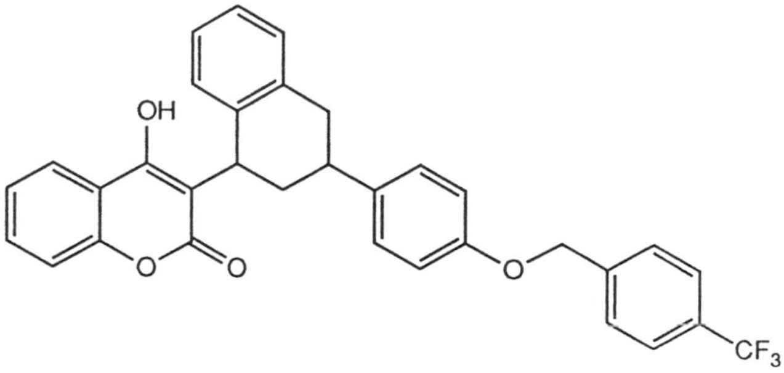

(30)

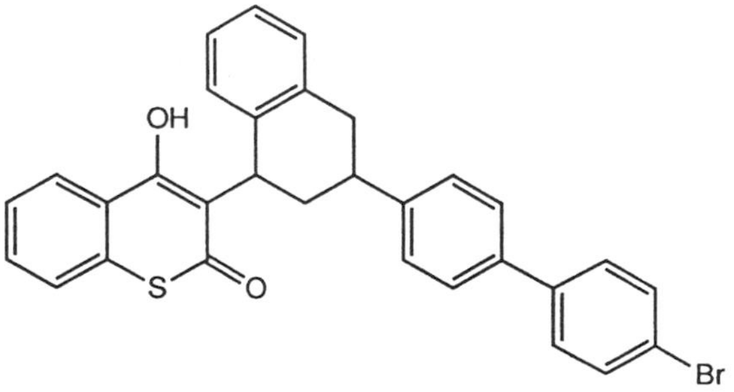

(31)

ticoagulants active against warfarin-resistant species, but they also possess higher intrinsic activity against fully susceptible rodents.

2. Rodenticidal Efficacy of the Anticoagulants

A large number of studies have been carried out on the toxicity of rodenticides, using many different species of rats and mice, and it is often difficult to deduce from these the relative toxicities of the major anticoagulants. The most straightforward comparisons of the toxicities of different anticoagulant rodenticides and of the toxicity of one anticoagulant against different species can be made from measurements of the acute oral rat toxicity. There is, however, considerable variation in the quoted LD_{50} values even for a given species, as this value also depends on factors such as the strain and health of the experimental animal and its diet.

Laboratory measurements of the LD_{50} (mg/kg) have been determined on *R. norvegicus* and *M. musculus* for most of the important anticoagulant rodenticides and these are summarized in Table 2 [60, 61]. These values of LD_{50} illustrate that, in general, the second-generation rodenticides, brodifacoum, difenacoum, bromadiolone, flocoumafen, and difethialone, are intrinsically more potent than the first-generation products. In practice this means that a rodent can ingest a lethal dose from a single application of either of the two more powerful second-generation compounds, such as brodifacoum or flocoumafen: these are, therefore, known as "Single Feed Anticoagulants." In contrast, five daily intakes of warfarin are typically required to achieve a lethal effect, and if the sequence is interrupted, the affected animal rapidly excretes the absorbed dose and is unharmed.

The higher potency of the more recent anticoagulants has also enabled the bait to be effective with a much lower concentration of the active ingredient. Typical bait concentrations are listed in Table 2.

3. *Mode of Action*

The anticoagulant rodenticides act as indirect antagonists of vitamin K_1 during blood clotting.

Table 2 Acute Oral LD_{50} of Active Ingredients of Some Anticoagulants Against *R. norvegicus* and *M. musculus*

Compound	Concentration a.i. in bait (ppm)	LD_{50} (mg/kg) *R. norvegicus*[a]	*M. musculus*[a]
Warfarin	250	186 [61]	374 [62]
Coumafuryl	250–500	400 [63]	N/A
Coumatetralyl	375	16.5 [64]	N/A
Coumachlor	300	900–1200 [1]	900–1200 [1]
Diphacinone	50	2.3 [65]	141 [61]
Chlorophacinone	50	20 [64]	N/A
Brodifacoum	50	0.26 [64]	0.4 [65]
Difenacoum	50	1.8 [65]	0.8 [65]
Bromadiolone	50	1.12 [65]	1.75 [65]
Flocoumafen	50	0.25 [64]	0.8–2.4 [64]
Difethialone	25	0.56 [65]	1.29 [65]

[a]Reference numbers are given in brackets.

Vitamin K_1 (32) (hereafter simply vitamin K) is essential for normal blood coagulation as a cofactor in the postribosomal synthesis of clotting factors II, VII, IX, and X [66]. During the vitamin K-dependent step, γ-carboxylation of glutamic residues in the clotting factors takes place. The reaction, catalyzed by a vitamin K-dependent carboxylase, takes place in the rough endoplasmic reticulum (the microsomal fraction) in the cell and requires vitamin K in its reduced, hydroquinone, form (33), molecular oxygen, carbon dioxide, and a peptide-bound glutamyl substrate (Fig. 1).

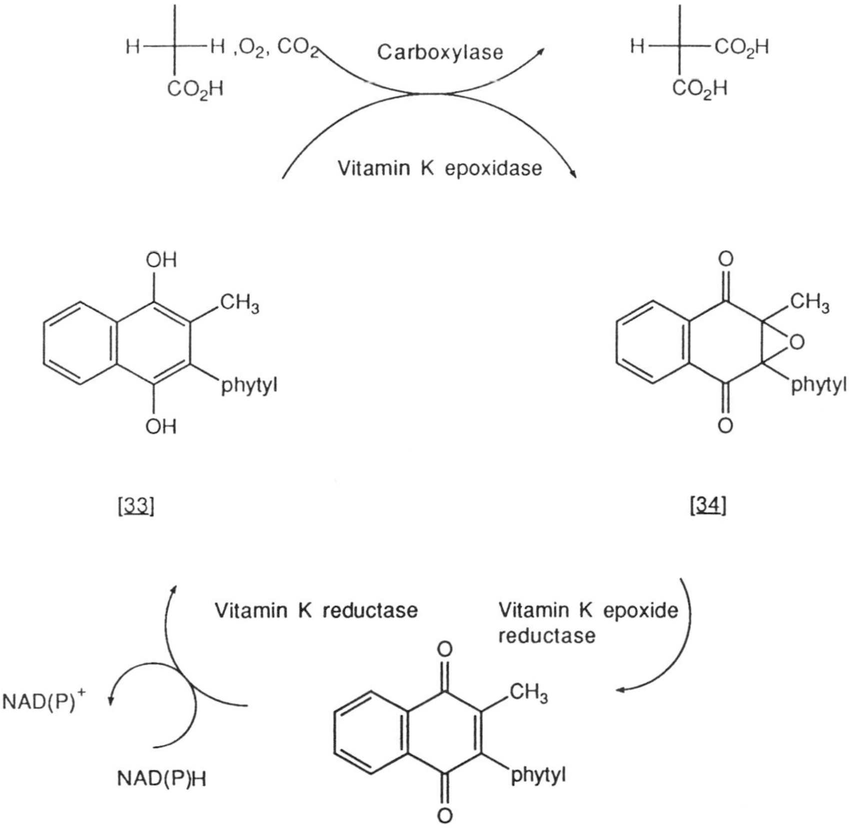

Figure 1 The mechanism of action of vitamin K_1.

During the reaction, vitamin K is converted by the enzyme vitamin K epoxidase into vitamin K 2,3-epoxide (34), which is inactive [67]. There is evidence that the carboxylation and epoxidation reactions are coupled [68, 69], although the mechanism of this coupling is not yet known.

Continued synthesis of clotting factors requires the regeneration of vitamin K hydroquinone (33) from the inactive epoxide (34). This takes place in two stages: The epoxide is first converted to the quinone form of the vitamin (32) by the enzyme vitamin K epoxide reductase, and then further reduced to the hydroquinone form (33). This second reduction can be carried out by two types of enzyme system: a vitamin K reductase or microsomal pyridine nucleotide-linked dehydrogenases.

Vitamin K reductase, in common with the epoxide reductase, can be assayed using dithiothreitol (DTT) as the reductant (and these enzymes are, therefore, described as DTT dependent). It is possible that epoxide reduction and quinone reduction occur at the same enzymic site [67, 70]; for example, there is evidence for the existence of a reducible disulphide at the active sites of both enzymes, because either substrate will protect the enzymes against sulphhydryl modification by N-ethylmaleimide (NEM) [71]. Mechanisms for the reductions have been proposed by Preusch [67] and Silverman [72], based on the participation of an active-site disulphide.

The alternative method of quinone/hydroquinone reduction is by pyridine nucleotide-linked dehydrogenases, one of which appears to be identical with the cytosolic enzyme DT-diaphorase [73, 74]. Although DT-diaphorase can reduce the quinone to hydroquinone *in vitro* [75], it is unlikely that the same enzyme could also carry out the epoxide reduction. The relative importance of the two reduction pathways is a subject of continuing debate.

Current evidence suggests that the coumarin (and indanedione) anticoagulants inhibit both the epoxide reductase and the DTT-dependent quinone reductase.

Early studies showed that administration of these compounds increased the ratio of vitamin K epoxide to vitamin K [76], and it was then demonstrated that the microsomal epoxide reductase was inhibited by these compounds [77]. The discovery of resistance to warfarin facilitated study of the mode of action of the coumarins: comparison of the *in vitro* effects of warfarin and difenacoum on rat liver microsomal vitamin K-dependent carboxylase, vitamin K epoxidase, vitamin K epoxide reductase, and cytosolic DT-diaphorase from warfarin-resistant and warfarin-sensitive rats showed that the epoxide reductase was the most sensitive to coumarin inhibition [78], and furthermore, that of those studied this enzyme was the only one in the warfarin-resistant animals to show differential sensitivity to warfarin and difenacoum. In this study, the DTT-dependent quinone reductase was not assayed.

By comparing the effect of warfarin on the epoxide reductase and quinone reductase, it was shown that the quinone reductase is also sensitive to coumarin anticoagulants [79] and that more warfarin (threefold to fourfold) was required to produce 50% inhibition of this enzyme from resistant rats than that from sensitive rats [80]. Thus, it is likely that the mode of action of the coumarin and indanedione anticoagulants involves inhibition of both the epoxide and DTT-dependent quinone reductases.

The coumarin (and indanedione) anticoagulants are also inhibitors of DT-diaphorase, although it appears that this is not important physiologically [78]; the inhibitor binding sites of the two enzymes may share some common features.

Although the mechanism of inhibition by the coumarin anticoagulants has not been established, two hypotheses have been proposed: Preusch [67] has suggested, on the basis of the structural similarity between the 4-hydroxycoumarins and a postulated enzyme-bound hydroxyvitamin K intermediate (35), that these compounds act as analogs of this transition state. In contrast, Silverman [72,81], based on his proposed mechanism, suggested that the 4-hydroxycoumarins act as suicide inhibitors of the enzyme by the mechanism outlined in Figure 2. It is also possible that the site of coumarin action is not at the active site, in which case the structural similarity may be coincidental [66].

$$(35)$$

4. *Resistance to Anticoagulant Rodenticides*

a. REPORTS OF RESISTANCE TO ANTICOAGULANTS In the United Kingdom, resistance to warfarins and to diphacinone in brown rats, *R. norvegicus*, was first noticed in 1958 [82]; in *M. musculus*, the first cases were detected in 1961. By the mid-1960s, many brown rat and house mouse populations had become resistant to warfarin. Populations

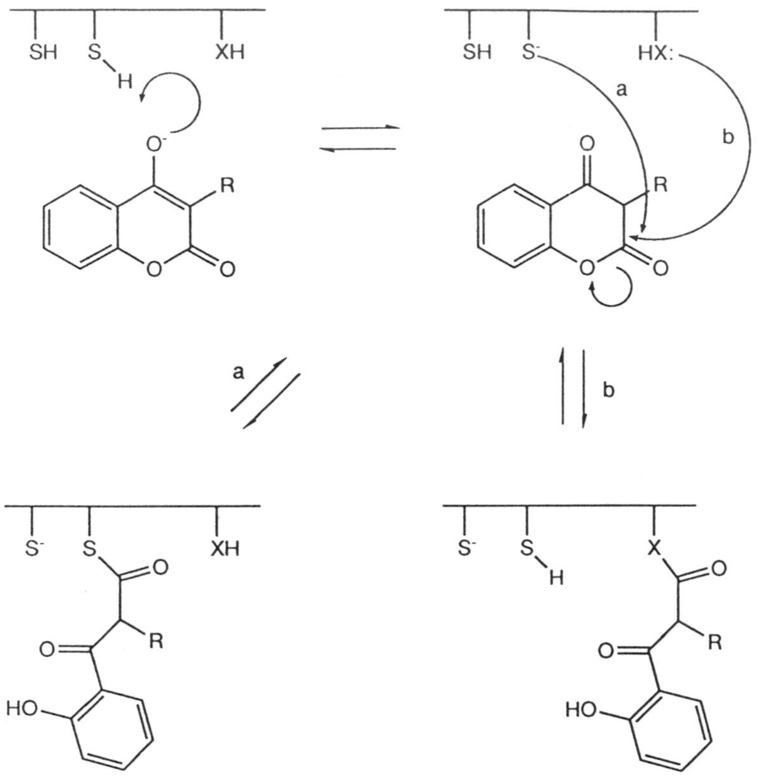

Figure 2 The mechanism of inhibition of epoxide reductase by coumarin anticoagulants proposed by Silverman [72].

of warfarin resistant black rats, *R. rattus*, a relatively rare pest in the United Kingdom, have also been found. Resistance to the first-generation anticoagulants has now been reported to occur in many other countries [83, 84]. In view of their chemical similarity and common mode of action, it is not surprising that resistant populations show cross-resistance with other first-generation compounds.

In contrast, resistance to the second-generation anticoagulants does not appear to be widespread. Although the first reports of the failure of difenacoum adequately to control brown rat populations in some areas appeared within a year of its introduction [85], resistance is still limited to a few discrete regions, and resistance to brodifacoum and flocoumafen has not been established unequivocally; in some reported cases of "resistance" [86], other factors may be the cause of the poor levels of control [87]. The exception is bromadiolone, to which resistance in *M. musculus* is now widespread [88].

The reasons for the ability of these second-generation compounds to retain their activity on warfarin-resistant rodents, despite their common mode of action, has not been clarified.

b. BIOCHEMICAL MECHANISM OF RESISTANCE It was established in a number of early studies that the metabolism of warfarin or vitamin K are apparently not altered in resistant rodents [89–91]. Work by Zimmerman and Matschiner in 1972 [92] demonstrated that the vitamin K epoxide reductase of warfarin-resistant rats is much less susceptible to inhibition *in vitro* than is the reductase of warfarin-susceptible animals. On the basis of these and other results [78, 93], it was proposed that resistance is due to a mutation that alters the vitamin K epoxide reductase, rendering it less susceptible to inhibition by warfarin. The discovery that, in addition to the epoxide reductase, the quinone reductase from resistant rats is less sensitive to warfarin than that from sensitive rats suggests that genetic alteration of this enzyme may also be involved in the mechanism of resistance [80].

The effectiveness of the second-generation anticoagulants against warfarin-resistant rats is reflected in their retained activity on the epoxide reductase; the I_{50} value is only slightly higher in the resistant animals [78]. Thus, it appears that the second-generation compounds display very similar affinities for the epoxide reductases from resistant and sensitive rats. Further rationalization of the differences in behavior of the first- and second-generation anticoagulants is not possible without detailed knowledge of the molecular interactions of these compounds with the enzyme(s).

5. *Environmental Safety*

All of the second-generation anticoagulants, and some of the earlier compounds, are persistent in animal tissues [61, 94]. The most important potential consequence of the persistence of anticoagulant rodenticides might be the secondary effect on nontarget animals, i.e., predators feeding on rodenticide-poisoned rodents. However, practical experience indicates that incidents of secondary poisoning with rodenticides are, in fact, rare and that the danger of secondary poisoning has often been exaggerated on the basis of laboratory studies carried out under conditions in which the test animals were given no choice but to eat the poisoned rodents. Although many of the anticoagulants have been used extensively over recent years, there is no evidence that the occasional incident of secondary poisoning has had any effect on the populations of the nontarget animals, any losses having been remedied by breeding [95, 96].

B. Analogs of Vitamin K as Anticoagulant Rodenticides

The synthesis of the 2-chloro analog of vitamin K, chloro-K (36), as a vitamin K antagonist, was reported in 1970 [97]. It was then noted [98] that the mechanism by which chloro-K inhibited blood coagulation differed from that shown by warfarin and 2-phenyl-1,3-indanedione, and that chloro-K was effective against warfarin-resistant rodents.

(36)

The mode of action of chloro-K was identified as inhibition of vitamin K epoxidase [99]; *in vitro* assays showed that 3 µM chloro-K inhibited the epoxidase by about 75%. It was also shown in this study that this inhibition could be reversed *in vivo* by vitamin K. Further experiments on rabbits showed that chloro-K decreased the vitamin K epoxide : vitamin K ratio in plasma from 0.44 to 0.13, consistent with the hypothesis that chloro-K inhibits the epoxidase [100].

Direct inhibition of vitamin K-dependent carboxylation by chloro-K has also been demonstrated [101]; analysis of the inhibition of carboxylation over a range of antagonist concentrations gave an I_{50} value of 285 µM. Direct inhibition of carboxylation would be expected because the carboxylase and epoxidase activities are closely coupled and, may, in fact, both be catalyzed by the same enzyme [102, 103].

The use of chloro-K as a rodenticide was patented by the Wisconsin Alumni Research Foundation in 1977 [104]; the patent described a composition consisting of chloro-K and warfarin which would be effective against warfarin-resistant rats.

In 1987, the rodenticidal properties of the related 2-halo-1,4-naphthoquinones were disclosed, and compositions containing them were patented as rodenticides [105]. 2-Chloro-1,4-naphthoquinone was quoted as the preferred compound; the rodenticidal activity of compositions containing 0.005–0.05% active ingredient was claimed to be greater than that of a composition containing warfarin and chloro-K. However, none of these compounds has been used commercially.

REFERENCES

1. A.P. Meehan, *Rats and Mice*, Rentokil Ltd., East Grinstead, 1984.
2. J.T. Marshall, *The Mouse in Biomedical Research*, (H.L. Foster, J.D. Small, and J.G. Fox, eds.), Vol. 1, Academic Press, New York, 1981, pp. 17–26.
3. J.R. Ellerman, *The Families and Genera of Living Rodents*, Vol. 2, British Museum (Natural History), London, 1941.
4. J.H. Greaves, *EPPO Bull.*, *18*:203 (1988).
5. A.P. Buckle and R.H. Smith, *Rodent Pests and Their Control*, CAB International, Wallingford, 1994.
6. L.J. Bruce-Chwatt, *Pestic. Sci.*, *3*:467 (1972).
7. J.E. Brooks, *CRC Crit. Rev. Environ. Control*, *3*:405 (1973).
8. E.W. Bentley and J.H. Greaves, *J. Hyg.*, *58*:125 (1960).
9. J.E. Brooks, *Int. Pest Control*, *5*:21 (1963).
10. E.W. Bentley, L.E. Hammond, A.H. Bathard, and J.H. Greaves, *J. Hyg.*, *59*:413 (1961).
11. K.P. DuBois, K.W. Cochran, and J.F. Thomson, *Proc. Soc. Exp. Biol. Med.*, *67*:169 (1948).
12. D.R. Maddock and H.F. Schoof, *Pest Control*, *35*:22 (1967).
13. B.D. Rennison, L.E. Hammond, and G.L. Jones, *J. Hyg.*, *66*:147 (1968).
14. H.F. Schoof, *Pest Control*, *38*:38 (1970).
15. M. Balasubramanyam and K.R. Purushotham, *Indian Rev. Life Sci.*, *8*:205 (1988).
16. B.D. Rennison and D.C. Drummond, *Environ. Health*, *92*:287 (1985).
17. D.G. Crabtree, *Econ. Bot.*, *1*:394 (1947).
18. R.E. Marsh and W.E. Howard, *Proc. Third Int. Biodegrad. Symp.* (J.M. Sharpley and A.M. Kaplan, eds.), Applied Science, London, 1975, pp. 317–329.
19. A. von Wartburg and J. Renz, *Helv. Chim. Acta*, *42*: 1620 (1959).
20. A. von Wartburg, A. Stoll, and J. Renz, *Helv. Chim. Acta*, *49*:30 (1966).
21. A.J. Verbascar, T.F. Banigan, and H.S. Gentry, *Proc. 12th Verteb. Pest Conf.* (1986), p. 51
22. A.J. Verbascar, J. Patel, T.F. Banigan, and R.A. Schatz, *J. Agric. Food Chem.*, *34*:973 (1986).
23. E. Rothlin and W.R. Schalch, *Helv. Physiol. Acta*, *10*:427 (1952).
24. A. Stoll and J. Renz, *Helv. Chim. Acta*, *25*:43 (1942).
25. D. Chitty, "Control of Rats and Mice," Oxford University Press, London, 1954, pp. 62–100.
26. A.J. Verbascar, T.F. Banigan, and R.A. Schatz, *J. Agric. Food Chem.*, *35*:365 (1987).
27. A.P. Meehan, *Pest. Sci.*, *11*:555 (1980).
28. R.B. Woodward, F.E. Bader, H. Bickel, A.J. Frey, and R.W. Kierstead, *Tetrahedron*, *2*:1 (1958).
29. J.J. Lewis, *Physiological Pharmacology*, (W.S. Root and F.G. Hofmann, eds.), Vol. 1, Academic Press, New York, 1963, pp. 479–536.
30. A.P. Meehan, *Pest. Sci.*, *11*:562 (1980).

31. J.O. Bull, Rentokil Ltd., Patent GB 1364418, priority date 21 January 1971.
32. A.P. Meehan, Rentokil Ltd., Patent GB 1511924, priority date 9 February 1976.
33. A.F. Pelfrene, *Handbook of Pesticide Toxicology*, (W.J. Hayes and E.R. Laws, eds.), Vol. 3, Academic Press, New York, 1991, pp. 1271–1316.
34. J.H. Greaves, R. Redfern, and R.E. King, *J. Hyg*, 73:341 (1974).
35. E.F. Marshall, *Proc. 11th Verteb. Pest Conf.* (1984), p. 95.
36. L.D. Brown and E.F. Marshall, *Proc. 13th Verteb. Pest Conf.* (1988), p. 70.
37. Anon, "A New Rodenticide", Ministry of Agriculture, Fisheries, and Food, Technical Circular No. 30, 1973.
38. F.P. Rowe, F.J. Smith, and T.J. Swinney, *J. Hyg*, 73:353 (1974).
39. E.R. Yendt, *International Encyclopedia of Pharmacology and Therapeutics*, Vol. 1, Pergamon Press, Oxford, 1970, Sect. 51, pp. 139–195.
40. I.W. Southon and J. Buckingham (eds.), *Dictionary of Alkaloids* Chapman and Hall, New York, 1989, pp. 1006–1007.
41. R. Robinson, *Experientia*, 2: 28 (1946).
42. R.B. Woodward, M.P. Cava, W.D. Ollis, A. Hunger, H.U. Daeniker, and K. Schenker, *Tetrahedron*, 19:247 (1963).
43. P. Magnus, M. Giles, R. Bonnert, G. Johnson, L. McQuire, M. Deluca, A. Merritt, C.S. Kim, and N. Vicker, *J. Am. Chem. Soc*, 15:8116 (1993), and references therein.
44. J. Mann, *Secondary Metabolism*, 2nd ed., Oxford University Press, Oxford, 1987.
45. S.H. Diecke and C.P. Richter, *Public Health Rep*, 61:672 (1946).
46. N.J.A. Gutteridge, *Chem. Soc. Rev*, 1:381 (1972).
47. B.A. Colvin, P.L. Hedgal, and W.B. Jackson, *EPPO Bull*, 18:301 (1988).
48. W.W. Jacobs, *Proc. 14th Verteb. Pest Conf.* (1990), p. 36.
49. "The Biologic and Economic Assessment of Strychnine and Strychnine Sulfate 1080/1081", U.S. Dept. Agric. Tech. Bull., No. 1626 (1980).
50. L.M. Roderick, *J. Am. Vet. Med. Assoc*, 74:314 (1929).
51. K.P. Link, *The Harvey Lectures*, 39:162 (1944).
52. K.P. Link, *Circulation*, 19:95 (1959).
53. E.M. Mills, *Pest Control*, 23:14 (1955).
54. M.R. Hadler, R. Redfern, and F.P. Rowe, *J. Hyg*, 74:441 (1975).
55. A.C. Dubock and D.E. Kaukeinen, *Proc. 8th Verteb. Pest Conf.* (1978), p. 127.
56. R.E. Marsh, *EPPO Bull*, 7:495 (1977).
57. A.P. Meehan, *Proc. 8th Verteb. Pest Conf*, (1978), p. 122.
58. R. Redfern and J.E. Gill, *J. Hyg*, 84:263 (1980).
59. R.A. Johnson and R.M. Scott, *Proc. 7th Brit. Pest Control Conf.* (1986).
60. A.C. Dubock, *Proc. 5th Br. Vert. Pest Control Conf.* (1979).
61. M.R. Hadler and A.P. Buckle, *Proc. 15th Verteb. Pest Conf.* (1992), p. 149.
62. E.C. Hagen and J.L. Radomoski, *J. Am. Pharm. Assoc*, 42:379 (1953).
63. W.J. Wiswesser, *Pesticide Index*, 5th ed., Entomology Society of America, College Park, MD, 1976.

64. B. Garforth and R.A. Johnson, *Stored Products Pest Control*, British Crop Protection Council, Farnham, 1987, p. 115.
65. C.R. Worthington (ed.), *Pesticide Manual*, 9th ed., British Crop Protection Council, Farnham, 1991.
66. J.W. Suttie, *Annu. Rev. Biochem.*, 54:459 (1985).
67. P.C. Preusch and D.M. Smalley, *Free Radical Research Communications*, 8:401 (1990).
68. J.W. Suttie, A.E. Larson, L.M. Canfield, and T.L. Carlisle, *Fed. Proc., Fed. Am. Soc. Exp. Biol.*, 37:2605 (1978).
69. A.E. Larson, P.A. Friedman, and J.W. Suttie, *J. Biol. Chem.*, 256:11032 (1981).
70. P.C. Preusch and J.W. Suttie, *Biochem. Biophys. Acta*, 798:141 (1984).
71. J.J. Lee and M.J. Fasco, *Biochemistry*, 23:2246 (1984).
72. R.B. Silverman, *J. Am. Chem. Soc.*, 102:5421 (1980).
73. R. Wallin, O. Gebhardt, and H. Prydz, *Biochem. J.*, 169:95 (1978).
74. R. Wallin and J. W. Suttie, *Biochem. J.*, 194:983 (1981).
75. C. Martius, R. Ganser, and A. Viviani, *FEBS Lett.*, 59:13 (1975).
76. R.G. Bell, J.A. Sadowski, and J.T. Matschiner, *Biochemistry*, 11:1959 (1972).
77. A.K. Willingham and J.T. Matschiner, *Biochem. J.*, 140:435 (1974).
78. E.F. Hildebrandt and J.W. Suttie, *Biochemistry*, 21:2406 (1982).
79. M.J. Fasco and L.M. Principe, *J. Biol. Chem.*, 257:4894 (1982).
80. M.J. Fasco, E.F. Hildebrandt, and J.W. Suttie, *J. Biol. Chem.*, 257:11210 (1982).
81. R.B. Silverman and D.L. Nandi, *Biochem. Biophys. Res. Commun.*, 155:1248 (1988).
82. C.M. Boyle, *Nature*, 188:517 (1960).
83. A.D. MacNicoll, *EPPO Bull.*, 18:223 (1988).
84. M. Lund, *Proc. 11th Verteb. Pest Conf.*, (1984), p. 89.
85. R. Redfern and J.E. Gill, *J. Hyg.*, 81:427 (1978).
86. J.H. Greaves, D.S. Shepherd, and R. Quy, *J. Hyg.*, 89:295 (1982).
87. R.J. Quy, D.S. Shepherd, and I.R. Inglis, *Crop Protection*, 11:14 (1992).
88. F.P. Rowe, C.J. Plant, and A. Bradfield, *J. Hyg.*, 87:171 (1981).
89. J.G. Pool, R.A. O'Reilly, and L. J. Schneiderman, *Am. J. Physiol.*, 215:627 (1968).
90. M.A. Hermodson, J.W. Suttie, and K.P. Link, *Am. J. Physiol.*, 217:1316 (1969).
91. M.G. Townsend, E.M. Odam, and J.M. Page, *J. Biochem. Pharmacol.*, 24:729 (1975).
92. A. Zimmerman and J.T. Matschiner, *Fed. Proc., Fed. Am. Soc. Exp. Biol.*, 31:714 (1972).
93. R.G. Bell and P.T. Caldwell, *Biochemistry*, 12:1759 (1973).
94. G. Parmar, H. Bratt, R. Moore, and P.L. Batten, *Human Toxicol.*, 6:431 (1987).
95. D.E. Kaukeinen, *Proc. 10th Verteb. Pest Conf.*, (1982), p. 151.
96. S.M. Percival, *Br. Wildlife*, 2:131 (1990).
97. J. Lowenthal and M.N.R. Chowdbury, *Can. J. Chem.*, 48:3957 (1970).

98. P. Ren, R.E. Laliberte, and R.G. Bell, *Mol. Pharmacol,* 10:373 (1974).

99. A.K. Willingham, R.E. Laliberte, R.G. Bell, and J.T. Matschiner, *Biochem. Pharmacol,* 25:1063 (1976).

100. A.M. Breckenridge, J.B. Leck, B.K. Park, M.J. Serlin, and A. Wilson, *Br. J. Pharmacol,* 64:399P (1978).

101. D.M. Cocchetto and T.D. Bjornsson, *Haemostasis,* 16:321 (1986).

102. J.J. McTigue and J.W. Suttie, *FEBS Lett,* 200:71 (1986).

103. R. Wallin and J.W. Suttie, *Arch. Biochem. Biophys.* 214:155 (1982).

104. Wisconsin Alumni Res. Foundation, U.S. Patent 4021568 (14 May 1973).

105. T. Bjornsson and D.M. Cocchetto, Duke University, Patent EP 244309 (23 April 1986).

9

Biological Control Agents

Jane Louise Faull

*Birkbeck College, University of London,
London, England*

Keith A. Powell

*Zeneca Agrochemicals, Bracknell,
Berkshire, England*

I. INTRODUCTION

This chapter will not be an extensive review of biological control agents (BCAs) but will attempt to outline the current practical uses of microbial biological control agents and to examine in detail those where the mode of action is at least partially connected to the production of natural products. In the area of plant pathology alone, there have been more than 500 publications on biological control in the last 5 years. We, therefore, will restrict our observations to those areas where products have been launched or where there is some indication of commercial activity, in order to avoid a phenomenological collection of microbes.

Pressures are increasingly being applied to reduce the usage of chemical pesticides and instead to consider alternatives including the use of beneficial microbes to control plant pests, diseases, and weeds. However, in reality, there are very few products that are commercially available in this sphere. The market share for these products is small [1]; in 1991 the only significant commercial biological control agents (BCAs) were bioinsecticides and their sales were estimated at $120 million, less than 1% of the total crop protection market [2]. Sales of *Bacillus thuringiensis* (BT) accounted for 92% of the biopesticide market, and as BT is really a protein toxin rather than a living organism, its description as a biological control agent is not strictly accurate [2].

Optimistic projections of markets for BCAs estimate that sales will have doubled by 1995 but that bioinsecticides (predominantly BT) will still account for over 80% of these sales [3]. This is still far short of the earlier market projections [4].

The reasons behind this shortfall are complex and are partly due to an imperfect understanding of the mechanisms by which these microbes have their effects [5]. However, this situation is currently being remedied as mode-of-action studies clarify both the way in which biological control agents work and the external (or environmental) influences that significantly affect them.

Interactions between target species and microbes generally fall into three categories: competition, mycoparasitism, and antibiosis. Competition as a mechanism for biological control has been reviewed recently [6] as has mycoparasitism [7–9], and, therefore, this review will restrict itself to a consideration of those BCAs known to have modes of action that include the production and effects of antibiotics.

Commercial success of BCAs for plant disease control is very limited. Total sales of these agents amounted to only $1 million per annum in 1992. Those agents that have achieved significant success in commercial terms are restricted to a few genera of bacteria and fungi, and their probable modes of action are well documented.

II. BACTERIAL AGENTS

A. Agrobacterium radiobacter

By far the most successful commercial BCA for plant diseases is *Agrobacterium radiobacter* K84. This strain is currently being replaced by K1026. The bacterium controls Crown Gall disease of stone fruits, nut trees, and ornamentals [10] caused by the bacterium *Agrobacterium tumefasciens*. The disease symptoms of tumor formation on wound sites are mediated by the transfer of a tumor-inducing plasmid (Ti) from *A. tumefasciens* into the cells of the host plant. The plasmid codes for a number of products when integrated into the plant genome confers autonomy of plant hormone production of both auxin [11, 12] and cytokinin [13]. This, in turn, deregulates cell division and expansion, giving rise to tumors. Other plasmid-coded products include the opines, unique amino acids that provide the vector bacteria with substrates on which to grow.

Agrobacterium radiobacter is taxonomically very closely related to *A. tumefasciens*, differing only in the type of plasmid that it contains. *A. radiobacter* pAgK84, is a 47 kb conjugative plasmid [14] that codes

$$(1)$$

for agrocin 84 (1), an antibiotic that is an analog of the opine agrocinopine. Agrocin 84 competitively binds to sites normally used for the transport of opines into *A. tumefasciens* cells [15]. Pathogenic strains are thus starved of substrates by the irreversible binding of agrocin K84, whereas nonpathogenic strains are unaffected. In this way, *A. radiobacter* is able to effectively compete with and antagonize with *A. tumefasciens* and act as a BCA [10]. Furthermore, as agrocin K84 is unstable, accumulation of the product in the environment, and therefore selection for resistance to the antibiotic, is considered unlikely to occur [10].

However, recently there have been reports of the failure of this biological control system due to the appearance of strains of the pathogens that are resistant to agrocin 84 [16]. This resistance can be acquired by mutation or by genetic dissemination. Mutational resistance is associated with the Ti plasmid which in some cases may be lost or contain a large deletion that is accompanied by an inability to utilize agrocinopine. Some of these mutants become avirulent, but a significant number retain tumorigenic and these retain a full-sized Ti plasmid. Resistance occurs because of point or small deletion in *acc*, the locus for agrocinopine catabolism and agrocin 84 sensitivity [17]. As the *vir* region maps quite close to *acc*, large mutations in *acc* can extend into the *vir* region and thereby cause avirulence [17].

Genetic dissemination has also been shown to occur. pAgK84 is a self-conjugal plasmid that also confers immunity to the antibiotic [18, 19]. It has become apparent that transfer of pAgK84 has occurred at some geographical sites, conferring resistance to agrocin 84 of the pathogen. A deletion mutant has been constructed removing the *tra* region from pAgK84, and the new strain of *A. radiobacter*, K1026, shows no conjugative transfer of the plasmid and controls Crown Gall [20].

A limitation to the use of *Agrobacterium radiobacter* is its high degree of specificity. It is unable to control Crown Gall on vine, due partly to the pathovar of the bacterium and partly to the internal infections by *Agrobacterium tumefasciens* of the vascular system of the plant [10].

There is also some evidence that the success of *A. radiobacter* may not solely be due to the activity of agrocin 84. Studies by Moore [21], Moore and Warren [22], and Cooksey and Moore [23] have shown that *A. radiobacter* K84 can successfully control pathogenic strains resistant to the antibiotic in greenhouse experiments, although they conceed that control was not as effective as when antibiotic-sensitive strains were used. Agrocin 84 production has also been transferred to other strains of *Agrobacterium*, but these strains failed to give good control of Crown Gall [19, 24]. In these experiments, root colonization and binding to infection sites by the successful antagonist was shown to have an important role to play in disease control. The preinoculation of K84 on to infection sites has been shown to successfully exclude pathogenic strains [25].

B. Pseudomonas

There are many reports in the literature that strains of *Pseudomonas* will suppress infection of plants by soil-borne plant pathogens [26–28] and many other reports that they can promote plant growth [29, 30]. At least two strains of *Pseudomonas* are, or were, available as commercial products, DAGGER G (Ecogen Inc., Langhorne, PA) for the control of seedling blight of cotton and *Pseudomonas fluorescens* for the control of *P. toloassii*, mushroom blotch. The modes of action of *Pseudomonads* have been ascribed to a number of natural products that they produce. These include siderophores, antibiotics, and the production of plant growth regulators.

1. *Siderophores*

Siderophores are low-molecular-weight compounds produced by microbes under iron-limiting conditions [31]. These microbial ion transport compounds are capable of chelating ferric iron (Fe^{3+}) with high affinity, and they act as transport systems for the movement of Fe^{3+} into microbial cells [32]. Although their structures vary greatly, they fall into two main groups, those with hydroxamate and those with catechol groups that are involved in the iron chelation.

Evidence supporting the importance of siderophores in competition includes the poor performance of siderophore-minus TN_5 mutants of pseudomonads as biological control agents [33–35], the inhibition of the biological control effect of pseudomonads by the application of dissolved ferric iron to soils [36], and the inhibition of pathogen growth by purified siderophore under low-iron conditions [37, 38].

Pyoverdine siderophores, such as (2) produced by different strains of *Pseudomonas*, show structural differences in the peptide chain [39, 40] that may correspond to the differences in siderophore receptor sites on various strains of *Pseudomonas fluorescens* [41]. Specificity of siderophores for membrane receptors is the basis for their biological control activity [42]. Competition for iron and iron chelate appears to occur between both different strains of pseudomonad [43] and between pseudomonads and other rhizosphere bacteria and fungi. Such competition in an iron-depleted habitat would depend on the differing affinities for iron that microbial siderophores possess, and on their ability either to utilize ferric siderophores directly or to obtain iron by ligand exchange. When siderophores chelate iron they lose protons, trihydroxamate siderophores losing 3, and tri-catechols losing 6. Pyoverdine, with three bidentate chelating groups, a catecholate, a hydroxamate and either an α-hydroxy acid or another hydroxamate, is intermediate in proton-loss characteristics. The ratio among free iron, protonated siderophore, and the iron-siderophore complex will, therefore, be influenced by pH, and ligand exchange among siderophores at any given pH will be influenced by their relative equilibrium constants for iron, their relative concentrations and the concentration of iron, and the kinetics of both the initial chelation reaction and that of the ligand exchange.

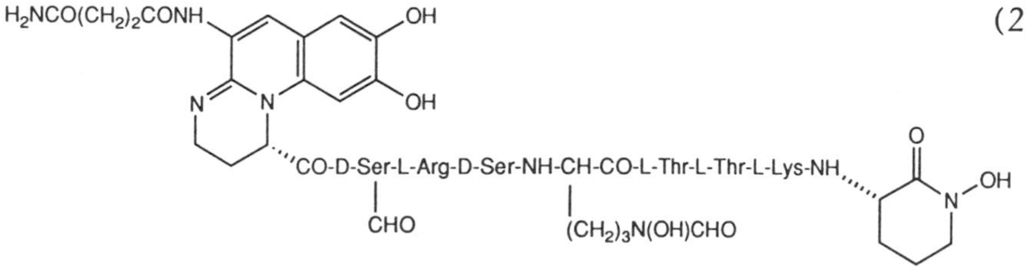

(2)

The iron complex of pyoverdine from *P. fluorescens* has a stability constant of 10^{32} [44] and the hydroxamate siderophore of *P. cepacia*, cepabactin (3), a stability constant around 10^{27} [45], and the expected result of competition for iron by these two siderophores would be that the pyoverdine compound would bind the iron [41, 42]. In experimental

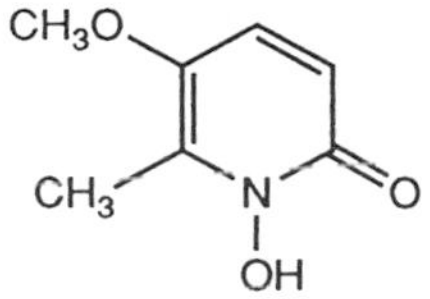

(3)

systems designed to test the effects of complexed and noncomplexed siderophores Kloepper et al. [38] found that pseudobactin would inhibit growth of *Fusarium oxysporum f.* sp. *lini* and *Gaeumannomyces graminis* var. *tritici* on agar plates, whereas ferric pseudobactin did not.

Plants may also obtain their iron from these microbial siderophores by ligand exchange or by direct uptake of ferric siderophores. However, microbial siderophores have greater iron-chelate stability constants than phytosiderophores, and it is possible that under certain conditions ferric pyoverdines are able to inhibit iron uptake by plants [46].

The importance of siderophore production and iron competition in all systems has been further challenged by the findings of Keel and Defago [47] using a wheat *Pythium ultimum* or *Gaeumannomyces graminis* var. *tritici* system. They found that strains of *P. fluorescens* either negative or positive for pyoverdine production did not differ in their influence on plant growth, colonization of roots or their biocontrol ability. It appears that siderophores are important in some systems and not in others. Furthermore, as many fungal hydroxamate siderophores (see Fig. 1) have stability constants between 10^{22} and 10^{32}, similar to

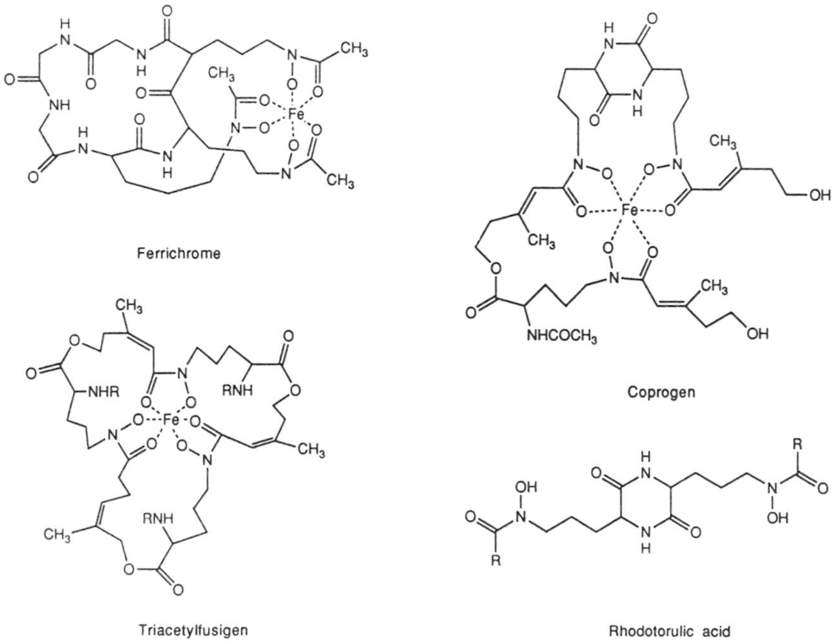

Figure 1 Some representative hydroxamate siderophores.

those of bacterial siderophores, it is difficult to imagine how iron sequestration could be a mode of antagonism [48].

2. *Antibiotics*

In 1979 when Leisinger and Magraff [49] reviewed the secondary products of the fluorescent pseudomonads they listed more than 78 different compounds belonging to six major chemical classes. Many of the antibiotics isolated from pseudomonads are nitrogen-containing heterocycles that are either intermediates or end products of aromatic amino acid biosynthesis. Unusual amino acids and peptides form a second major group. Since 1979 many new compounds from pseudomonads have been described and much is known about the regulation of their production *in vitro* [50, 51]. However, detailed evidence for their role in biocontrol in soil has been hard to find.

The strongest evidence that antibiotics do have a role in biological control by *Pseudomonads* has come from studies with *Pythium ultimum* and *Gaeumannomyces graminis* var. *tritici,* where biotechnological approaches have been used. Control of these pathogens by *Pseudomonas fluorescens* seems to be mediated by the production of phenazine antibiotics, such as (4) [51–57], and mutants of *Pseudomonas fluorescens* that were unable to produce phenazines failed to control *Gaeumannomyces,* whereas those that produced phenazines controlled disease [55, 58]. Phenazines are bacterial pigments that have been known since the mid-nineteenth century. The way in which they affect target organisms is unproven, but both phenazine (4) and pyocyanin (5) have been used as electron acceptors in metabolic studies of photosynthesis and electron transport, and they have been ob-

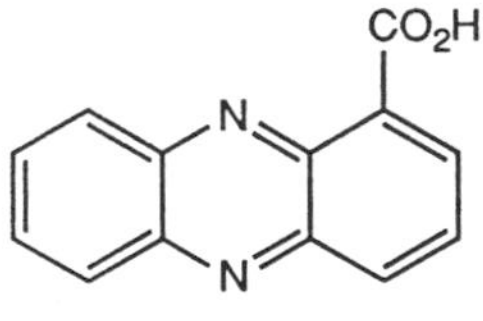

(4)

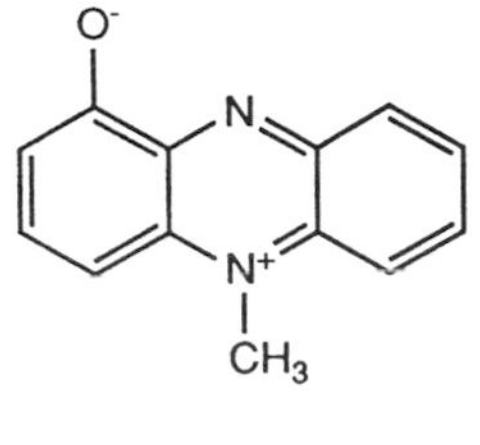

(5)

served to increase respiration rates of cells [59]. A more likely mode of action for these compounds may be because of their ability to bind to RNA and DNA by intercalation between bases or base pairs.

Hollstein and Van-Gemaert [60] found that all phenazine derivatives that they studied inhibited DNA-template-controlled RNA synthesis in an *in vitro* system. Interactions were strongly affected by ionic strengths of buffers used and completely absent at an ionic strength above 0.2. These observations are confirmed by the findings of Brisbane et al. [53] who found that pH strongly influenced the activity of phenazine-1-carboxylic acid, the carboxylate anion being biologically inactive toward indicator strains of fungi. With an ionization constant (pKa) of 4.4 at neutral or alkaline pH, the active species would represent only 0.17% of applied loadings. This may explain the poor performance of these BCAs in some field situations. New genetic approaches to elucidating the effect of environmental factors on phenazine production may yield more information on this subject [61]. Furthermore, it is by no means sure that phenazines represent the only mode of action of these pseudomonads; not all of the phenazine minus TN_5 mutants produced by Thomashow and Weller [55] failed to control disease, implying that there may well be more than one mode of action for these BCAs.

Pyocyanin is another phenazine-based antibiotic isolated from *Pseudomonas aeruginosa* [6]. It is a water-soluble blue-green pigment with antibacterial properties. Its inhibitory action may be as a result of its unique redox potential. During respiration, pyocyanin becomes reduced and univalently reduces oxygen to superoxide radicals [62]. Susceptibility of different bacterial strains to pyocyanin would be dependent on levels of superoxide dismutase and catalase that the organism possessed and on the presence of oxygen. Alternatively, pyocyanin could be a competitor for electrons at several levels of respiration, diverting electron flows at specific points in the respiratory chain [59]. However, as with other phenazines, intercalation with DNA and RNA base pairs may also be a mode of action [60].

Other compounds have been implicated as having a role in biological control activity of the pseudomonads. These include the pseudomonic acids A (6) and B (7) [63–65], responsible for broad-spectrum antibacterial activity and 2,4 diacetyl phloroglucinol (8) [66]. This antibiotic was demonstrated as being a significant factor in the suppression of Take-All disease of wheat [67], in the control of Black Rot of Tobacco [68], causes inhibition of *Septoria tritici* [69] and has also been shown to be active against gram-positive bacteria including the actinomycetes. Keel et al. (67) demonstrated that a mutant *Pseudomonas* lacking the ability to produce 2,4 diacetyl phloroglucinol was less able to

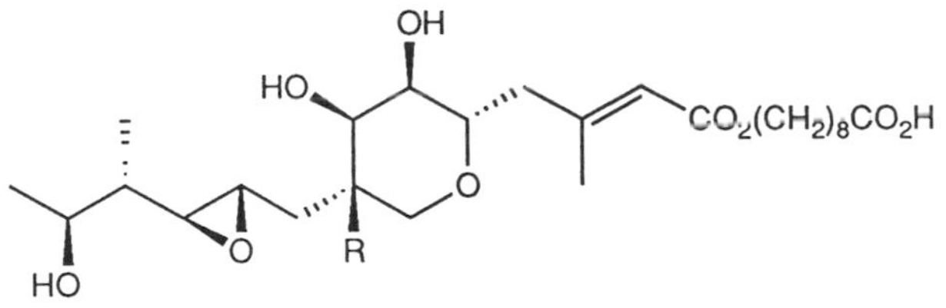

R=H   (6)

R=OH   (7)

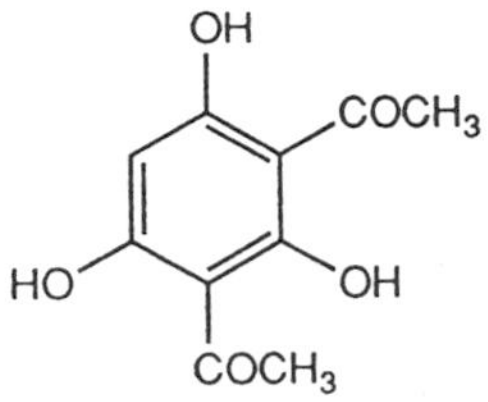

(8)

control Take-All than wild-type strains, and that the control ability could be restored by inserting into the mutant strain a cosmid containing the wild-type gene which was shown to restore 2,4 diacetyl phloroglucinol production by the bacterium.

However, phytotoxic activity of this compound has also been demonstrated [28, 70] with a mode of action similar to that of the chemical herbicide 2-4D(dichlorophenoxyacetate) [71]. Pyrrolnitrin (9), also produced by *Pseudomonas* strains, has broad-spectrum antibiotic activity [72]. HCN has been implicated in both the biological control of plant pathogens [73–75] and with phytotoxic symptoms [34]. It appears that there is competition between HCN minus and HCN plus *Pseudomonas* strains in the field, and that soil type, variety of crop plant, pathogen pressure, and saprophyte population will influence whether or not the pseudomonads are acting as BCAs, PGPRs, or pathogens [28]. There seems to be a correlation between iron availability and production of HCN by pseudomonads; the more available iron, the higher the levels of HCN [35].

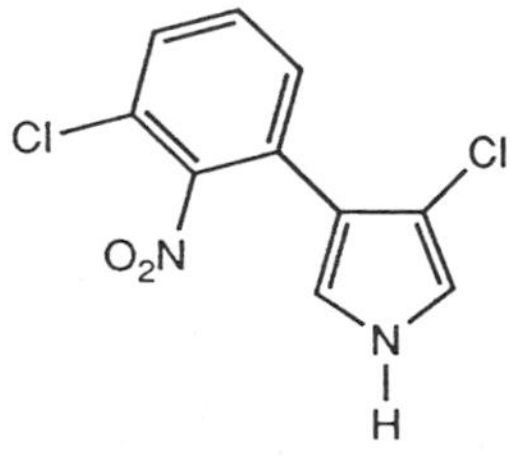

(9)

There are undoubtedly more antibiotics from pseudomonads to be discovered and identified. For example, oomycin A is a chemically unidentified antibiotic substance produced in small quantities with activity against *Pythium ultimum* [76]. The chemical structure of this antibiotic is unknown, although considerable detail is known about the regulation of its production.

C. Actinomycetes

Despite the widespread long-term use of *Actinomycetes* as biological control agents of cotton diseases in the Peoples Republic of China [77, 78], there is only one commercially available actinomycete, *Streptomyces griseoviridis* (Mycostop, Kemira) [79]. The *Streptomyces* species used in this product was one of many isolated from Finnish sphagnum peat known to suppress plant root disease [80]. Mycostop is described as a biofungicide with its main target *Fusarium* spp. on ornamentals, vegetables, and field crops [81]. It is also reported as controlling *Rhizoctonia solani* and *Alternaria brassicicola* in experimental systems [82–85].

Actinomycetes have been reported to produce many antibiotics with applications to human, animal, and plant diseases, but their role in biological control is only now being elucidated. The mode of action of Mycostop (Kemira) has not been published in detail. However, Raatikainen [86] identified a heptaene complex antibiotic isolated from one of the *Streptomyces griseoviridis* strains from Finnish sphagnum peat. Heptaenes are multicomponent polyenic macrolide antibiotics, and in this case the major heptaene had a candicidin type of HPLC profile. Candicidin appears to be synonymous with levorin A (*10*), a complex antibiotic possessing a macrolide ring with conjugated double bonds and a number of hydroxyl, keto, and methyl groups substituents [87]. It also contains the amino sugar mycosamine. This isolate also produced many other components dissimilar to candicidin. There are also many other

(*10*)

antibiotics isolated from strains of *S. griseoviridis*, including the peptidolactones neoviriogrisein [88] and etamycin [89], actinobolin [90], and roseophilin, an antibiotic containing pyrrole and furan moieties [91].

Other members of the streptomycete family also produce antibiotics with herbicidal and antibacterial activity. The Racemomycins belong to the streptothricin group of antibiotics [92] with a very broad spectrum of activity as antimicrobials, but due to their delayed nephrotoxicity in mammalian systems, they have not been introduced for medical use. These antibiotics are active against *Fusarium oxysporum* and *Pseudomonas syryngae* pv. *tabaci*, but they are also inhibitors of *Brassica rapa* root growth.

III. CONTROL OF FUNGI

In 1989 there were no fungal agents for plant disease control available in the United States, and only two were available in Europe [93]. The situation has not changed in the last 3 years. The two products that are available account for a small percentage of the less than $1 million per annum market for biofungicides [2]. *Peniophora gigantea* and *Trichoderma viride* are the two fungi and they will be considered in turn.

A. Peniophora gigantea

The basidiomycete *Peniophora gigantea*, a bracket fungus, is an aggressive colonizer of dying and dead pine tree wood. Its target as a BCA is another bracket fungus, *Heterobasidion* (*Fomes*) *annosum*, which colonizes the freshly cut tree stumps and wounds created during tree harvesting and other forestry practices. *Heterobasidion annosum* is able to colonize intact trees via roots of the felled trees, and it is, hence, a serious problem. Part of the mode of action of this BCA is undoubtedly by its vigorous competition for substrate on the cut stumps, but the lysis of *H. annosum* mycelium by *P. gigantea* ensures eradication of the pathogen from wood that has already been colonized [94]. Lysis appears to be by hyphal interference at the point of contact between the two fungi [95], but the nature of the interference substance is unknown. The effect is very localized, and the substance appears not to diffuse through media to any detectable extent.

B. Trichoderma sp.

Despite the many reports of the antagonistic properties of *Trichoderma* sp. to plant pathogens [96, 97 and references therein], there are very few registered products with *Trichoderma* as the active ingredient.

Trichoderma viride has been registered in Europe for use against *Chondrostereum purpureum* and *Armillaria mellea* in trees, and *Trichoderma harzianum* has been registered in Israel for use against damping-off diseases. In many cases, the basis of the antagonism appears to be by antibiosis, and several different classes of antibiotic have been isolated, both volatile and nonvolatile [98, 99]. These include Trichodermin (*11*), a sesquiterpenoid [100], peptides like suzakacillin A (*12*), and alamethicin I (*13*) [101, 102], pyrones [103], and isocyanides (*14–17*) [104]. However, the two products registered for use as BCAs are not reported to be toxin producers. A second genus, *Gliocladium*, is often reported to have biocontrol properties [97]. Interestingly, work by Howell and Stipancovic [105] showed that mutants of *G. virens* unable to produce the diketopiperazine antibiotic gliovirin (*18*) were also unable to protect plants from the pathogen *Pythium ultimum*, whereas wild-type *G. virens* was effective in laboratory tests.

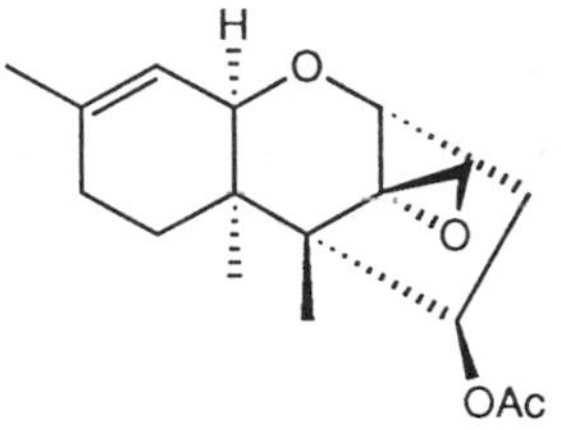

(11)

Ac-Aib-Ala-Aib-Ala-Aib-Ala(Aib)-Gln-Aib-Aib(Leu)-Aib-Gly-Leu-Aib-Pro-   (12)
Val-Aib-Aib-(D-Iva)-Gln-Gln-Pheol

Ac-Aib-Pro-Aib-Ala-Aib-L-Ala-Gln-Aib-Val-Aib-Gly-Leu-Aib-Pro-Val-Aib-   (13)
Aib-Glu(OH)-Gln-Phenylalaninol

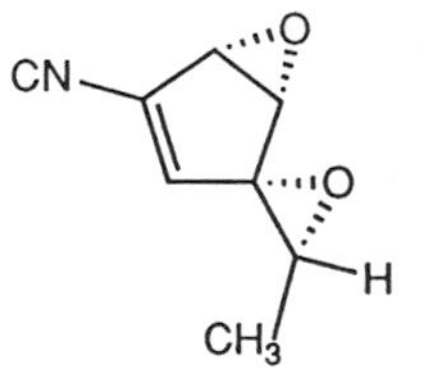

(14)

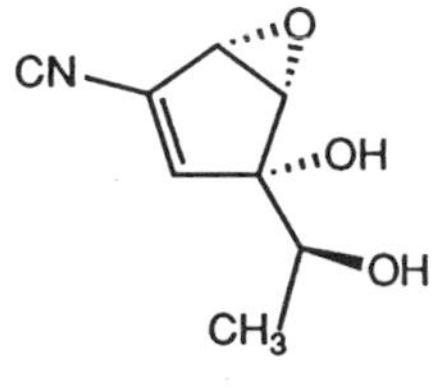

(15)

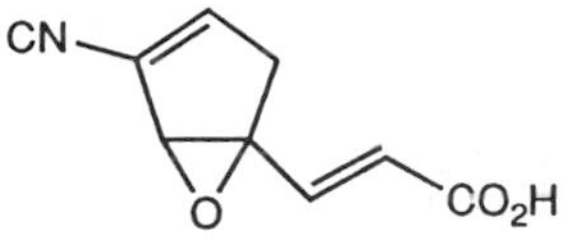

(16)

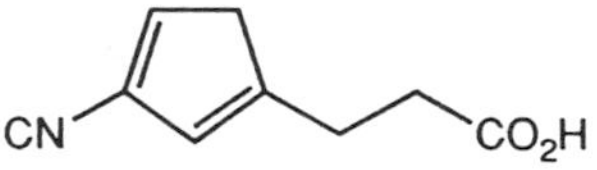

(17)

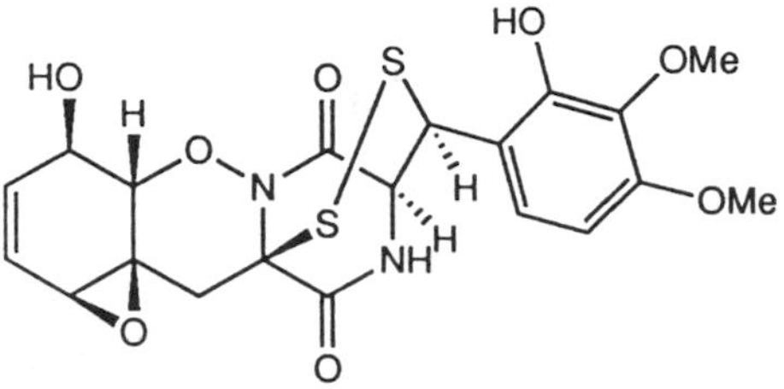

(18)

IV. CONTROL OF INSECT PESTS

The total market size for insecticides in 1991 was estimated at over $7600 million, of which bioinsecticides accounted for less than 1.5% of the market, with other organisms such as fungi, viruses, and nematodes accounting for only $10 million of sales. Of the bioinsecticides, BT was by far the most successful, but with characteristics more akin to chemical pesticides, it is doubtful if it is a true BCA.

A. Bacterial Insecticides (Bacillus thuringiensis)

Bacillus thuringiensis is a gram-positive spore-forming bacillus that produces a crystalline protein inclusion at sporulation. This is toxic to certain insect pests on injestion and can be used as an insecticide. In many cases, the injestion of the crystal itself is sufficient to cause paralysis and death of the insect, although there are circumstances in which the presence of the spores may enhance toxin kill [106] or they may even be essential for kill. There are many excellent reviews of this vast subject area [107–111] and the current commercial situation is well summarized by Currier and Gawron-Burke [112]. BT will be dealt with in outline only in this review.

BT has over 30 recognized subspecies based on serotyping and biochemical tests [112]. They produce several insecticidal toxins, two of which are used in agriculture, the δ endotoxin and the β exotoxin. By far the greater research effort has gone into the δ endotoxin, and there are over 20 commercial products based on strains of BT that produce it. Currently, strains are being used to produce products for Lepidopteran larvae (var. *kurstaki*), for mosquito and blackfly (var. *isrealiensis*), for wax moth larvae (var. *aizawa*); and for colorado beetle (var. *berliner*). The recent discoveries of beetle-specific toxins have excited considerable interest and are under development [113, 114]. The structure of the δ endotoxin has been reviewed by Fast [115] and advances in its molecular biology summarized by Currier and Gawron-Burke [112]. In essence, BT δ endotoxin is a bipyramidal aggregate of dumbbell-shaped protein molecules with a molecular weight of 230,000 D [115]. The crystal is actually a protoxin, which must be solubilized and activated by proteolysis. These two events occur in the insect mid-gut, where alkaline pH and the presence of reductant enhance solubilization [116]. The mode of action of the BT toxin appears to be via the creation of potassium ion channels across brush border membranes [117–120], resulting in the rapid metabolic breakdown of these cells, followed by paralysis and death of the insect from ionic imbalance. Septiceamia from invasion of the hemolymph by gut commensals contributes to death in around 50% of affected individuals.

The β exotoxin thuringiensin (*19*) is a structural ATP analog (reviewed in Ref. 121) which specifically inhibits both prokaryotic and eukaryotic DNA-dependent RNA polymerases. Although it is present in preparations of BT used in the former Soviet Union (Bitoxibacillin), its toxicity and teratogenic effects have prevented its useage in the United States and Canada [106].

BT strains are known to produce other toxins with lesser effects on insects, including a phospholipase C, the γ exotoxin [122].

(*19*)

B. Fungal Insecticides

Despite the existence of many reports of fungal parasitism of insects in the scientific literature, commercial success in this group has been limited to only three species of imperfect fungi, *Verticillium*, *Metarhizium*, and *Beauvaria*.

1. *Verticillium lecanii*

Verticillium lecanii strains are the active ingredients of Mycotal and Vertalec (Koppert). These products were developed by Tate and Lyle, and then Microbial Resources Ltd. for greenhouse use to control white-fly and aphids, respectively. Although the toxic depsipeptide bassiano-lide (*20*) has been isolated from *V. lecanii* mycelium, its importance in pathogenicity of strains is unproven [123]. Toxicity is more probably based on a combination of factors such as proliferation of the fungus within the insect tissues [124] and the production of high levels of ex-tracellular proteolytic enzymes [125, 126] as well as toxin production.

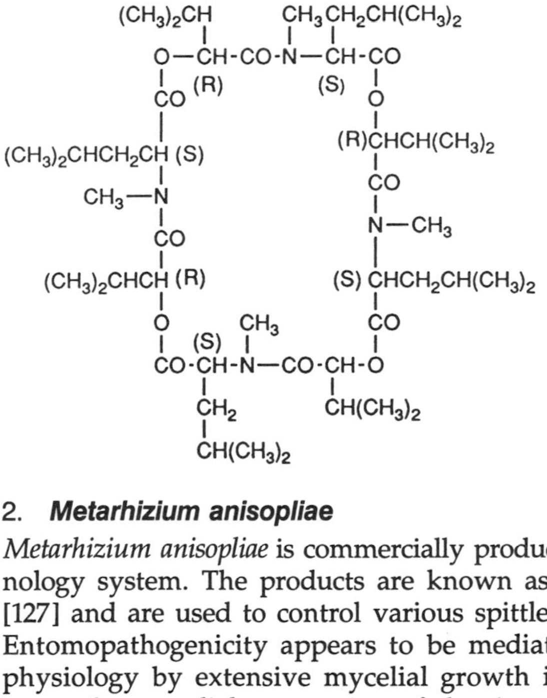

$$
\begin{array}{cc}
(CH_3)_2CH & CH_3CH_2CH(CH_3)_2 \\
| & | \\
O-CH\text{-}CO\text{-}N-CH\text{-}CO \\
| \ (R) & (S) \ | \\
CO & O
\end{array}
$$

(CH₃)₂CHCH₂CH (S)

CH₃—N

CO

(CH₃)₂CHCH (R)

O   CH₃

(S) |

CO·CH-N—CO·CH-O

CH₂   CH(CH₃)₂

CH(CH₃)₂

(R)CHCH(CH₃)₂

CO

N—CH₃

(S) CHCH₂CH(CH₃)₂

CO

CH(CH₃)₂

(20)

2. *Metarhizium anisopliae*

Metarhizium anisopliae is commercially produced in Brasil in a low-technology system. The products are known as Metaquino and Metabiol [127] and are used to control various spittlebug species on sugarcane. Entomopathogenicity appears to be mediated by disruption of host physiology by extensive mycelial growth in the hemolymph of the insect, by mycelial penetration of the tissues, and by the production of toxins known as destruxins [e.g., destruxin A (*21*)] [128]. These tox-ins cause paralysis and death in lepidopteran larvae and adult diptera

[129–131]. Fifteen related destruxins have been isolated, destruxins A and B being the most widely distributed, E being the most toxic.

$$\text{CO} - \text{L-Pro-L-Ile-Me-L-Val-Me-L-Ala-}\beta\text{-Ala} \qquad (21)$$

Mode of action studies of destruxins have shown that they cause depolarization of muscle membranes by activating calcium ion channels [131], but these toxins also have cytotoxic effects on other tissues, albeit less dramatic [128]. The importance of the destruxins in pathogenesis remains uncertain however, as the susceptibility to the muscle paralyzing effects of destruxins in the Insecta is restricted, whereas the host range of *Metarhizium* is wide. Some of this difference can be accounted for by the ability of some insects to detoxify the destruxins [130], and it may be that the myotoxic effects may potentiate other aspects of pathogenicity. Other fungal metabolites produced by *Metarhizium* include the cytochalasins. These compounds are known to inhibit cell division and are also potential candidates in potentiating effects of other aspects of entomopathogenicity.

3. *Beauvaria bassiana*

Beauvaria bassiana has been commercially produced as Boverin in the CIS for control of codling moth and colorado beetle on potato. Again, the role of toxins like bassianolide (*20*) and beauvericin (*22*) that this fungi is known to produce is uncertain [132] and it has been suggested that they may be important instead in allowing extensive colonisation of the insect by *Beauvaria* unhindered by secondary invaders [123]. In this way, these toxins would be less pathogenicity factors than substrate possession factors.

$$(22)$$

V. CONTROL OF WEEDS

Currently sales of bioherbicides amount to less than $1 million per annum [2] and two products, Collego (Ecogen) and Devine (Abbott), account for the largest part of this market. These are both products where fungi are the active ingredient; as yet there are no commercially available bioherbicides based on bacteria or virus. As Devine (*Phytophthora palmivora*) is not thought to have a mode of action mediated by a toxin, it will not be further discussed.

A. Colletotrichum sp.

There are a number of *Colletotrichum* sp. that are commercial bioherbicides, the most successful in economic terms being Collego (*Colletotrichum gloeosporiodes f.* sp. *aeschynomene*, used to control the Northern Jointvetch (*Aeshcynomene virginica*) in rice and soya [133]. Metabolites from the fungus have been found to be toxic to the host plants [134], and one toxin has been identified as gloeosporone (23), the final configuration of which was not fully assigned until 1987 [135]. The importance of gloeosporone in the pathogenicity of *C. gloeosporioides f.* sp. *aeschynomene* on its target plant is not proven, but it is known to act as a self-inhibitor of germination. It is possible that gloeosporone may complex with certain metal ions and that their transport into the cell may be affected, inhibiting germination [136]. Other species of *Colletotrichum* that were developed as commercial bioherbicides include *C. gloeosporioides f.* sp. *malvae* (Biomal) used to control round leaved mallow (*Malva pusilla*), *C. coccodes* (Velgo) against velvetleaf (*Abutilon thoephrasti*), and *C. orbiculare* against Bathurst burr (*Xanthium theophrasti*). *C. gloeosporiodes f.* sp. *cuscutae* (Luboa 2) is used in the Peoples Republic of China to control dodder (*Cuscuta chinensis* and *C. australis*) on soybean. The symptoms that these species of *Colletotrichum*

(23)

produce on their target plants are strongly reminiscent to those produced by Collego, and it seems likely that their modes of action will include toxin production. It is interesting to note that *C. gloeosporioides* was reported to produce the polyamino acid phytotoxin Aspegillomarasmin A (24) [137].

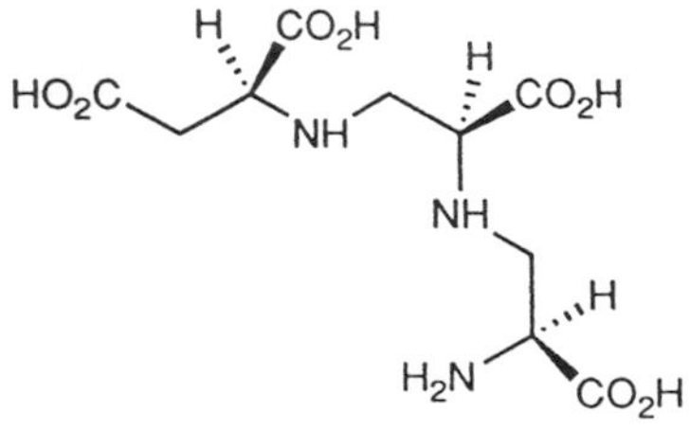

(24)

B. Alternaria sp.

Alternaria cassiae is the active ingredient of Casst (Mycogen), a mycoherbicide which was developed for control of sicklepod (*Cassia obtusifolia*) and coffee senna (*Cassia occidentalis*) in soybean and peanut crops. The product has been withdrawn because of technical difficulties. Several toxins have been isolated from species of *Alternaria*, including the diketopiperazines maculosin (25) [138] and tenuazonic acid (26) [139] from *A. alternata*, the quinone alteichin (27) [140] from *A. eichornia*, and zinniol (28) from at least four different species of *Alternaria* [141]. Like the *Colletotrichum* species group, it seems likely that the *Alternaria* sp. will have modes of action that are mediated by toxin production.

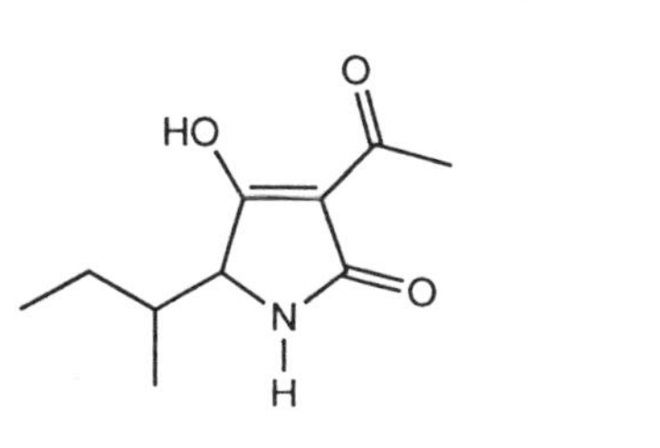

(25)

(26)

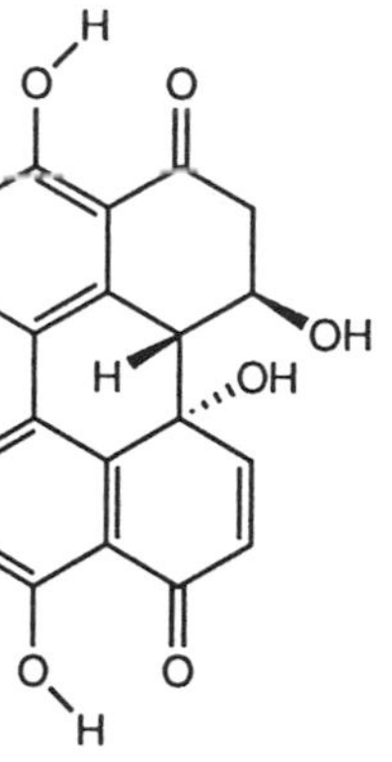

(27)

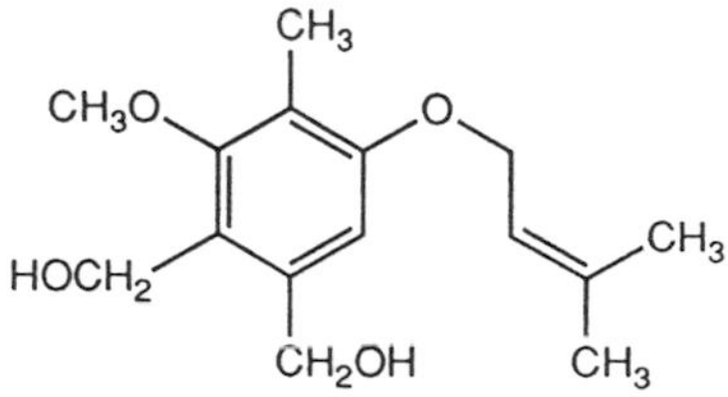

(28)

REFERENCES

1. K.A. Powell, *Chem. Ind.,* 5:168. (1992).
2. K.A. Powell and A.R. Jutsum, *Chem. Ind.,* 5:168 (1992).
3. Wood Mackenzie, *Agrochemical Monitor,* August 1991, p. 9.
4. S.G. Lisansky, *Perform. Chem.* Feb. 1989, p. 14.
5. K.A. Powell, J.L. Faull, and A.R. Renwick, *Biological Control of Soil-borne Plant Pathogens* (D. Hornby, ed.), Commonwealth Agricultural Bureau International, Wallingford, Oxfordshire U.K., 1990, p. 445.
6. J.L. Faull, *Fungi in Biological Control Systems* (M.N. Burge, ed.), Manchester University Press, Manchester, 1988, p. 125.
7. J. Singh and J.L. Faull, *Biocontrol of Plant Disease, Vol. 2* (K.G. Mukerji and K.L. Garg, eds.), CRC Press, Boca Raton, FL, 1988, p. 167.
8. J.M. Whipps, K. Lewis, and R.C. Cook, *Fungi in Biological Control Systems* (M.N. Burge, ed.), Manchester University Press, Manchster, 1988, p. 161.
9. K. Lewis, J.M. Whipps, and R.C. Cook, *Biotechnology of Fungi for Improving Plant Growth,* (J.M. Whipps and R.D. Lumsden, eds.), Cambridge University Press, Cambridge, 1990, p. 111.
10. S.K. Farrand, *New Directions in Biological Control: Alternatives for Suppressing Agricultural Pests and Diseases* (R.R. Baker and P.E. Dunn, eds.), Alan Liss, Inc., New York. 1990, p. 679.
11. D. Inze, A. Follin, M. Van Lijsebettens, C. Simoens, G. Genetello, M. Van Montague, and J. Schell, *Mol. Genet.,* 194:265 (1984).

12. G. Schroeder, S. Waffenschmidt, E. W. Weiler, and J. Schroeder, *Eur. J. Biochem.,* 139:387 (1984).
13. D.E. Akiyoshi, H. Klee, R.M. Amasina, E.W. Nester, and M.P. Gordon, *Proc. Natl. Acad. Sci., USA,* 81:5994 (1984).
14. S.K. Farrand, J.E. Slota, J.S. Shim, and A. Kerr, *Plasmid,* 13:106. (1985).
15. P.J. Murphy and W.P. Roberts, *J. Gen. Micro.,* 114:207. (1979).
16. C.G. Panagopoulos, P.G. Psallidas, and A.S. Alivizatos, *Biology and Control of Soil Borne Plant Pathogens* (B. Schippers and W. Gams, eds.), Academic Press, New York, 1979, p. 569.
17. G.T. Hayman and S.K. Farrand, *J. Bacteriol.,* 170:1759 (1988).
18. J.E. Slota and S.K. Farrand, *Plasmid,* 8:175 (1982).
19. J.G. Ellis, A. Kerr, M. Van Montague, and J. Schell, *Physiol. Plant Pathol.* 15:311 (1979).
20. D.A. Jones, M.H. Ryder, B.G. Clara, S.K. Farrand, and A. Kerr, *Mol. Gen. Genet,* 212:207 (1988).
21. L.W. Moore, *Biology and Control of Soil-Borne Plant Pathogens* (B. Schippers and W. Gams, eds.), Academic Press, New York, 1978.
22. L.W. Moore and G. Warren, *Annu. Rev. Phytopathol.,* 17:163 (1979).
23. D.A. Cooksey and L.W. Moore, *Phytopathology,* 70:506 (1980).
24. J-S. Shim, S.K. Farrand, and A. Kerr, *Phytopathol.,* 77:463 (1987).
25. D.A. Cooksey and L.W. Moore, *Phytopathology,* 72:919 (1982).
26. R.J. Cook and A.D. Rovira, *Soil Biol. Biochem.,* 8:269 (1976).
27. D.M. Weller, W.J. Howie, and R.J. Cook, *Phytopathology,* 78:1094 (1988).
28. G. Defago and D. Haas, *Soil Biochemistry* 6 (J.M. Bollag and G. Stotsky, eds.), Marcel Dekker, New York, 1990, p. 249.
29. M.N. Schroth and J.G. Handcock, *Annu. Rev. Microbiol.,* 35:453 (1981).
30. M.N. Schroth and J.G. Handcock, *Science,* 216:1376 (1982).
31. B.C. Hemming, *New Directions in Biological Control, Alternatives for Suppressing Agricultural Pests and Diseases* (R. Baker and P.E. Dunn, eds.), Alan Liss Inc. New York, 1990, p. 223.
32. J.B. Neilands, *Annu. Rev. Biochem.,* 50:715 (1981).
33. P.A.H.M. Bakker, J.G. Lamers, A.W. Bakker, J.D. Marugg, P.J. Wiesbeek, and B. Schippers, *Neth. J. Plant Pathol.,* 92:249 (1986).
34. A.W. Bakker and B. Schippers, *Soil Biol. Biochem.,* 19:451 (1987).
35. A.W. Bakker, W.L.M. Punte, and B. Schippers, *Biotic Interactions and Soilborne Disease* (A.B.R. Beemster, G.J. Bollen, M.A. Ruissen, B. Schippers, and A. Tempel, eds.), Elsevier, Amsterdam, 1991, p. 297.
36. J. Leong, *Annu. Rev. Plant Pathol.,* 26:187 (1986).
37. J.W. Kloepper, J. Leong, M. Teintze, and M.N. Schroth, *Nature (London),* 286:885 (1980).
38. J.W. Kloepper, J. Leong, M. Teintze, and M.N. Schroth, *Curr. Microbiol.* 4:317 (1980).
39. D. Hohnadel and J.M. Meyer, *Iron, Siderophores and Plant Diseases* (T.R. Swinburn, ed.), Plenum Press, New York, 1986, p. 119.
40. D. Hohnadel and J.M. Meyer, *J. Bact.,* 170:4865 (1986).

41. J. Leong, W. Bitter, M. Koster, J.D. Marugg, V. Venturi, and P.J. Weisbeek, *Proceedings of the Second International Workshop on Plant Growth Promoting Rhizobacteria 1990*, Bulletin of the International Control of Noxious Animals and Plants, West Palearctic Regional Section, (C. Keel, B. Koller and G. Defago, eds.), 1991, pp. 127–185.

42. J.E. Loper, *New Directions in Biological Control. Alternatives for Suppressing Agricultural Pests and Diseases*, Alan Liss Inc, New York, 1990, p. 735.

43. P.A.H.M. Bakker, R.V. Peer, and B. Schippers, *Biological Control of Soil Borne Plant Pathogens* (D. Hornby, ed.), Common Agricultural Bureau International, Wallingford, 1990, p. 131.

44. J.M. Meyer and M.A. Abdallah, *J. Gen. Microbiol.,* 107:319 (1978).

45. J.M. Meyer, D. Hohnadel, and F. Halle, *J. Gen. Microbiol.,* 135:1479 (1989).

46. J.O. Becker, R.W. Hedges, and E. Hessens, *Appl. Environm. Microbiol.,* 49:1090 (1985).

47. C. Keel and G. Defago, *Proceedings of the Second International Workshop on Plant Growth Promoting Rhizobacteria 1990*, (C. Keel, B. Koller and G. Defago, eds.), 1991, pp. 136–142.

48. G. Winkelman, *Mycol. Res.,* 96:529 (1992).

49. T. Leisinger and R. Magraff, *Microbiol. Rev.,* 43:422 (1979).

50. N. Gutterson, J.S. Seigle, G.J. Warren, and T.J. Layton, *J. Bacteriol.,* 170:380 (1988).

51. L.S. Pierson and L.S. Thomashow, *Proceedings of the Second International Workshop on Plant Growth Promoting Rhizobacteria, 1990*, (C. Keel, B. Koller and G. Defago, eds.), (1991), pp. 119–121.

52. S. Gurusiddiaih, D.M. Weller, A. Sarkar, and R.J. Cook, *Antimicrob. Agents Chemother.,* 29:488 (1986).

53. P.G. Brisbane, L.J. Janik, M.E. Tate, and R.F.O. Warren, *Antimicrobial Agents Chemother.,* 31:1967 (1987).

54. P.G. Brisbane and A.D. Rovira, *Plant Pathol.,* 37:104 (1988).

55. L.S. Thomashow and D.M. Weller, *J. Bacteriol.,* 178:3499 (1988).

56. D.M. Weller and R.J. Cook, *Phytopathology,* 73: 463 (1983).

57. W.J. Howie and T.V. Suslow, *Molec. Plant Microb. Interactions,* 4:393 (1991).

58. L.S. Thomashow and D.M. Weller, *Biological Control of Soil-borne Plant Pathogens* (D. Hornby, ed.), Common Agricultural Bureau International, Wallinford, 1990, p. 109.

59. S.S. Baron and J.J. Rowe, *Antimicrob. Agents Chemother.,* 20:814 (1981).

60. U. Hollstein and R.J. Van-Gemaert, *Biochemistry,* 10:497 (1971)

61. L.S. Thomashow and L.S. Pierson, *Advances in Molecular Genetics of Plant–Microbe Interactions* (H. Hennecke and D.P.S. Verma, eds.), Kluwer Academic, Boston, 1991.

62. H.M. Hassan and I. Fridovich, *J. Bacteriol.,* 141:156 (1980).

63. A.T. Fuller, G. Mellows, M. Woolford, G.T. Banks, K.D. Barrow, and E.B. Chain, *Nature (London),* 234:416 (1971).

64. E.B. Chain and G. Mellows, *J. Chem Soc. Perkin I* 294 (1977).

65. E.B. Chain and G. Mellows, *J. Chem. Soc. Perkin I*, 318 (1977).
66. A. Dutrecq, P. Debras, J. Stevaux, and M. Marlier, *Biotic Interactions and Soil-Borne Diseases* (A.B.R. Beemster, G.J. Bollen, M.A. Ruissen, B. Schippers, and A. Tempel, eds.), Elsevier, Amsterdam, 1991, p. 252.
67. C. Keel, M. Maurhofer, T. Oberhansli, C. Voisard, D. Haas, and G. Defago, *Biotic Interactions and Soil-Borne Diseases* (A.B.R. Beemster, G.J. Bollen, M.A. Ruissen, B. Schippers, and A. Tempel, eds.), Elsevier, Amsterdam, 1991, p. 335.
68. G. Defago, C.H. Berling, U. Burger, D. Haas, G. Kahr, C. Keel, C. Voisard, P. Wirthner, and B. Wuttrich, *Biological Control of Soil-Borne Plant Pathogens* (D. Hornby, ed.), Common Agricultural Bureau International, Wallingford, 1990, p. 93.
69. E. Levy, F.J. Gough, K.D. Berlin, P.W. Guiana, and J.T. Smith, *Plant Pathol.*, 41:335 (1992).
70. D. Haas, C. Keel, J. Laville, M. Maurhofer, Th. Oberhansli, V. Schnider, and G. Defago, *Biotic Interactions and Plant Diseases* (A.B.R. Beemster, G.J. Bollen, M. Gerlagh, M.A. Ruissen, B. Schippers, and R. Tempel, eds.), Elsevier, Amsterdam, 1991, p. 450.
71. B.T. Kataryan and G.C. Torgashova, *Dokl. Akad. Nauk. Arm. SSR.*, 63:109 (1976).
72. K. Arima, I. Hiroshi, K. Masanobu, A. Fukuda, and G. Tamura, *J. Antibiot. Ser A*, 18:201 (1965).
73. C.J. Knowles and A.W. Bunch, *Adv. Microbiol. Physiol.*, 27:73 (1986).
74. R.A. Askeland and S.M. Morrison, *Appl. Env. Microbiol.*, 45:1802 (1983).
75. C. Voisard, C. Keel, D. Haas, and G. Defago, *EMBO J.*, 8:351 (1989).
76. N. Gutterson, *Crit. Rev. Biotech.*, 10:69 (1990).
77. S.Y. Yin, D.C. Keng, K.Y. Yang, and D. Cheu, *Acta Phytopathol. Sin.*, 3:55 (1957).
78. S.Y. Yin, J.K. Chang, and P.C. Xun, *Acta Microbiol. Sin.*, 11:259 (1965).
79. O. Mohammadi, *Biological Control of Plant Diseases* (E.S. Tjamos, G. Papavisas, and R.J. Cook, eds.), Plenum Press, New York, 1992, p. 207.
80. M-L. Lahdenpera, *Acta. Hort.*, 216:85 (1987).
81. J.G. White, L.A. Linfield, M-L. Lahdenpera, and J. Uoti, *Proceedings of the BCPC Pests and Diseases*, Levenham Press, Farnham, Surrey, U.K., 1990, pp. 221–226.
82. R. Tahvonen and H. Avikainen, *J. Agric. Sci. Finl.*, 59:199 (1987).
83. R. Tahvonen, *J. Agric. Sci. Finl.*, 54:357 (1982).
84. R. Tahvonen, *EPPO Bull.*, 18:55 (1988).
85. M-L. Lahdenpera, E. Simon, and J. Uoti, *Biotic Interactions and Soil-Borne Diseases* (A.B.R. Beemster, G.J. Bollen, M.A. Ruissen, B. Schippers, and A. Tempel, eds.), Elsevier, Amsterdam, 1991, p. 258.
86. O. Raatikainen, *J. Chrom.*, 588:356 (1991).
87. V.V. Belakov, A.Y. Vlasenko, A.A. Levina, Y.D. Shenin, N.G. Rhozhkova, and G.K. Smirnova, *Antibiotiki I Khimioterapiya*, 36:11 (1991).
88. N. Nagato, Y. Okumura, R. Okamato, and T. Ishikura, T. *Agric. Biol. Chem.*, 48:3041 (1984).

89. E. Katz, F. Kamal, and K. Mason, *J. Biol. Chem.,* 254:6684 (1979).
90. R.S. Garagipati, D.M. Tschaen, and S.M. Weinreb, *J. Am. Chem. Soc.,* 107:7790 (1985).
91. Y. Hayakawa, K. Kawakami, H. Seto, and K. Furihati, *Tetrahedron Lett,* 33:2701 (1992).
92. Y. Inamori, H. Tominaga, M. Okuno, H. Sato, and H. Tsujibo, *Chem. Pharm. Bull,* 36:1577 (1988).
93. R.D. Lumsden and J.A. Lewis, *Biotechnology of Fungi for Improving Plant Growth* (J.M. Whipps and R.D. Lumsden, eds.), Cambridge University Press, Cambridge, 1989.
94. D.S. Meredith, *Ann. Bot,* 24:63 (1960).
95. C.E.O. Ikediugwu, C. Dennis, and J. Webster, *Trans. Br. Mycol. Soc.,* 54:307 (1970).
96. I. Chet, *Biological Control of Soil-Borne Plant Pathogens* (D. Hornby, ed.), Common Agricultural Bureau International, Wallingford, 1990, p. 15.
97. G.C. Papavisas, *Biological Control of Plant Diseases, Prospects and Challenges for the Future* (E.C. Tjamos, G.C. Papavisas, and R.J. Cook, eds.), Plenum Press, New York, 1992, p. 211.
98. C. Dennis and J. Webster, *Trans. Br. Mycol. Soc.,* 57:25 (1971).
99. C. Dennis and J. Webster, *Trans. Br. Mycol. Soc.,* 57:41 (1971).
100. W.O. Gotfredsen and S. Vandegal, *Acta Chemica Scand,* 19:1088 (1965).
101. T. Ooka, Y. Shimojima, T. Akimoto, I. Takeda, S. Senoh, and J. Abe, *Agric. Biol. Chem.,* 30:700 (1966).
102. C.E. Meyer and F. Reusser, *Experientia,* 23:85 (1967).
103. M.O. Moss, R.M. Jackson, and D. Rogers, *Phytochemistry,* 14:2706 (1975).
104. M.S. Edenborough and R.B. Herbert, *Natural Product Reports.,* 5:229 (1988).
105. C.R. Howell and R.D. Stipancovic, *Can. J. Microbiol,* 29:231 (1983).
106. A.M. Heimpel and T.A. Angus, *J. Insect Pathol,* 1:152 (1959).
107. H.T. Dulmage and Cooperators, *Microbial Control of Pests and Plant Diseases 1970–1980* (H.D. Burges, ed.), Academic Press, New York, 1981, p. 193.
108. B.C. Carlton, *Biotechnology of Crop Protection* (P.A. Hedrin, J.J. Menn, and R.M. Hollingworth, eds.), American Chemical Society, Washington, D.C., 1988, p. 261.
109. A.I. Aaronson, W. Beckman, and P. Dunn, P. *Microbiol. Rev,* 50:1 (1986).
110. R.M. Faust and L.A. Bulla, *Microbial and Viral Pesticides* (E. Kurstak, ed.), Marcel Dekker, New York, 1982, p. 209.
111. P. Luthy, J.L. Cordier, and H-M. Fischer, *Microbial and Viral Pesticides* (E. Kurstak, ed.), Marcel Dekker, New York, 1982, p. 35.
112. T.C. Currier and C. Gawron-Burke, *Biotechnology of Plant–Microbe Interactions* (J.P. Nakas and C. Hagedorn, eds.), McGraw-Hill, New York, 1990, p. 111.
113. A. Kreig, A.M. Huger, G.A. Langenbruch, and W. Schnetter, *Z. Agnew. Entomol,* 96:500 (1983).

114. C. Hernstadt, G.C. Soares, E.R. Wilcox, and D.L. Edwards, *Biotechnology,* 4:305 (1986).
115. P.G. Fast, *Microbial Control of Pests and Plant Diseases 1970–1980* (H.D. Burges, ed.), Academic Press, New York, 1981, p. 223.
116. M. Lilley, R.N. Ruffel, and H.J. Somerville, *J. Gen. Microbiol,* 118:1 (1980).
117. B.L. Gupta, J.A.T. Dow, T.A. Hall, and W.R. Harvey, *J. Cell Sci,* 74:137 (1985).
118. V.F. Sacci, P. Parenti, G.M. Hanozet, B. Giordana, P. Luthy, and M.G. Wolfersburger, *FEBS Letts,* 204:213 (1986).
119. D.N. Crawford and W.R. Harvey, *J. Exp. Biol,* 137:277 (1988).
120. C. Hofman, P. Luthy, R. Hutter, and V. Pliska, *Eur. J. Biochem,* 173:85 (1988).
121. K. Sebesta, J. Farkas, and K. Horska, *Microbial Control of Pests and Plant Diseases 1970–1980* (H.D. Burges, ed.), Academic Press, New York, 1981, p. 249.
122. S. Toumanoff, *Annales Institute Pasteur 85*:90 (1953).
123. C.W. McCoy, R.A. Samson, and D.G. Boucias, *Microbial Insecticides, 5* (C.M. Ignoffo and N.B. Mandore, eds.), CRC Press, Boca Raton, FL, 1988, p. 151.
124. R.J. Quinlan, *Fungi in Biological Control* (M.N. Burge, ed.), Manchester University Press, Manchester, 1988, p. 19.
125. R.J. St Leger, A.K. Charnley, and R.M. Cooper, *Journal of Invertebrate Pathology,* 48:85 (1986).
126. R.J. St Leger, A.K. Charnley, and R.M. Cooper, *J. Inv. Pathol.,* 47:167 (1986).
127. A.T. Gillespie and E.R. Moorhouse, *Biotechnology of Fungi for Improving Plant Growth* (J.M. Whipps and R.D. Lumsden, eds.), Cambridge University Press, Cambridge, 1989, p. 55.
128. A.K. Charnley, *Biotechnology of Fungi for Improving Plant Growth* (J.M. Whipps and R.D. Lumsden, eds.), Cambridge University Press, Cambridge, 1989, p. 85.
129. D.W. Roberts, *Microbial Control of Pests and Diseases 1970–1980* (H.D. Burges, ed.), Academic Press, New York, 1981, p. 441.
130. R.I. Samuels, S.E. Reynolds, and A.K. Charnley, *Comp. Biochem. Physiol.,* 90C:403 (1988).
131. R.I. Samuels, A.K. Charnley, and S.E. Reynolds, *Mycopathologia,* 104:51 (1988).
132. F.R. Champlin and E.A. Grula, *Appl. Environ. Micro,* 37:1122 (1968).
133. G.E. Templeton and D.K. Heiny, *Biotechnology of Fungi for Improving Plant Growth* (J.R. Whipps and R.D. Lumsden, eds.), Cambridge University Press, Cambridge, 1989, p. 127.
134. H.L. Walker and G.E. Templeton, *Plant Sci. Lett.* 13:91 (1978).
135. G. Adam, R. Zibuck, and D.J. Seebach, *Am. Chem. Soc,* 109:6176 (1987).
136. W.L. Meyer, W.B. Schweizer, A.K. Beck, W. Scheifle, D. Seebach, and S.E. Kelly, *Helv. Chim. Acta,* 70:281 (1987).

137. A. Ballio, A. Bottalico, V. Buonocore, A. Carilli, V. Di Vittorio, and A. Graniti, *Phytopathol. Mediterr.*, *8*:187 (1969).

138. A.C. Stierle, J.H. Cardellina, and G.A. Strobel, *Proc. Natl. Acad. Sci., USA*, *85*:8008 (1988).

139. T. Kinoshita, Y. Renbutsu, I.D. Khan, K. Kohmoto, and S. Nishimura, *Ann. Phytothol. Soc. Japan*, *38*:397 (1972).

140. G. Strobel, F. Sugawara, and J. Clardy, *Allelochemicals: Role in Agriculture and Forestry* (G.R. Waller, ed.), American Chemical Society, Washington, D.C., 1987, p. 516.

141. D.J. Robeson and G.A. Strobel, *Phytochemistry*, *23*:1597 (1984).

Index